KREYSZIG'S
Mathematical Methods

KREYSZIG'S Mathematical Methods

Professor Erwin Kreyszig
Professor of Mathematics
Ohio State University
Columbus, Ohio

Dr. A. Ramakrishna Prasad
M.Sc., M.Phil., M.Tech. (CSDP, IIT), Ph.D.,
MCSI, MIETE, MHRDN, MFLNLP

Professor of Mathematics
JNTU, Kukatpally, Hyderabad

WILEY-
INDIA

KREYSZIG'S
Mathematical Methods

Other Wiley Editorial Offices:
John Wiley & Sons, Inc. 111 River Street, Hoboken, NJ 07030, USA
Wiley-VCH Verlag GmbH, Pappellaee 3, D-69469 Weinheim, Germany
John Wiley & Sons Australia Ltd, 42 McDougall Street, Milton, Queensland 4064, Australia
John Wiley & Sons (Asia) Pte Ltd, 2 Clementi Loop #02-01, Jin Xing Distripark, Singapore 129809
John Wiley & Sons Canada Ltd, 22 Worcester Road, Etobicoke, Ontario, Canada, M9W 1L1

First Edition: 2009
ISBN: 978-81-265-2138-8
www.wileyindia.com
www.wileyprecisetextbook.com

Printed at: *Magic International, UP.*

JAWAHARLAL NEHRU TECHNOLOGICAL UNIVERSITY, HYDERABAD

Prof. D.N. Reddy
B.E., M.Tech., Ph.D., FIE., MISME., MSESI., MIIPE., MISTE.,
VICE-CHANCELLOR

Kukatpally
Hyderabad - 500 085
Andhra Pradesh (India)
Phone: 040-23156109 (O)
Fax: 040-23156112
E-mail: vcjntu@yahoo.com
www.jntu.ac.in

Foreword

To benefit the student community and to enrich the standards of curriculum, Jawaharlal Nehru Technological University Hyderabad undertakes the rigorous exercise of streamlining and modifying the syllabi every two years. The University is quality conscious with regard to choice of textbooks prescribed for the students. In this connection, I am happy to know that Prof. A. Ramakrishna Prasad, Head of the Department of Mathematics, authored a book titled **Kreyszig Mathematical Methods**.

I am glad to note that the well-known book Advanced Engineering Mathematics by Kreyszig is fine tuned to the requirements of UG Curriculum of JNTU. The book, I understand, is written in strict adherence to the required specifications and keeping in view the current trends in the field of Engineering. The significant feature of the book is the emphasis laid on the application part, which I think is the key for grasping the basics of Engineering subjects. I appreciate the sincere efforts of Prof. A. Ramakrishna Prasad in bringing out such a comprehensive and student - friendly textbook. I am hopeful that this textbook will be well received by students, researchers and teachers. I wish Prof. A. Ramakrishna Prasad all success in his future academic endeavours.

20/6/08

Prof.D.N.Reddy

About the Author

Dr. A. Ramakrishna Prasad is presently Professor and Head of the Department of Mathematics, Jawaharlal Nehru Technological University College of Engineering, Hyderabad and Chairman Board of Studies (Mathematics) of Jawaharlal Nehru Technological University, Hyderabad.

Having a passion for teaching, he made a modest beginning as a teaching assistant in 1980 and has been serving the Jawaharlal Nehru Technological University in various capacities since then. He has a vast teaching experience of over two decades and has been teaching varied subjects such as Engineering Mathematics, Information Systems, Database Management Systems, etc. and continues to explore different domains in Sciences and Engineering. He played a pioneering role in elevating the image of the university during his stint (2004–2008) as the Professor-in-charge of Training and Placement for United Jawaharlal Nehru Technological University, Hyderabad. He was instrumental in developing a healthy rapport with the industry and in helping the students in their career search. To cite a few, he was the co-coordinator for the BHEL-JNTU project and business associate for Wipro.

Besides vast experience in teaching, he is an able administrator and has held several key positions. He has published qualitative research papers in reputed journals. He is on the expert panel for many academic bodies like IGNOU and BROAU. He has been actively involved in efficient organization of numerous seminars, workshops and orientation programs. In collaboration with Honeywell, he successfully organized Nobel Laureates' visits to JNTUH University. After taking over as Head, Department of Mathematics at JNTUHCE, an exclusive website of the department was launched. To update the scholars and teachers of Mathematics, the department also started a newsletter. Sensitive to the needs and innovative trends in the academic arena, he continues to put in solemn efforts to make teaching as well as learning a meaningful and productive experience.

Dr. Prasad is an M.Tech (1992) in Computer Science and Data Processing from one of the most prestigious premier institutions, IIT Kharagpur. Subsequently, in 1995, he obtained PhD in Mathematics from SK University, Anantapur. He is also a life member of ISTE, CSI, NHRDN, IETE and Neuro-Linguistic Program Practitioners Club Florida (USA).

Dr. Prasad's areas of interest are Business Intelligence, Multidiscipline Optimization, Software Engineering and Social Networks.

Preface

It is a widely accepted notion that Mathematics is the foundation for all the disciplines of Engineering and Technology. Conceptual strength in Mathematics is crucial in understanding the basics of engineering. It is observed that the majority of textbooks available in the market lack clarity in explanation of the concepts and their applications and are confined to mere discussion of a little theory and problems. **Advanced Engineering Mathematics by Kreyszig** is the best-known prescribed textbook in almost all the universities in India and abroad. An attempt is made to fine-tune the components of this book, keeping in view the requirements of the undergraduate curriculum of JNTU.

This book is a modest effort to enhance the conceptual understanding of the learners and to improve the application capability in their respective branches of Engineering and Technology. This user-friendly textbook consists of relevant illustrations, concepts and additional information. It is also helpful for the teachers and researchers to develop qualitative mathematical ability. In tune with the needs of the students, adequate care has been taken to incorporate all the necessary topics. I sincerely hope that the student community and teaching fraternity would get the maximum benefit from this endeavor.

Comments and criticism are welcome from academicians and students for enrichment of the book in subsequent editions.

Acknowledgments

I am indebted to many of my teachers, colleagues and students who helped me directly or indirectly in preparing this book.

My first and foremost thanks are due to Prof. Kreyszig for granting me permission to utilize his book – Advanced Engineering Mathematics – as a source. My sincere thanks are due to Prof. D. N. Reddy, Vice-Chancellor, JNTUH and Prof. N. V. Ramana Rao, Principal, JNTUH College of Engineering, Hyderabad, for their continuous support and encouragement. Special cordial thanks go to my inspirer Prof. B. Krishna Gandhi, Vice-Chancellor, JNTUA.

My heartfelt thanks are due to my beloved wife, Maj. Anuha Prasad and son, Master Anish Prasad for their cooperation and love during the preparation of this book, without which I could not have completed this project.

Thanks are also due to my colleagues in the Department of Mathematics, and special thanks to Mr. Deepak, Mr. Naidu and Ms. Nirmala for their valuable assistance in this effort.

Furthermore, I wish to thank Wiley India Pvt. Ltd, Mr. Vikas Gupta, MD and Mr. Paras Bansal, Publisher and in particular their editorial team – Ms. Meenakshi, Ms. Mamta, Mr. Chandrabhan, Ms. Nupoor and all team members – for their sustained cooperation and great care in getting this book published within a record time.

A. Ramakrishna Prasad

Preface of the Ninth Edition

Goal of the Book. Arrangement of Material

This new edition continues the tradition of providing instructors and students with a comprehensive and up-to-date resource for teaching and learning ***engineering mathematics,*** that is, ***applied mathematics*** for engineers and physicists, mathematicians and computer scientists, as well as members of other disciplines. A course in elementary calculus is the sole ***prerequisite.***

The subject matter is arranged into seven parts A–G:

A. Ordinary Differential Equations (ODEs) (Chaps. 1–6)
B. Linear Algebra. Vector Calculus (Chaps. 7–9)
C. Fourier Analysis. Partial Differential Equations (PDEs) (Chaps. 11–12)
D. Complex Analysis (Chaps. 13–18)
E. Numeric Analysis (Chaps. 19–21)
F. Optimization, Graphs (Chaps. 22–23)
G. Probability, Statistics (Chaps. 24–25)

This is followed by five appendices:

App. 1. References (ordered by parts)
App. 2. Answers to Odd-Numbered Problems
App. 3. Auxiliary Material (see also inside covers)
App. 4. Additional Proofs
App. 5. Tables of Functions.

This book has helped to pave the way for the present development of engineering mathematics. By a modern approach to those areas A–G, this new edition will prepare the student for the tasks of the present and of the future. The latter can be predicted to some extent by a judicious look at the present trend. Among other features, this trend shows the appearance of more complex production processes, more extreme physical conditions (in space travel, high-speed communication, etc.), and new tasks in robotics and communication systems (e.g., fiber optics and scan statistics on random graphs) and elsewhere. This requires the refinement of existing methods and the creation of new ones.

It follows that students need solid knowledge of basic principles, methods, and results, and a clear view of what engineering mathematics is all about, and that it requires proficiency in all three phases of problem solving:

- **Modeling,** that is, translating a physical or other problem into a mathematical form, into a mathematical *model;* this can be an algebraic equation, a differential equation, a graph, or some other mathematical expression.
- **Solving** the model by selecting and applying a suitable mathematical method, often requiring numeric work on a computer.
- **Interpreting** the mathematical result in physical or other terms to see what it practically means and implies.

It would make no sense to overload students with all kinds of little things that might be of occasional use. Instead they should recognize that mathematics rests on relatively few basic concepts and involves powerful unifying principles. This should give them a firm grasp on the ***interrelations among theory, computing, and*** (physical or other) ***experimentation.***

PARTS AND CHAPTERS OF THE BOOK

PART A
Chaps. 1–6
Ordinary Differential Equations (ODEs)

- Chaps. 1–4 Basic Material
  - Chap. 5 Series Solutions
  - Chap. 6 Laplace Transforms

PART B
Chaps. 7–10
Linear Algebra. Vector Calculus

- Chap. 7 Matrices, Linear Systems
  - Chap. 8 Eigenvalue Problems
- Chap. 9 Vector Differential Calculus
  - Chap. 10 Vector Integral Calculus

PART C
Chaps. 11–12
Fourier Analysis. Partial Differential Equations (PDEs)

- Chap. 11 Fourier Analysis
  - Chap. 12 Partial Differential Equations

PART D
Chaps. 13–18
Complex Analysis, Potential Theory

- Chap. 13–17 Basic Material
  - Chap. 18 Potential Theory

PART E
Chaps. 19–21
Numeric Analysis

- Chap. 19 Numerics in General
- Chap. 20 Numeric Linear Algebra
- Chap. 21 Numerics for ODEs and PDEs

PART F
Chaps. 22–23
Optimization, Graphs

- Chap. 22 Linear Programming
- Chap. 23 Graphs, Optimization

PART G
Chaps. 24–25
Probability, Statistics

- Chap. 24 Data Analysis. Probability Theory
- Chap. 25 Mathematical Statistics

GUIDES AND MANUALS
Maple Computer Guide
Mathematica Computer Guide

- Student Solutions Manual
- Instructor's Manual

General Features of the Book Include:

- **Simplicity of examples,** to make the book teachable—why choose complicated examples when simple ones are as instructive or even better?
- **Independence of chapters,** to provide ***flexibility*** in tailoring courses to special needs.
- **Self-contained presentation,** except for a few clearly marked places where a proof would exceed the level of the book and a reference is given instead.
- **Modern standard notation,** to help students with other courses, *modern* books, and mathematical and engineering journals.

Many sections were rewritten in a more detailed fashion, to make it a ***simpler book***. This also resulted in a ***better balance between theory and applications***.

Use of Computers

The presentation is ***adaptable to various levels of technology and use of a computer or graphing calculator***: very little or no use, medium use, or intensive use of a graphing calculator or of an unspecified *CAS* (Computer Algebra System, ***Maple, Mathematica, or Matlab*** being popular examples). In either case texts and problem sets form an entity without gaps or jumps. And many problems can be solved by hand or with a computer or both ways. (For ***software,*** see the beginnings of *Part E* on Numeric Analysis and *Part G* on Probability and Statistics.)

More specifically, this new edition on the one hand gives more prominence to tasks the computer ***cannot*** do, notably, modeling and interpreting results. On the other hand, it includes ***CAS projects, CAS problems,*** and ***CAS experiments,*** which ***do require*** a computer and show its power in solving problems that are difficult or impossible to access otherwise. Here our goal is the combination of *intelligent* computer use with high-quality mathematics. This has resulted in a change from a formula-centered teaching and learning of engineering mathematics to a more quantitative, project-oriented, and visual approach. CAS experiments also exhibit the computer as an instrument for observations and experimentations that may become the beginnings of new research, for "proving" or disproving conjectures, or for formalizing empirical relationships that are often quite useful to the engineer as working guidelines. These changes will also help the student in discovering the ***experimental aspect of modern applied mathematics***.

Some ***routine and drill work*** is retained as a necessity for keeping firm contact with the subject matter. In some of it the computer can (but must not) give the student a hand, but there are plenty of problems that are more suitable for pencil-and-paper work.

Major Changes

1. **New Problem Sets.** Modern engineering mathematics is mostly ***teamwork***. It usually combines analytic work in the process of modeling and the use of computer algebra and numerics in the process of solution, followed by critical evaluation of results. Our problems—some straightforward, some more challenging, some "thinking problems" not accessible by a CAS, some open-ended—reflect this modern situation with its increased emphasis on qualitative methods and applications, and the problem sets take care of this novel situation by including team projects, CAS projects, and writing projects. The latter will also help the student in writing general reports, as they are required in engineering work quite frequently.
2. **Computer Experiments,** using the computer as an instrument of **"experimental mathematics"** for exploration and research (see also above). These are mostly open-ended experiments,

demonstrating the use of computers in experimentally finding results, which may be provable afterward or may be valuable heuristic qualitative guidelines to the engineer, in particular in complicated problems.

3. **More on modeling and selecting methods,** tasks that usually cannot be automated.
4. **Student Solutions Manual and Study Guide enlarged,** upon explicit requests of the users. This Manual contains worked-out solutions to carefully selected odd-numbered problems (to which App. 1 gives only the final answers) as well as general comments and hints on studying the text and working further problems, including explanations on the significance and character of concepts and methods in the various sections of the book.

Further Changes, New Features

- Electric circuits moved entirely to Chap. 2, to avoid duplication and repetition
- Second-order ODEs and Higher Order ODEs placed into two separate chaps. (2 and 3)
- In Chap. 2, applications presented before variation of parameters
- Series solutions somewhat shortened, without changing the order of sections
- Material on Laplace transforms brought into a better logical order: partial fractions used earlier in a more practical approach, unit step and Dirac's delta put into separate subsequent sections, differentiation and integration of transforms (not of functions!) moved to a later section in favor of practically more important topics
- Second- and third-order determinants made into a separate section for reference throughout the book
- Complex matrices made optional
- Three sections on curves and their application in mechanics combined in a single section
- First two sections on Fourier series combined to provide a better, more direct start
- Discrete and Fast Fourier Transforms included
- Conformal mapping presented in a separate chapter and enlarged
- Numeric analysis updated
- Backward Euler method included
- Stiffness of ODEs and systems discussed
- List of software (in Part E) updated; another list for statistics software added (in Part G)
- References updated, now including about 75 books published or reprinted after 1990

Suggestions for Courses: A Four-Semester Sequence

The material, when taken in sequence, is suitable for four consecutive semester courses, meeting 3–4 hours a week:

1st Semester. *ODEs* (Chaps. 1–5 or 6)
2nd Semester. *Linear Algebra. Vector Analysis* (Chaps. 7–10)
3rd Semester. *Complex Analysis* (Chaps. 13–18)
4th Semester. *Numeric Methods* (Chaps. 19–21)

Suggestions for Independent One-Semester Courses

The book is also suitable for various independent one-semester courses meeting 3 hours a week. For instance:

Introduction to ODEs (Chaps. 1–2, Sec. 21.1)
Laplace Transforms (Chap. 6)
Matrices and Linear Systems (Chaps. 7–8)
Vector Algebra and Calculus (Chaps. 9–10)
Fourier Series and PDEs (Chaps. 11-12, Secs. 21.4–21.7)
Introduction to Complex Analysis (Chaps. 13–17)
Numeric Analysis (Chaps. 19, 21)
Numeric Linear Algebra (Chap. 20)
Optimization (Chaps. 22–23)
Graphs and Combinatorial Optimization (Chap. 23)
Probability and Statistics (Chaps. 24–25)

Acknowledgments

I am indebted to many of my former teachers, colleagues, and students who helped me directly or indirectly in preparing this book, in particular, the present edition. I profited greatly from discussions with engineers, physicists, mathematicians, and computer scientists, and from their written comments. I want to mention particularly Y. Antipov, D. N. Buechler, S. L. Campbell, R. Carr, P. L. Chambre, V. F. Connolly, Z. Davis, J. Delany, J. W. Dettman, D. Dicker, L. D. Drager, D. Ellis, W. Fox, A. Goriely, R. B. Guenther, J. B. Handley, N. Harbertson, A. Hassen, V. W. Howe, H. Kuhn, G. Lamb, M. T. Lusk, H. B. Mann, I. Marx, K. Millet, J. D. Moore, W. D. Munroe, A. Nadim, B. S. Ng, J. N.Ong, Jr., D. Panagiotis, A. Plotkin, P. J. Pritchard, W. O. Ray, J. T. Scheick, L. F. Shampine, H. A. Smith, J. Todd, H. Unz, A. L. Villone, H. J. Weiss, A. Wilansky, C. H. Wilcox, H. Ya Fan, and A. D. Ziebur, all from the United States, Professors E. J. Norminton and R. Vaillancourt from Canada, and Professors H. Florian and H. Unger from Europe. I can offer here only an inadequate acknowledgment of my gratitude and appreciation.

Special cordial thanks go to Privatdozent Dr. M. Kracht and to Mr. Herbert Kreyszig, MBA, the coauthor of the Student Solutions Manual, who both checked the manuscript in all details and made numerous suggestions for improvements and helped me proofread the galley and page proofs.

Furthermore, I wish to thank John Wiley and Sons (see the list on p. iv) as well as GGS Information Services, in particular Mr. K. Bradley and Mr. J. Nystrom, for their effective cooperation and great care in preparing this new edition.

Suggestions of many readers worldwide were evaluated in preparing this edition. Further comments and suggestions for improving the book will be gratefully received

Erwin Kreyszig

Contents

1 Solution for Linear Systems

LEARNING OBJECTIVES

After reading this chapter, you will be able to understand the following:

- Basic concepts related to matrices: addition, multiplication, transposition.
- Various forms of matrices.
- Basic concepts of determinants and Cramer's rule.
- Gauss elimination and back substitution.
- Echelon form and row-reduced Echelon form.
- Rank of a matrix: Existence and uniqueness.
- Inverse of a matrix: Gauss–Jordan method and adjoint method.
- Consistent and inconsistent nonhomogeneous linear equations.
- Gauss–Seidel iteration method and Jacobi iteration.
- ILL conditioning.
- LU decomposition and various other methods.
- Tridiagonal systems and QR factorization.

In this chapter we introduce the basic concepts and rules of matrix and vector algebra as also the main application to **linear systems** (systems of linear equations). Calculations with matrices handle matrices as single objects, denote them by single letters, and calculate with them in a very compact form, almost as with numbers, so that matrix calculations constitute a powerful "mathematrical shorthand." Calculations with matrices and vectors are defined and explained. The discussion centers around linear systems, with a thorough discussion of Gauss elimination, the role of rank, the existence and uniqueness problem for solutions, and matrix inversion. This also includes determinants (Cramer's rule). Applications are considered throughout this chapter.

1.1 Introduction to Matrices

A **matrix** is a rectangular array of numbers (or functions) enclosed in brackets. These numbers (or functions) are called the **entries** (or sometimes the *elements*) of the matrix. For example,

$$\begin{bmatrix} 0.3 & 1 & -5 \\ 0 & -0.2 & 16 \end{bmatrix}, \quad \begin{bmatrix} a_{11} & a_{12} & a_{13} \\ a_{21} & a_{22} & a_{23} \\ a_{31} & a_{32} & a_{33} \end{bmatrix}, \quad \begin{bmatrix} e^{-x} & 2x^2 \\ e^{6x} & 4x \end{bmatrix}, \quad [a_1 \quad a_2 \quad a_3], \quad \begin{bmatrix} 4 \\ 1/2 \end{bmatrix} \tag{1.1}$$

are matrices. The first matrix has two **rows** (horizontal lines of entries) and three **columns** (vertical lines). The second and third matrices are **square matrices,** that is, each has as many rows as columns (3 and 2, respectively). The entries of the second matrix have two indices giving the location of the entry. The first index is the number of the row and the second is the number of the column in which the entry stands. Thus, a_{23} (read *a two three*) is in Row 2 and Column 3, etc. This notation is standard, regardless of whether a matrix is square or not.

Matrices having just a single row or column are called **vectors.** Thus the fourth matrix in (1.1) has just one row and is called a **row vector.** The last matrix in (1.1) has just one column and is called a **column vector.**

We shall see that matrices are practical in various applications for storing and processing data. As a first illustration let us consider two simple but typical examples.

Example 1 Linear Systems, A Major Application of Matrices

In a system of linear equations, briefly called a **linear system**, such as

$$\begin{aligned} 4x_1 + 6x_2 + 9x_3 &= 6 \\ 6x_1 \qquad - 2x_3 &= 20 \\ 5x_1 - 8x_2 + x_3 &= 10 \end{aligned}$$

the coefficients of the **unknowns** x_1, x_2, x_3 are the entries of the **coefficient matrix**, say **A**,

$$\mathbf{A} = \begin{bmatrix} 4 & 6 & 9 \\ 6 & 0 & -2 \\ 5 & -8 & 1 \end{bmatrix}$$

The matrix

$$\tilde{\mathbf{A}} = \begin{bmatrix} 4 & 6 & 9 & 6 \\ 6 & 0 & -2 & 20 \\ 5 & -8 & 1 & 10 \end{bmatrix}$$

is obtained by augmenting **A** by the right sides of the linear system and is called the **augmented matrix** of the system (also denoted as [A|B]). In **A** the coefficients of the system are displayed in the pattern of the equations. That is, their position in **A** corresponds to that in the system when written as shown. The same is true for $\tilde{\mathbf{A}}$.

We shall see that the augmented matrix $\tilde{\mathbf{A}}$ contains all the information about the solutions of a system, so that we can solve a system just by calculations on its augmented matrix. Meanwhile you may verify by substitution that the solution is $x_1 = 3$, $x_2 = 1/2$, $x_3 = -1$. The notation x_1, x_2, x_3 for the unknowns is practical but not essential; we could choose x, y, z or some other letters.

Example 2 Sales Figures in Matrix Form

Sales figures for three products I, II, III in a store on Monday (M), Tuesday (T),... may for each week be arranged in a matrix

$$\mathbf{A} = \begin{array}{c} \\ \\ \end{array}\begin{array}{cccccc} M & T & W & Th & F & S \end{array}$$

$$\mathbf{A} = \begin{bmatrix} 400 & 330 & 810 & 0 & 210 & 470 \\ 0 & 120 & 780 & 500 & 500 & 960 \\ 100 & 0 & 0 & 270 & 430 & 780 \end{bmatrix} \begin{matrix} \text{I} \\ \text{II} \\ \text{III} \end{matrix}$$

If the company has 10 stores, we can set up 10 such matrices, one for each store. Then by adding corresponding entries of these matrices we can get a matrix showing the total sales of each product on each day. Can you think of other data for which matrices are feasible? For instance, in transportation or storage problems? Or in recording phone calls, or in listing distances in a network of roads?

General Concepts and Notations

We shall denote matrices by capital boldface letters **A**, **B**, **C**, …, or by writing the general entry in brackets; thus $\mathbf{A} = [a_{jk}]$, and so on. By an **$m \times n$ matrix** (read *m by n matrix*) we mean a matrix with m rows and n columns – rows come always first! $m \times n$ is called the **size** of the matrix. Thus an $m \times n$ matrix is of the form

$$\mathbf{A} = [a_{jk}] = \begin{bmatrix} a_{11} & a_{12} & \cdots & a_{1n} \\ a_{21} & a_{22} & \cdots & a_{2n} \\ \cdot & \cdot & \cdots & \cdot \\ a_{m1} & a_{m2} & \cdots & a_{mn} \end{bmatrix} \tag{1.2}$$

The matrices in (1.1) are of sizes 2×3, 3×3, 2×2, 1×3, and 2×1, respectively. Each entry in (1.2) has two subscripts. The first is the *row number* and the second is the *column number*. Thus a_{21} is the entry in Row 2 and Column 1. If $m = n$, we call **A** an $n \times n$ **square matrix.** Then its diagonal containing the entries $a_{11}, a_{22}, \ldots, a_{nn}$ is called the **main diagonal** of **A.** Thus the main diagonals of the two square matrices in (1.1) are a_{11}, a_{22}, a_{33} and e^{-x}, $4x$, respectively. Square matrices are particularly important, as we shall see. A matrix that is not square is called a **rectangular matrix.**

Vectors

A **vector** is a matrix with only one row or column. Its entries are called the **components** of the vector. We shall denote vectors by ***lowercase*** boldface letters **a**, **b**, … or by its general component in brackets, $\mathbf{a} = [a_j]$, and so on. Our special vectors in (1.1) suggest that a (general) **row vector** is of the form

$$\mathbf{a} = [a_1 \quad a_2 \quad \cdots \quad a_n]$$

For instance,

$$\mathbf{a} = [-2 \quad 5 \quad 0.8 \quad 0 \quad 1]$$

A **column vector** is of the form

$$\mathbf{b} = \begin{bmatrix} b_1 \\ b_2 \\ \vdots \\ b_m \end{bmatrix}$$

For instance,

$$\mathbf{b} = \begin{bmatrix} 4 \\ 0 \\ -7 \end{bmatrix}$$

Matrix Addition and Scalar Multiplication

What makes matrices and vectors really useful and particularly suitable for computers is the fact that we can calculate with them almost as easily as with numbers. Indeed, we now introduce rules for addition and for scalar multiplication (multiplication by numbers) that were suggested by practical applications. We first need the concept of equality.

DEFINITION **(Equality of Matrices):** Two matrices $\mathbf{A} = [a_{jk}]$ and $\mathbf{B} = [b_{jk}]$ are **equal,** written $\mathbf{A} = \mathbf{B}$, if and only if they have the same size and the corresponding entries are equal, that is, $a_{11} = b_{11}$, $a_{12} = b_{12}$, and so on. Matrices that are not equal are called **different**. Thus, matrices of different sizes are always different.

Example 3 Equality of Matrices

Let

$$\mathbf{A} = \begin{bmatrix} a_{11} & a_{12} \\ a_{21} & a_{22} \end{bmatrix} \quad \text{and} \quad \mathbf{B} = \begin{bmatrix} 4 & 0 \\ 3 & -1 \end{bmatrix}$$

Then $\mathbf{A} = \mathbf{B}$ if and only if $a_{11} = 4$, $a_{12} = 0$, $a_{21} = 3$, $a_{22} = -1$.

The following matrices are all different. Explain!

$$\begin{bmatrix} 1 & 3 \\ 4 & 2 \end{bmatrix} \quad \begin{bmatrix} 4 & 2 \\ 1 & 3 \end{bmatrix} \quad \begin{bmatrix} 4 & 1 \\ 2 & 3 \end{bmatrix} \quad \begin{bmatrix} 1 & 3 & 0 \\ 4 & 2 & 0 \end{bmatrix} \quad \begin{bmatrix} 0 & 1 & 3 \\ 0 & 4 & 2 \end{bmatrix}$$

DEFINITION **(Addition of Matrices):** The **sum** of two matrices $\mathbf{A} = [a_{jk}]$ and $\mathbf{B} = [b_{jk}]$ ***of the same size*** is written $\mathbf{A} + \mathbf{B}$ and has the entries $a_{jk} + b_{jk}$ obtained by adding the corresponding entries of $\mathbf{A}$ and $\mathbf{B}$. Matrices of different sizes cannot be added.

As a special case, the **sum a + b** of two row vectors or two column vectors, which must have the same number of components, is obtained by adding the corresponding components.

Example 4 Addition of Matrices and Vectors

If

$$\mathbf{A} = \begin{bmatrix} -4 & 6 & 3 \\ 0 & 1 & 2 \end{bmatrix} \quad \text{and} \quad \mathbf{B} = \begin{bmatrix} 5 & -1 & 0 \\ 3 & 1 & 0 \end{bmatrix}$$

then

$$\mathbf{A} + \mathbf{B} = \begin{bmatrix} 1 & 5 & 3 \\ 3 & 2 & 2 \end{bmatrix}$$

$\mathbf{A}$ in Example 3 and our present $\mathbf{A}$ cannot be added. If $\mathbf{a} = [5 \quad 7 \quad 2]$ and $\mathbf{b} = [-6 \quad 2 \quad 0]$, then $\mathbf{a} + \mathbf{b} = [-1 \quad 9 \quad 2]$. An application of matrix addition was suggested in Example 2. Many others will follow.

DEFINITION **[Scalar Multiplication (Multiplication by a Number)]:** The **product** of any $m \times n$ matrix $\mathbf{A} = [a_{jk}]$ and any **scalar** c (number c) is written $c\mathbf{A}$ and is the $m \times n$ matrix $c\mathbf{A} = [ca_{jk}]$ obtained by multiplying each entry of $\mathbf{A}$ by c.

Here $(-1)\mathbf{A}$ is simply written $-\mathbf{A}$ and is called the **negative** of $\mathbf{A}$. Similarly, $(-k)\mathbf{A}$ is written $-k\mathbf{A}$. Also, $\mathbf{A} + (-\mathbf{B})$ is written $\mathbf{A} - \mathbf{B}$ and is called the **difference** of $\mathbf{A}$ and $\mathbf{B}$ (which must have the same size!).

Example 5 Scalar Multiplication

If $\mathbf{A} = \begin{bmatrix} 2.7 & -1.8 \\ 0 & 0.9 \\ 9.0 & -4.5 \end{bmatrix}$, then $-\mathbf{A} = \begin{bmatrix} -2.7 & 1.8 \\ 0 & -0.9 \\ -9.0 & 4.5 \end{bmatrix}$, $\frac{10}{9}\mathbf{A} = \begin{bmatrix} 3 & -2 \\ 0 & 1 \\ 10 & -5 \end{bmatrix}$, $0\mathbf{A} = \begin{bmatrix} 0 & 0 \\ 0 & 0 \\ 0 & 0 \end{bmatrix}$

If a matrix **B** shows the distances between some cities in miles, 1.609**B** gives these distances in kilometers.

Rules for Matrix Addition and Scalar Multiplication

From the familiar laws for the addition of numbers we obtain similar laws for the addition of matrices of the same size $m \times n$, namely,

$$\begin{aligned} &\text{(a)} \quad \mathbf{A}+\mathbf{B}=\mathbf{B}+\mathbf{A} \\ &\text{(b)} \quad (\mathbf{A}+\mathbf{B})+\mathbf{C}=\mathbf{A}+(\mathbf{B}+\mathbf{C}) \\ &\text{(c)} \quad \mathbf{A}+\mathbf{0}=\mathbf{A} \qquad \text{(written } \mathbf{A}+\mathbf{B}+\mathbf{C}) \\ &\text{(d)} \quad \mathbf{A}+(-\mathbf{A})=\mathbf{0} \end{aligned} \tag{1.3}$$

Here **0** denotes the **zero matrix** (of size $m \times n$), that is, the $m \times n$ matrix with all entries zero. (The last matrix in Example 5 is a zero matrix.) Hence matrix addition is *commutative* and *associative* [by (1.3a) and (1.3b)]. Similarly, for scalar multiplication we obtain the rules

$$\begin{aligned} &\text{(a)} \quad c(\mathbf{A}+\mathbf{B})=c\mathbf{A}+c\mathbf{B} \\ &\text{(b)} \quad (c+k)\mathbf{A}=c\mathbf{A}+k\mathbf{A} \\ &\text{(c)} \quad c(k\mathbf{A})=(ck)\mathbf{A} \qquad \text{(written } ck\mathbf{A}) \\ &\text{(d)} \quad 1\mathbf{A}=\mathbf{A} \end{aligned} \tag{1.4}$$

Matrix Multiplication

Matrix multiplication means multiplication of matrices by matrices. This is the last algebraic operation to be defined (except for *transposition,* which is of lesser importance). Now matrices are *added* by *adding* corresponding entries. In *multiplication,* do we *multiply* corresponding entries? The answer is no. Why not? Such an operation would not be of much use in applications. The standard definition of multiplication looks artificial, but will be fully motivated later in this section by the use of matrices in "linear transformations," by which this multiplication is suggested.

DEFINITION **(Multiplication of a Matrix by a Matrix):** The **product C** = **AB** (in this order) of an $m \times n$ matrix $\mathbf{A} = [a_{jk}]$ times an $r \times p$ matrix $\mathbf{B} = [b_{jk}]$ is defined if and only if $r = n$ and is then the $m \times p$ matrix $\mathbf{C} = [c_{jk}]$ with entries

$$c_{jk} = \sum_{l=1}^{n} a_{jl}b_{lk} = a_{j1}b_{1k} + a_{j2}b_{2k} + \cdots + a_{jn}b_{nk} \quad (j=1,\ldots,m;\ k=1,\ldots,p) \tag{1.5}$$

The condition $r = n$ means that the second factor, **B**, must have as many rows as the first factor has columns, namely n. As a diagram of sizes (denoted as shown):

$$\begin{matrix} \mathbf{A} & \mathbf{B} & = & \mathbf{C} \\ [m \times n] & [n \times r] & & [m \times r] \end{matrix}$$

c_{jk} in (1.5) is obtained by multiplying each entry in the jth row of **A** by the corresponding entry in the kth column of **B** and then adding these n products. For instance, $c_{21} = a_{21}b_{11} + a_{22}b_{21} + \cdots + a_{2n}b_{n1}$, and so on. One calls this briefly a "***multiplication of rows into columns.***" See the illustration in Figure 1 where $n = 3$.

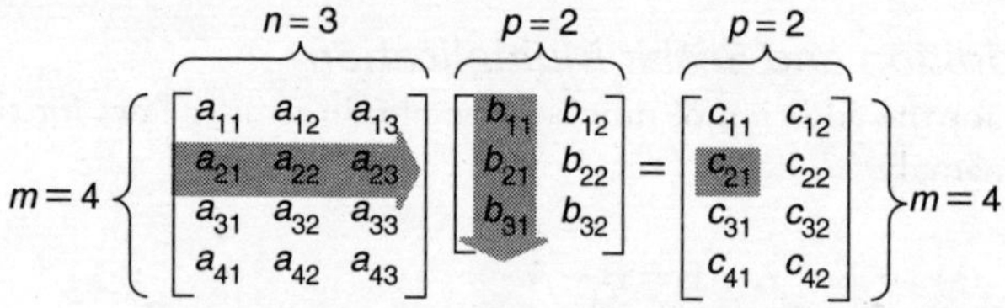

Figure 1 Notations in a product **AB** = **C**.

Example 6 Matrix Multiplication

$$\mathbf{AB} = \begin{bmatrix} 3 & 5 & -1 \\ 4 & 0 & 2 \\ -6 & -3 & 2 \end{bmatrix} \begin{bmatrix} 2 & -2 & 3 & 1 \\ 5 & 0 & 7 & 8 \\ 9 & -4 & 1 & 1 \end{bmatrix} = \begin{bmatrix} 22 & -2 & 43 & 42 \\ 26 & -16 & 14 & 6 \\ -9 & 4 & -37 & -28 \end{bmatrix}$$

Here $c_{11} = 3 \times 2 + 5 \times 5 + (-1) \times 9 = 22$, and so on. The entry in the box is $c_{23} = 4 \times 3 + 0 \times 7 + 2 \times 1 = 14$. The product **BA** is not defined.

Example 7 Multiplication of a Matrix and a Vector

$$\begin{bmatrix} 4 & 2 \\ 1 & 8 \end{bmatrix} \begin{bmatrix} 3 \\ 5 \end{bmatrix} = \begin{bmatrix} 4 \times 3 + 2 \times 5 \\ 1 \times 3 + 8 \times 5 \end{bmatrix} = \begin{bmatrix} 22 \\ 43 \end{bmatrix} \quad \text{whereas} \quad \begin{bmatrix} 3 \\ 5 \end{bmatrix} \begin{bmatrix} 4 & 2 \\ 1 & 8 \end{bmatrix} \text{ is undefined}$$

Example 8 Products of Row and Column Vectors

$$[3 \quad 6 \quad 1] \begin{bmatrix} 1 \\ 2 \\ 4 \end{bmatrix} = [19], \quad \begin{bmatrix} 1 \\ 2 \\ 4 \end{bmatrix} [3 \quad 6 \quad 1] = \begin{bmatrix} 3 & 6 & 1 \\ 6 & 12 & 2 \\ 12 & 24 & 4 \end{bmatrix}$$

Example 9 **CAUTION! Matrix Multiplication is not Commutative, AB ≠ BA in General**

This is illustrated by Examples 6 and 7, where one of the two products is not even defined, and by Example 8, where the two products have different sizes. But it also holds for square matrices. For instance,

$$\begin{bmatrix} 1 & 1 \\ 100 & 100 \end{bmatrix}\begin{bmatrix} -1 & 1 \\ 1 & -1 \end{bmatrix}=\begin{bmatrix} 0 & 0 \\ 0 & 0 \end{bmatrix} \quad \text{but} \quad \begin{bmatrix} -1 & 1 \\ 1 & -1 \end{bmatrix}\begin{bmatrix} 1 & 1 \\ 100 & 100 \end{bmatrix}=\begin{bmatrix} 99 & 99 \\ -99 & -99 \end{bmatrix}$$

It is interesting that this also shows that $\mathbf{AB} = \mathbf{0}$ does ***not*** necessarily imply $\mathbf{BA} = \mathbf{0}$ or $\mathbf{A} = \mathbf{0}$ or $\mathbf{B} = \mathbf{0}$.

Our examples show that the ***order of factors*** in matrix products ***must always be observed very carefully.*** Otherwise matrix multiplication satisfies rules similar to those for numbers, namely.

$$\begin{aligned}
&\text{(a)} \quad (k\mathbf{A})\mathbf{B} = k(\mathbf{AB}) = \mathbf{A}(k\mathbf{B}) \quad &&\text{written } k\mathbf{AB} \text{ or } \mathbf{A}k\mathbf{B} \\
&\text{(b)} \quad \mathbf{A}(\mathbf{BC}) = (\mathbf{AB})\mathbf{C} \quad &&\text{written } \mathbf{ABC} \\
&\text{(c)} \quad (\mathbf{A}+\mathbf{B})\mathbf{C} = \mathbf{AC}+\mathbf{BC} \\
&\text{(d)} \quad \mathbf{C}(\mathbf{A}+\mathbf{B}) = \mathbf{CA}+\mathbf{CB}
\end{aligned} \tag{1.6}$$

provided **A**, **B**, and **C** are such that the expressions on the left are defined; here, k is any scalar. Equation (1.6b) is called the **associative law.** Equations (1.6c) and (1.6d) are called the **distributive laws.** Since matrix multiplication is a multiplication of rows into columns, we can write the defining formula (1.5) more compactly as

$$c_{jk} = \mathbf{a}_j \mathbf{b}_k, \quad j = 1, \dots, m, \quad k = 1, \dots, p \tag{1.7}$$

where $\mathbf{a}_j$ is the jth row vector of **A** and $\mathbf{b}_k$ is the kth column vector of **B**, so that in agreement with (1.5),

$$\mathbf{a}_j \mathbf{b}_k = [a_{j1} \quad a_{j2} \quad \cdots \quad a_{jn}]\begin{bmatrix} b_{1k} \\ \vdots \\ b_{nk} \end{bmatrix} = a_{j1}b_{1k} + a_{j2}b_{2k} + \cdots + a_{jn}b_{nk}$$

Example 10 **Product in Terms of Row and Column Vecto**

If $\mathbf{A} = [a_{jk}]$ is of size 3×3 and $\mathbf{B} = [b_{jk}]$ is of size 3×4, then

$$\mathbf{AB} = \begin{bmatrix} \mathbf{a}_1\mathbf{b}_1 & \mathbf{a}_1\mathbf{b}_2 & \mathbf{a}_1\mathbf{b}_3 & \mathbf{a}_1\mathbf{b}_4 \\ \mathbf{a}_2\mathbf{b}_1 & \mathbf{a}_2\mathbf{b}_2 & \mathbf{a}_2\mathbf{b}_3 & \mathbf{a}_2\mathbf{b}_4 \\ \mathbf{a}_3\mathbf{b}_1 & \mathbf{a}_3\mathbf{b}_2 & \mathbf{a}_3\mathbf{b}_3 & \mathbf{a}_3\mathbf{b}_4 \end{bmatrix} \tag{1.8}$$

Taking $\mathbf{a}_1 = [3 \quad 5 \quad -1]$, $\mathbf{a}_2 = [4 \quad 0 \quad 2]$, etc., verify (1.8) for the product in Example 6.

Parallel processing of products on the computer is facilitated by a variant of (1.7) for computing $\mathbf{C} = \mathbf{AB}$, which is used by standard algorithms (such as in Lapack). In this method, **A** is used as given, **B** is taken in terms of its column vectors, and the product is computed columnwise; thus,

$$\mathbf{AB} = \mathbf{A}[\mathbf{b}_1 \quad \mathbf{b}_2 \quad \cdots \quad \mathbf{b}_p] = [\mathbf{Ab}_1 \quad \mathbf{Ab}_2 \quad \cdots \quad \mathbf{Ab}_p] \tag{1.9}$$

Columns of **B** are then assigned to different processors (individually or several to each processor), which simultaneously compute the columns of the product matrix $\mathbf{Ab}_1$, $\mathbf{Ab}_2$, etc.

Example 11 Computing Products Columnwise by (1.9)

To obtain

$$\mathbf{AB} = \begin{bmatrix} 4 & 1 \\ -5 & 2 \end{bmatrix} \begin{bmatrix} 3 & 0 & 7 \\ -1 & 4 & 6 \end{bmatrix} = \begin{bmatrix} 11 & 4 & 34 \\ -17 & 8 & -23 \end{bmatrix}$$

from (1.9), calculate the columns

$$\begin{bmatrix} 4 & 1 \\ -5 & 2 \end{bmatrix} \begin{bmatrix} 3 \\ -1 \end{bmatrix} = \begin{bmatrix} 11 \\ -17 \end{bmatrix}, \quad \begin{bmatrix} 4 & 1 \\ -5 & 2 \end{bmatrix} \begin{bmatrix} 0 \\ 4 \end{bmatrix} = \begin{bmatrix} 4 \\ 8 \end{bmatrix}, \quad \begin{bmatrix} 4 & 1 \\ -5 & 2 \end{bmatrix} \begin{bmatrix} 7 \\ 6 \end{bmatrix} = \begin{bmatrix} 34 \\ -23 \end{bmatrix}$$

of **AB** and then write them as a single matrix, as shown in the first formula on the right.

Motivation of Multiplication by Linear Transformations

Let us now motivate the "unnatural" matrix multiplication by its use in **linear transformations.** For $n = 2$ variables these transformations are of the form

$$\begin{aligned} y_1 &= a_{11}x_1 + a_{12}x_2 \\ y_2 &= a_{21}x_1 + a_{22}x_2 \end{aligned} \tag{1.10}$$

and suffice to explain the idea. For instance, (1.10) may relate an x_1x_2-coordinate system to a y_1y_2-coordinate system in the plane. In vectorial form we can write (1.10) as

$$\mathbf{y} = \begin{bmatrix} y_1 \\ y_2 \end{bmatrix} = \mathbf{Ax} = \begin{bmatrix} a_{11} & a_{12} \\ a_{21} & a_{22} \end{bmatrix} \begin{bmatrix} x_1 \\ x_2 \end{bmatrix} = \begin{bmatrix} a_{11}x_1 + a_{12}x_2 \\ a_{21}x_1 + a_{22}x_2 \end{bmatrix} \tag{1.11}$$

Now suppose further that the x_1x_2-system is related to a w_1w_2-system by another linear transformation, say,

$$\mathbf{x} = \begin{bmatrix} x_1 \\ x_2 \end{bmatrix} = \mathbf{Bw} = \begin{bmatrix} b_{11} & b_{12} \\ b_{21} & b_{22} \end{bmatrix} \begin{bmatrix} w_1 \\ w_2 \end{bmatrix} = \begin{bmatrix} b_{11}w_1 + b_{12}w_2 \\ b_{21}w_1 + b_{22}w_2 \end{bmatrix} \tag{1.12}$$

Then the y_1y_2-system is related to the w_1w_2-system indirectly via the x_1x_2-system, and we wish to express this relation directly. Substitution will show that this direct relation is a linear transformation, too, say,

$$\mathbf{y} = \mathbf{Cw} = \begin{bmatrix} c_{11} & c_{12} \\ c_{21} & c_{22} \end{bmatrix} \begin{bmatrix} w_1 \\ w_2 \end{bmatrix} = \begin{bmatrix} c_{11}w_1 + c_{12}w_2 \\ c_{21}w_1 + c_{22}w_2 \end{bmatrix} \tag{1.13}$$

Indeed, substituting (1.12) into (1.11), we obtain

$$\begin{aligned} y_1 &= a_{11}(b_{11}w_1 + b_{12}w_2) + a_{12}(b_{21}w_1 + b_{22}w_2) \\ &= (a_{11}b_{11} + a_{12}b_{21})w_1 + (a_{11}b_{12} + a_{12}b_{22})w_2 \\ y_2 &= a_{21}(b_{11}w_1 + b_{12}w_2) + a_{22}(b_{21}w_1 + b_{22}w_2) \end{aligned}$$

$$= (a_{21}b_{11} + a_{22}b_{21})w_1 + (a_{21}b_{12} + a_{22}b_{22})w_2$$

Comparing this with (1.13), we see that

$$c_{11} = a_{11}b_{11} + a_{12}b_{21} \qquad c_{12} = a_{11}b_{12} + a_{12}b_{22}$$

$$c_{21} = a_{21}b_{11} + a_{22}b_{21} \qquad c_{22} = a_{21}b_{12} + a_{22}b_{22}$$

This proves that $\mathbf{C} = \mathbf{AB}$ with the product defined as in (1.5). For larger matrix sizes the idea and result are exactly the same. Only the number of variables changes. We then have m variables y and n variables x and p variables w. The matrices **A**, **B**, and $\mathbf{C} = \mathbf{AB}$ then have sizes $m \times n$, $n \times p$, and $m \times p$, respectively. And the requirement that **C** be the product **AB** leads to formula (1.5) in its general form. *This motivates matrix multiplication completely.*

Transposition

Transposition provides a transition from row vectors to column vectors and conversely. More generally, it gives us a choice to work either with a matrix or with its transpose, whatever will be more practical in a specific situation.

DEFINITION **(Transposition of Matrices and Vectors):** The transpose of an $m \times n$ matrix $\mathbf{A} = [a_{jk}]$ is the $n \times m$ matrix $\mathbf{A}^\mathrm{T}$ (read *A transpose*) that has the first *row* of **A** as its first *column*, the second *row* of **A** as its second *column*, and so on. Thus the transpose of **A** in (1.6) is $\mathbf{A}^\mathrm{T} = [a_{kj}]$, written out

$$\mathbf{A}^\mathrm{T} = [a_{kj}] = \begin{bmatrix} a_{11} & a_{21} & \cdots & a_{m1} \\ a_{12} & a_{22} & \cdots & a_{m2} \\ \cdot & \cdot & \cdots & \cdot \\ a_{1n} & a_{2n} & \cdots & a_{mn} \end{bmatrix} \tag{1.14}$$

As a special case, transposition converts row vectors to column vectors and conversely.

Example 12 Transposition of Matrices and Vectors

If

$$\mathbf{A} = \begin{bmatrix} 5 & -8 & 1 \\ 4 & 0 & 0 \end{bmatrix}$$

then

$$\mathbf{A}^\mathrm{T} = \begin{bmatrix} 5 & 4 \\ -8 & 0 \\ 1 & 0 \end{bmatrix}$$

A little more compactly, we can write

$$\begin{bmatrix} 5 & -8 & 1 \\ 4 & 0 & 0 \end{bmatrix}^\mathrm{T} = \begin{bmatrix} 5 & 4 \\ -8 & 0 \\ 1 & 0 \end{bmatrix}, \quad \begin{bmatrix} 3 & 0 \\ 8 & -1 \end{bmatrix}^\mathrm{T} = \begin{bmatrix} 3 & 8 \\ 0 & -1 \end{bmatrix}$$

$$[6 \quad 2 \quad 3]^{T} = \begin{bmatrix} 6 \\ 2 \\ 3 \end{bmatrix}, \qquad \begin{bmatrix} 6 \\ 2 \\ 3 \end{bmatrix}^{T} = [6 \quad 2 \quad 3]$$

Note that for a square matrix, the transpose is obtained by interchanging entries that are symmetrically positioned with respect to the main diagonal, e.g., a_{12} and a_{21}, and so on.

Rules for transposition are

$$\begin{aligned} &\text{(a)} && (\mathbf{A}^{T})^{T} = \mathbf{A} \\ &\text{(b)} && (\mathbf{A}+\mathbf{B})^{T} = \mathbf{A}^{T}+\mathbf{B}^{T} \\ &\text{(c)} && (c\mathbf{A})^{T} = c\mathbf{A}^{T} \\ &\text{(d)} && (\mathbf{AB})^{T} = \mathbf{B}^{T}\mathbf{A}^{T} \end{aligned} \tag{1.15}$$

CAUTION! Note that in (1.15) the transposed matrices are ***in reversed order.*** We leave the proofs to the student.

Special Matrices

Certain kinds of matrices will occur quite frequently in our work, and we now list the most important ones of them.

Symmetric and Skew-Symmetric Matrices

Transposition gives rise to two useful classes of matrices, as follows. *Symmetric matrices* and *skew-symmetric matrices* are ***square*** matrices whose transpose equals the matrix itself or minus the matrix, respectively:

Symmetric Matrix

$$\mathbf{A}^{T} = \mathbf{A} \qquad (\text{thus } a_{kj} = a_{jk}) \tag{1.16}$$

Skew-Symmetric Matrix

$$\mathbf{A}^{T} = -\mathbf{A} \qquad (\text{thus } a_{kj} = -a_{jk}, \text{ hence } a_{jj} = 0) \tag{1.17}$$

Example 13 Symmetric and Skew-Symmetric Matrices

$$\mathbf{A} = \begin{bmatrix} 20 & 120 & 200 \\ 120 & 10 & 150 \\ 200 & 150 & 30 \end{bmatrix} \text{ is symmetric and } \mathbf{B} = \begin{bmatrix} 0 & 1 & -3 \\ -1 & 0 & -2 \\ 3 & 2 & 0 \end{bmatrix} \text{ is skew-symmetric.}$$

For instance, if a company has three building supply centers C_1, C_2, C_3, then **A** could show costs, say, a_{jj} for handling 1000 bags of cement on center C_j, and a_{jk} $(j \neq k)$ the cost of shipping 1000 bags from C_j to C_k. Clearly, $a_{jk} = a_{kj}$ because shipping in the opposite direction will usually cost the same.

Symmetric matrices have several general properties which make them important. This will be seen as we proceed.

Triangular Matrices

Upper triangular matrices are square matrices that can have nonzero entries only on and ***above*** the main diagonal, whereas any entry below the diagonal must be zero. Similarly, **lower triangular matrices** can have nonzero entries only on and ***below*** the main diagonal. Any entry on the main diagonal of a triangular matrix may be zero or not.

Orthogonal Matrices

These are those matrices where transposition gives the inverse, that is

$$A^T = A^{-1}$$

Example 14 Orthogonal Matrix

$$\begin{bmatrix} -2/3 & 1/3 & 2/3 \\ 2/3 & 2/3 & 1/3 \\ 1/3 & -2/3 & 2/3 \end{bmatrix}$$

Example 15 Upper and Lower Triangular Matrices

$$\begin{bmatrix} 1 & 3 \\ 0 & 2 \end{bmatrix}, \begin{bmatrix} 1 & 4 & 2 \\ 0 & 3 & 2 \\ 0 & 0 & 6 \end{bmatrix}, \begin{bmatrix} 2 & 0 & 0 \\ 8 & -1 & 0 \\ 7 & 6 & 8 \end{bmatrix}, \begin{bmatrix} 3 & 0 & 0 & 0 \\ 9 & -3 & 0 & 0 \\ 1 & 0 & 2 & 0 \\ 1 & 9 & 3 & 6 \end{bmatrix}$$

Upper triangular Lower triangular

Diagonal Matrices

These are square matrices that can have nonzero entries only on the main diagonal. Any entry above or below the main diagonal must be zero. If all the diagonal entries of a diagonal matrix S are equal, say, c, we call S a **scalar matrix** because multiplication of any square matrix **A** of the same size by S has the same effect as the multiplication by a scalar, that is,

$$AS = SA = cA \tag{1.18}$$

In particular, a scalar matrix whose entries on the main diagonal are all 1 is called a **unit matrix** (or **identity matrix**) and is denoted by $\mathbf{I}_n$ or simply by **I**. For **I**, formula (1.18) becomes

$$AI = IA = A \tag{1.19}$$

Example 16 Diagonal Matrix D. Scalar Matrix S. Unit Matrix I

$$D = \begin{bmatrix} 2 & 0 & 0 \\ 0 & -3 & 0 \\ 0 & 0 & 0 \end{bmatrix}, \quad S = \begin{bmatrix} c & 0 & 0 \\ 0 & c & 0 \\ 0 & 0 & c \end{bmatrix}, \quad I = \begin{bmatrix} 1 & 0 & 0 \\ 0 & 1 & 0 \\ 0 & 0 & 1 \end{bmatrix}$$

Applications of Matrix Multiplication

Matrix multiplication will play a crucial role in connection with linear systems of equations, beginning in the next section. For the time being we mention some other simple applications that need no lengthy explanations.

Example 17 Computer Production. Matrix Times Matrix

Supercomp Ltd. produces two computer models PC1086 and PC1186. The matrix **A** shows the cost per computer (in thousands of dollars) and **B** the production figures for the year 2005 (in multiples of 10000 units). Find a matrix **C** that shows the shareholders the cost per quarter (in millions of dollars) for raw material, labor, and miscellaneous.

$$\begin{array}{c} \\ \mathbf{A} = \end{array} \begin{array}{c} \text{PC1086} \quad \text{PC1186} \\ \begin{bmatrix} 1.2 & 1.6 \\ 0.3 & 0.4 \\ 0.5 & 0.6 \end{bmatrix} \end{array} \begin{array}{l} \\ \text{Raw Components} \\ \text{Labor} \\ \text{Miscellaneous} \end{array} \qquad \mathbf{B} = \overset{\text{Quarter}}{\overset{1 \quad 2 \quad 3 \quad 4}{\begin{bmatrix} 3 & 8 & 6 & 9 \\ 6 & 2 & 4 & 3 \end{bmatrix}}} \begin{array}{l} \text{PC1086} \\ \text{PC1186} \end{array}$$

Solution:

$$\mathbf{C} = \mathbf{AB} = \overset{\text{Quarter}}{\overset{1 \qquad 2 \qquad 3 \qquad 4}{\begin{bmatrix} 13.2 & 12.8 & 13.6 & 15.6 \\ 3.3 & 3.2 & 3.4 & 3.9 \\ 5.1 & 5.2 & 5.4 & 6.3 \end{bmatrix}}} \begin{array}{l} \text{Raw Components} \\ \text{Labor} \\ \text{Miscellaneous} \end{array}$$

Since cost is given in multiples of \$1000 and production in multiples of 10000 units, the entries of **C** are multiples of \$10 millions; thus $c_{11} = 13.2$ means \$132 million, etc.

Example 18 Weight Watching. Matrix Times Vector

Suppose that in a weight-watching program, a person of 185 lb burns 350 cal/hr in walking (3 mph), 500 in bycycling (13 mph) and 950 in jogging (5.5 mph). Bill, weighing 185 lb, plans to exercise according to the matrix shown. Verify the calculations (W = Walking, B = Bicycling, J = Jogging).

$$\begin{array}{l} \text{MON} \\ \text{WED} \\ \text{FRI} \\ \text{SAT} \end{array} \overset{\text{W} \quad \text{B} \quad \text{J}}{\begin{bmatrix} 1.0 & 0 & 0.5 \\ 1.0 & 1.0 & 0.5 \\ 1.5 & 0 & 0.5 \\ 2.0 & 1.5 & 1.0 \end{bmatrix}} \begin{bmatrix} 350 \\ 500 \\ 950 \end{bmatrix} = \begin{bmatrix} 825 \\ 1325 \\ 1000 \\ 2400 \end{bmatrix} \begin{array}{l} \text{MON} \\ \text{WED} \\ \text{FRI} \\ \text{SAT} \end{array}$$

Example 19 Markov Process. Powers of a Matrix. Stochastic Matrix

Suppose that the 2004 state of land use in a city of 60 mi^2 of built-up area is

C: Commercially Used 25% I: Industrially Used 20% R: Residentially Used 55%

Find the states in 2009, 2014, and 2019, assuming that the transition probabilities for five-year intervals are given by the matrix **A** and remain practically the same over the time considered.

$$\begin{array}{cccccc} & \text{From C} & \text{From I} & \text{From R} & \\ \mathbf{A} = & \left[\begin{array}{c} 0.7 \\ 0.2 \\ 0.1 \end{array}\right. & \begin{array}{c} 0.1 \\ 0.9 \\ 0 \end{array} & \left.\begin{array}{c} 0 \\ 0.2 \\ 0.8 \end{array}\right] & \begin{array}{l} \text{To C} \\ \text{To I} \\ \text{To R} \end{array} \end{array}$$

A is a **stochastic matrix**, that is, a square matrix with all entries nonnegative and all column sums equal to 1. Our example concerns a **Markov process**, that is, a process for which the probability of entering a certain state depends only on the last state occupied (and the matrix **A**), not on any earlier state.

Solution: From the matrix **A** and the 2004 state we can compute the 2009 state

$$\begin{array}{c} \text{C} \\ \text{I} \\ \text{R} \end{array} \begin{bmatrix} 0.7\times 25+0.1\times 20+\ \ 0\times 55 \\ 0.2\times 25+0.9\times 20+0.2\times 55 \\ 0.1\times 25+\ \ 0\times 20+0.8\times 55 \end{bmatrix} = \begin{bmatrix} 0.7 & 0.1 & 0 \\ 0.2 & 0.9 & 0.2 \\ 0.1 & 0 & 0.8 \end{bmatrix} \begin{bmatrix} 25 \\ 20 \\ 55 \end{bmatrix} = \begin{bmatrix} 19.5 \\ 34.0 \\ 46.5 \end{bmatrix}$$

To explain: The 2009 figure for C equals 25% times the probability 0.7 that C goes into C, plus 20% times the probability 0.1 that I goes into C, plus 55% times the probability 0 that R goes into C. Together, $25\times 0.7+20\times 0.1+55\times 0=19.5$ (%). Also $25\times 0.2+20\times 0.9+55\times 0.2=34$ (%). Similarly, the new R is 46.5%. We see that the 2009 state vector is the column vector

$$\mathbf{y}=[19.5 \quad 34.0 \quad 46.5]^{\mathrm{T}}=\mathbf{Ax}=\mathbf{A}\,[25 \quad 20 \quad 55]^{\mathrm{T}}$$

where the column vector $\mathbf{x}=[25 \quad 20 \quad 55]^{\mathrm{T}}$ is the given 2004 state vector. Note that the sum of the entries of **y** is 100 (%). Similarly, you may verify that for 2014 and 2019 we get the state vectors

$$\mathbf{z}=\mathbf{Ay}=\mathbf{A}(\mathbf{Ax})=\mathbf{A}^2\mathbf{x}=[17.05 \quad 43.80 \quad 39.15]^{\mathrm{T}}$$
$$\mathbf{u}=\mathbf{Az}=\mathbf{A}^2\mathbf{y}=\mathbf{A}^3\mathbf{x}=[16.315 \quad 50.660 \quad 33.025]^{\mathrm{T}}$$

Answer: In 2009 the commercial area will be 19.5% (11.7 mi^2), the industrial 34% (20.4 mi^2) and the residential 46.5% (27.9 mi^2). For 2014 the corresponding figures are 17.05%, 43.80%, 39.15%. For 2019 they are 16.315%, 50.660%, 33.025%.

Determinants

Second- and Third-Order Determinants

A **determinant of second order** is denoted and defined by

$$D=\det \mathbf{A}=\begin{vmatrix} a_{11} & a_{12} \\ a_{21} & a_{22} \end{vmatrix}=a_{11}a_{22}-a_{12}a_{21} \tag{1.20}$$

So here we have ***bars*** (whereas a matrix has ***brackets***). We now explain the **Cramer's rule** for solving linear systems of two equations in two unknowns. Consider the equations

$$\begin{aligned}&\text{(a)}\ a_{11}x_1 + a_{12}x_2 = b_1\\ &\text{(b)}\ a_{21}x_1 + a_{22}x_2 = b_2\end{aligned} \tag{1.21}$$

Then Cramer's rule gives

$$x_1 = \frac{\begin{vmatrix} b_1 & a_{12} \\ b_2 & a_{22} \end{vmatrix}}{D} = \frac{b_1a_{22} - a_{12}b_2}{D}$$

$$x_2 = \frac{\begin{vmatrix} a_{11} & b_1 \\ a_{21} & b_2 \end{vmatrix}}{D} = \frac{a_{11}b_2 - b_1a_{21}}{D} \tag{1.22}$$

with D as in (1.20), provided $D \neq 0$. The value $D = 0$ appears for inconsistent nonhomogeneous systems and for homogeneous systems with nontrivial solutions.

Example 20 Cramer's Rule for Two Equations

If

$$\begin{aligned}4x_1 + 3x_2 &= 12\\ 2x_1 + 5x_2 &= -8\end{aligned}$$

then

$$x_1 = \frac{\begin{vmatrix} 12 & 3 \\ -8 & 5 \end{vmatrix}}{\begin{vmatrix} 4 & 3 \\ 2 & 5 \end{vmatrix}} = \frac{84}{14} = 6, \qquad x_2 = \frac{\begin{vmatrix} 4 & 12 \\ 2 & -8 \end{vmatrix}}{\begin{vmatrix} 4 & 3 \\ 2 & 5 \end{vmatrix}} = \frac{-56}{14} = -4$$

Third-Order Determinants

A determinant of third order can be defined by

$$D = \begin{vmatrix} a_{11} & a_{12} & a_{13} \\ a_{21} & a_{22} & a_{23} \\ a_{31} & a_{32} & a_{33} \end{vmatrix} = a_{11}\begin{vmatrix} a_{22} & a_{23} \\ a_{32} & a_{33} \end{vmatrix} - a_{21}\begin{vmatrix} a_{12} & a_{13} \\ a_{32} & a_{33} \end{vmatrix} + a_{31}\begin{vmatrix} a_{12} & a_{13} \\ a_{22} & a_{23} \end{vmatrix} \tag{1.23}$$

Note the following. The signs on the right are $+ - +$. Each of the three terms on the right is an entry in the first column of D times its minor, that is, the second-order determinant obtained from D by deleting the row and column of that entry; thus, for a_{11} delete the first row and first column, and so on.

If we write out the minors in (1.23), we obtain

$$D = a_{11}a_{22}a_{33} - a_{11}a_{23}a_{32} + a_{21}a_{13}a_{32} - a_{21}a_{12}a_{33} + a_{31}a_{12}a_{23} - a_{31}a_{13}a_{22} \tag{1.24}$$

Cramer's Rule for Linear Systems of Three Equations

$$\begin{aligned}a_{11}x_1 + a_{12}x_2 + a_{13}x_3 &= b_1\\ a_{21}x_1 + a_{22}x_2 + a_{23}x_3 &= b_2\\ a_{31}x_1 + a_{32}x_2 + a_{33}x_3 &= b_3\end{aligned} \tag{1.25}$$

is

$$x_1 = \frac{D_1}{D}, \quad x_2 = \frac{D_2}{D}, \quad x_3 = \frac{D_3}{D} \qquad (D \neq 0) \tag{1.26}$$

with the *determinant D of the system* given by (1.23) and

$$D_1 = \begin{vmatrix} b_1 & a_{12} & a_{13} \\ b_2 & a_{22} & a_{23} \\ b_3 & a_{32} & a_{33} \end{vmatrix}, \quad D_2 = \begin{vmatrix} a_{11} & b_1 & a_{13} \\ a_{21} & b_2 & a_{23} \\ a_{31} & b_3 & a_{33} \end{vmatrix}, \quad D_3 = \begin{vmatrix} a_{11} & a_{12} & b_1 \\ a_{21} & a_{22} & b_2 \\ a_{31} & a_{32} & b_3 \end{vmatrix}$$

Note that D_1, D_2, D_3 are obtained by replacing Columns 1, 2, 3, respectively, by the column of the right sides of (1.25). Cramer's rule (1.26) can be derived by eliminations similar to those for (1.22), but it also follows from the general case in the next section.

Determinants and Cramer's Rule

A determinant of order n is a scalar associated with an $n \times n$ (hence ***square***!) matrix $\mathbf{A} = [a_{jk}]$, which is written as

$$D = \det \mathbf{A} = \begin{vmatrix} a_{11} & a_{12} & \cdots & a_{1n} \\ a_{21} & a_{22} & \cdots & a_{2n} \\ \cdot & \cdot & \cdots & \cdot \\ \cdot & \cdot & \cdots & \cdot \\ a_{n1} & a_{n2} & \cdots & a_{nn} \end{vmatrix} \tag{1.27}$$

and is defined for $n = 1$ by

$$D = a_{11} \tag{1.28}$$

and for $n \geqq 2$ by

$$D = a_{j1}C_{j1} + a_{j2}C_{j2} + \cdots + a_{jn}C_{jn} \qquad (j = 1, 2, \ldots, n) \tag{1.29a}$$

or

$$D = a_{1k}C_{1k} + a_{2k}C_{2k} + \cdots + a_{nk}C_{nk} \qquad (k = 1, 2, \ldots, n) \tag{1.29b}$$

Here,

$$C_{jk} = (-1)^{j+k} M_{jk}$$

and M_{jk} is a determinant of order $n - 1$, namely, the determinant of the submatrix of **A** obtained from **A** by omitting the row and column of the entry a_{jk}, that is, the jth row and the kth column. In this way, D is defined in terms of n determinants of order $n - 1$, each of which is, in turn, defined in terms of $n - 1$ determinants of order $n - 2$, and so on; we finally arrive at second-order determinants, in which those submatrices consist of single entries whose determinant is defined to be the entry itself.

From the definition it follows that *we may **expand D** by any row or column,* that is, choose in (1.29) the entries in any row or column, similarly when expanding the C_{jk}'s in (1.29), and so on. ***This definition is unambiguous***, that is, it yields the same value for D no matter which columns or rows we choose in expanding. Terms used in connection with determinants are taken from matrices. In D we have n^2 **entries** a_{jk}, also n **rows** and n **columns,** and a **main diagonal** on which $a_{11}, a_{22}, \ldots, a_{nn}$ stand. The following two terms are new:

▶ M_{jk} is called the **minor** of a_{jk} in D, and C_{jk} the **cofactor** of a_{jk} in D.

For later use we note that (1.22) may also be written in terms of minors as

$$D=\sum_{k=1}^{n}(-1)^{j+k}a_{jk}M_{jk} \quad (j=1,2,\ldots,n) \tag{1.30a}$$

$$D=\sum_{j=1}^{n}(-1)^{j+k}a_{jk}M_{jk} \quad (k=1,2,\ldots,n) \tag{1.30b}$$

Example 21 Minors and Cofactors of a Third-Order Determinant

In (1.23) the minors and cofactors of the entries in the first column can be seen directly. For the entries in the second row the minors are

$$M_{21}=\begin{vmatrix} a_{12} & a_{13} \\ a_{32} & a_{33} \end{vmatrix}, \quad M_{22}=\begin{vmatrix} a_{11} & a_{13} \\ a_{31} & a_{33} \end{vmatrix}, \quad M_{23}=\begin{vmatrix} a_{11} & a_{12} \\ a_{31} & a_{32} \end{vmatrix}$$

and the cofactors are $C_{21}=-M_{21}$, $C_{22}=+M_{22}$, and $C_{23}=-M_{23}$. Similarly for the third row – write these down yourself. And verify that the signs in C_{jk} form a checkerboard pattern

$$\begin{matrix} + & - & + \\ - & + & - \\ + & - & + \end{matrix}$$

Example 22 Expansions of a Third-Order Determinant

$$D=\begin{vmatrix} 1 & 3 & 0 \\ 2 & 6 & 4 \\ -1 & 0 & 2 \end{vmatrix} = 1\begin{vmatrix} 6 & 4 \\ 0 & 2 \end{vmatrix} - 3\begin{vmatrix} 2 & 4 \\ -1 & 2 \end{vmatrix} + 0\begin{vmatrix} 2 & 6 \\ -1 & 0 \end{vmatrix}$$

$$=1(12-0)-3(4+4)+0(0+6)=-12$$

This is the expansion by the first row. The expansion by the third column is

$$D=0\begin{vmatrix} 2 & 6 \\ -1 & 0 \end{vmatrix} - 4\begin{vmatrix} 1 & 3 \\ -1 & 0 \end{vmatrix} + 2\begin{vmatrix} 1 & 3 \\ 2 & 6 \end{vmatrix} = 0-12+0=-12$$

Verify that the other four expansions also give the value −12.

Example 23 Determinant of a Triangular Matrix

$$\begin{vmatrix} -3 & 0 & 0 \\ 6 & 4 & 0 \\ -1 & 2 & 5 \end{vmatrix} = -3\begin{vmatrix} 4 & 0 \\ 2 & 5 \end{vmatrix} = -3\times 4\times 5=-60$$

Inspired by this, can you formulate a little theorem on determinants of triangular matrices? Of diagonal matrices?

General Properties of Determinants

To obtain the value of a determinant (1.27), we can first simplify it systematically by elementary row operations, as follows.

THEOREM 1 (Behavior of an *n*th-Order Determinant under Elementary Row Operations): (*a*) *Interchange of two rows multiplies the value of the determinant by* -1. (*b*) *Addition of a multiple of a row to another row does not alter the value of the determinant.* (*c*) *Multiplication of a row by a nonzero constant c multiplies the value of the determinant by c. (This holds also when* $c = 0$, *but gives no longer an elementary row operation.)*

Example 24 Evaluation of Determinants by Reduction to Triangular Form

Because of Theorem 1 we may evaluate determinants by reduction to triangular form, as in the Gauss elimination for a matrix. For instance

$$D = \begin{vmatrix} 2 & 0 & -4 & 6 \\ 4 & 5 & 1 & 0 \\ 0 & 2 & 6 & -1 \\ -3 & 8 & 9 & 1 \end{vmatrix}$$

$$= \begin{vmatrix} 2 & 0 & -4 & 6 \\ 0 & 5 & 9 & -12 \\ 0 & 2 & 6 & -1 \\ 0 & 8 & 3 & 10 \end{vmatrix} \begin{matrix} \\ \\ \text{Row } 2 \to \text{Row } 2 - 2 \text{ Row } 1 \\ \text{Row } 4 \to \text{Row } 4 + 1.5 \text{ Row} \end{matrix}$$

$$= \begin{vmatrix} 2 & 0 & -4 & 6 \\ 0 & 5 & 9 & -12 \\ 0 & 0 & 2.4 & 3.8 \\ 0 & 0 & -11.4 & 29.2 \end{vmatrix} \begin{matrix} \\ \\ \text{Row}\,3 \to \text{Row}\,3 - 0.4\,\text{Row}\,2 \\ \text{Row}\,4 \to \text{Row}\,4 - 1.6\,\text{Row}\,2 \end{matrix}$$

$$= \begin{vmatrix} 2 & 0 & -4 & 6 \\ 0 & 5 & 9 & 12 \\ 0 & 0 & 2.4 & 3.8 \\ 0 & 0 & -0 & 47.25 \end{vmatrix} \quad \text{Row } 4 \to \text{Row } 4 + 4.75 \text{ Row } 3$$

$$= 2 \times 5 \times 2.4 \times 47.25 = 1134$$

THEOREM 2 (Further Properties of *n*th-Order Determinants): (*a*)–(*c*) *in Theorem 1 hold also for columns.* (***d***) ***Transposition*** *leaves the value of a determinant unaltered.* (***e***) ***A zero row or column*** *renders the value of a determinant zero.* (***f***) ***Proportional rows or columns*** *render the value of a determinant zero. In particular, a determinant with two identical rows or columns has the value zero.*

THEOREM 3 [Rank in Terms of Determinants (Explained Later)]: *An* $m \times n$ *matrix* $\mathbf{A} = [a_{jk}]$ *has rank* $r \geqq 1$ *if and only if* $\mathbf{A}$ *has an* $r \times r$ *submatrix with nonzero determinant, whereas every square submatrix with more than r rows that* $\mathbf{A}$ *has (or does not have!) has determinant equal to zero. In particular, if* $\mathbf{A}$ *is square,* $n \times n$, *it has rank n if and only if*

$$\det \mathbf{A} \neq 0$$

Cramer's Rule

Theorem 3 opens the way to the classical solution formula for linear systems known as ***Cramer's rule***, which gives solutions as quotients of determinants. ***Cramer's rule is not practical in computations***, but is of ***theoretical interest*** in differential equations and other theories that have engineering applications.

THEOREM 4 [Cramer's Theorem (Solution of Linear Systems by Determinants)]: (*a*) *If a linear system of n equations in the same number of unknowns* $x_1, \ldots, x_n$

$$\begin{aligned} a_{11}x_1 + a_{12}x_2 + \cdots + a_{1n}x_n &= b_1 \\ a_{21}x_1 + a_{22}x_2 + \cdots + a_{2n}x_n &= b_2 \\ \cdots\cdots\cdots\cdots\cdots\cdots\cdots\cdots \\ a_{n1}x_1 + a_{n2}x_2 + \cdots + a_{nn}x_n &= b_n \end{aligned} \tag{1.31}$$

has a nonzero coefficient determinant $D = \det \mathbf{A}$, *the system has precisely one solution. This solution is given by the formulas*

$$x_1 = \frac{D_1}{D}, \quad x_2 = \frac{D_2}{D}, \quad \ldots, \quad x_n = \frac{D_n}{D} \qquad \textbf{(Cramer's rule)} \tag{1.32}$$

where D_k *is the determinant obtained from D by replacing in D the kth column by the column with the entries* $b_1, \ldots, b_n$.

(*b*) *Hence if the system* (1.31) *is* ***homogeneous*** *and* $D \neq 0$, *it has only the trivial solution* $x_1 = 0, x_2 = 0, \ldots, x_n = 0$. *If* $D = 0$, *the homogeneous system also has nontrivial solutions.*

PROOF: The augmented matrix $\tilde{\mathbf{A}}$ of the system (1.31) is of size $n \times (n+1)$. Hence its rank can be at most n. Now if

$$D = \det \mathbf{A} = \begin{vmatrix} a_{11} & \cdots & a_{1n} \\ \cdot & \cdots & \cdot \\ \cdot & \cdots & \cdot \\ a_{n1} & \cdots & a_{nn} \end{vmatrix} \neq 0 \tag{1.33}$$

then rank $\mathbf{A} = n$ by Theorem 3. Thus rank $\tilde{\mathbf{A}}$ = rank $\mathbf{A}$. Hence, by the Fundamental Theorem, the system (1.31) has a unique solution.

Let us now prove (1.32). Expanding D by its kth column, we obtain

$$D = a_{1k}C_{1k} + a_{2k}C_{2k} + \cdots + a_{nk}C_{nk} \tag{1.34}$$

where C_{ik} is the cofactor of entry a_{ik} in D. If we replace the entries in the kth column of D by any other numbers, we obtain a new determinant, say, $\hat{D}$. Clearly, its expansion by the kth column will be of the form (1.35), with $a_{1k}, \ldots, a_{nk}$ replaced by those new numbers and the cofactors C_{ik} as before. In particular, if we choose as new numbers the entries $a_{1l}, \ldots, a_{nl}$ of the lth column of D (where $l \neq k$), we have a new determinant $\hat{D}$ which has twice the column $[a_{1l}, \ldots, a_{nl}]^T$, once as its lth column, and once as its kth because of the replacement. Hence $\hat{D} = 0$ by Theorem 2(f). If we now expand $\hat{D}$ by the column that has been replaced (the kth column), we thus obtain

$$a_{1l}C_{1k} + a_{2l}C_{2k} + \cdots + a_{nl}C_{nk} = 0 \quad (l \neq k) \tag{1.35}$$

We now multiply the first equation in (1.31) by C_{1k} on both sides, the second by C_{2k}, …, the last by C_{nk}, and add the resulting equations. This gives

$$C_{1k}(a_{11}x_1 + \cdots + a_{1n}x_n) + \cdots + C_{nk}(a_{n1}x_1 + \cdots + a_{nn}x_n) = b_1C_{1k} + \cdots + b_nC_{nk} \tag{1.36}$$

Collecting terms with the same x_j, we can write the left side as

$$x_1(a_{11}C_{1k} + a_{21}C_{2k} + \cdots + a_{n1}C_{nk}) + \cdots + x_n(a_{1n}C_{1k} + a_{2n}C_{2k} + \cdots + a_{nn}C_{nk})$$

From this we see that x_k is multiplied by

$$a_{1k}C_{1k} + a_{2k}C_{2k} + \cdots + a_{nk}C_{nk}$$

Equation (1.34) shows that this equals D. Similarly, x_l is multiplied by

$$a_{1l}C_{1k} + a_{2l}C_{2k} + \cdots + a_{nl}C_{nk}$$

Equation (1.35) shows that this is zero when $l \neq k$. Accordingly, the left side of (1.36) equals simply x_kD, so that (1.36) becomes

$$x_kD = b_1C_{1k} + b_2C_{2k} + \cdots + b_nC_{nk}$$

Now the right side of this is D_k as defined in the theorem, expanded by its kth column, so that division by D gives (1.32). This proves Cramer's rule.

If (1.31) is homogeneous and $D \neq 0$, then each D_k has a column of zeros, so that $D_k = 0$ by Theorem 2(e), and (1.32) gives the trivial solution. Finally, if (1.31) is homogeneous and $D = 0$, then rank $\mathbf{A} < n$ by Theorem 3, so that nontrivial solutions exist. ■

CONCEPT CHECK

1. Let

$$\mathbf{A} = \begin{bmatrix} 3 & 0 & 4 \\ -1 & 2 & 2 \\ 6 & 5 & -4 \end{bmatrix}, \quad \mathbf{B} = \begin{bmatrix} 0 & -5 & -3 \\ -5 & 2 & 4 \\ -3 & 4 & 0 \end{bmatrix},$$

$$\mathbf{C} = \begin{bmatrix} 0 & 2 \\ 2 & 4 \\ 1 & 3 \end{bmatrix}, \quad \mathbf{D} = \begin{bmatrix} 6 & 1 \\ -4 & 7 \\ -8 & 3 \end{bmatrix},$$

$$\mathbf{u} = \begin{bmatrix} 2 \\ 0 \\ -1 \end{bmatrix}, \quad \mathbf{v} = \begin{bmatrix} -4.5 \\ 0.8 \\ 1.2 \end{bmatrix}$$

Find the following expressions or give reasons why they are undefined.

(a) $\mathbf{C} + \mathbf{D}$, $\mathbf{D} + \mathbf{C}$, $6(\mathbf{D} - \mathbf{C})$, $6\mathbf{C} - 6\mathbf{D}$

(b) $4\mathbf{C}$, $2\mathbf{D}$, $4\mathbf{C} + 2\mathbf{D}$, $8\mathbf{C} - 0\mathbf{D}$

(c) $\mathbf{A} + \mathbf{C} - \mathbf{D}$, $\mathbf{C} - \mathbf{D}$, $\mathbf{D} - \mathbf{C}$, $\mathbf{B} + 2\mathbf{C} + 4\mathbf{D}$

(d) $2(\mathbf{A} + \mathbf{B})$, $2\mathbf{A} + 2\mathbf{B}$, $5\mathbf{A} - (1/2)\mathbf{B}$, $\mathbf{A} + \mathbf{B} + \mathbf{C}$

(e) $3\mathbf{C} - 8\mathbf{D}$, $4(3\mathbf{A})$, $(4 \times 3)\mathbf{A}$, $\mathbf{B} - (1/10)\mathbf{A}$

(f) $5\mathbf{A} - 3\mathbf{C}$, $\mathbf{A} - \mathbf{B} + \mathbf{D}$, $4(\mathbf{B} - 6\mathbf{A})$, $4\mathbf{B} - 24\mathbf{A}$

(g) $33\mathbf{u}$, $4\mathbf{v} + 9\mathbf{u}$, $4(\mathbf{v} + 2.25\mathbf{u})$, $\mathbf{u} - \mathbf{v}$

(h) $\mathbf{A} + \mathbf{u}$, $12\mathbf{u} + 10\mathbf{v}$, $0(\mathbf{B} - \mathbf{v})$, $0\mathbf{B} + \mathbf{u}$

2. **(Linear system)** Write down a linear system (as in Example 1) whose augmented matrix is the matrix **B** in this problem set.

3. **(Scalar multiplication)** The matrix **A** in Example 2 shows the numbers of items sold. Find the matrix showing the number of units sold if a unit consists of (a) 5 items, (b) 10 items?

4. **(Double subscript notation)** Write the entries of **A** in Example 2 in the general notation shown in (1.2).

5. **(Sizes, diagonal)** What sizes do **A**, **B**, **C**, **D**, **u**, **v** in this problem set have? What are the main diagonals of **A** and **B**, and what about **C**?

6. **(Equality)** Give reasons why the five matrices in Example 3 are different.

7. **(Addition of vectors)** Can you add (a) row vectors whose numbers of components are different, (b) a row and a column vector with the same number of components, (c) a vector and a scalar?

8. **(General rules)** Prove (1.3) and (1.4) for general 3×2 matrices and scalars c and k.

9. **TEAM PROJECT. Matrices in Modeling Networks.** Matrices have various applications, as we shall see, in a form that these problems can be efficiently handled on the

computer. For instance, they can be used to characterize connections in electrical networks, in nets of roads, in production processes, etc., as follows.

(a) **Nodal incidence matrix.** The network in Figure 2 consists of 5 *branches* or *edges* (connections, numbered 1, 2, …, 5) and 4 *nodes* (points where two or more branches come together), with one node being grounded. We number the nodes and branches and give each branch a direction shown by an arrow. This we do arbitrarily. The network can now be described by a *"nodal incidence matrix"* $\mathbf{A} = [a_{jk}]$, where

$$a_{jk} = \begin{cases} +1 \text{ if branch } k \text{ leaves node } (j) \\ -1 \text{ if branck } k \text{ enters node } (j) \\ 0 \text{ if branch } k \text{ does not touch } (j) \end{cases}$$

Show that for the network in Figure 2 the matrix $\mathbf{A}$ has the given form

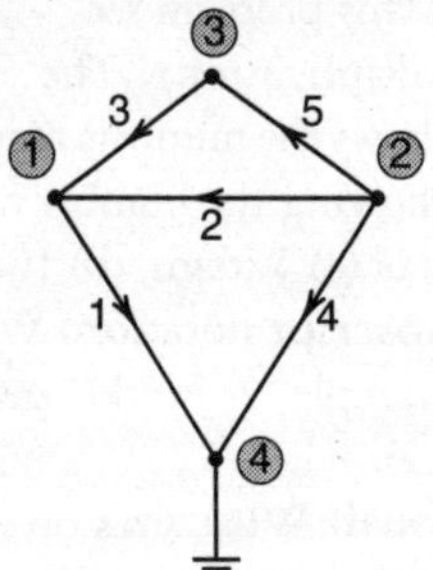

Branch	1	2	3	4	5
Node ①	1	−1	−1	0	0
Node ②	0	1	0	1	1
Node ③	0	0	1	0	−1
Node ④	−1	0	0	−1	0

Figure 2 Network and nodal incidence matrix in Team Project 16(a).

(b) Find the nodal incidence matrices of the networks in Figure 3.

(c) Graph the three networks corresponding to the nodal incidence matrices.

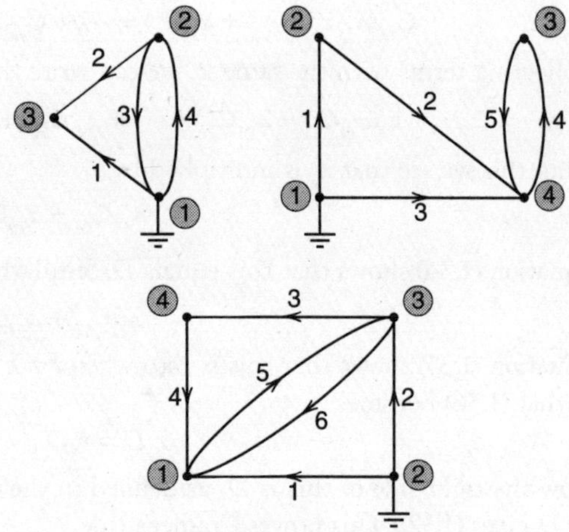

Figure 3 Networks in Team Project 16(b).

$$\begin{bmatrix} 1 & -1 & 1 & -1 \\ 0 & 0 & -1 & 1 \\ -1 & 1 & 0 & 0 \end{bmatrix}, \quad \begin{bmatrix} 1 & 0 & 0 \\ 0 & -1 & 1 \\ -1 & 1 & 0 \\ 0 & 0 & -1 \end{bmatrix},$$

$$\begin{bmatrix} 1 & 1 & 1 & 0 & 0 & 0 \\ 0 & -1 & 0 & 0 & -1 & 1 \\ -1 & 0 & 0 & 1 & 1 & 0 \\ 0 & 0 & -1 & -1 & 0 & -1 \end{bmatrix}$$

(d) **Mesh incidence matrix.** A network can also be characterized by the *mesh incidence matrix* $\mathbf{M} = [m_{jk}]$, where

$$m_{jk} = \begin{cases} +1 \text{ if branch } k \text{ is in mesh } \boxed{j} \\ \text{and has the same orientation} \\ -1 \text{ if branch } k \text{ is in mesh } \boxed{j} \\ \text{and has the opposite orientation} \\ 0 \text{ if branch } k \text{ is not in mesh } \boxed{j} \end{cases}$$

and a mesh is a loop with no branch in its interior (or in its exterior). Here, the meshes are numbered and directed (oriented) in an arbitrary fashion. Show that in Figure 4 the matrix $\mathbf{M}$ corresponds to the given figure, where Row 1 corresponds to mesh 1, etc.

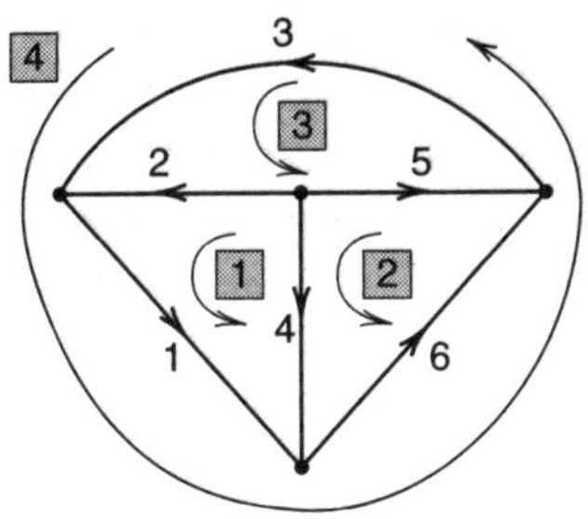

$$\mathbf{M} = \begin{bmatrix} 1 & 1 & 0 & -1 & 0 & 0 \\ 0 & 0 & 0 & 1 & -1 & 1 \\ 0 & -1 & 1 & 0 & 1 & 0 \\ 1 & 0 & 1 & 0 & 0 & 1 \end{bmatrix}$$

Figure 4 Network and matrix **M** in Team Project 16(d).

(e) Number the nodes in Figure 4 from left to right 1, 2, 3 and the low node by 4. Find the corresponding nodal incidence matrix.

10. Let

$$\mathbf{A} = \begin{bmatrix} 6 & -2 & -2 \\ 10 & -3 & 1 \\ -10 & 5 & 1 \end{bmatrix}, \quad \mathbf{B} = \begin{bmatrix} 9 & 4 & -4 \\ 4 & 7 & 0 \\ -4 & 0 & 11 \end{bmatrix}$$

$$\mathbf{C} = \begin{bmatrix} 3 & 1 \\ 0 & -2 \\ 4 & 0 \end{bmatrix}, \quad \mathbf{a} = \begin{bmatrix} 5 \\ 1 \\ 2 \end{bmatrix}, \quad \mathbf{b} = [3 \;\; 0 \;\; 8]$$

Calculate the following products and sums or give reasons why they are not defined. (Show all intermediate results.)

(a) $\mathbf{Aa}$, $\mathbf{Ab}$, $\mathbf{Ab}^T$, $\mathbf{AB}$

(b) $\mathbf{Ab}^T + \mathbf{Bb}^T$, $(\mathbf{A} + \mathbf{B})\mathbf{b}^T$, $\mathbf{bA}$, $\mathbf{B} - \mathbf{B}^T$

(c) $\mathbf{AB}$, $\mathbf{BA}$, $\mathbf{AA}^T$, $\mathbf{A}^T\mathbf{A}$

(d) $\mathbf{A}^2$, $\mathbf{B}^2$, $(\mathbf{A}^T)^2$, $(\mathbf{A}^2)^T$

(e) $\mathbf{a}^T\mathbf{A}$, $\mathbf{bA}$, $5\mathbf{B}(3\mathbf{a} + 2\mathbf{b}^T)$, $15\mathbf{Ba} + 10\mathbf{Bb}^T$

(f) $\mathbf{A}^T\mathbf{b}$, $\mathbf{b}^T\mathbf{B}$, $(3\mathbf{A} - 2\mathbf{B})^T\mathbf{a}$, $\mathbf{a}^T(3\mathbf{A} - 2\mathbf{B})$

(g) $\mathbf{ab}$, $\mathbf{ba}$, $(\mathbf{ab})\mathbf{A}$, $\mathbf{a}(\mathbf{bA})$

(h) $\mathbf{ab} - \mathbf{ba}$, $-(4\mathbf{b})(7\mathbf{a})$, $-28\mathbf{ba}$, $5\mathbf{abB}$

(i) $(\mathbf{A} + \mathbf{B})^2$, $\mathbf{A}^2 + \mathbf{AB} + \mathbf{BA} + \mathbf{B}^2$, $\mathbf{A}^2 + 2\mathbf{AB} + \mathbf{B}^2$

(j) $(\mathbf{A} + \mathbf{B})(\mathbf{A} - \mathbf{B})$, $\mathbf{A}^2 - \mathbf{AB} + \mathbf{BA} - \mathbf{B}^2$, $\mathbf{A}^2 - \mathbf{B}^2$

(k) $\mathbf{A}^2\mathbf{B}$, $\mathbf{A}^3$, $(\mathbf{AB})^2$, $\mathbf{A}^2\mathbf{B}^2$

(l) $\mathbf{B}^3$, $\mathbf{BC}$, $(\mathbf{BC})^2$, $(\mathbf{BC})(\mathbf{BC})^T$

(m) $\mathbf{a}^T\mathbf{Aa}$, $\mathbf{a}^T(\mathbf{A} + \mathbf{A}^T)\mathbf{a}$, $\mathbf{bBb}^T$, $\mathbf{b}(\mathbf{B} - \mathbf{B}^T)\mathbf{b}^T$

(n) $\mathbf{a}^T\mathbf{CC}^T\mathbf{a}$, $\mathbf{a}^T\mathbf{C}^2\mathbf{a}$, $\mathbf{bC}^T\mathbf{Cb}^T$, $\mathbf{bCC}^T\mathbf{b}^T$

11. **(Commutativity)** Find all 2×2 matrices $\mathbf{A} = [a_{jk}]$ that commute with $\mathbf{B} = [b_{jk}]$, where $b_{jk} = j + k$.

12. **(Product)** Write **AB** in Probs. 10(a)–(n) in terms of row and column vectors.

13. **(Product)** Calculate **AB** in Prob. 10(a) columnwise.

14. **TEAM PROJECT. Symmetric and Skew-Symmetric Matrices.** These matrices occur quite frequently in applications, so it is worthwhile to study some of their most important properties.

(a) Verify the claims in (1.16) that $a_{kj} = a_{jk}$ for a symmetric matrix, and $a_{kj} = -a_{jk}$ for a skew-symmetric matrix. Give examples.

(b) Show that for every square matrix **C** the matrix $\mathbf{C} + \mathbf{C}^T$ is symmetric and $\mathbf{C} - \mathbf{C}^T$ is skew-symmetric. Write **C** in the form $\mathbf{C} = \mathbf{S} + \mathbf{T}$, where **S** is symmetric and **T** is skew-symmetric and find **S** and **T** in terms of **C**. Represent **A** and **B** in Probs. 10(a)–(n) in this form.

(c) A **linear combination** of matrices **A**, **B**, **C**, …, **M** of the same size is an expression of the form

$$a\mathbf{A} + b\mathbf{B} + c\mathbf{C} + \cdots + m\mathbf{M} \qquad (1.37)$$

where a, …, m are any scalars. Show that if these matrices are square and symmetric, so is (1.37); similarly, if they are skew-symmetric, so is (1.37).

(d) Show that **AB** with symmetric **A** and **B** is symmetric if and only if **A** and **B** **commute**, that is, $\mathbf{AB} = \mathbf{BA}$.

(e) Under what condition is the product of skew-symmetric matrices skew-symmetric?

15. **(Idempotent and nilpotent matrices)** By definition, **A** is **idempotent** if $\mathbf{A}^2 = \mathbf{A}$, and **B** is **nilpotent** if $\mathbf{B}^m = \mathbf{0}$ for some positive integer m. Give examples (different from **0** or **I**). Also give examples such that $\mathbf{A}^2 = \mathbf{I}$ (the unit matrix).

16. **(Triangular matrices)** Let $\mathbf{U}_1$, $\mathbf{U}_2$ be upper triangular and $\mathbf{L}_1$, $\mathbf{L}_2$ lower triangular. Which of the following are triangular? Give examples. How can you save half of your work by transposition?
$$\mathbf{U}_1+\mathbf{U}_2,\ \mathbf{U}_1\mathbf{U}_2,\ \mathbf{U}_1^2,\ \mathbf{U}_1+\mathbf{L}_1,\ \mathbf{U}_1\mathbf{L}_1,$$
$$\mathbf{L}_1+\mathbf{L}_2,\ \mathbf{L}_1\mathbf{L}_2,\ \mathbf{L}_1^2$$
17. **(Transposition of products)** Prove (1.15a)–(1.15c). Illustrate the basic formula (1.15d) by examples of your own. Then prove it.
18. **(Markov process)** If the transition matrix $\mathbf{A}$ has the entries $a_{11}=0.5$, $a_{12}=0.3$, $a_{21}=0.5$, $a_{22}=0.7$ and the initial state is $[1\quad 1]^T$, what will be the next three states?
19. **(Concert subscription)** In a community of 300000 adults, subscribers to a concert series tend to renew their subscription with probability 90% and persons presently not subscribing will subscribe for the next season with probability 0.1%. If the present number of subscribers is 2000, can one predict an increase, decrease, or no change over each of the next three seasons?
20. **(Production)** In a production process, let N mean "no trouble" and T "trouble." Let the transition probabilities from one day to the next be 0.9 for $N\rightarrow N$, hence 0.1 for $N\rightarrow T$, and 0.5 for $T\rightarrow N$, hence 0.5 for $T\rightarrow T$. If today there is no trouble, what is the probability of N two days after today? Three days after today?
21. **(Profit vector)** Two factory outlets F_1 and F_2 in New York and Los Angeles sell sofas (S), chairs (C), and tables (T) with a profit of \$110, \$45, and \$80, respectively. Let the sales in a certain week be given by the matrix
$$\mathbf{A}=\begin{array}{c} \begin{matrix} \text{S} & \text{C} & \text{T} \end{matrix} \\ \begin{bmatrix} 600 & 400 & 100 \\ 300 & 820 & 205 \end{bmatrix} \end{array}\begin{matrix} F_1 \\ F_2 \end{matrix}$$
Introduce a "profit vector" $\mathbf{p}$ such that the components of $\mathbf{v}=\mathbf{Ap}$ give the total profits of F_1 and F_2.
22. **(Second-order determinant)** Expand a general second order determinant in four possible ways and show that the results agree.
23. **(Minors, cofactors)** Complete the list of minors and cofactors in Example 21.
24. **(Third-order determinant)** Do the task indicated in Example 22. Also evaluate D by reduction to triangular form.
25. **(Scalar multiplication)** Show that $\det(k\mathbf{A})=k^n\det\mathbf{A}$ (not $k\det\mathbf{A}$), where $\mathbf{A}$ is any $n\times n$ matrix. Give an example.
26. Evaluate, showing the details of your work.

(a) $\begin{vmatrix} 13 & 8 \\ -2 & 7 \end{vmatrix}$ (b) $\begin{vmatrix} \cos n\theta & \sin n\theta \\ -\sin n\theta & \cos n\theta \end{vmatrix}$

(c) $\begin{vmatrix} \cos\alpha & \sin\alpha \\ \sin\beta & \cos\beta \end{vmatrix}$ (d) $\begin{vmatrix} 14 & 2 & 5 \\ 2 & 0 & 8 \\ 5 & 8 & 2 \end{vmatrix}$

(e) $\begin{vmatrix} 70.4 & 0.3 & 0.8 \\ 0 & 0.5 & 2.6 \\ 0 & 0 & -1.9 \end{vmatrix}$ (f) $\begin{vmatrix} 2 & 1 & 2 \\ -2 & 2 & 1 \\ 1 & 2 & -2 \end{vmatrix}$

(g) $\begin{vmatrix} 0 & 3 & 1 \\ -3 & 0 & -4 \\ 1 & 4 & 0 \end{vmatrix}$ (h) $\begin{vmatrix} 0 & a & b \\ -a & 0 & c \\ -b & -c & 0 \end{vmatrix}$

(i) $\begin{vmatrix} u & v & w \\ w & u & v \\ v & w & u \end{vmatrix}$ (j) $\begin{vmatrix} 1 & -2 & 0 & 0 \\ 4 & 3 & 5 & 0 \\ 0 & 2 & 7 & 5 \\ 0 & 0 & 2 & 4 \end{vmatrix}$

(k) $\begin{vmatrix} 1 & 2 & 0 & 0 \\ 3 & 4 & 0 & 0 \\ 0 & 0 & 5 & 6 \\ 0 & 0 & 7 & 8 \end{vmatrix}$ (l) $\begin{vmatrix} 0 & -2 & 1 & 0 \\ 2 & 0 & -2 & 4 \\ -1 & 2 & 0 & 1 \\ 0 & -4 & -1 & 0 \end{vmatrix}$

27. Solve by Cramer's rule. (Show details.)

(a) $2x-5y=23$
$4x+6y=-2$

(b) $3y+4z=14.8$
$4x+2y-1z=-6.3$
$x-1y+5z=13.5$

(c) $w+2x+0y-3z=30$
$0w+4x-5y+2z=13$
$2w+8x-4y+z=42$
$3w+0x+1y-5z=35$

1.2 Systems of Linear Algebraic Equations

The most important use of matrices occurs in the solution of *systems of linear equations,* briefly called **linear systems**. Such systems model various problems, for instance, in frameworks, electrical networks, traffic flow, economics, statistics, and many others. In this section we show an important solution method, the *Gauss elimination.*

Linear System, Coefficient Matrix, Augmented Matrix

A linear system of m equations in n unknowns $x_1, \ldots, x_n$ is a set of equations of the form

$$\begin{aligned} a_{11}x_1 + \cdots + a_{1n}x_n &= b_1 \\ a_{21}x_1 + \cdots + a_{2n}x_n &= b_2 \\ \cdots\cdots\cdots\cdots\cdots\cdots \\ a_{m1}x_1 + \cdots + a_{mn}x_n &= b_m \end{aligned} \tag{1.38}$$

The system is called *linear* because each variable x_j appears in the first power only, just as in the equation of a straight line. $a_{11}, \ldots, a_{mn}$ are given numbers, called the **coefficients** of the system. $b_1, \ldots, b_m$ on the right are also given numbers. If all the b_j are zero, then (1.38) is called a **homogeneous system.** If at least one b_j is not zero, then (1.38) is called a **nonhomogeneous system.**

A **solution** of (1.38) is a set of numbers $x_1, \ldots, x_n$ that satisfies all the m equations. A **solution vector** of (1.38) is a vector **x** whose components form a solution of (1.38). If the system (1.38) is homogeneous, it has at least the **trivial solution** $x_1 = 0, \ldots, x_n = 0$.

Matrix Form of the Linear System (1.38)

From the definition of matrix multiplication we see that the m equations of (1.38) may be written as a single vector equation

$$\mathbf{Ax} = \mathbf{b} \tag{1.39}$$

where the **coefficient matrix** $\mathbf{A} = [a_{jk}]$ is the $m \times n$ matrix

$$\mathbf{A} = \begin{bmatrix} a_{11} & a_{12} & \cdots & a_{1n} \\ a_{21} & a_{22} & \cdots & a_{2n} \\ \cdot & \cdot & \cdots & \cdot \\ a_{m1} & a_{m2} & \cdots & a_{mn} \end{bmatrix}, \quad \mathbf{x} = \begin{bmatrix} x_1 \\ \cdot \\ \cdot \\ \cdot \\ x_n \end{bmatrix}, \quad \text{and} \quad \mathbf{b} = \begin{bmatrix} b_1 \\ \vdots \\ b_m \end{bmatrix}$$

are column vectors. We assume that the coefficients a_{jk} are not all zero, so that **A** is not a zero matrix. Note that **x** has n components, whereas **b** has m components. The matrix

$$\tilde{\mathbf{A}} = [\mathbf{A} \mid \mathbf{O}] = \begin{bmatrix} a_{11} & \cdots & a_{1n} & b_1 \\ \cdot & \cdots & \cdot & \cdot \\ \cdot & \cdots & \cdot & \cdot \\ a_{m1} & \cdots & a_{mn} & b_m \end{bmatrix}$$

is called the **augmented matrix** of the system (1.38). The dashed vertical line could be omitted (as we shall do later); it is merely a reminder that the last column of $\tilde{\mathbf{A}}$ does not belong to **A**. ***The augmented matrix*** $\tilde{\mathbf{A}}$ ***determines the system* (1.38) *completely*** because it contains all the given numbers appearing in (1.38).

Example 25 Geometric Interpretation. Existence and Uniqueness of Solutions

If $m = n = 2$, we have two equations in two unknowns x_1, x_2

$$a_{11}x_1 + a_{12}x_2 = b_1$$
$$a_{21}x_1 + a_{22}x_2 = b_2$$

If we interpret x_1, x_2 as coordinates in the x_1x_2-plane, then each of the two equations represents a straight line, and (x_1, x_2) is a solution if and only if the point P with coordinates x_1, x_2 lies on both lines. Hence there are three possible cases:

(a) Precisely one solution if the lines intersect.
(b) Infinitely many solutions if the lines coincide.
(c) No solution if the lines are parallel

For instance,

$x_1 + x_2 = 1$ $2x_1 - x_2 = 0$	$x_1 + x_2 = 1$ $2x_1 + 2x_2 = 2$	$x_1 + x_2 = 1$ $x_1 + x_2 = 0$
Case (*a*)	Case (*b*)	Case (*c*)

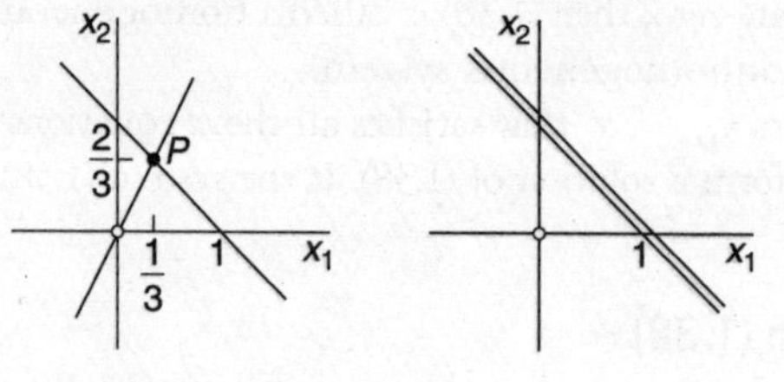

If the system is homogenous, Case (c) cannot happen, because then those two straight lines pass through the origin, whose coordinates 0, 0 constitute the trivial solution. If you wish, consider three equations in three unknowns as representations of three planes in space and discuss the various possible cases in a similar fashion. See Figure 5.

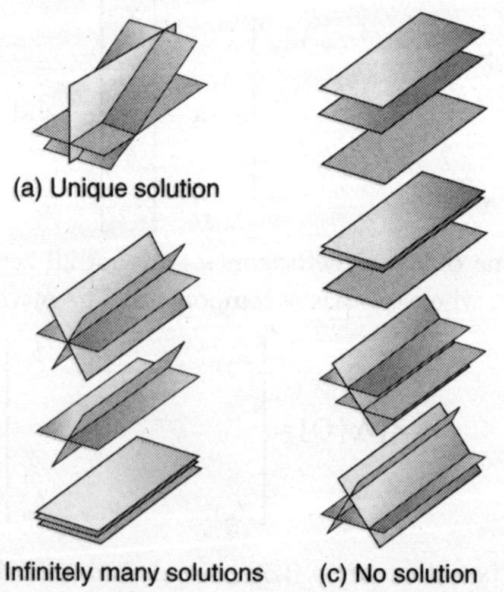

Figure 5 Three equations in three unknowns interpreted as planes in space.

Our simple example illustrates that a system (1.38) may perhaps have no solution. This poses the following problem. Does a given system (1.38) have a solution? Under what conditions does it have precisely one solution? If it has more than one solution, how can we characterize the set of all solutions? How can we actually obtain the solutions? Perhaps the last question is the most immediate one from a practical viewpoint.

Gauss Elimination and Back Substitution

This is a standard elimination method for solving linear systems that proceeds systematically irrespective of particular features of the coefficients. It is a method of great practical importance and is reasonable with respect to computing time and storage demand. We begin by motivating the method. If a system is in "**triangular form**," say,

$$\begin{aligned} 2x_1 + 5x_2 &= 2 \\ 13x_2 &= -26 \end{aligned}$$

we can solve it by "**back substitution**," that is, solve the last equation for the variable, $x_2 = -26/13 = -2$, and then work backward, substituting $x_2 = -2$ into the first equation and solve it for x_1, obtaining $x_1 = 1/2(2 - 5x_2) = 1/2[2 - 5 \times (-2)] = 6$. This gives us the idea of first reducing a general system to triangular form. For instance, let the given system be

$$\begin{aligned} 2x_1 + 5x_2 &= 2 \\ -4x_1 + 3x_2 &= -30 \end{aligned}$$

Its augmented matrix is

$$\begin{bmatrix} 2 & 5 & 2 \\ -4 & 3 & -30 \end{bmatrix}$$

We leave the first equation as it is. We eliminate x_1 from the second equation, to get a triangular system. For this we add twice the first equation to the second, and we do the same operation on the ***rows*** of the augmented matrix. This gives $-4x_1 + 4x_1 + 3x_2 + 10x_2 = -30 + 2 \times 2$, that is,

$$\begin{aligned} 2x_1 + 5x_2 &= 2 \\ 13x_2 &= -26 \quad \text{Row 2} \to \text{Row 2} + 2\ \text{Row 1} \end{aligned} \qquad \begin{bmatrix} 2 & 5 & 2 \\ 0 & 13 & -26 \end{bmatrix}$$

where Row 2 + 2 Row 1 means "Add twice Row 1 to Row 2" in the original matrix. This is the **Gauss elimination** (for 2 equations in 2 unknowns) giving the triangular form, from which back substitution now yields $x_2 = -2$ and $x_1 = 6$, as before.

Since a linear system is completely determined by its augmented matrix, ***Gauss elimination can be done by merely considering the matrices***, as we have just indicated. We do this again in the next example, emphasizing the matrices by writing them first and the equations behind them, just as a help in order not to lose track.

Example 26 Gauss Elimination. Electrical Network

Solve the linear system

$$\begin{aligned} x_1 - x_2 + x_3 &= 0 \\ -x_1 + x_2 - x_3 &= 0 \\ 10x_2 + 25x_3 &= 90 \\ 20x_1 + 10x_2 &= 80 \end{aligned}$$

Derivation from the circuit in Figure 6. This is the system for the unknown currents $x_1 = i_1$, $x_2 = i_2$, $x_3 = i_3$ in the electrical network in Figure 6. To obtain it, we label the currents as shown, choosing directions arbitrarily; if a current will come out negative, this will simply mean that the current flows against the direction of our arrow. The current entering each battery will be the same as the current leaving it. The equations for the currents result from Kirchhoff's laws:

Kirchhoff's current law **(KCL).** *At any point of a circuit, the sum of the inflowing currents equals the sum of the outflowing currents.*

Kirchhoff's voltage law **(KVL).** *In any closed loop, the sum of all voltage drops equals the impressed electromotive force.*

Node P gives the first equation, node Q the second, the right loop the third, and the left loop the fourth, as indicated in the figure.

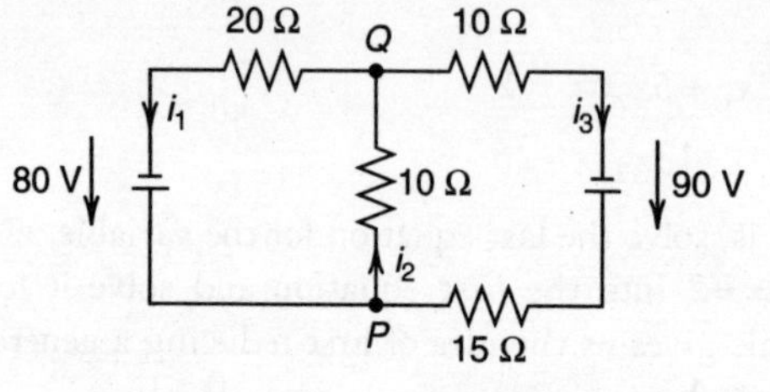

Node P: $i_1 - i_2 + i_3 = 0$

Node Q: $-i_1 + i_2 - i_3 = 0$

Right loop: $10i_2 + 25i_3 = 90$

Left loop: $20i_1 + 10i_2 = 80$

Figure 6 Network in Example 26 and equations relating the currents.

Solution by Gauss Elimination: This system could be solved rather quickly by noticing its particular form. But this is not the point. The point is that the Gauss elimination is systematic and will work in general, also for large systems. We apply it to our system and then do back substitution. As indicated let us write the augmented matrix of the system first and then the system itself:

Augmented Matrix $\tilde{\mathbf{A}}$

$$\begin{array}{l} \text{Pivot 1} \longrightarrow \\ \\ \text{Eliminate} \longrightarrow \\ \\ \end{array} \left[\begin{array}{rrr|r} \boxed{1} & -1 & 1 & 0 \\ -1 & 1 & -1 & 0 \\ 0 & 10 & 25 & 90 \\ 20 & 10 & 0 & 80 \end{array}\right]$$

Equations

$$\begin{array}{l} \text{Pivot 1} \longrightarrow \\ \\ \text{Eliminate} \longrightarrow \\ \\ \end{array} \begin{array}{rrrl} \boxed{x_1} & - \; x_2 & + \; x_3 & = 0 \\ -x_1 & + \; x_2 & - \; x_3 & = 0 \\ & 10x_2 & + 25x_3 & = 90 \\ 20x_1 & +10x_2 & & = 80 \end{array}$$

Step **1.** ***Elimination of*** x_1

Call the first row of A the **pivot row** and the first equation the **pivot equation.** Call the coefficient 1 of its x_1-term the **pivot** in this step. Use this equation to eliminate x_1 (get rid of x_1) in the other equations. For this, do:

1. Add 1 times the pivot equation to the second equation.
2. Add –20 times the pivot equation to the fourth equation.

This corresponds to **row operations** on the augmented matrix. So the operations are performed on the ***preceding matrix***. The result is

$$\left[\begin{array}{rrr|r} 1 & -1 & 1 & 0 \\ 0 & 0 & 0 & 0 \\ 0 & 10 & 25 & 90 \\ 0 & 30 & -20 & 80 \end{array}\right] \begin{array}{l} \\ \text{Row 2} \to \text{Row 2} + \text{Row 1} \\ \text{Row 4} \to \text{Row 4} - 20 \text{ Row 1} \\ \\ \end{array} \begin{array}{rl} x_1 - x_2 + x_3 & = 0 \\ 0 & = 0 \\ 10x_2 - 25x_3 & = 90 \\ 30x_2 - 20x_3 & = 80 \end{array} \tag{1.40}$$

Step **2.** ***Elimination of*** x_2

The first equation remains as it is. We want the new second equation to serve as the next pivot equation. But since it has no x_2-term (in fact, it is $0 = 0$), we must first change the order of the equations and the corresponding rows of the new matrix. We put 0 = 0 at the end and move the third equation and the fourth equation one place up. This is called **partial pivoting** (as opposed to the rarely used *total pivoting*, in which also the order of the unknowns is changed). It gives

$$\begin{array}{ll} \text{Pivot } 10 \longrightarrow \\ \text{Eliminate } 30 \longrightarrow \end{array} \left[\begin{array}{ccc|c} 1 & 1 & 1 & 0 \\ 0 & \textcircled{10} & 25 & 90 \\ 0 & \boxed{30} & -20 & 80 \\ 0 & 0 & 0 & 0 \end{array}\right] \quad \begin{array}{ll} \text{Pivot } 10 \longrightarrow \\ \text{Eliminate } 30x_2 \longrightarrow \end{array} \quad \begin{aligned} x_1 - x_2 + x_3 &= 0 \\ \textcircled{10x_2} + 25x_3 &= 90 \\ \boxed{30x_2} - 20x_3 &= 80 \\ 0 &= 0 \end{aligned}$$

To eliminate x_2, add –4 times the pivot equation to the third equation. The result is

$$\left[\begin{array}{ccc|c} 1 & -1 & 1 & 0 \\ 0 & 10 & 25 & 90 \\ 0 & 0 & -95 & -190 \\ 0 & 0 & 0 & 0 \end{array}\right] \quad \text{Row } 3 \to \text{Row } 3 - 3 \text{ Row } 2 \quad \begin{aligned} x_1 - x_2 + x_3 &= 0 \\ 10x_2 + 25x_3 &= 90 \\ -95x_3 &= -190 \\ 0 &= 0 \end{aligned} \tag{1.41}$$

Back Substitution ***[Determination of x_3, x_2, x_1 (in this order)]***: Working backward from the last to the first equation of this "triangular" system (1.41), we can now readily find x_3, then x_2, and then x_1:

$$\begin{aligned} x_3 &= -190 \\ 10x_2 + 25x_3 &= 90 \\ x_1 - x_2 + x_3 &= 0 \end{aligned} \qquad \begin{aligned} x_3 &= i_3 = 2 \text{ [A]} \\ x_2 &= \frac{90 - 25x_3}{10} = i_2 = 4 \text{ [A]} \\ x_1 &= x_2 - x_3 = i_1 = 2 \text{ [A]} \end{aligned}$$

where A stands for "amperes." This is the answer to our problem. The solution is unique.

Example 26 illustrates the operations of the Gauss elimination. These are the first two of three operations, which are called as follows:

1. **Elementary Row Operations for Matrices:** These operations are for rows, ***not for columns***!
   - *Interchange of two rows.*
   - *Addition of a constant multiple of one row to another row.*
   - *Multiplication of a row by a* ***nonzero*** *constant c.*
2. **Elementary Operations for Equations:**
   - *Interchange of two equations.*
   - *Addition of a constant multiple of one equation to another equation.*
   - *Multiplication of an equation by a* ***nonzero*** *constant c.*

Clearly, the interchange of two equations does not alter the solution set. Neither does that addition because we can undo it by a corresponding subtraction. Similarly for that multiplication, which we can undo by multiplying the new equation by $1/c$ (since $c \neq 0$), producing the original equation.

We now call a linear system S_1 **row-equivalent** to a linear system S_2 if S_1 can be obtained from S_2 by (finitely many!) row operations. Thus we have proved the following result, which also justifies the Gauss elimination.

THEOREM 5 (Row-Equivalent Systems): *Row-equivalent linear systems have the same set of solutions.*

Because of this theorem, systems having the same solution sets are often called *equivalent systems.* But note well that we are dealing with ***row operations.*** No column operations on the augmented matrix are permitted in this context because they would generally alter the solution set.

A linear system (1.38) is called **overdetermined** if it has more equations than unknowns, as in Example 26, **determined** if $m = n$, as in Example 25, and **underdetermined** if it has fewer equations than unknowns. Furthermore, a system (1.38) is called **consistent** if it has at least one solution (thus, one solution or infinitely many solutions), but **inconsistent** if it has no solutions at all, as $x_1 + x_2 = 1$, $x_1 + x_2 = 0$ in Example 25.

Gauss Elimination: The Three Possible Cases of Systems

The Gauss elimination can take care of linear systems with a unique solution (see Example 26), with infinitely many solutions (Example 27), and without solutions (inconsistent systems; see Example 28).

Example 27 Gauss Elimination if Infinitely Many Solutions Exist

Solve the following linear systems of three equations in four unknowns whose augmented matrix is

$$\left[\begin{array}{cccc|c} 3.0 & 2.0 & 2.0 & -5.0 & 8.0 \\ 0.6 & 1.5 & 1.5 & -5.4 & 2.7 \\ 1.2 & -0.3 & -0.3 & 2.4 & 2.1 \end{array}\right]$$

Thus,

$$\begin{aligned} 3.0x_1 + 2.0x_2 + 2.0x_3 - 5.0x_4 &= 8.0 \\ \boxed{0.6x_1} + 1.5x_2 + 1.5x_3 - 5.4x_4 &= 2.7 \\ \boxed{1.2x_1} - 0.3x_2 - 0.3x_3 + 2.4x_4 &= 2.1 \end{aligned} \tag{1.42}$$

Solution: As in the previous example, we circle pivots and box terms of equations and corresponding entries to be eliminated. We indicate the operations in terms of equations and operate on both equations and matrices.

Step **1.** ***Elimination of*** x_1 from the second and third equations by adding $-0.6/3.0 = -0.2$ times the first equation to the second equation, $-1.2/3.0 = -0.4$ times the first equation to the third equation. This gives the following, in which the pivot of the next step is circled.

$$\left[\begin{array}{cccc|c} 3.0 & 2.0 & 2.0 & -5.0 & 8.0 \\ 0 & 1.1 & 1.1 & -4.4 & 1.1 \\ 0 & -1.1 & -1.1 & 4.4 & -1.1 \end{array}\right] \quad \begin{array}{l} \text{Row 2} \rightarrow \text{Row 2} - 0.2 \text{ Row 1} \\ \text{Row 3} \rightarrow \text{Row 3} - 0.4 \text{ Row 1} \end{array} \quad \begin{aligned} 3.0x_3 + 2.0x_2 + 2.0x_3 - 5.0x_4 &= 8.0 \\ 1.1x_2 + 1.1x_3 - 4.4x_4 &= 1.1 \\ \boxed{-1.1x_2} - 1.1x_3 + 4.4x_4 &= -1.1 \end{aligned} \tag{1.43}$$

Step **2.** ***Elimination of*** x_2 from the third equation of (1.43) by adding $1.1/1.1 = 1$ times the second equation to the third equation. This gives

$$\left[\begin{array}{cccc|c} 3.0 & 2.0 & 2.0 & -5.0 & 8.0 \\ 0 & 1.1 & 1.1 & -4.4 & 1.1 \\ 0 & 0 & 0 & 0 & 0 \end{array}\right] \quad \text{Row 3} \rightarrow \text{Row 3} + \text{Row 2} \quad \begin{aligned} 3.0x_1 + 2.0x_2 + 2.0x_3 - 5.0x_4 &= 8.0 \\ 1.1x_2 + 1.1x_3 - 4.4x_4 &= 1.1 \\ 0 &= 0 \end{aligned} \tag{1.44}$$

Back Substitution. From the second equation, $x_2 = 1 - x_3 + 4x_4$. From this and the first equation, $x_1 = 2 - x_4$. Since x_3 and x_4 remain arbitrary, we have infinitely many solutions. If we choose a value of x_3 and a value of x_4, then the corresponding values of x_1 and x_2 are uniquely determined.

On Notation. If unknowns remain arbitrary, it is also customary to denote them by other letters $t_1, t_2, \ldots$. In this example we may thus write $x_1 = 2 - x_4 = 2 - t_2$, $x_2 = 1 - x_3 + 4x_4 = 1 - t_1 + 4t_2$, $x_3 = t_1$ (first arbitrary unknown), $x_4 = t_2$ (second arbitrary unknown).

Example 28 Gauss Elimination if no Solution Exists

What will happen if we apply the Gauss elimination to a linear system that has no solution? The answer is that in this case the method will show this fact by producing a contradiction. For instance, consider

$$\left[\begin{array}{ccc|c} 3 & 1 & 1 & 3 \\ 2 & 1 & 1 & 0 \\ 6 & 2 & 4 & 6 \end{array}\right] \qquad \begin{aligned} \boxed{3x_1} + 2x_2 + x_3 &= 3 \\ \boxed{2x_1} + x_2 + x_3 &= 0 \\ \boxed{6x_1} + 2x_2 + 4x_3 &= 6 \end{aligned}$$

Step 1. ***Elimination of*** x_1 from the second and third equations by adding −2/3 times the first equation to the second equation, −6/3 = −2 times the first equation to the third equation. This gives

$$\left[\begin{array}{ccc|c} 3 & 2 & 1 & 3 \\ 0 & -1/3 & 1/3 & -2 \\ 0 & -2 & 2 & 0 \end{array}\right] \quad \begin{array}{l} \text{Row 2} \to \text{Row 2} - \frac{2}{3}\,\text{Row 1} \\ \text{Row 3} \to \text{Row 3} - 2\,\text{Row 1} \end{array} \quad \begin{aligned} 3x_1 + 2x_2 + x_3 &= 3 \\ -\frac{1}{3}x_2 + (1/3)x_3 &= -2 \\ \boxed{-2x_2} + 2x_3 &= 0 \end{aligned}$$

Step 2. ***Elimination of*** x_2 from the third equation gives

$$\left[\begin{array}{ccc|c} 3 & 2 & 1 & 3 \\ 0 & -1/3 & -1/3 & -2 \\ 0 & 0 & 0 & 12 \end{array}\right] \quad \text{Row 3} \to \text{Row 3} - 6\,\text{Row 2} \quad \begin{aligned} 3x_1 + 2x_2 + x_3 &= 3 \\ -(1/3)x_2 + (1/3)x_3 &= -2 \\ 0 &= 12 \end{aligned}$$

The false statement 0 = 12 shows that the system has no solution.

Echelon Form and Row-Reduced Echelon Form

At the end of the Gauss elimination the form of the coefficient matrix, the augmented matrix, and the system itself are called the **row echelon form.** In it, rows of zeros, if present, are the last rows, and in each nonzero row the leftmost nonzero entry is farther to the right than in the previous row. For instance, in Example 28 the coefficient matrix and its augmented in row echelon form are

$$\begin{bmatrix} 3 & 2 & 1 \\ 0 & -1/3 & 1/3 \\ 0 & 0 & 0 \end{bmatrix} \quad \text{and} \quad \left[\begin{array}{ccc|c} 3 & 2 & 1 & 3 \\ 0 & -1/3 & 1/3 & -2 \\ 0 & 0 & 0 & 12 \end{array}\right]$$

Note that we do not require that the leftmost nonzero entries be 1 since this would have no theoretic or numeric advantage. (The so-called *reduced echelon form*, in which those entries *are* 1, will be discussed later.)

At the end of the Gauss elimination (before the back substitution) the row echelon form of the augmented matrix will be

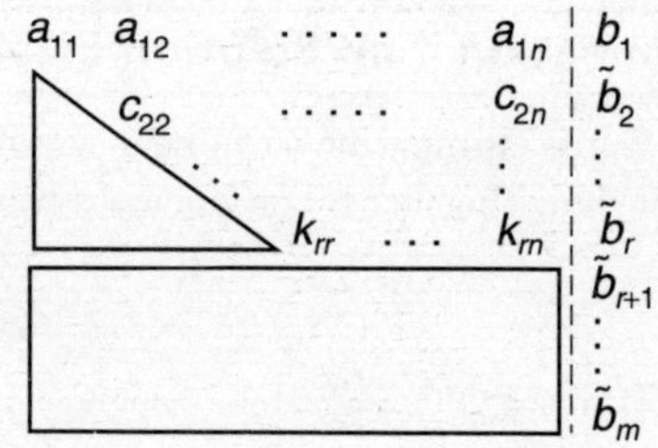

(1.45)

Here, $r \leqq m$ and $a_{11} \neq 0$, $c_{22} \neq 0$, …, $k_{rr} \neq 0$, and all the entries in the blue triangle as well as in the blue rectangle are zero. From this we see that with respect to solutions of the system with augmented matrix (1.45) (and thus with respect to the originally given system) there are three possible cases:

(a) ***Exactly one solution*** if $r = n$ and $\tilde{b}_{r+1}, \dots, \tilde{b}_m$, if present, are zero. To get the solution, solve the nth equation corresponding to (1.45) (which is $k_{nn}x_n = \tilde{b}_n$) for x_n, then the $(n-1)$th equation for x_{n-1}, and so on up the line. See Example 26, where $r = n = 3$ and $m = 4$.

(b) ***Infinitely many solutions*** if $r < n$ and $\tilde{b}_{r+1}, \dots, \tilde{b}_m$, if present, are zero. To obtain any of these solutions, choose values of $x_{r+1}, \dots, x_n$ arbitrarily. Then solve the rth equation for x_r, then the $(r-1)$th equation for x_{r-1}, and so on up the line. See Example 27.

(c) ***No solution*** if $r < m$ and one of the entries $\tilde{b}_{r+1}, \dots, \tilde{b}_m$ is not zero. See Example 28, where $r = 2 < m = 3$ and $\tilde{b}_{r+1} = \tilde{b}_3 = 12$.

CONCEPT CHECK

1. Solve the following systems or indicate the nonexistence of solutions. (Show the details of your work.)

(a) $\begin{aligned} 5x - 2y &= 20.9 \\ -x + 4y &= 19.3 \end{aligned}$

(b) $\begin{aligned} 3.0x + 6.2y &= 0.2 \\ 2.1x + 8.5y &= 4.3 \end{aligned}$

(c) $\begin{aligned} 0.5x + 3.5y &= 5.7 \\ -x + 5.0y &= 7.8 \end{aligned}$

(d) $\begin{aligned} 4y - 2z &= 2 \\ 6x - 2y + z &= 29 \\ 4x + 8y - 4z &= 24 \end{aligned}$

(e) $\begin{aligned} 0.8x + 1.2y - 0.6z &= -7.8 \\ 2.6x + 1.7z &= 15.3 \\ 4.0x - 7.3y - 1.5z &= 1.1 \end{aligned}$

(f) $\begin{aligned} 14x - 2y - 4z &= 0 \\ 18x - 2y - 6z &= 0 \\ 4x + 8y - 14z &= 0 \end{aligned}$

(g) $\begin{aligned} y + z &= -2 \\ 4y + 6z &= -12 \\ x + y + z &= 2 \end{aligned}$

(h) $\begin{aligned} 2x + y - 3z &= 8 \\ 5x + 2z &= 3 \\ 8x - y + 7z &= 0 \end{aligned}$

(i) $\begin{aligned} 4y + 4z &= 24 \\ 3x - 11y - 2z &= -6 \\ 6x - 17y + z &= 18 \end{aligned}$

(j) $\begin{aligned} 0.6x + 0.3y - 0.4z &= -1.9 \\ -4.6x + 0.5y + 1.2z &= -1.3 \end{aligned}$

(k) $\begin{aligned} 2x - y + 3z &= -1 \\ -4x + 2y - 6z &= 2 \end{aligned}$

(l) $\begin{aligned} -2y - 2z &= -8 \\ 3x + 4y - 5z &= 13 \end{aligned}$

(m) $\begin{aligned} x + y - 2z &= 0 \\ -4w - x - y + 2z &= -4 \\ -2w + 3x + 3y - 6z &= -2 \end{aligned}$

(n) $\begin{aligned} w - 2x + 5y - 3z &= 0 \\ -3w + 6x + y + z &= 0 \\ 2w - 4x + 3y - z &= 3 \end{aligned}$

(o)
$$\begin{aligned} 3x + 7y - 4z &= -46 \\ 5w + 4x + 8y + z &= 7 \\ 8w + 4y - 2z &= 0 \\ -w + 6x + 2z &= 13 \end{aligned}$$

(p)
$$\begin{aligned} -2w - 17x + 4y + 3z &= 0 \\ 7w + 3y - 2z &= 0 \\ 2x + 8y - 6z &= -20 \\ 5w - 13x - y + 5z &= 16 \end{aligned}$$

2. Using Kirchhoff's laws (see Example 26), find the currents. (Show the details of your work.)

(a)

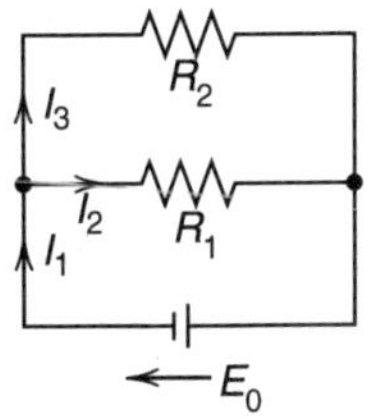

(b)

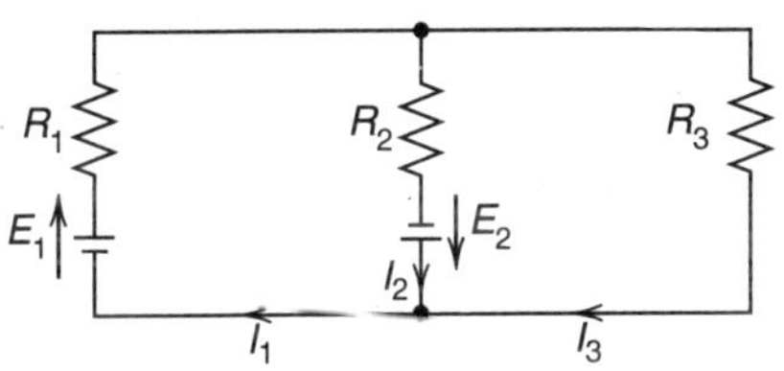

(c)

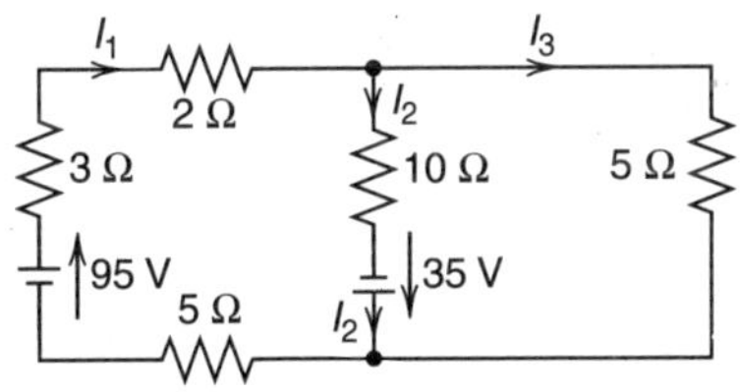

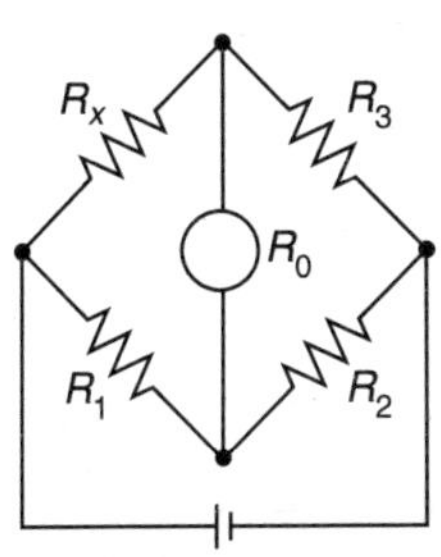

Figure 7 Wheatstone bridge.

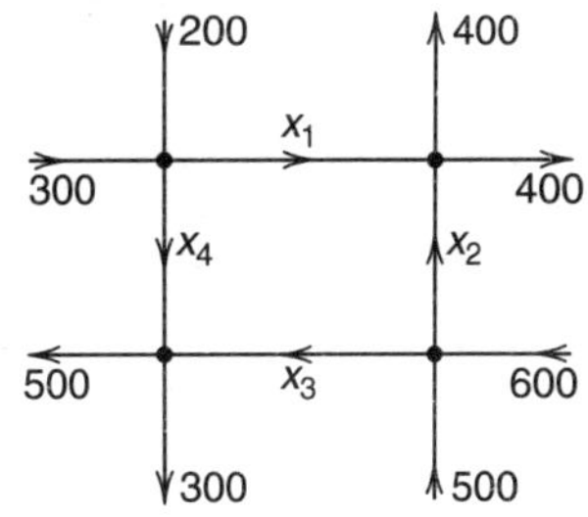

Figure 8 Net of one-way streets.

3. **(Wheatstone bridge, Figure 7)** Show that if $R_x/R_3 = R_1/R_2$ in the figure, then $I = 0$. (R_0 is the resistance of the instrument by which I is measured.) This bridge is a method for determining R_x, R_1, R_2, R_3 are known. R_3 is variable. To get R_x, make $I = 0$ by varing R_3. Then calculate $R_x = R_3R_1/R_2$.

4. **(Traffic flow, Figure 8)** Methods of electrical circuit analysis have applications to other fields. For instance, applying the analog of Kirchhoff's current law, find the traffic flow (cars per hour) in the net of one-way streets (in the directions indicated by the arrows) shown in the figure. Is the solution unique?

5. **(Models of markets)** Determine the equilibrium solution ($D_1 = S_1$, $D_2 = S_2$) of the two-commodity market with linear model (D, S, P = demand, supply, price; index 1 = first commodity, index 2 = second commodity)

$$D_1 = 60 - 2P_1 - P_2, \quad S_1 = 4P_1 - 2P_2 + 14$$
$$D_2 = 4P_1 - P_2 + 10, \quad S_2 = 5P_2 - 2$$

6. **(Equivalence relation)** By definition, an *equivalence relation* on a set is a relation satisfying three conditions (named as indicated):
   (a) Each element A of the set is equivalent to itself ("*Reflexivity*").
   (b) If A is equivalent to B, then B is equivalent to A ("*Symmetry*").
   (c) If A is equivalent to B and B is equivalent to C, then A is equivalent to C ("*Transitivity*").

   Show that row equivalence of matrices satisfies these three conditions. *Hint.* Show that for each of the three elementary row operations these conditions hold.

7. **PROJECT. Elementary Matrices.** The idea is that elementary operations can be accomplished by matrix multiplication. If **A** is an $m \times n$ matrix on which we want to do an elementary operation, then there is a matrix **E** such that **EA** is the new matrix after the operation. Such an **E** is called an **elementary matrix.** This idea can be helpful, for instance, in the design of algorithms. (*Computationally,* it is generally preferable to do row operations *directly,* rather than by multiplication by **E**.)
   (a) Show that the following are elementary matrices, for interchanging Rows 2 and 3, for adding −5 times the first row to the third, and for multiplying the fourth row by 8.

$$\mathbf{E}_1 = \begin{bmatrix} 1 & 0 & 0 & 0 \\ 0 & 0 & 1 & 0 \\ 0 & 1 & 0 & 0 \\ 0 & 0 & 0 & 1 \end{bmatrix}$$

$$\mathbf{E}_2 = \begin{bmatrix} 1 & 0 & 0 & 0 \\ 0 & 1 & 0 & 0 \\ -5 & 0 & 1 & 0 \\ 0 & 0 & 0 & 1 \end{bmatrix}$$

$$\mathbf{E}_3 = \begin{bmatrix} 1 & 0 & 0 & 0 \\ 0 & 1 & 0 & 0 \\ 0 & 0 & 1 & 0 \\ 0 & 0 & 0 & 8 \end{bmatrix}$$

   Apply $\mathbf{E}_1, \mathbf{E}_2, \mathbf{E}_3$ to a vector and to a 4×3 matrix of your choice. Find $\mathbf{B} = \mathbf{E}_3\mathbf{E}_2\mathbf{E}_1\mathbf{A}$, where $\mathbf{A} = [a_{jk}]$ is the general 4×2 matrix. Is **B** equal to $\mathbf{C} = \mathbf{E}_1\mathbf{E}_2\mathbf{E}_3\mathbf{A}$?
   (b) Conclude that $\mathbf{E}_1$, $\mathbf{E}_2$, $\mathbf{E}_3$ are obtained by doing the corresponding elementary operations on the 4×4 unit matrix. Prove that if **M** is obtained from **A** by an elementary row operation, then

$$\mathbf{M} = \mathbf{EA}$$

   where **E** *is* obtained from the $n \times n$ unit matrix $\mathbf{I}_n$ by the same row operation.

1.3 Rank of a Matrix

In the last section we explained the Gauss elimination with back substitution, the most important numeric solution method for linear systems of equations. It appeared that such a system may have a unique solution or infinitely many solutions, or it may be inconsistent, that is, have no solution at all. Hence we are confronted with the questions of **existence and uniqueness of solutions.** As the key concept for this (and other questions) we introduce the ***rank of a matrix***. To define rank, we first need the following concepts, which are of general importance.

Linear Independence and Dependence of Vectors

Given any set of m vectors $\mathbf{a}_{(1)}, \ldots, \mathbf{a}_{(m)}$ (with the same number of components), a **linear combination** of these vectors is an expression of the form

$$c_1\mathbf{a}_{(1)} + c_2\mathbf{a}_{(2)} + \cdots + c_m\mathbf{a}_{(m)}$$

where $c_1, c_2, \ldots, c_m$ are any scalars. Now consider the equation

$$c_1\mathbf{a}_{(1)} + c_2\mathbf{a}_{(2)} + \cdots + c_m\mathbf{a}_{(m)} = \mathbf{0} \tag{1.46}$$

Clearly, this vector equation (1.46) holds if we choose all c_j's zero, because then it becomes $\mathbf{0} = \mathbf{0}$. If this is the only m-tuple of scalars for which (1.46) holds, then our vectors $\mathbf{a}_{(1)}, \ldots, \mathbf{a}_{(m)}$ are said to form a *linearly independent set* or, more briefly, we call them **linearly independent**. Otherwise, if (1.46) also holds with scalars not all zero, we call these vectors **linearly dependent**, because then we can express (at least) one of them as a linear combination of the others. For instance, if (1.46) holds with, say, $c_1 \neq 0$, we can solve (1.46) for $\mathbf{a}_{(1)}$:

$$\mathbf{a}_{(1)} = k_2\mathbf{a}_{(2)} + \cdots + k_m\mathbf{a}_{(m)} \quad \text{where } k_j = -c_j/c_1$$

(Some k_j's may be zero. Or even all of them, namely, if $\mathbf{a}_{(1)} = \mathbf{0}$.)

Why is this important? Well, in the case of linear dependence we can get rid of some of the vectors until we arrive at a linearly independent set that is optimal to work with because it is smallest possible in the sense that it consists only of the "really essential" vectors, which can no longer be expressed linearly in terms of each other. This motivates the idea of a "basis" used in various contexts, notably later in our present section.

Example: Linear Independence and Dependence

The three vectors

$$\begin{aligned} \mathbf{a}_{(1)} &= [3 \quad 0 \quad 2 \quad 2] \\ \mathbf{a}_{(2)} &= [-6 \quad 42 \quad 24 \quad 54] \\ \mathbf{a}_{(3)} &= [21 \quad -21 \quad 0 \quad -15] \end{aligned}$$

are linearly dependent because

$$6\mathbf{a}_{(1)} - \frac{1}{2}\mathbf{a}_{(2)} - \mathbf{a}_{(3)} = \mathbf{0}$$

Although this is easily checked (do it!), it is not so easy to discover. However, a systematic method for finding out about linear independence and dependence follows below.

The first two of the three vectors are linearly independent because $c_1\mathbf{a}_{(1)} + c_2\mathbf{a}_{(2)} = \mathbf{0}$ implies $c_2 = 0$ (from the second components) and then $c_1 = 0$ (from any other component of $\mathbf{a}_{(1)}$).

DEFINITION The **rank** of a matrix **A** is the maximum number of linearly independent row vectors of **A**. It is denoted by rank **A**.

Our further discussion will show that the rank of a matrix is an important key concept for understanding general properties of matrices and linear systems of equations.

Example 29 Rank

The matrix

$$\mathbf{A} = \begin{bmatrix} 3 & 0 & 2 & 2 \\ -6 & 42 & 24 & 54 \\ 21 & -21 & 0 & -15 \end{bmatrix} \tag{1.47}$$

has rank 2, because the example in box shows that the first two row vectors are linearly independent, whereas all three row vectors are linearly dependent.

Note further that rank $\mathbf{A} = 0$ if and only if $\mathbf{A} = \mathbf{0}$. This follows directly from the definition.

We call a matrix $\mathbf{A}_1$ **row-equivalent** to a matrix $\mathbf{A}_2$ if $\mathbf{A}_1$ can be obtained from $\mathbf{A}_2$ by (finitely many!) elementary row operations. Now the maximum number of linearly independent row vectors of a matrix does not change if we change the order of rows or multiply a row by an nonzero c or take a linear combination by adding a multiple of a row to another row. This proves that rank is **invariant** under elementary row operations:

THEOREM 6 (Row-Equivalent Matrices): *Row-equivalent matrices have the same rank.*

Hence we can determine the rank of a matrix by reduction to row-echelon form and then see the rank directly.

Example 30 Determination of Rank

For the matrix in Example 29 we obtain successively

$$\mathbf{A} = \begin{bmatrix} 3 & 0 & 2 & 2 \\ -6 & 42 & 24 & 54 \\ 21 & -21 & 0 & -15 \end{bmatrix} \quad \text{(given)}$$

$$= \begin{bmatrix} 3 & 0 & 2 & 2 \\ 0 & 42 & 28 & 58 \\ 0 & -21 & -14 & -29 \end{bmatrix} \quad \begin{matrix} \text{Row } 2 \to \text{Row } 2 + 2 \text{ Row } 1 \\ \text{Row } 3 \to \text{Row } 3 - 7 \text{ Row } 1 \end{matrix}$$

$$= \begin{bmatrix} 3 & 0 & 2 & 2 \\ 0 & 42 & 28 & 58 \\ 0 & 0 & 0 & 0 \end{bmatrix} \quad \text{Row } 3 \to \text{Row } 3 + (1/2) \text{ Row } 2$$

Since rank is defined in terms of two vectors, we immediately have the useful theorem below.

THEOREM 7 (Linear Independence and Dependence of Vectors): *p vectors with n components each are linearly independent if the matrix with these vectors as row vectors has rank p, but they are linearly dependent if that rank is less than p.*

Further important properties will result from the following basic theorem.

THEOREM 8 (Rank in Terms of Column Vectors): *The rank r of a matrix* **A** *equals the maximum number of linearly independent* ***column*** *vectors of* **A**. *Hence* **A** *and its transpose* $\mathbf{A}^T$ *have the same rank.*

PROOF: In this proof we write simply "rows" and "columns" for row and column vectors. Let **A** be an $m \times n$ matrix of rank $\mathbf{A} = r$. Then by definition of rank, **A** has r linearly independent rows which we denote by $\mathbf{v}_{(1)}, \ldots, \mathbf{v}_{(r)}$ (regardless of their position in **A**), and all the rows $\mathbf{a}_{(1)}, \ldots, \mathbf{a}_{(m)}$ of **A** are linear combinations of those, say,

$$\begin{aligned} \mathbf{a}_{(1)} &= c_{11}\mathbf{v}_{(1)} + c_{12}\mathbf{v}_{(2)} + \cdots + c_{1r}\mathbf{v}_{(r)} \\ \mathbf{a}_{(2)} &= c_{21}\mathbf{v}_{(1)} + c_{22}\mathbf{v}_{(2)} + \cdots + c_{2r}\mathbf{v}_{(r)} \\ &\vdots \\ \mathbf{a}_{(m)} &= c_{m1}\mathbf{v}_{(1)} + c_{m}\mathbf{v}_{(2)} + \cdots + c_{mr}\mathbf{v}_{(r)} \end{aligned} \tag{1.48}$$

These are vector equations for rows. To switch to columns, we write (1.48) in terms of components as n such systems, with $k = 1, \ldots, n$,

$$\begin{aligned} a_{1k} &= c_{11}v_{1k} + c_{12}v_{2k} + \cdots + c_{1r}v_{rk} \\ a_{2k} &= c_{21}v_{1k} + c_{22}v_{2k} + \cdots + c_{2r}v_{rk} \\ &\vdots \\ a_{mk} &= c_{m1}v_{1k} + c_{m2}v_{2k} + \cdots + c_{mr}v_{rk} \end{aligned} \tag{1.49}$$

and collect components in columns. Indeed, we can write (1.49) as

$$\begin{bmatrix} a_{1k} \\ a_{2k} \\ \vdots \\ a_{mk} \end{bmatrix} = v_{1k}\begin{bmatrix} c_{11} \\ c_{21} \\ \vdots \\ c_{m1} \end{bmatrix} + v_{2k}\begin{bmatrix} c_{12} \\ c_{22} \\ \vdots \\ c_{m2} \end{bmatrix} + \cdots + v_{rk}\begin{bmatrix} c_{1r} \\ c_{2r} \\ \vdots \\ c_{mr} \end{bmatrix} \tag{1.50}$$

where $k = 1, \ldots, n$. Now the vector on the left is the kth column vector of **A**. We see that each of these n columns is a linear combination of the same r columns on the right. Hence **A** cannot have more linearly independent columns than rows, whose number is rank $\mathbf{A} = r$. Now rows of **A** are columns of the transpose $\mathbf{A}^T$. For $\mathbf{A}^T$ our conclusion is that $\mathbf{A}^T$ cannot have more linearly independent columns than rows, so that **A** cannot have more linearly independent rows than columns. Together, the number of linearly independent columns of **A** must be r, the rank of **A**. This completes the proof. ■

Example 31 **Illustration of Theorem 8**

The matrix in (1.47) has rank 2. From Example 30 we see that the first two row vectors are linearly independent and by "working backward" we can verify that Row 3 = 6 Row 1 − (1 /2) Row 2. Similarly, the first two columns are linearly independent, and by reducing the last matrix in Example 30 by columns we find that

$$\text{Column } 3 = \frac{2}{3}\text{ Column } 1 + \frac{2}{3}\text{ Column } 2 \quad \text{and} \quad \text{Column } 4 = \frac{2}{3}\text{ Column } 1 + \frac{29}{21}\text{ Column } 2$$

Combining Theorems 7 and 8 we obtain Theorem 9.

THEOREM 9 (Linear Dependence of Vectors): *p vectors with $n < p$ components are always linearly dependent.*

PROOF: The matrix **A** with those p vectors as row vectors has p rows and $n < p$ columns; hence by Theorem 8 it has rank $\mathbf{A} \leqq n < p$, which implies linear dependence by Theorem 7. ■

Vector Space

The following related concepts are of general interest in linear algebra. In the present context they provide a clarification of essential properties of matrices and their role in connection with linear systems.

DEFINITION A **vector space** is a (nonempty) set V of vectors such that with any two vectors **a** and **b** in V all their linear combinations $\alpha\mathbf{a} + \beta\mathbf{b}$ (α, β any real numbers) are elements of V, and these vectors satisfy the laws (1.3) and (1.4) (written in lowercase letters **a**, **b**, **u**, …, which is our notation for vectors).

DEFINITION The maximum number of linearly independent vectors in V is called the **dimension** of V and is denoted by dim V.

DEFINITION A linearly independent set in V consisting of a maximum possible number of vectors in V is called a **basis** for V. Thus the number of vectors of a basis for V equals dim V.

DEFINITION The set of all linear combinations of given vectors $\mathbf{a}_{(1)}, \ldots, \mathbf{a}_{(p)}$ with the same number of components is called the **span** of these vectors. Obviously, a span is a vector space.

DEFINITION By a **subspace** of a vector space V we mean a nonempty subset of V (including V itself) that forms itself a vector space with respect to the two algebraic operations (addition and scalar multiplication) defined for the vectors of V.

THEOREM 10 (Vector Space R^n): *The vector space R^n consisting of all vectors with n components (n real numbers) has dimension n.*

PROOF: A basis of n vectors is $\mathbf{a}_{(1)} = [1 \quad 0 \quad \ldots \quad 0]$, $\mathbf{a}_{(2)} = [0 \quad 1 \quad 0 \quad \ldots \quad 0], \ldots,$ $\mathbf{a}_{(n)} = [0 \quad \ldots \quad 0 \quad 1]$. ■

In the case of a matrix **A** we call the span of the row vectors the **row space** of A and the span of the column vectors the **column space** of A. Now, Theorem 8 shows that a matrix **A** has as many linearly independent rows as columns. By the definition of dimension, their number is the dimension of the row space or the column space of A. This proves the following theorem.

THEOREM 11 (Row Space and Column Space): *The row space and the column space of a matrix* **A** *have the same dimension, equal to rank* **A**.

Finally, for a given matrix **A** the solution set of the homogeneous system $\mathbf{Ax} = \mathbf{0}$ is a vector space, called the **null space** of A, and its dimension is called the **nullity** of A. In the next section we motivate and prove the basic relation

$$\text{Rank } \mathbf{A} + \text{Nullity } \mathbf{A} = \text{Number of columns of } \mathbf{A} \tag{1.51}$$

Solutions of Linear Systems: Existence, Uniqueness

Rank as just defined gives complete information about existence, uniqueness, and general structure of the solution set of linear systems as follows.

A linear system of equations in n unknowns has a unique solution if the coefficient matrix and the augmented matrix have the same rank n, and infinitely many solution if that common rank is less than n. The system has no solution if those two matrices have different rank.

To state this precisely and prove it, we shall use the (generally important) concept of a **submatrix** of A. By this we mean any matrix obtained from **A** by omitting some rows or columns (or both). By definition this includes **A** itself (as the matrix obtained by omitting no rows or columns); this is practical.

THEOREM 12 (Fundamental Theorem for Linear Systems):

(a) **Existence.** *A linear system of m equations in n unknowns $x_1, \ldots, x_n$*

$$\begin{aligned} a_{11}x_1 + a_{12}x_2 + \cdots + a_{1n}x_n &= b_1 \\ a_{21}x_1 + a_{22}x_2 + \cdots + a_{2n}x_n &= b_2 \\ \cdots\cdots\cdots\cdots\cdots\cdots\cdots\cdots\cdots \\ a_{m1}x_1 + a_{m2}x_2 + \cdots + a_{mn}x_n &= b_m \end{aligned} \tag{1.52}$$

*is **consistent**, that is, has solutions, if and only if the coefficient matrix* **A** *and the augmented matrix* $\tilde{\mathbf{A}}$ *have the same rank. Here,*

$$\mathbf{A} = \begin{bmatrix} a_{11} & \cdots & a_{1n} \\ \cdot & \cdots & \cdot \\ \cdot & \cdots & \cdot \\ a_{m1} & \cdots & a_{mn} \end{bmatrix} \quad \text{and} \quad \tilde{\mathbf{A}} = \begin{bmatrix} a_{11} & \cdots & a_{1n} & b_1 \\ \cdot & \cdots & \cdot & \cdot \\ \cdot & \cdots & \cdot & \cdot \\ a_{m1} & \cdots & a_{mn} & b_m \end{bmatrix}$$

(b) **Uniqueness.** *The system* (1.52) *has precisely one solution if and only if this common rank r of* **A** *and* $\tilde{\mathbf{A}}$ *equals n.*

(c) **Infinitely many solutions.** *If this common rank r is less than n, the system* (1.52) *has infinitely many solutions. All of these solutions are obtained by determining r suitable unknowns (whose submatrix of coefficients must have rank r) in terms of the remaining* $n - r$ *unknowns, to which arbitrary values can be assigned.*

(d) **Gauss elimination.** *If solutions exist, they can all be obtained by the Gauss elimination.*

Homogeneous Linear System

Recall that a linear system is called **homogeneous** if all the b_j's are zero, and **nonhomogeneous** if one or several b_j's are not zero. For the homogeneous system we obtain from the Fundamental Theorem the following results.

THEOREM 13 (Homogeneous Linear System): *A homogeneous linear system*

$$\begin{aligned} a_{11}x_1 + a_{12}x_2 + \cdots + a_{1n}x_n &= 0 \\ a_{21}x_1 + a_{22}x_2 + \cdots + a_{2n}x_n &= 0 \\ \cdots\cdots\cdots\cdots\cdots\cdots\cdots\cdots & \\ a_{m1}x_1 + a_{m2}x_2 + \cdots + a_{mn}x_n &= 0 \end{aligned} \tag{1.53}$$

*always is consistent if it has unique or **trivial solution*** $x_1 = 0, \ldots, x_n = 0$ *that is if* $\det \mathbf{A} \neq 0$ *or* $\rho(\mathbf{A})$ = *number of unknowns. Nontrivial solutions exist if and only if rank* $\mathbf{A} < n$. *If rank* $\mathbf{A} = r < n$, *these solutions, together with* $\mathbf{x} = \mathbf{0}$, *form a vector space of dimension* $n - r$, *called the **solution space** of* (1.53).

In particular, if $\mathbf{x}_{(1)}$ *and* $\mathbf{x}_{(2)}$ *are solution vectors of* (1.53), *then* $\mathbf{x} = c_1\mathbf{x}_{(1)} + c_2\mathbf{x}_{(2)}$ *with any scalars* c_1 *and* c_2 *is a solution vector of* (1.53). *(This **does not hold** for nonhomogeneous systems. Also, the term **solution space** is used for **homogeneous** systems only.)*

THEOREM 14 (Homogeneous Linear System with Fewer Equations than Unknowns): *A homogeneous linear system with fewer equations than unknowns has always nontrivial solutions.*

Nonhomogeneous Linear Systems

The characterization of all solutions of the linear system (1.52) is now quite simple, as follows.

THEOREM 15 (Nonhomogeneous Linear System): *If a nonhomogeneous linear system* (1.52) *is consistent, then all of its solutions are obtained as*

$$\mathbf{x} = \mathbf{x}_0 + \mathbf{x}_h \tag{1.54}$$

where $\mathbf{x}_0$ *is any (fixed) solution of* (1.52) *and* $\mathbf{x}_h$ *runs through all the solutions of the corresponding homogeneous system* (1.54).

CONCEPT CHECK

1. Find the rank and a basis for the row space and for the column space. *Hint.* Row-reduce the matrix and its transpose. (You may omit obvious factors from the vectors of these bases.)

(a) $\begin{bmatrix} 1 & -2 \\ 0 & 0 \\ -3 & 6 \end{bmatrix}$ (b) $\begin{bmatrix} 8 & 2 & 5 \\ 16 & 6 & 29 \\ 4 & 0 & -7 \end{bmatrix}$

(c) $\begin{bmatrix} 0 & -2 & 1 & 3 \\ 1 & 4 & 0 & 7 \\ 5 & 5 & 5 & 5 \end{bmatrix}$ (d) $\begin{bmatrix} a & b & c \\ b & a & c \end{bmatrix}$

(e) $\begin{bmatrix} 0 & 3 & 4 \\ -3 & 0 & -5 \\ -4 & 5 & 0 \end{bmatrix}$ (f) $\begin{bmatrix} 1 & 1 & a \\ 1 & a & 1 \\ a & 1 & 1 \end{bmatrix}$

(g) $\begin{bmatrix} 8 & 0 & 4 \\ 0 & 2 & 0 \\ 4 & 0 & 2 \\ 0 & 4 & 0 \end{bmatrix}$ (h) $\begin{bmatrix} 1 & 2 & 3 & 4 \\ 2 & 3 & 4 & 5 \\ 3 & 4 & 5 & 6 \\ 4 & 5 & 6 & 7 \end{bmatrix}$

(i) $\begin{bmatrix} 1 & 0 & 3 & 0 \\ 0 & 5 & 8 & -37 \\ 3 & 8 & 7 & 0 \\ 0 & -37 & 0 & 37 \end{bmatrix}$

(j) $\begin{bmatrix} 1 & -2 & 3 & -4 \\ 2 & -3 & 4 & -1 \\ 3 & -4 & 1 & -2 \\ 4 & -1 & 2 & -3 \end{bmatrix}$

(k) $\begin{bmatrix} 2 & 4 & 8 & 16 \\ 16 & 8 & 4 & 2 \\ 4 & 8 & 16 & 2 \\ 2 & 16 & 8 & 4 \end{bmatrix}$

(l) $\begin{bmatrix} 0 & 0 & -7 & 1 \\ 0 & 0 & 5 & 0 \\ -7 & 5 & 0 & 2 \\ 1 & 0 & 2 & 0 \end{bmatrix}$

2. Are the following sets of vectors linearly independent? (Show the details.)

(a) $[3 \quad -2 \quad 0 \quad 4]$, $[5 \quad 0 \quad 0 \quad 1]$, $[-6 \quad 1 \quad 0 \quad 1]$, $[2 \quad 0 \quad 0 \quad 3]$

(b) $[1 \quad 1 \quad 0]$, $[1 \quad 0 \quad 0]$, $[1 \quad 1 \quad 1]$

(c) $[6 \quad 0 \quad 3 \quad 1 \quad 4 \quad 2]$, $[0 \quad -1 \quad 2 \quad 7 \quad 0 \quad 5]$, $[12 \quad 3 \quad 0 \quad -19 \quad 8 \quad -11]$

(d) $[3 \; 4 \; 7]$, $[2 \; 0 \; 3]$, $[8 \; 2 \; 3]$, $[5 \; 5 \; 6]$

(e) $[0.2 \quad 1.2 \quad 5.3 \quad 2.8 \quad 1.6]$, $[4.3 \quad 3.4 \quad 0.9 \quad 2.0 \quad -4.3]$

(f) $[3 \quad 2 \quad 1]$, $[0 \quad 0 \quad 0]$, $[4 \quad 3 \quad 6]$

(g) $[1 \quad 2 \quad 3 \quad 4]$, $[2 \quad 3 \quad 4 \quad 5]$, $[3 \quad 4 \quad 5 \quad 6]$, $[4 \quad 5 \quad 6 \quad 7]$

3. Show the following.

(a) rank $\mathbf{B}^T\mathbf{A}^T$ = rank $\mathbf{AB}$. (Note the order!)

(b) rank $\mathbf{A}$ = rank $\mathbf{B}$ does *not* imply rank $\mathbf{A}^2$ = rank $\mathbf{B}^2$. (Give a counterexample.)

(c) If $\mathbf{A}$ is not square, either the row vectors or the column vectors of $\mathbf{A}$ are linearly dependent.

(d) If the row vectors of a square matrix are linearly independent, so are the column vectors, and conversely.

(e) Give examples showing that the rank of a product of matrices cannot exceed the rank of either factor.

4. Is the given set of vectors a vector space? (Give reason.) If your answer is yes, determine the dimension and find a basis. (v_1, v_2, … denote components.)

(a) All vectors in R^3 such that $v_1 + v_2 = 0$

(b) All vectors in R^4 such that $2v_2 - 3v_4 = k$

(c) All vectors in R^3 with $v_1 \geqq 0$, $v_2 = -4v_3$

(d) All vectors in R^2 with $v_1 \leqq v_2$

(e) All vectors in R^3 with $4v_1 + v_3 = 0, 3v_2 = v_3$

(f) All vectors in R^4 with $v_1 - v_2 = 0, v_3 = 5v_1, v_4 = 0$

(g) All vectors in R^n with $|v_j| \leqq 1$ for $j = 1, \ldots, n$

(h) All ordered quadruples of positive real numbers

(i) All vectors in R^5 with $v_1 = 2v_2 = 3v_3 = 4v_4 = 5v_5$

(j) All vectors in R^4 with $3v_1 - v_3 = 0, 2v_1 + 3v_2 - 4v_4 = 0$

1.4 Consistent and Inconsistent Nonhomogeneous Linear Equations

DEFINITION **(Consistent Linear Equations):** A system of equations is said to be **consistent** if it has one or more solutions.

DEFINITION **(Inconsistent Linear Equations):** If a system of equations has no solution at all, it is said to be **inconsistent.**

Consistency of a Set of m Equations (Nonhomogeneous) in "n" Unknowns

Let

$$\begin{array}{c} a_{11} x_1 + a_{12}x_2 + \cdots + a_{1n}x_n = b_1 \\ a_{21}x_1 + a_{22}x_2 + \cdots + a_{2n}x_n = b_2 \\ a_{31}x_1 + a_{32}x_2 + \cdots + a_{3n}x_n = b_3 \\ \cdots\cdots\cdots\cdots\cdots\cdots\cdots\cdots \\ a_{m1}x_1 + a_{m2}x_2 + \cdots + a_{mn}x_n = b_m \end{array}$$

In the matrix form, the system of equations can be written as

$$\begin{bmatrix} a_{11} & a_{12} & a_{13} & \cdot & \cdot & \cdot & a_{1m} \\ a_{21} & a_{22} & a_{23} & \cdot & \cdot & \cdot & a_{2n} \\ \cdot & \cdot & \cdot & \cdot & \cdot & \cdot & \cdot \\ \cdot & \cdot & \cdot & \cdot & \cdot & \cdot & \cdot \\ \cdot & \cdot & \cdot & \cdot & \cdot & \cdot & \cdot \\ \cdot & \cdot & \cdot & \cdot & \cdot & \cdot & \cdot \\ a_{m1} & a_{m2} & a_{m3} & \cdot & \cdot & \cdot & a_{mn} \end{bmatrix} \begin{bmatrix} x_1 \\ x_2 \\ \cdot \\ \cdot \\ \cdot \\ \cdot \\ x_n \end{bmatrix} = \begin{bmatrix} b_1 \\ b_2 \\ \cdot \\ \cdot \\ \cdot \\ \cdot \\ b_n \end{bmatrix}_{m\times 1}$$

Let this system be denoted as

$$\mathbf{AX = B}$$

where **A** is the coefficient matrix; **X** the column matrix of unknowns; **B** the column of constants. In this case the augmented matrix is

$$[\mathbf{A}\,|\,\mathbf{B}] = \begin{bmatrix} a_{11} & a_{12} & a_{13} & \cdot & \cdot & a_{1n} & b_1 \\ a_{21} & a_{22} & a_{23} & \cdot & \cdot & a_{2n} & b_2 \\ a_{31} & a_{32} & a_{33} & \cdot & \cdot & a_{3n} & b_3 \\ \cdot & \cdot & \cdot & \cdot & \cdot & \cdot & \cdot \\ \cdot & \cdot & \cdot & \cdot & \cdot & \cdot & \cdot \\ \cdot & \cdot & \cdot & \cdot & \cdot & \cdot & \cdot \\ a_{m1} & a_{m2} & a_{m3} & \cdot & \cdot & q_{mn} & b_m \end{bmatrix}$$

To find the solution of **AX** = **B** we have to find the rank of the coefficient matrix **A** and the rank of the augmented matrix. Now there are two possibilities:

Case 1: Rank of **A** ≠ Rank of [**A** | **B**], that is $\rho(\mathbf{A}) \neq \rho[\mathbf{A} \mid \mathbf{B}]$

In this case the equations are said to be inconsistent, that is they have no solution.

Case 2: If Rank of **A** = Rank of [**A** | **B**] = r (say), that is $\rho(\mathbf{A}) = \rho[\mathbf{A} \mid \mathbf{B}]$

In this case the equations are said to be consistent, that is they have a solution.

If r is the rank of **A** and [**A** | **B**] then the following three cases are possible:

Case 1: If $r = n$, then $n - r = 0$. In this case there will be a unique solution of the set of equations.

Case 2: If $r < n$, then $n - r > 0$. In this case the arbitrary value will be given to the $n - r$ variables. Thus in this case there will be an infinite number of solutions.

Case 3: If $m < n$, where $r \leqq m < n$, then $n - r > 0$. In this case there will be infinite number of solutions.

Example 32 Consistency Test

Test for consistency and find the solution of the equation

$$\begin{aligned} x + y + z &= 6 \\ x - y + 2z &= 5 \\ 3x + y + z &= 8 \\ 2x - 2y + 3z &= 7 \end{aligned}$$

Solution: The system of equation can be written as **AX** = **B** where

$$\mathbf{A} = \begin{bmatrix} 1 & -1 & 1 \\ 1 & -1 & 2 \\ 3 & 1 & 1 \\ 2 & -2 & 3 \end{bmatrix}, \quad \mathbf{X} = \begin{bmatrix} x \\ y \\ z \end{bmatrix}, \quad \mathbf{B} = \begin{bmatrix} 6 \\ 5 \\ 8 \\ 7 \end{bmatrix}$$

The augmented matrix [**A** | **B**] is

$$[\mathbf{A} \mid \mathbf{B}] = \begin{bmatrix} 1 & 1 & 1 & 6 \\ 1 & -1 & 2 & 5 \\ 3 & 1 & 1 & 8 \\ 2 & -2 & 3 & 7 \end{bmatrix}$$

$$= \begin{bmatrix} 1 & 1 & 1 & 6 \\ 0 & -2 & 1 & -1 \\ 0 & -2 & -2 & -10 \\ 0 & -4 & 1 & -5 \end{bmatrix} \quad \begin{aligned} &\text{Row } 2 \to \text{Row } 2 - \text{Row } 1 \\ &\text{Row } 3 \to \text{Row } 3 - 3\,\text{Row } 1 \\ &\text{Row } 4 \to \text{Row } 4 - 2\,\text{Row } 1 \end{aligned}$$

$$= \begin{bmatrix} 1 & 1 & 1 & 6 \\ 0 & -2 & 1 & -1 \\ 0 & 0 & -3 & -9 \\ 0 & 0 & 5 & 15 \end{bmatrix} \quad \begin{aligned} &\text{Row } 3 \to \text{Row } 3 - \text{Row } 2 \\ &\text{Row } 4 \to \text{Row } 4 - 2\,\text{Row } 3 \end{aligned}$$

$$[\mathbf{A}\,|\,\mathbf{B}] = \begin{bmatrix} 1 & 1 & 1 & 6 \\ 1 & -2 & 1 & -1 \\ 0 & 0 & -3 & -9 \\ 0 & 0 & 0 & 0 \end{bmatrix} \text{ Row } 4 \rightarrow \text{Row } 4 + (5/3)\text{ Row}3$$

Therefore, Rank of $[\mathbf{A}\,|\,\mathbf{B}] = 3$ and Rank of $(\mathbf{A}) = 3$. Since $\rho(\mathbf{A}) = \rho\,[\mathbf{A}\,|\,\mathbf{B}] = 3$. Hence the given system of equations is consistent. Also since the rank is equal to the number of unknowns, the system possess a unique solution. We now have

$$\begin{bmatrix} 1 & 1 & 1 \\ 0 & -1 & 1 \\ 0 & 0 & -3 \end{bmatrix} \begin{bmatrix} x \\ y \\ z \end{bmatrix} = \begin{bmatrix} 6 \\ -1 \\ -9 \end{bmatrix}$$

Expanding we get

$$\begin{aligned} x + y + z &= 6 \\ -2y + z &= -1 \\ -3z &= -9 \end{aligned}$$

Therefore, the solution is $z = 3$, $y = 2$, $x = 1$.

Example 33

For what values of λ and μ the given system of equations

$$\begin{aligned} x + y + z &= 6 \\ x + 2y + 3z &= 10 \\ x + 2y + \lambda z &= \mu \end{aligned}$$

has

(a) no solution
(b) a unique solution
(c) infinite number of solutions

Solution: Here

$$\mathbf{A} = \begin{bmatrix} 1 & 1 & 1 \\ 1 & 2 & 3 \\ 1 & 2 & \lambda \end{bmatrix}, \quad \mathbf{X} = \begin{bmatrix} x \\ y \\ z \end{bmatrix}, \quad \mathbf{B} = \begin{bmatrix} 6 \\ 10 \\ \mu \end{bmatrix}$$

The augmented matrix is given by

$$[\mathbf{A}|\mathbf{B}] = \begin{bmatrix} 1 & 1 & 1 & 6 \\ 1 & 2 & 3 & 10 \\ 1 & 2 & \lambda & \mu \end{bmatrix}$$

Now

$$[A|B]=\begin{bmatrix}1 & 1 & 1 & 6\\0 & 1 & 2 & 4\\0 & 1 & \lambda-1 & \mu-6\end{bmatrix}\begin{matrix}\text{Row 2}\to\text{Row 2 - Row 1}\\\text{Row 3}\to\text{Row 3 - Row 1}\end{matrix}$$

Therefore

$$[A|B]=\begin{bmatrix}1 & 1 & 1 & 6\\0 & 1 & 2 & 4\\0 & 0 & \lambda-3 & \mu-10\end{bmatrix}\quad\text{Row 3}\to\text{Row 3 - Row 2}$$

There exist the following cases:

Case 1: If $\lambda = 3$ and $\mu = 10$ then $\rho(A) = 2 = \rho\,[A|B] < 3$. Since Rank of A = Rank of $[A|B]$ and is less than the number of unknown, the system is consistent and hence infinite number of solutions.

Case 2: If $\lambda = 3$ and $\mu \neq 10$ then Rank of (A) = 2 and Rank of $[A|B] = 3$. Since $\rho(A) \neq \rho\,[A|B]$, the system of equation is inconsistent and there is no solution.

Case 3: If $\lambda \neq 3$ and μ has any value then Rank of (A) = 3 = Rank of $[A|B]$ and therefore a unique solution.

CONCEPT CHECK

1. Test for consistency and hence solve the system:

$$\begin{aligned}x + y + z &= 6\\x - y + 2z &= 5\\3x + y + z &= 8\\2x - 2y + 3z &= 7\end{aligned}$$

2. Test for the consistency the following system of equations:

$$\begin{aligned}x + y + z &= 1\\x - y + 2z &= 1\\x - y + 2z &= 5\\2x - 2y + 3z &= 1\\3x - y + z &= 2\end{aligned}$$

1.5 Inverse of a Matrix

In this section we consider **square** ***matrices exclusively***. The **inverse** of an $n \times n$ matrix $\mathbf{A} = [a_{jk}]$ is denoted by $\mathbf{A}^{-1}$ and is an $n \times n$ matrix such that

$$AA^{-1} = A^{-1}A = I \tag{1.55}$$

where **I** is the $n \times n$ unit matrix. If **A** has an inverse, then **A** is called a **nonsingular matrix.** If **A** has no inverse, then **A** is called a **singular matrix.**

THEOREM 16 (Uniqueness of Inverse): *Every invertible matrix possesses a unique inverse.*

OR

If **A** *has an inverse then the inverse is unique.* [JNTU 2003]

PROOF: Let is possible **B** and **C** be the inverses of **A**. Then

$$AB = BA = I \text{ and } AC = CA = I$$

Now $\mathbf{B} = \mathbf{BI} = \mathbf{B(AC)} = \mathbf{(BA)C} = \mathbf{IC} = \mathbf{C}$. Therefore

$$\mathbf{B} = \mathbf{C}$$

Hence there is only one inverse of $\mathbf{A}$ ■

We prove next that $\mathbf{A}$ has an inverse (is nonsingular) if and only if it has maximum possible rank n. The proof will also show that $\mathbf{Ax} = \mathbf{b}$ implies $\mathbf{x} = \mathbf{A}^{-1}\mathbf{b}$ provided $\mathbf{A}^{-1}$ exists, and will thus give a motivation for the inverse as well as a relation to linear systems.

THEOREM 17 (Existence of the Inverse): *The inverse of an $n \times n$ matrix $\mathbf{A}$ exists if and only if rank $\mathbf{A} = n$, thus* (by Theorem 3) *if and only if* det $\mathbf{A} \neq 0$. *Hence $\mathbf{A}$ is nonsingular if rank $\mathbf{A} = n$, and is singular if rank $\mathbf{A} < n$.*

PROOF: Let $\mathbf{A}$ be a given $n \times n$ matrix and consider the linear system

$$\mathbf{Ax} = \mathbf{b} \tag{1.56}$$

If the inverse $\mathbf{A}^{-1}$ exists, then multiplication from the left on both sides and use of (1.56) gives

$$\mathbf{A}^{-1}\mathbf{Ax} = \mathbf{x} = \mathbf{A}^{-1}\mathbf{b}$$

This shows that (1.56) has a unique solution $\mathbf{x}$. Hence $\mathbf{A}$ must have rank n by the Fundamental Theorem in the previous section.

Conversely, let rank $\mathbf{A} = n$. Then by the same theorem, the system (1.56) has a unique solution $\mathbf{x}$ for any $\mathbf{b}$. Now the back substitution following the Gauss elimination shows that the components x_j of $\mathbf{x}$ are linear combinations of those of $\mathbf{b}$. Hence we can write

$$\mathbf{x} = \mathbf{Bb} \tag{1.57}$$

with $\mathbf{B}$ to be determined. Substitution into (1.57) gives

$$\mathbf{Ax} = \mathbf{A(Bb)} = \mathbf{(AB)b} = \mathbf{Cb} = \mathbf{b} \qquad (\mathbf{C} = \mathbf{AB})$$

for any $\mathbf{b}$. Hence $\mathbf{C} = \mathbf{AB} = \mathbf{I}$, the unit matrix. Similarly, if we substitute (1.56) into (1.57) we get

$$\mathbf{x} = \mathbf{Bb} = \mathbf{B(Ax)} = \mathbf{(BA)x}$$

for any $\mathbf{x}$ (and $\mathbf{b} = \mathbf{Ax}$). Hence $\mathbf{BA} = \mathbf{I}$. Together, $\mathbf{B} = \mathbf{A}^{-1}$ exists. ■

Determination of the Inverse by the Gauss–Jordan Method

For the practical determination of the inverse $\mathbf{A}^{-1}$ of a nonsingular $n \times n$ matrix $\mathbf{A}$ we can use the Gauss elimination, actually a variant of it, called the **Gauss–Jordan elimination**. The idea of the method is as follows.

Using $\mathbf{A}$, we form n linear systems

$$\mathbf{Ax}_{(1)} = \mathbf{e}_{(1)}, \ldots, \mathbf{Ax}_{(n)} = \mathbf{e}_{(n)}$$

where $\mathbf{e}_{(1)}, \ldots, \mathbf{e}_{(n)}$ are the columns of the $n \times n$ unit matrix $\mathbf{I}$; thus, $\mathbf{e}_{(1)} = [1 \quad 0 \quad \ldots \quad 0]^T$, $\mathbf{e}_{(2)} = [0 \quad 1 \quad 0 \quad \ldots \quad 0]^T$, etc. These are n vector equations in the unknown vectors $\mathbf{x}_{(1)}, \ldots, \mathbf{x}_{(n)}$. We combine them into a single matrix equation $\mathbf{AX} = \mathbf{I}$, with the unknown matrix $\mathbf{X}$ having the columns $\mathbf{x}_{(1)}, \ldots, \mathbf{x}_{(n)}$. Correspondingly, we combine the n augmented matrices $[\mathbf{A} \quad \mathbf{e}_{(1)}], \ldots, [\mathbf{A} \quad \mathbf{e}_{(n)}]$ into one $n \times 2n$ "augmented matrix" $\tilde{\mathbf{A}} = [\mathbf{A} \quad \mathbf{I}]$. Now multiplication of $\mathbf{AX} = \mathbf{I}$ by $\mathbf{A}^{-1}$ from the left gives $\mathbf{X} = \mathbf{A}^{-1}\mathbf{I} = \mathbf{A}^{-1}$. Hence, to solve $\mathbf{AX} = \mathbf{I}$ for $\mathbf{X}$, we can apply the Gauss elimination to $\tilde{\mathbf{A}} = [\mathbf{A} \quad \mathbf{I}]$. This gives a matrix of the form $[\mathbf{U} \quad \mathbf{H}]$ with upper triangular $\mathbf{U}$ because the Gauss elimination triangularizes systems. The Gauss–Jordan method reduces $\mathbf{U}$ by further elementary row operations to diagonal form, in fact to the unit matrix $\mathbf{I}$. This is done by eliminating the entries of $\mathbf{U}$ above the main diagonal and making the diagonal entries all 1 by multiplication (see the example below). Of course, the method operates on the entire matrix [U H], transforming **H** into some matrix **K**, hence the entire [U H] to [I K]. This is the "augmented matrix" of **IX** = **K**. Now **IX** = **X** = $\mathbf{A}^{-1}$, as shown before. By comparison, **K** = $\mathbf{A}^{-1}$, so that we can read $\mathbf{A}^{-1}$ directly from [I K]. The following example illustrates the practical details of the method.

Example 34 Inverse of a Matrix. Gauss–Jordan Elimination

Determine the inverse $\mathbf{A}^{-1}$ of

$$\mathbf{A} = \begin{bmatrix} -1 & 1 & 2 \\ 3 & -1 & 1 \\ -1 & 3 & 4 \end{bmatrix}$$

Solution: We apply the Gauss elimination to the following $n \times 2n = 3 \times 6$ matrix:

$$[\mathbf{A} \mid \mathbf{I}] = \left[\begin{array}{ccc|ccc} -1 & 1 & 2 & 1 & 0 & 0 \\ 3 & -1 & 1 & 0 & 1 & 0 \\ -1 & 3 & 4 & 0 & 0 & 1 \end{array}\right]$$

$$= \left[\begin{array}{ccc|ccc} -1 & 1 & 2 & 1 & 0 & 0 \\ 0 & 2 & 7 & 3 & 1 & 0 \\ 0 & 2 & 2 & -1 & 0 & 1 \end{array}\right] \quad \begin{array}{l} \text{Row 2} \to \text{Row 2} + 3\,\text{Row 1} \\ \text{Row 3} \to \text{Row 3} - \text{Row 1} \end{array}$$

$$= \left[\begin{array}{ccc|ccc} -1 & 1 & 2 & 1 & 0 & 0 \\ 0 & 2 & 7 & 3 & 1 & 0 \\ 0 & 0 & -5 & -4 & -1 & 1 \end{array}\right] \quad \text{Row 3} \to \text{Row 3} - \text{Row 2}$$

This is [U H] as produced by the Gauss elimination. Now follow the additional Gauss–Jordan steps, reducing **U** to **I**, that is, to diagonal form with entries 1 on the main diagonal. We let

$$[\mathbf{A} \mid \mathbf{I}] = \left[\begin{array}{ccc|ccc} 1 & -1 & -2 & -1 & 0 & 0 \\ 0 & 1 & 3.5 & 1.5 & 0.5 & 0 \\ 0 & 0 & 1 & 0.8 & 0.2 & -0.2 \end{array}\right] \quad \begin{array}{l} \text{Row 1} \to -\ \text{Row 1} \\ \text{Row 2} \to 0.5\ \text{Row 2} \\ \text{Row 3} \to -0.2\ \text{Row 3} \end{array}$$

$$= \left[\begin{array}{ccc|ccc} 1 & -1 & 0 & 0.6 & 0.4 & -0.4 \\ 0 & 1 & 0 & -1.3 & -0.2 & 0.7 \\ 0 & 0 & 1 & 0.8 & 0.2 & -0.2 \end{array}\right] \quad \begin{array}{l} \text{Row 1} \to \text{Row 1} + 2\ \text{Row 3} \\ \text{Row 2} \to \text{Row 2} - 3.5\ \text{Row 3} \end{array}$$

$$= \left[\begin{array}{ccc|ccc} 1 & 0 & 0 & -0.7 & 0.2 & 0.3 \\ 0 & 1 & 0 & -1.3 & -0.2 & 0.7 \\ 0 & 0 & 1 & 0.8 & 0.2 & -0.2 \end{array}\right] \quad \text{Row 1} \to \text{Row 1} + \text{Row 2}$$

The last three columns constitute $\mathbf{A}^{-1}$. Check:

$$\begin{bmatrix} -1 & 1 & 2 \\ 3 & -1 & 1 \\ -1 & 3 & 4 \end{bmatrix} \begin{bmatrix} -0.7 & 0.2 & 0.3 \\ -1.3 & -0.2 & 0.7 \\ 0.8 & 0.2 & -0.2 \end{bmatrix} = \begin{bmatrix} 1 & 0 & 0 \\ 0 & 1 & 0 \\ 0 & 0 & 1 \end{bmatrix}$$

Hence $\mathbf{AA}^{-1} = \mathbf{I}$. Similarly, $\mathbf{A}^{-1}\mathbf{A} = \mathbf{I}$.

Determination of Inverse by Adjoint Method

The explicit formula (1.58) in the following theorem is often useful in theoretical studies (as opposed to *computing* inverses). In fact, the special case $n = 2$ occurs quite frequently in geometrical and other applications.

THEOREM 18 (Inverse of a Matrix): *The inverse of a nonsingular $n \times n$ matrix $\mathbf{A} = [a_{jk}]$ is given by*

$$\mathbf{A}^{-1} = \frac{1}{\det \mathbf{A}}[C_{jk}]^{\mathrm{T}} = \frac{1}{\det \mathbf{A}} \begin{bmatrix} C_{11} & C_{21} & \cdots & C_{n1} \\ C_{12} & C_{12} & \cdots & C_{n1} \\ \cdot & \cdot & \cdots & \cdot \\ C_{1n} & C_{2n} & \cdots & C_{nn} \end{bmatrix} \tag{1.58}$$

where C_{jk} is the cofactor of a_{jk} in det **A**. [CAUTION! Note well that in $\mathbf{A}^{-1}$, the cofactor C_{jk} occupies the same place as a_{kj} (not a_{jk}) does in **A**.]

In particular, the inverse of

$$\mathbf{A} = \begin{bmatrix} a_{11} & a_{12} \\ a_{21} & a_{22} \end{bmatrix}$$

is

$$\mathbf{A}^{-1} = \frac{1}{\det \mathbf{A}} \begin{bmatrix} a_{22} & -a_{12} \\ -a_{21} & -a_{11} \end{bmatrix} \tag{1.59}$$

PROOF: We denote the right side of (1.58) by **B** and show that $\mathbf{BA} = \mathbf{I}$. We first write

$$\mathbf{BA} = \mathbf{G} = [g_{kl}] \tag{1.60}$$

and then show that $\mathbf{G} = \mathbf{I}$. Now by the definition of matrix multiplication and because of the form of **B** in (1.58), we obtain [CAUTION! C_{sk}, not C_{ks}]

$$g_{kl} = \sum_{s=1}^{n} \frac{C_{sk}}{\det \mathbf{A}} a_{sl} = \frac{1}{\det \mathbf{A}}(a_{1l}C_{1k} + \cdots + a_{nl}C_{nk}) \tag{1.61}$$

Now (1.34) and (1.35) show that the sum $(\cdots)$ on the right is $D = \det \mathbf{A}$ when $l = k$, and is zero when $l \neq k$. Hence

$$g_{kk} = \frac{1}{\det \mathbf{A}} \det \mathbf{A} = 1$$

$$g_{kl} = 0 \quad (l \neq k)$$

In particular, for $n = 2$ we have in (1.58) in the first row $C_{11} = a_{22}$, $C_{21} = -a_{12}$ and in the second row $C_{12} = -a_{21}$, $C_{22} = a_{11}$. This gives (1.59). ■

Example 35 **Inverse of a 2 × 2 Matrix**

$$\mathbf{A}=\begin{bmatrix}3 & 1\\ 2 & 4\end{bmatrix},\qquad \mathbf{A}^{-1}=\frac{1}{10}\begin{bmatrix}4 & -1\\ -2 & 3\end{bmatrix}=\begin{bmatrix}0.4 & -0.1\\ -0.2 & 0.3\end{bmatrix}$$

Example 36 **Further Illustration of Theorem 18**

Using (1.50), find the inverse of

$$\mathbf{A}=\begin{bmatrix}-1 & 1 & 2\\ 3 & -1 & 1\\ -1 & 3 & 4\end{bmatrix}$$

Solution: We obtain det $\mathbf{A} = -1(-7) - 1\times 13 + 2\times 8 = 10$, and in (1.58),

$$C_{11}=\begin{vmatrix}-1 & 1\\ 3 & 4\end{vmatrix}=-7,\qquad C_{21}=-\begin{vmatrix}1 & 2\\ 3 & 4\end{vmatrix}=2,\qquad C_{31}=\begin{vmatrix}1 & 2\\ -1 & 1\end{vmatrix}=3$$

$$C_{12}=-\begin{vmatrix}3 & 1\\ -1 & 4\end{vmatrix}=-13,\quad C_{22}=\begin{vmatrix}-1 & 2\\ -1 & 4\end{vmatrix}=-2,\quad C_{32}=-\begin{vmatrix}-1 & 2\\ 3 & 1\end{vmatrix}=7$$

$$C_{13}=\begin{vmatrix}3 & -1\\ -1 & 3\end{vmatrix}=8,\qquad C_{23}=-\begin{vmatrix}-1 & 1\\ -1 & 3\end{vmatrix}=2,\qquad C_{33}=\begin{vmatrix}-1 & 1\\ 3 & 1\end{vmatrix}=2$$

so that by (1.58), in agreement with Example 34,

$$\mathbf{A}^{-1}=\begin{bmatrix}-0.7 & 0.2 & 0.3\\ -1.3 & -0.2 & 0.7\\ 0.8 & 0.2 & -0.2\end{bmatrix}$$

THEOREM 19 (Diagonal Matrices): $\mathbf{A} = [a_{jk}]$, $a_{jk} = 0$ *when* $j \neq k$, *have an inverse if and only if all* $a_{jj} \neq 0$. *Then* $\mathbf{A}^{-1}$ *is diagonal too, with entries* $1/a_{11}, \ldots, 1/a_{nn}$.

PROOF: For a diagonal matrix we have in (1.50)

$$\frac{C_{11}}{D}=\frac{a_{22}\cdots a_{nn}}{a_{11}a_{22}\cdots a_{nn}}=\frac{1}{a_{11}},\ \text{etc.}$$

■

Example 37 **Inverse of a Diagonal Matrix**

Let

$$\mathbf{A}=\begin{bmatrix}-0.5 & 0 & 0\\ 0 & 4 & 0\\ 0 & 0 & 1\end{bmatrix}$$

Then the inverse is

$$A^{-1} = \begin{bmatrix} -2 & 0 & 0 \\ 0 & 0.25 & 0 \\ 0 & 0 & 1 \end{bmatrix}$$

THEOREM 20: *Products can be inverted by taking the inverse of each factor and multiplying these inverses* ***in reverse order***,

$$(\mathbf{AC})^{-1} = \mathbf{C}^{-1}\mathbf{A}^{-1} \tag{1.62}$$

Hence for more than two factors,

$$(\mathbf{AC} \cdots \mathbf{PQ})^{-1} = \mathbf{Q}^{-1}\mathbf{P}^{-1} \cdots \mathbf{C}^{-1}\mathbf{A}^{-1} \tag{1.63}$$

PROOF: The idea is to start from (1.55) for $\mathbf{AC}$ instead of $\mathbf{A}$, that is, $\mathbf{AC}(\mathbf{AC})^{-1} = \mathbf{I}$, and multiply it on both sides from the left, first by $\mathbf{A}^{-1}$, which because of $\mathbf{A}^{-1}\mathbf{A} = \mathbf{I}$ gives

$$\mathbf{A}^{-1}\mathbf{AC}(\mathbf{AC})^{-1} = \mathbf{C}(\mathbf{AC})^{-1} = \mathbf{A}^{-1}\mathbf{I} = \mathbf{A}^{-1}$$

and then multiplying this on both sides from the left, this time by $\mathbf{C}^{-1}$ and by using $\mathbf{C}^{-1}\mathbf{C} = \mathbf{I}$,

$$\mathbf{C}^{-1}\mathbf{C}(\mathbf{AC})^{-1} = (\mathbf{AC})^{-1} = \mathbf{C}^{-1}\mathbf{A}^{-1}$$

This proves (1.62), and from it, (1.63) follows by induction. ■

THEOREM 21: *The inverse of the inverse is the given matrix, that is,*

$$(\mathbf{A}^{-1})^{-1} = \mathbf{A}$$

PROOF: The proof is left as an exercise for the readers. ■

CONCEPT CHECK

1. Find the inverse by Gauss–Jordan [or by (1.51) if $n = 2$] or state that it does not exist. Check by using (1.47).

(a) $\begin{bmatrix} 1.20 & 4.64 \\ 0.50 & 3.60 \end{bmatrix}$

(b) $\begin{bmatrix} 0.6 & 0.8 \\ 0.8 & -0.6 \end{bmatrix}$

(c) $\begin{bmatrix} \cos 2\theta & \sin 2\theta \\ -\sin 2\theta & \cos 2\theta \end{bmatrix}$

(d) $\begin{bmatrix} 2/3 & 1/3 & 2/3 \\ -2/3 & 2/3 & 1/3 \\ 1/3 & 2/3 & -2/3 \end{bmatrix}$

(e) $\begin{bmatrix} 3 & -1 & 1 \\ -15 & 6 & -5 \\ 5 & -2 & 2 \end{bmatrix}$

(f) $\begin{bmatrix} 29 & -11 & 10 \\ -160 & 61 & -55 \\ 55 & -21 & 19 \end{bmatrix}$

(g) $\begin{bmatrix} 1 & 0 & 0 \\ 2 & 1 & 0 \\ 5 & 4 & 1 \end{bmatrix}$

(h) $\begin{bmatrix} 1 & 2 & 5 \\ 0 & -1 & 2 \\ 0 & 4 & 11 \end{bmatrix}$

(i) $\begin{bmatrix} 0 & 1 & 0 \\ 1 & 0 & 0 \\ 0 & 0 & 1 \end{bmatrix}$

(j) $\begin{bmatrix} 0 & 8 & 0 \\ 0 & 0 & 4 \\ 2 & 0 & 0 \end{bmatrix}$

(k) $\begin{bmatrix} 1 & 2 & 5 \\ 0 & -1 & 2 \\ 2 & 4 & 10 \end{bmatrix}$

(l) $\begin{bmatrix} 1 & 2 & -9 \\ -2 & -4 & 19 \\ 0 & -1 & 2 \end{bmatrix}$

2. **(Triangular matrix)** Is the inverse of a triangular matrix always triangular [as in Prob. 1(g)]? Give reason.
3. **(Rotation)** Give an application of the matrix in Prob. 1(c) that makes the form of its inverse obvious.
4. **(Inverse of the square)** Verify $(\mathbf{A}^2)^{-1} = (\mathbf{A}^{-1})^2$ for **A** in Prob. 1(e).
5. Prove the formula in Prob. 4.
6. **(Inverse of the transpose)** Verify $(\mathbf{A}^T)^{-1} = (\mathbf{A}^{-1})^T$ for **A** in Prob. 1(e).
7. Prove the formula in Prob. 6.
8. **(Inverse of the inverse)** Prove that $(\mathbf{A}^{-1})^{-1} = \mathbf{A}$.
9. **(Row interchange)** Same question as in Prob. 4 for the matrix in Prob. 1(i).
10. Formula (1.50) is generally not very practical. To understand its use, apply it:
    (a) To Prob. 1(i)
    (b) To Prob. 1(d)
    (c) To Prob. 1(g)

1.6 Reduction to Normal Form

THEOREM 22: *Every $m \times n$ matrix of rank r can be reduced to the form*

$$\begin{bmatrix} \mathbf{I}_r & 0 \\ 0 & 0 \end{bmatrix}$$

by a finite number of elementary row and column operations where $\mathbf{I}_r$ is the r-rowed unit matrix.

COROLLARY A: The rank of an $m \times n$ matrix **A** is r if and only if it can be reduced to the form

$$\begin{bmatrix} \mathbf{I}_r & 0 \\ 0 & 0 \end{bmatrix}$$

by a finite number of elementary operations.

COROLLARY B: If A be an $m \times n$ matrix of rank, there exists non-singular matrix **P** and **Q** such that

$$\mathbf{PAQ} = \begin{bmatrix} \mathbf{I}_r & 0 \\ 0 & 0 \end{bmatrix}$$

PROOF: If A be an $m \times n$ matrix of rank "r" it can be transformed into the form

$$\begin{bmatrix} \mathbf{I}_r & 0 \\ 0 & 0 \end{bmatrix}$$

by elementary operations since elementary row a column operations are equivalent to pre- or post-multiplication by the corresponding elementary matrices we have the following result:

If **A** be an $m \times n$ matrix of rank r, there exist elementary matrices

$$\mathbf{P}_1, \mathbf{P}_2, \ldots, \mathbf{P}_n, \mathbf{A}, \mathbf{Q}_1, \mathbf{Q}_2, \ldots, \mathbf{Q}_n = \begin{bmatrix} \mathbf{I}_r & 0 \\ 0 & 0 \end{bmatrix}$$

Now each elementary matrix is either a nonsingular matrix or the product of nonsingular matrices. Therefore if

$$\mathbf{P} = \mathbf{P}_n, \mathbf{P}_{n-1}, \ldots, \mathbf{P}_1 \text{ and } \mathbf{Q} = \mathbf{Q}_1, \mathbf{Q}_2, \ldots, \mathbf{Q}_n$$

then **P** and **Q** are nonsingular matrices. Hence

$$\mathbf{PAQ} = \begin{bmatrix} \mathbf{I}_r & 0 \\ 0 & 0 \end{bmatrix}$$

Example 38

Reduce the matrix

$$A = \begin{bmatrix} 1 & -1 & 2 & -3 \\ 4 & 1 & 0 & 2 \\ 0 & 3 & 0 & 4 \\ 0 & 1 & 0 & 2 \end{bmatrix}$$

to the normal form $\begin{bmatrix} \mathbf{I}_r & 0 \\ 0 & 0 \end{bmatrix}$ and hence determine its rank.

Solution: Let

$$A = \begin{bmatrix} 1 & -1 & 2 & -3 \\ 4 & 1 & 0 & 2 \\ 0 & 3 & 0 & 4 \\ 0 & 1 & 0 & 2 \end{bmatrix}$$

$$= \begin{bmatrix} 1 & 0 & 0 & 0 \\ 4 & 5 & -8 & 14 \\ 0 & 3 & 0 & 4 \\ 0 & 1 & 0 & 2 \end{bmatrix} \quad \begin{matrix} \text{Column 2} \to \text{Column 2} + \text{Column 1} \\ \text{Column 3} \to \text{Column 3} - 2\,\text{Column 1} \\ \text{Column 4} \to \text{Column 4} + 3\,\text{Column 1} \end{matrix}$$

$$= \begin{bmatrix} 1 & 0 & 0 & 0 \\ 0 & 5 & -8 & 14 \\ 0 & 3 & 0 & 4 \\ 0 & 1 & 0 & 2 \end{bmatrix} \quad \text{Row 2} \to \text{Row 2} - 4\,\text{Row 1}$$

$$= \begin{bmatrix} 1 & 0 & 0 & 0 \\ 0 & 1 & 0 & 2 \\ 0 & 3 & 0 & 4 \\ 0 & 5 & -8 & 14 \end{bmatrix} \quad \text{Row 2} \leftrightarrow \text{Row 4}$$

$$= \begin{bmatrix} 1 & 0 & 0 & 0 \\ 0 & 1 & 0 & 0 \\ 0 & 3 & 0 & -2 \\ 0 & 5 & -8 & 4 \end{bmatrix} \quad \text{Column 4} \to \text{Column 4} - 2\,\text{Column 2}$$

$$= \begin{bmatrix} 1 & 0 & 0 & 0 \\ 0 & 1 & 0 & 0 \\ 0 & 0 & 0 & -2 \\ 0 & 0 & -8 & 4 \end{bmatrix} \quad \begin{matrix} \text{Row 3} \to \text{Row 3} - 3\,\text{Row 2} \\ \text{Row 4} \to \text{Row 4} - 5\,\text{Row 2} \end{matrix}$$

$$= \begin{bmatrix} 1 & 0 & 0 & 0 \\ 0 & 1 & 0 & 0 \\ 0 & 0 & -2 & 0 \\ 0 & 0 & 4 & -8 \end{bmatrix} \quad \text{Column } 3 \leftrightarrow \text{Column } 4$$

$$= \begin{bmatrix} 1 & 0 & 0 & 0 \\ 0 & 1 & 0 & 0 \\ 0 & 0 & 1 & 0 \\ 0 & 0 & -2 & 1 \end{bmatrix} \quad \begin{matrix} \text{Column } 3 \to \text{Column } 3 \times (-1/2) \\ \text{Column } 4 \to \text{Column } 4 \times (-1/8) \end{matrix}$$

$$= \begin{bmatrix} 1 & 0 & 0 & 0 \\ 0 & 1 & 0 & 0 \\ 0 & 0 & 1 & 0 \\ 0 & 0 & 0 & 1 \end{bmatrix} \quad \text{Row } 4 \to \text{Row } 4 + 2\,\text{Row } 3$$

$$= [\mathbf{I}_4]$$

Hence the rank of matrix **A** is 4.

Example 39

Find the nonsingular matrices **P** and **Q** such that **PAQ** is in the normal form where

$$\mathbf{A} = \begin{bmatrix} 1 & 1 & 1 \\ 1 & -1 & -1 \\ 3 & 1 & 1 \end{bmatrix}$$

Also find the rank of the matrix **A**.

Solution: Let $\mathbf{A} = \mathbf{I}_3\mathbf{A}\mathbf{I}_3$. That is

$$\begin{bmatrix} 1 & 1 & 1 \\ 1 & -1 & 1 \\ 3 & 1 & 1 \end{bmatrix} = \begin{bmatrix} 1 & 0 & 0 \\ 0 & 1 & 0 \\ 0 & 0 & 1 \end{bmatrix} \mathbf{A} \begin{bmatrix} 1 & 0 & 0 \\ 0 & 1 & 0 \\ 0 & 0 & 1 \end{bmatrix}$$

Apply elementary operations on the matrix A until it is reduced to the normal form. Every elementary row operation will also be applied to the pre-factor $\mathbf{I}_3$ of the product on the right-hand member of the above equation and every elementary column operation to the post-factor $\mathbf{I}_3$.

$$\begin{bmatrix} 1 & 1 & 1 \\ 0 & -2 & -2 \\ 0 & -2 & -2 \end{bmatrix} = \begin{bmatrix} 1 & 0 & 0 \\ -1 & 1 & 0 \\ -3 & 0 & 1 \end{bmatrix} \mathbf{A} \begin{bmatrix} 1 & 0 & 0 \\ 0 & 1 & 0 \\ 0 & 0 & 1 \end{bmatrix} \quad \begin{matrix} \text{Row } 2 \to \text{Row } 2 - \text{Row } 1 \\ \text{Row } 3 \to \text{Row } 3 - 3\,\text{Row } 1 \end{matrix}$$

$$\begin{bmatrix} 1 & 0 & 0 \\ 0 & -2 & -2 \\ 0 & -2 & -2 \end{bmatrix} = \begin{bmatrix} 1 & 0 & 0 \\ -1 & 1 & 0 \\ -3 & 0 & 1 \end{bmatrix} \mathbf{A} \begin{bmatrix} 1 & -1 & -1 \\ 0 & 1 & 0 \\ 0 & 0 & 1 \end{bmatrix} \quad \begin{matrix} \text{Column } 2 \to \text{Column } 2 - \text{Column } 1 \\ \text{Column } 3 \to \text{Column } 3 - \text{Column } 1 \end{matrix}$$

$$\begin{bmatrix} 1 & 0 & 0 \\ 0 & 1 & 0 \\ 0 & -2 & -2 \end{bmatrix} = \begin{bmatrix} 1 & 0 & 0 \\ 1/2 & -1/2 & 0 \\ -3 & 0 & 1 \end{bmatrix} \mathbf{A} \begin{bmatrix} 1 & -1 & -1 \\ 0 & 1 & 0 \\ 0 & 0 & 1 \end{bmatrix} \quad \text{Row } 2 \to \text{Row } 2 \times (-1/2)$$

$$\begin{bmatrix} 1 & 0 & 0 \\ 0 & 1 & 0 \\ 0 & 0 & 0 \end{bmatrix} = \begin{bmatrix} 1 & 0 & 0 \\ 1/2 & -1/2 & 0 \\ -2 & -1 & 1 \end{bmatrix} \mathbf{A} \begin{bmatrix} 1 & -1 & -1 \\ 0 & 1 & 0 \\ 0 & 0 & 1 \end{bmatrix} \quad \text{Row } 3 \to \text{Row } 3 + 2\,\text{Row } 2$$

$$\begin{bmatrix} 1 & 0 & 0 \\ 0 & 1 & 0 \\ 0 & 0 & 0 \end{bmatrix} = \begin{bmatrix} 1 & 0 & 0 \\ 1/2 & -1/2 & 0 \\ -2 & -1 & 1 \end{bmatrix} \mathbf{A} \begin{bmatrix} 1 & -1 & -1 \\ 0 & 1 & -1 \\ 0 & 0 & 1 \end{bmatrix} \quad \text{Column } 3 \to \text{Column } 3 - \text{Column } 2$$

Finally we have

$$\mathbf{PAQ} = \begin{bmatrix} \mathbf{I}_2 & 0 \\ 0 & 0 \end{bmatrix}$$

where

$$\mathbf{P} = \begin{bmatrix} 1 & 0 & 0 \\ 1/2 & -1/2 & 0 \\ -2 & -1 & 1 \end{bmatrix}, \quad \mathbf{Q} = \begin{bmatrix} 1 & -1 & -1 \\ 0 & 1 & -1 \\ 0 & 0 & 1 \end{bmatrix}$$

Since

$$\mathbf{A} = \begin{bmatrix} \mathbf{I}_2 & 0 \\ 0 & 0 \end{bmatrix}$$

Hence rank of $(\mathbf{A}) = 2$.

Example 40

Find the rank of the matrix

$$\mathbf{A} = \begin{bmatrix} 1 & 2 & -1 & 3 \\ 4 & 1 & 2 & 1 \\ 3 & -1 & 1 & 2 \\ 1 & 2 & 0 & 1 \end{bmatrix}$$

by reducing it to the normal form.

Solution: $\rho(\mathbf{A}) = 3$. Left to readers as an exercise.

CONCEPT CHECK

Reduce the following matrices to normal form and hence find the rank:

1. $\begin{bmatrix} 2 & -4 & 3 & -1 & 0 \\ 1 & -2 & -1 & -4 & 2 \\ 0 & 1 & -1 & 3 & 1 \\ 4 & -7 & 4 & -4 & 5 \end{bmatrix}$

2. $\begin{bmatrix} 1 & 4 & 3 & -2 & 1 \\ -2 & -3 & -1 & 4 & 3 \\ -1 & 6 & 7 & 2 & 9 \\ -3 & 3 & 6 & 2 & 12 \end{bmatrix}$

3. $\begin{bmatrix} 0 & 1 & -3 & -1 \\ 1 & 0 & 1 & 1 \\ 3 & 1 & 0 & 2 \\ 1 & 1 & -2 & 0 \end{bmatrix}$

1.7 Gauss Elimination Method

Theory

Recall that a **linear system of *n* equations** in *n unknowns* $x_1, \ldots, x_n$ is a set of equations $E_1, \ldots, E_n$ of the form

$$\begin{aligned} E_1&: a_{11}x_1 + \cdots + a_{1n}x_n = b_1 \\ E_2&: a_{21}x_1 + \cdots + a_{2n}x_n = b_2 \\ &\cdots\cdots\cdots\cdots\cdots\cdots\cdots\cdots \\ E_n&: a_{n1}x_1 + \cdots + a_{nn}x_n = b_n \end{aligned} \tag{1.64}$$

where the **coefficients** a_{jk} and the b_j are given numbers. The system is called **homogeneous** if all the b_j are zero; otherwise it is called **nonhomogeneous.** Using matrix multiplication, we can write (1.64) as a single vector equation

$$\mathbf{Ax} = \mathbf{b} \tag{1.65}$$

where the **coefficient matrix** $\mathbf{A} = [a_{jk}]$ is the $n \times n$ matrix

$$\mathbf{A} = \begin{bmatrix} a_{11} & a_{12} & \cdots & a_{1n} \\ a_{21} & a_{22} & \cdots & a_{2n} \\ \cdot & \cdot & \cdots & \cdot \\ a_{n1} & a_{n2} & \cdots & a_{nn} \end{bmatrix}, \mathbf{x} = \begin{bmatrix} x_1 \\ \vdots \\ x_n \end{bmatrix} \text{ and } \mathbf{b} = \begin{bmatrix} b_1 \\ \vdots \\ b_n \end{bmatrix}$$

are column vectors. The following matrix $\tilde{\mathbf{A}}$ is called the **augmented matrix** of the system (1.56):

$$\tilde{\mathbf{A}} = [\mathbf{A}|\mathbf{b}] = \left[\begin{array}{ccc|c} a_{11} & \cdots & a_{1n} & b_1 \\ a_{21} & \cdots & a_{2n} & b_2 \\ \cdot & \cdots & \cdot & \cdot \\ a_{n1} & \cdots & a_{nn} & b_n \end{array}\right]$$

A **solution** of (1.64) is a set of numbers $x_1, \ldots, x_n$ that satisfy all the n equations, and a **solution vector** of (1.64) is a vector **x** whose components constitute a solution of (1.64).

The method of solving such a system by determinants (Cramer's rule) is not practical, even with efficient methods for evaluating the determinants. A practical method for the solution of a linear system is the so-called *Gauss elimination,* which we shall now discuss.

The Process

This standard method for solving linear systems (1.64) is a systematic process of elimination that reduces (1.64) to **"triangular form"** because the system can then be easily solved by **"back substitution."** For instance, a triangular system is

$$\begin{aligned} 3x_1 + 5x_2 + 2x_3 &= 8 \\ 8x_2 + 2x_3 &= -7 \\ 6x_3 &= 3 \end{aligned}$$

and back substitution gives $x_3 = 3/6 = 1/2$ from the third equation, then

$$x_2 = \frac{1}{8}(-7 - 2x_3) = -1$$

from the second equation, and finally from the first equation

$$x_1 = \frac{1}{3}(8 - 5x_2 - 2x_3) = 4$$

How do we reduce a given system (1.64) to triangular form? In the first step we ***eliminate*** $\boldsymbol{x_1}$ from equation E_2 to E_n in (1.64). We do this by adding (or subtracting) suitable multiples of E_1 from equations $E_2, \ldots, E_n$ and taking the resulting equations, call them $E_2^*, \ldots, E_n^*$ as the new equations. The first equation, E_1, is called the **pivot equation** in this step, and a_{11} is called the **pivot.** This equation is left unaltered. In the second step we take the new second equation E_2^* (which no longer contains x_1) as the pivot equation and use it to ***eliminate*** $\boldsymbol{x_2}$ from E_3^* to E_n^*. And so on. After $n - 1$ steps this gives a triangular system that can be solved by back substitution as just shown. In this way we obtain precisely all solutions of the *given* system.

The pivot a_{kk} (in step k) ***must be*** different from zero and ***should be*** large in absolute value, to avoid roundoff magnification by the multiplication in the elimination. For this we choose as our pivot equation one that has the absolutely largest a_{jk} in column k on or below the main diagonal (actually, the uppermost if there are several such equations). This popular method is called **partial pivoting.**

Partial pivoting distinguishes it from **total pivoting,** which involves both row and column interchanges but is hardly used in practice. Let us illustrate this method with a simple example.

Example 41 Gauss Elimination. Partial Pivoting

Solve the system

$$\begin{aligned} E_1&: \qquad\quad 8x_2 + 2x_3 = -7 \\ E_2&: 3x_1 + 5x_2 + 2x_3 = 8 \\ E_3&: 6x_1 + 2x_2 + 8x_3 = 26 \end{aligned}$$

Solution: We must pivot since E_1 has no x_1-term. In Column 1, equation E_3 has the largest coefficient. Hence we interchange E_1 and E_3,

$$\begin{aligned} 6x_1 + 2x_2 + 8x_3 &= 26 \\ 3x_1 + 5x_2 + 2x_3 &= \ 8 \\ 8x_2 + 2x_3 &= -7 \end{aligned}$$

***Step* 1. *Elimination of* x_1**

It would suffice to show the augmented matrix and operate on it. We show both the equations and the augmented matrix. In the first step, the first equation is the pivot equation. Thus

$$\begin{array}{ll} \text{Pivot 6} & \textcircled{6x_1} + 2x_2 + 8x_3 = 26 \\ \text{Eliminate} & \boxed{3x_2} + 5x_2 + 2x_3 = \ 8 \\ & \qquad 8x_2 + 2x_3 = -7 \end{array} \qquad \left[\begin{array}{ccc|c} 6 & 2 & 8 & 26 \\ 3 & 5 & 2 & 8 \\ 0 & 8 & 2 & -7 \end{array}\right]$$

To eliminate x_1 from the other equations (here, from the second equation), subtract $3/6 = 1/2$ times the pivot equation from the second equation. The result is

$$\begin{aligned} 6x_1 + 2x_2 + 8x_3 &= 26 \\ 4x_2 + 2x_3 &= -5 \\ 8x_2 + 2x_3 &= -7 \end{aligned} \qquad \left[\begin{array}{ccc|c} 6 & 2 & 8 & 26 \\ 0 & 4 & -2 & -5 \\ 0 & 8 & 2 & -7 \end{array}\right]$$

***Step* 2. *Elimination of* x_2**

The largest coefficient in Column 2 is 8. Hence we take the *new* third equation as the pivot equation, interchanging equations 2 and 3,

$$\begin{array}{ll} & 6x_1 + 2x_2 + 8x_3 = 26 \\ \text{Pivot 8} & \qquad 8x_2 + 2x_3 = -7 \\ \text{Eliminate} & \qquad \boxed{4x_2} - 2x_3 = -5 \end{array} \qquad \left[\begin{array}{ccc|c} 6 & 2 & 8 & 26 \\ 0 & 8 & 2 & -7 \\ 0 & 4 & -2 & -5 \end{array}\right]$$

To eliminate x_2 from the third equation, subtract 1/2 times the pivot equation from the third equation. The resulting triangular system is shown below. This is the end of the forward elimination. Now comes the back substitution.

***Back substitution. Determination of* x_3, x_2, x_1**

The triangular system obtained in Step 2 is

$$\begin{aligned} 6x_1 + 2x_2 + 8x_3 &= 26 \\ 8x_2 + 2x_3 &= -7 \\ -3x_3 &= -3/2 \end{aligned} \qquad \left[\begin{array}{ccc|c} 0 & 2 & 8 & 26 \\ 0 & 8 & 2 & -7 \\ 0 & 4 & -2 & -3/2 \end{array}\right]$$

From this system, taking the last equation, then the second equation, and finally the first equation, we compute the solution

$$x_3 = \frac{1}{2}$$

$$x_2 = \frac{1}{8}(-7 - 2x_3) = -1$$

$$x_1 = \frac{1}{6}(26 - 2x_2 - 8x_3) = 4$$

This agrees with the values given above, before the beginning of the example.

The general algorithm for the Gauss elimination is shown in the algorithm on the next page. To help explain the algorithm, we have numbered some of its lines. b_j is denoted by $a_{j,\,n+1}$, for uniformity. In lines 1 and 2 we look for a possible pivot. [For $k = 1$ we can always find one; otherwise x_1 would not occur in (1.64).] In line 2 we do pivoting if necessary, picking an a_{jk} of greatest absolute value (the one with the smallest j if there are several) and interchange the corresponding rows. If $|a_{kk}|$ is greatest, we do no pivoting. m_{jk} in line 3 suggests *multiplier,* since these are the factors by which we have to multiply the pivot equation E_k^* in Step k before subtracting it from an equation E_j^* below E_k^* from which we want to eliminate x_k. Here we have written E_k^* and E_j^* to indicate that after Step 1 these are no longer the given equations in (1.64), but these underwent a change in each step, as indicated in line 4. Accordingly, a_{jk} etc. in lines 1–4 refer to the most recent equations, and $j \geqq k$ in line 1 indicates that we leave untouched all the equations that have served as pivot equations in previous steps. For $p = k$ in line 4 we get 0 on the right, as it should be in the elimination,

$$a_{jk} - m_{jk}a_{kk} = a_{jk} - \frac{a_{jk}}{a_{kk}}a_{kk} = 0$$

In line 5, if the last equation in the *triangular* system is $0 = b_n^* \neq 0$, we have no solution. If it is $0 = b_n^* = 0$, we have no unique solution because we then have fewer equations than unknowns.

Example 42 Gauss Elimination, Sample Computation

In Example 41 we had $a_{11} = 0$, so that pivoting was necessary. The greatest coefficient in Column 1 was a_{31}. Thus $\tilde{j} = 3$ in line 2, and we interchanged E_1 and E_3. Then in lines 3 and 4 we computed $m_{21} = 3/6 = 1/2$ and

$$a_{22} = 5 - (1/2)\times 2 = 4, \qquad a_{23} = 2 - (1/2)\times 8 = -2, \qquad a_{24} = 8 - (1/2)\times 26 = -5$$

and then $m_{31} = 0/6 = 0$, so that the third equation $8x_2 + 2x_3 = -7$ did not change in Step 1. In Step 2 ($k = 2$) we had 8 as the greatest coefficient in Column 2, hence $\tilde{j} = 3$. We interchanged equations 2 and 3, computed $m_{32} = -4/8 = -(1/2)$ in line 4, and the $a_{33} = -2 - (1/2)\times 2 = -3$, $a_{34} = -5 - (1/2), (-7) = -(3/2)$. This produced the triangular form used in the back substitution.

If $a_{kk} = 0$ in Step k, ***we must pivot.*** If $|a_{kk}|$ is small, ***we should pivot*** because of roundoff error magnification that may seriously affect accuracy or even produce nonsensical results.

Example 43 Difficulty with Small Pivots

The solution of the system

$$0.0004x_1 + 1.402x_2 = 1.406$$
$$0.4003x_1 - 1.502x_2 = 2.501$$

is $x_1 = 10$, $x_2 = 1$. We solve this system by the Gauss elimination, using four-digit floating-point arithmetic. (4D is for simplicity. Make an 8D-arithmetic example that shows the same.)

(a) Picking the first of the given equations as the pivot equation, we have to multiply this equation by $m = 0.4003/0.0004 = 1001$ and subtract the result from the second equation, obtaining

$$-1405x_2 = -1404$$

Hence $x_2 = -1404/(-1405) = 0.9993$, and from the first equation, instead of $x_1 = 10$, we get

$$x_1 = \frac{1}{0.0004}(1.406 - 1.402 \times 0.9993) = \frac{0.005}{0.0004} = 12.5$$

This failure occurs because $|a_{11}|$ is small compared with $|a_{12}|$, so that a small roundoff error in x_2 leads to a large error in x_1.

(b) Picking the second of the given equations as the pivot equation, we have to multiply this equation by $0.0004/0.4003 = 0.0009993$ and subtract the result from the first equation, obtaining

$$1.404x_2 = 1.404$$

Hence $x_2 = 1$, and from the pivot equation $x_1 = 10$. This success occurs because $|a_{21}|$ is not very small compared to $|a_{22}|$, so that a small roundoff error in x_2 would not lead to a large error in x_1. Indeed, for instance, if we had the value $x_2 = 1.002$, we would still have from the pivot equation the good value $x_1 = (2.501 + 1.505)/0.4003 = 10.01$.

ALGORITHM GAUSS ($\tilde{\mathbf{A}} = [a_{jk}] = [\mathbf{A} \quad \mathbf{b}]$)

This algorithm computes a unique solution $\mathbf{x} = [x_j]$ of the system (1.56) or indicates that (1.64) has no unique solution.

INPUT: Augmented $n \times (n+1)$ matrix $\tilde{\mathbf{A}} = [a_{jk}]$, where $a_{j,n+1} = b_j$

OUTPUT: Solution $\mathbf{x} = [x_j]$ of (1.56) or message that the system (1.56) has no unique solution

For $k = 1, \ldots, n-1$, do:

1 If $a_{jk} = 0$ for all $j \geqq k$ then OUTPUT "No unique solution exists." Stop

[*Procedure completed unsuccessfully;* **A** *is singular*]

2 Else exchange the contents of rows $\tilde{j}$ and k of $\tilde{\mathbf{A}}$ with $\tilde{j}$ the smallest $j \geqq k$ such that $|a_{jk}|$ is maximum in column k.

3 For $j = k+1, \ldots, n$, do:

$$m_{jk} := \frac{a_{jk}}{a_{kk}}$$

4 For $p = k+1, \ldots, n+1$, do:

$$a_{jp} := a_{jp} - m_{jk}a_{kp}$$

End

End

End

5 If $a_{nn} = 0$ then OUTPUT "No unique solution exists." Stop

Else

6 $x_n = \dfrac{a_{n,n+1}}{a_{nn}}$ [*Start back substitution*]

For $i = n - 1, \ldots, 1$, do:

7 $x_i = \dfrac{1}{a_{ii}}\left(a_{i,n+1} - \sum_{j=i+1}^{n} a_{ij}x_j\right)$

End

OUTPUT $\mathbf{x} = [x_j]$. Stop

End GAUSS

Row scaling means the multiplication of each Row j by a suitable scaling factor s_j. It is done in connection with partial pivoting to get more accurate solutions. Despite much research and the proposition of several principles, scaling is still not well understood. As a possibility, one can scale for pivot choice only (not in the calculation, to avoid additional roundoff) and take as first pivot the entry a_{j1} for which $|a_{j1}|/|A_j|$ is largest; here A_j is an entry of largest absolute value in Row j. Similarly in the further steps of the Gauss elimination. For instance, for the system

$$\begin{aligned} 4.0000x_1 + 14020x_2 &= 14060 \\ 0.4003x_1 - 1.502x_2 &= 2.501 \end{aligned}$$

we might pick 4 as pivot, but dividing the first equation by 10^4 gives the system in Example 43, for which the second equation is a better pivot equation.

Operation Count

Quite generally, important factors in judging the quality of a numeric method are

Amount of storage
Amount of time ($\equiv$ number of operations)
Effect of roundoff error

For the Gauss elimination, the operation count for a full matrix (a matrix with relatively many nonzero entries) is as follows. In Step k we eliminate x_k from $n - k$ equations. This needs $n - k$ divisions in computing the m_{jk} (line 3) and $(n - k)(n - k + 1)$ multiplications and as many subtractions (both in line 4). Since we do $n - 1$ steps, k goes from 1 to $n - 1$ and thus the total number of operations in this forward elimination is

$$f(n) = \sum_{k=1}^{n-1}(n-k) + 2\sum_{k=1}^{n-1}(n-k)(n-k+1) \quad (\text{write } n - k = s)$$

$$= \sum_{s=1}^{n-1} s + 2\sum_{s=1}^{n-1} s(s+1) = \frac{1}{2}(n-1)n + \frac{2}{3}(n^2-1)n \approx \frac{2}{3}n^3$$

where $2n^3/3$ is obtained by dropping lower powers of n. We see that $f(n)$ grows about proportional to n^3. We say that $f(n)$ is of *order* n^3 and write

$$f(n) = O(n^3)$$

where O suggests **order.** The general definition of O is as follows. We write

$$f(n) = O(h(n))$$

if the quotient $|f(n)/h(n)|$ remains bounded (does not trail off to infinity) as $n \to \infty$. In our present case, $h(n) = n^3$ and, indeed, $f(n)/n^3 \to 2/3$ because the omitted terms divided by n^3 go to zero as $n \to \infty$.

In the back substitution of x_i we make $n - i$ multiplications and as many subtractions, as well as 1 division. Hence the number of operations in the back substitution is

$$b(n) = 2\sum_{i=1}^{n}(n-i) + n = 2\sum_{s=1}^{n} s + n = n(n+1) + n = n^2 + 2n = O(n^2)$$

We see that it grows more slowly than the number of operations in the forward elimination of the Gauss algorithm, so that it is negligible for large systems because it is smaller by a factor n, approximately. For instance, if an operation takes 10^{-9} sec, then the times needed are:

Algorithm	$n = 1000$	$n = 10000$
Elimination	0.7 sec	11 min
Back substitution	0.001 sec	0.1 sec

CONCEPT CHECK

1. Solve the following linear systems by Gauss elimination, with partial pivoting if necessary (but without scaling). Show the intermediate steps. Check the result by substitution. If no solution or more than one solution exists, give a reason.

(a) $\begin{aligned} 6x_1 + x_2 &= -3 \\ 4x_1 - 2x_2 &= 6 \end{aligned}$

(b) $\begin{aligned} 2x_1 - 8x_2 &= -4 \\ 6x_1 + 2x_2 &= 14 \end{aligned}$

(c) $\begin{aligned} 25.38x_1 - 15.48x_2 &= 30.60 \\ -7.05x_1 + 4.30x_2 &= -8.50 \end{aligned}$

(d) $\begin{aligned} 6x_2 + 13x_3 &= 137.86 \\ 6x_1 - 8x_3 &= -85.88 \\ 13x_1 - 8x_2 &= 178.54 \end{aligned}$

(e) $\begin{aligned} 5x_1 + 3x_2 + x_3 &= 2 \\ -4x_2 + 8x_3 &= -3 \\ 10x_1 - 6x_2 + 26x_3 &= 0 \end{aligned}$

(f) $\begin{aligned} 4x_1 + 10x_2 - 2x_3 &= -20 \\ -x_1 - 15x_2 + 3x_3 &= 30 \\ 25x_2 - 5x_3 &= -50 \end{aligned}$

(g) $\begin{aligned} x_1 + 2x_2 - 3x_3 &= -11 \\ 10x_1 + x_2 + x_3 &= 8 \\ 10x_2 + 2x_3 &= 2 \end{aligned}$

(h) $\begin{aligned} 3.4x_1 - 6.12x_2 - 2.72x_3 &= 0 \\ -x_1 + 1.80x_2 + 0.80x_3 &= 0 \\ 2.7x_1 - 4.86x_2 - 2.16x_3 &= 0 \end{aligned}$

(i) $\begin{aligned} 3x_2 + 5x_3 &= 1.20736 \\ 3x_1 - 4x_2 &= -2.34066 \\ 5x_1 + 6x_3 &= -0.329193 \end{aligned}$

(j) $\begin{aligned} 6.4x_1 + 3.2x_2 &= -1.6 \\ 3.2x_1 - 1.6x_2 + 4.8x_3 &= 32.0 \\ 4.8x_2 - 9.6x_3 + 7.2x_4 &= -78.0 \\ 7.2x_3 + 4.8x_4 &= 20.4 \end{aligned}$

(k) $4.4x_2 + 3.0x_3 - 6.6x_4 = -4.65$
$0.4x_1 + 3.6x_2 + 8.4x_4 = 4.62$
$-2.0x_1 - 6.2x_2 + 5.0x_3 = -4.35$
$x_1 - 7.6x_3 + 3.0x_4 = 5.97$

2. **TEAM PROJECT. Linear Systems and Gauss Elimination. (a) Existence and uniqueness.** Find a and b such that $ax_1 + x_2 = b$, $x_1 + x_2 = 3$ has (i) a unique solution, (ii) infinitely many solutions, (iii) no solutions.

(b) **Gauss elimination and nonexistence.** Apply the Gauss elimination to the following two systems and compare the calculations step by step. Explain why the elimination fails if no solution exists.

$$x_1 + x_2 + x_3 = 3$$
$$4x_1 + 2x_2 - x_3 = 5$$
$$9x_1 + 5x_2 - x_3 = 13$$

$$x_1 + x_2 + x_3 = 3$$
$$4x_1 + 2x_2 - x_3 = 5$$
$$9x_1 + 5x_2 - x_3 = 12$$

(c) **Zero determinant.** Why may a computer program give you the result that a homogeneous linear system has only the trivial solution although you know its coefficient determinant to be zero?

(d) **Pivoting.** Solve System (A) (below) by the Gauss elimination first without pivoting. Show that for any fixed machine word length and sufficiently small $\epsilon > 0$ the computer gives $x_2 = 1$ and then $x_1 = 0$. What is the exact solution? If its limit is $\epsilon \to 0$ then solve the system by the Gauss elimination with pivoting. Compare and comment.

(e) **Pivoting.** Solve System (B) by the Gauss elimination and three-digit rounding arithmetic, choosing (i) the first equation, (ii) the second equation as pivot equation. Then use four-digit rounding arithmetic in those two calculations. Compare and comment.

(A) $\epsilon x_1 + x_2 = 1$
$x_1 + x_2 = 2$

(B) $4.03x_1 + 2.16x_2 = -4.61$
$6.21x_1 + 3.35x_2 = -7.19$

1.8 Gauss–Seidel Iteration Method

This is an iterative method of great practical importance, which we can simply explain in terms of an example.

Example 44 Gauss–Seidel Iteration

We consider the linear system

$$\begin{aligned} x_1 - 0.25x_2 - 0.25x_3 + 0x_4 &= 50 \\ -0.25x_1 + x_2 + 0x_3 - 0.25x_4 &= 50 \\ -0.25x_1 + 0x_2 + x_3 - 0.25x_4 &= 25 \\ 0x_1 - 0.25x_2 - 0.25x_3 + x_4 &= 25 \end{aligned} \tag{1.66}$$

(Equations of this form arise in the numeric solution of PDEs and in spline interpolation.) We write the system in the form

$$\begin{aligned} x_1 &= 0.25x_2 + 0.25x_3 + 50 \\ x_2 &= 0.25x_1 + 0.25x_4 + 50 \\ x_3 &= 0.25x_1 + 0.25x_4 + 25 \\ x_4 &= 0.25x_2 + 0.25x_3 + 25 \end{aligned} \tag{1.67}$$

These equations are now used for iteration; that is, we start from a (possibly poor) approximation to the solution, say $x_1^{(0)} = 100$, $x_2^{(0)} = 100$, $x_3^{(0)} = 100$, $x_4^{(0)} = 100$, and compute from (1.67) a perhaps better approximation

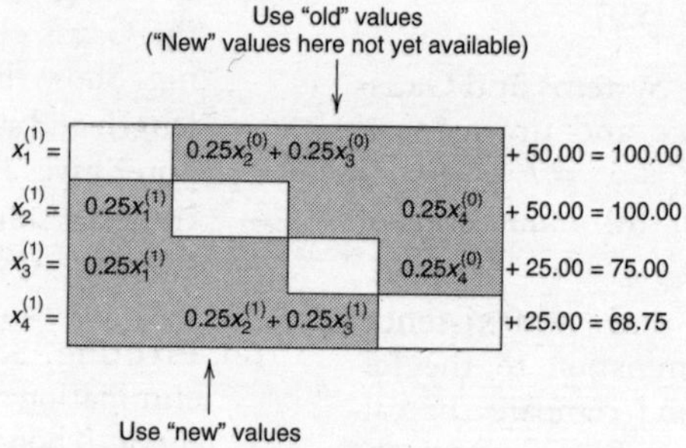

(1.68)

These equations (1.68) are obtained from (1.67) by substituting on the right the ***most recent*** approximation for each unknown. In fact, corresponding values replace previous ones as soon as they have been computed, so that in the second and third equations we use $x_1^{(1)}$ (not $x_1^{(0)}$), and in the last equation of (1.68) we use $x_2^{(1)}$ and $x_3^{(1)}$ (not $x_2^{(0)}$ and $x_3^{(0)}$). Using the same principle, we obtain in the next step

$$x_1^{(2)} = 0.25x_2^{(1)} + 0.25x_3^{(1)} + 50.00 = 93.750$$

$$x_2^{(2)} = 0.25x_1^{(2)} + 0.25x_4^{(1)} + 50.00 = 90.625$$

$$x_3^{(2)} = 0.25x_1^{(2)} + 0.25x_4^{(1)} + 25.00 = 65.625$$

$$x_4^{(2)} = 0.25x_2^{(2)} + 0.25x_3^{(2)} + 25.00 = 64.062$$

Further steps give the values

x_1	x_2	x_3	x_4
89.062	88.281	63.281	62.891
87.891	87.695	62.695	62.598
87.598	87.549	62.549	62.524
87.524	87.512	62.512	62.506
87.506	87.503	62.503	62.502

Hence convergence to the exact solution $x_1 = x_2 = 87.5$, $x_3 = x_4 = 62.5$ (verify!) seems rather fast.

An algorithm for the Gauss–Seidel iteration is shown on the next page. To obtain the algorithm, let us derive the general formulas for this iteration.

We assume that $a_{jj} = 1$ for $j = 1, \ldots, n$. (Note that this can be achieved if we can rearrange the equations so that no diagonal coefficient is zero; then we may divide each equation by the corresponding diagonal coefficient.) We now write

$$\mathbf{A} = \mathbf{I} + \mathbf{L} + \mathbf{U} \quad (a_{jj} = 1) \tag{1.69}$$

where $\mathbf{I}$ is the $n \times n$ unit matrix and $\mathbf{L}$ and $\mathbf{U}$ are respectively lower and upper triangular matrices with zero main diagonals. If we substitute (1.69) into $\mathbf{Ax} = \mathbf{b}$, we have

$$\mathbf{Ax} = (\mathbf{I} + \mathbf{L} + \mathbf{U})\mathbf{x} = \mathbf{b}$$

Taking $\mathbf{Lx}$ and $\mathbf{Ux}$ to the right, we obtain, since $\mathbf{Ix} = \mathbf{x}$,

$$\mathbf{x} = \mathbf{b} - \mathbf{Lx} - \mathbf{Ux} \tag{1.70}$$

Remembering from (1.68) in Example 44 that below the main diagonal we took "new" approximations and above the main diagonal "old" ones, we obtain from (1.68) the desired iteration formulas

"New" "Old"

$$\mathbf{x}^{(m+1)} = \mathbf{b} - \mathbf{L}\mathbf{x}^{(m+1)} - \mathbf{U}\mathbf{x}^{(m)} \quad (a_{jj} = 1) \tag{1.71}$$

where $\mathbf{x}^{(m)} = [x_j^{(m)}]$ is the mth approximation and $\mathbf{x}^{(m+1)} = [x_j^{(m+1)}]$ is the $(m + 1)$th approximation. In components this gives the formula in line 1 of the algorithm. The matrix $\mathbf{A}$ must satisfy $a_{jj} \neq 0$ for all j. In the algorithm below our assumption $a_{jj} = 1$ is no longer required, but is automatically taken care of by the factor $1/a_{jj}$ in line 1.

ALGORITHM GAUSS–SEIDEL ($\mathbf{A}$, $\mathbf{b}$, $\mathbf{x}^{(0)}$, ϵ, N)

This algorithm computes a solution $\mathbf{x}$ of the system $\mathbf{Ax} = \mathbf{b}$ given an initial approximation $\mathbf{x}^{(0)}$, where $\mathbf{A} = [a_{jk}]$ is an $n \times n$ matrix with $a_{jj} \neq 0$, $j = 1, \ldots, n$.

INPUT: $\mathbf{A}$, $\mathbf{b}$, initial approximation $\mathbf{x}^{(0)}$, tolerance $\epsilon > 0$, maximum number of iterations N

OUTPUT: Approximate solution $\mathbf{x}^{(m)} = [x_j^{(m)}]$ or failure message that $\mathbf{x}^{(N)}$ does not satisfy the tolerance condition

For $m = 0, \ldots, N - 1$, do:

For $j = 1, \ldots, n$, do:

1 $$x_j^{(m+1)} = \frac{1}{a_{jj}}\left(b_j - \sum_{k=1}^{j-1} a_{jk}x_k^{(m+1)} - \sum_{k=j+1}^{n} a_{jk}x_k^{(m)}\right)$$

End

2 If $\max_j |x_j^{(m+1)} - x_j^{(m)}| < \epsilon$ then OUTPUT $\mathbf{x}^{(m+1)}$. Stop

[Procedure completed successfully]

End

OUTPUT: "No solution satisfying the tolerance condition obtained after N iteration steps." Stop

[Procedure completed unsuccessfully]

End GAUSS–SEIDEL

Convergence and Matrix Norms

An iteration method for solving $\mathbf{Ax} = \mathbf{b}$ is said to **converge** for an initial $\mathbf{x}^{(0)}$ if the corresponding iterative sequence $\mathbf{x}^{(0)}, \mathbf{x}^{(1)}, \mathbf{x}^{(2)}, \ldots$ converges to a solution of the given system. Convergence depends on the relation between $\mathbf{x}^{(m)}$ and $\mathbf{x}^{(m+1)}$. To get this relation for the Gauss–Seidel method, we use (1.71). We first have

$$(\mathbf{I} + \mathbf{L})\mathbf{x}^{(m+1)} = \mathbf{b} - \mathbf{U}\mathbf{x}^{(m)} \tag{1.72}$$

and by multiplying by $(\mathbf{I}+\mathbf{L})^{-1}$ from the left,

$$\mathbf{x}^{(m+1)} = \mathbf{C}\mathbf{x}^{(m)} + (\mathbf{I}+\mathbf{L})^{-1}\mathbf{b} \qquad \text{where} \qquad \mathbf{C} = -(\mathbf{I}+\mathbf{L})^{-1}\mathbf{U} \tag{1.73}$$

The Gauss–Seidel iteration converges for every $\mathbf{x}^{(0)}$ if and only if all the eigenvalues explained in Chapter 2 of the "iteration matrix" $\mathbf{C} = [c_{jk}]$ have absolute value less than 1.

CAUTION! If you want to get **C**, first divide the rows of **A** by a_{jj} to have main diagonal 1, …, 1. If the *spectral radius* of **C** (= maximum of those absolute values) is small, then the convergence is rapid.

Sufficient Convergence Condition. A sufficient condition for convergence is

$$\|\mathbf{C}\| < 1 \tag{1.74}$$

Here $\|\mathbf{C}\|$ is some **matrix norm,** such as

$$\|\mathbf{C}\| = \sqrt{\sum_{j=1}^{n}\sum_{k=1}^{n} c_{jk}^{2}} \qquad \textbf{(Frobenius norm)} \tag{1.75}$$

or the greatest of the sums of the $|c_{jk}|$ in a *column* of **C**

$$\|\mathbf{C}\| = \max_{k} \sum_{j=1}^{n} |c_{jk}| \qquad \textbf{(Column "sum" norm)} \tag{1.76}$$

or the greatest of the sums of the $|c_{jk}|$ in a *row* of **C**

$$\|\mathbf{C}\| = \max_{j} \sum_{k=1}^{n} |c_{jk}| \qquad \textbf{(Row "sum" norm)} \tag{1.77}$$

These are the most frequently used matrix norms in numerics.

In most cases the choice of one of these norms is a matter of computational convenience. However, the following example shows that sometimes one of these norms is preferable to the others.

Example 45 Test of Convergence of the Gauss–Seidel Iteration

Test whether the Gauss–Seidel iteration converges for the system

$$\begin{aligned} 2x+y+z&=4 \\ x+2y+z&=4 \\ x+y+2z&=4 \end{aligned} \qquad \text{written as} \qquad \begin{aligned} x&=2-(1/2)y-(1/2)z \\ y&=2-(1/2)x-(1/2)z \\ z&=2-(1/2)x-(1/2)y \end{aligned}$$

Solution: The decomposition (multiply the matrix by 1/2 – why?) is

$$\begin{bmatrix} 1 & 1/2 & 1/2 \\ 1/2 & 1 & 1/2 \\ 1/2 & 1/2 & 1 \end{bmatrix} = \mathbf{I}+\mathbf{L}+\mathbf{U} = \mathbf{I} + \begin{bmatrix} 0 & 0 & 0 \\ 1/2 & 0 & 0 \\ 1/2 & 1/2 & 0 \end{bmatrix} + \begin{bmatrix} 0 & 1/2 & 1/2 \\ 0 & 0 & 1/2 \\ 0 & 0 & 0 \end{bmatrix}$$

It shows that

$$\mathbf{C} = -(\mathbf{I}+\mathbf{L})^{-1}\mathbf{U} = -\begin{bmatrix} 1 & 0 & 0 \\ -1/2 & 1 & 0 \\ -1/4 & -1/2 & 1 \end{bmatrix}\begin{bmatrix} 0 & 1/2 & 1/2 \\ 0 & 0 & 1/2 \\ 0 & 0 & 0 \end{bmatrix} = \begin{bmatrix} 0 & -1/2 & -1/2 \\ 0 & 1/4 & -1/4 \\ 0 & 1/8 & 3/8 \end{bmatrix}$$

We compute the Frobenius norm of **C**

$$\|\mathbf{C}\| = \left(\frac{1}{4}+\frac{1}{4}+\frac{1}{16}+\frac{1}{16}+\frac{1}{64}+\frac{9}{64}\right)^{1/2} = \left(\frac{50}{64}\right)^{1/2} = 0.884 < 1$$

and conclude from (1.75) that this Gauss–Seidel iteration converges. It is interesting that the other two norms would permit no conclusion, as you should verify. Of course, this points to the fact that (1.75) is sufficient for convergence rather than necessary.

Residual

Given a system $\mathbf{Ax} = \mathbf{b}$, the **residual r** of **x** with respect to this system is defined by

$$\mathbf{r} = \mathbf{b} - \mathbf{Ax} \tag{1.78}$$

Clearly, $\mathbf{r} = \mathbf{0}$ if and only if **x** is a solution. Hence $\mathbf{r} \neq \mathbf{0}$ for an approximate solution. In the Gauss–Seidel iteration, at each stage we modify or *relax* a component of an approximate solution in order to reduce a component of **r** to zero. Hence the Gauss–Seidel iteration belongs to a class of methods often called **relaxation methods.** More about the residual follows in the next section.

Jacobi Iteration

The Gauss–Seidel iteration is a method of **successive corrections** because for each component we successively replace an approximation of a component by a corresponding new approximation as soon as the latter has been computed. An iteration method is called a method of **simultaneous corrections** if no component of an approximation $\mathbf{x}^{(m)}$ is used until ***all*** the components of $\mathbf{x}^{(m)}$ have been computed. A method of this type is the **Jacobi iteration,** which is similar to the Gauss–Seidel iteration but involves *not* using improved values until a step has been completed and then replacing $\mathbf{x}^{(m)}$ by $\mathbf{x}^{(m+1)}$ at once, directly before the beginning of the next step. Hence if we write $\mathbf{Ax} = \mathbf{b}$ (*with* $a_{jj} = 1$ ***as before***!) in the form $\mathbf{x} = \mathbf{b} + (\mathbf{I} - \mathbf{A})\mathbf{x}$, the Jacobi iteration in matrix notation is

$$\mathbf{x}^{(m+1)} = \mathbf{b} + (\mathbf{I} - \mathbf{A})\mathbf{x}^{(m)} \qquad (a_{jj} = 1) \tag{1.79}$$

This method converges for every choice of $\mathbf{x}^{(0)}$ if and only if the spectral radius of $\mathbf{I} - \mathbf{A}$ is less than 1. It has recently gained greater practical interest since on parallel processors all n equations can be solved simultaneously at each iteration step.

CONCEPT CHECK

1. Verify the claim at the end of Example 44.
2. Show that for the system in Example 44 the Jacobi iteration diverges. [*Hint.* Use eigenvalues].
3. Do 5 steps, starting from $\mathbf{x}_0 = [1 \quad 1 \quad 1]^T$ and using 6S in the computation. [*Hint.* Make sure that you solve each equation for the variable

that has the largest coefficient (why?). Show the details].

(a) $x_1 + x_2 + 6x_3 = -61.3$
$x_1 + 9x_2 - 2x_3 = 49.1$
$8x_1 + 2x_2 - x_3 = 185.8$

(b) $x_2 + 7x_3 = 25.5$
$5x_1 + x_2 = 0$
$x_1 + 6x_2 + x_3 = 10.5$

(c) $5x_1 + x_2 + 2x_3 = 19$
$x_1 + 4x_2 - 2x_3 = -2$
$2x_1 + 3x_2 + 8x_3 = 39$

(d) $4x_1 - x_2 = 21$
$-x_1 + 4x_2 - x_3 = -45$
$-x_2 + 4x_3 = 33$

(e) $10x_1 + x_2 + x_3 = 6$
$x_1 + 10x_2 + x_3 = 6$
$x_1 + x_2 + 10x_3 = 6$

(f) $4x_1 + 5x_3 = 12.5$
$x_1 + 6x_2 + 2x_3 = 18.5$
$8x_1 + 2x_2 + x_3 = -11.5$

4. Apply the Gauss–Seidel iteration (3 steps) to the system in Prob. 2(e), starting from (a) 0, 0, 0, (b) 10, 10, 10. Compare and comment.

5. In Prob. 2(e), compute **C** (a) if you solve the first equation for x_1, the second for x_2, the third for x_3, proving convergence; (b) if you nonsensically solve the third equation for x_1, the first for x_2, the second for x_3, proving divergence.

6. Do 5 steps, starting from $\mathbf{x}_0 = [1 \quad 1 \quad 1]^T$. Compare with the Gauss–Seidel iteration. Which of the two seems to converge faster? (Show the details of your work.)
   (a) The system in Prob. 2(d)
   (b) The system in Prob. 2(c)
   (c) The system in Prob. 2(f)
   (d) Show convergence in Prob. 5(d) by verifying that **I** – **A**, where **A** is the matrix in Prob. 5(d) with the rows divided by the corresponding main diagonal entries, has the eigenvalues –0.519589 and 0.259795 ± 0.246603 *i*.

7. Compute the norms (1.67), (1.68), (1.69) for the following (square) matrices. Comment on the reasons for greater or smaller differences among the three numbers.
   (a) The matrix in Prob. 2(a)
   (b) The matrix in Prob. 2(e)
   (c) The matrix in Prob. 2(f)
   (d) $\begin{bmatrix} 2k & -k & -k \\ k & -2k & k \\ -k & -k & 2k \end{bmatrix}$
   (e) $\begin{bmatrix} 4 & -3 & 5 \\ -7 & 1 & 0 \\ 17 & -12 & 2 \end{bmatrix}$

1.9 Linear Systems: ILL-Conditioning, Norms

One does not need much experience to observe that some systems $\mathbf{Ax} = \mathbf{b}$ are good, giving accurate solutions even under roundoff or coefficient inaccuracies, whereas others are bad, so that these inaccuracies affect the solution strongly. We want to see what is going on and whether or not we can "trust" a linear system. Let us first formulate the two relevant concepts (ill- and well-conditioned) for general numeric work and then turn to linear systems and matrices.

A computational problem is called **ill-conditioned** (or *ill-posed*) if "small" changes in the data (the input) cause "large" changes in the solution (the output). On the other hand, a problem is called **well-conditioned** (or *well-posed*) if "small" changes in the data cause only "small" changes in the solution.

These concepts are qualitative. We would certainly regard a magnification of inaccuracies by a factor 100 as "large," but could debate where to draw the line between "large" and "small," depending on the kind of problem and on our viewpoint. Double precision may sometimes help, but if data are measured inaccurately, one should attempt ***changing the mathematical setting*** of the problem to a well-conditioned one.

Let us now turn to linear systems. Figure 9 explains that ill-conditioning occurs if and only if the two equations give two nearly parallel lines, so that their intersection point (the solution of the system) moves substantially if we raise or lower a line just a little. For larger systems the situation is similar in principle, although geometry no longer helps. We shall see that we may regard ill-conditioning as an approach to singularity of the matrix.

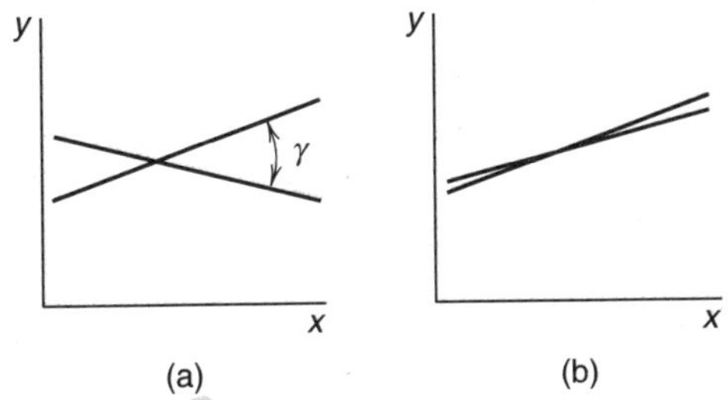

Figure 9 (a) Well-conditioned; (b) ill-conditioned linear system of two equations in two unknowns.

Example 46 An ILL-Conditioned System

You may verify that the system

$$\begin{aligned} 0.9999x - 1.0001y &= 1 \\ x - y &= 1 \end{aligned}$$

has the solution $x = 0.5$, $y = -0.5$, whereas the system

$$\begin{aligned} 0.9999x - 1.0001y &= 1 \\ x - y &= 1 + \epsilon \end{aligned}$$

has the solution $x = 0.5 + 5000.5\,\epsilon$, $y = -0.5 + 4999.5\,\epsilon$. This shows that the system is ill-conditioned because a change on the right of magnitude ϵ produces a change in the solution of magnitude 5000ϵ, approximately. We see that the lines given by the equations have nearly the same slope.

Well-conditioning can be asserted if the main diagonal entries of **A** have large absolute values compared to those of the other entries. Similarly if $\mathbf{A}^{-1}$ and **A** have maximum entries of about the same absolute value.

Ill-conditioning is indicated if $\mathbf{A}^{-1}$ has entries of large absolute value compared to those of the solution (about 5000 in Example 46) and if poor approximate solutions may still produce small residuals.

Residual: The *residual* **r** of an approximate solution $\tilde{\mathbf{x}}$ of $\mathbf{Ax} = \mathbf{b}$ is defined as

$$\mathbf{r} = \mathbf{b} - \mathbf{A}\tilde{\mathbf{x}} \tag{1.80}$$

Now $\mathbf{b} = \mathbf{Ax}$, so that

$$\mathbf{r} = \mathbf{A}(\mathbf{x} - \tilde{\mathbf{x}}) \tag{1.81}$$

Hence **r** is small if $\tilde{\mathbf{x}}$ has high accuracy, but the converse may be false:

Example 47 **Inaccurate Approximate Solution with a Small Residual**

The system

$$\begin{aligned} 1.0001x_1 + \quad\quad x_2 &= 2.0001 \\ x_1 + 1.0001x_2 &= 2.0001 \end{aligned}$$

has the exact solution $x_1 = 1, x_2 = 1$. Can you see this by inspection? The very inaccurate approximation $\tilde{x}_1 = 2.0000, \tilde{x}_2 = 0.0001$ has the very small residual (to 4D):

$$\mathbf{r} = \begin{bmatrix} 2.0001 \\ 2.0001 \end{bmatrix} - \begin{bmatrix} 1.0001 & 1.0000 \\ 1.0000 & 1.0001 \end{bmatrix} \begin{bmatrix} 2.0000 \\ 0.0001 \end{bmatrix} = \begin{bmatrix} 2.0001 \\ 2.0001 \end{bmatrix} - \begin{bmatrix} 2.0003 \\ 2.0001 \end{bmatrix} = \begin{bmatrix} -0.0002 \\ 0.0000 \end{bmatrix}$$

From this, a naive person might draw the false conclusion that the approximation should be accurate to 3 or 4 decimals.

Our result is probably unexpected, but we shall see that it has to do with the fact that the system is ill-conditioned.

Our goal is to show that ill-conditioning of a linear system and of its coefficient matrix **A** can be measured by a number, the *"condition number"* $\kappa(\mathbf{A})$. Other measures for ill-conditioning have also been proposed, but $\kappa(\mathbf{A})$ is probably the most widely used one. $\kappa(\mathbf{A})$ is defined in terms of norm, a concept of great general interest throughout numerics (and in modern mathematics in general!). We shall reach our goal in three steps, discussing:

1. **Vector norms.**
2. **Matrix norms.**
3. **Condition number** κ of a square matrix.

Vector Norms

A vector norm for column vectors $\mathbf{x} = [x_j]$ with n components (n fixed) is a generalized length or distance. It is denoted by $\|\mathbf{x}\|$ and is defined by four properties of the usual length of vectors in three-dimensional space, namely,

$$\begin{aligned} &\text{(a)} \quad \|\mathbf{x}\| \text{ is a nonnegative real number} \\ &\text{(b)} \quad \|\mathbf{x}\| = 0 \quad \text{if and only if} \quad \mathbf{x} = \mathbf{0} \\ &\text{(c)} \quad \|k\mathbf{x}\| = |k|\,\|\mathbf{x}\| \quad \text{where } k \text{ is a scalar} \\ &\text{(d)} \quad \|\mathbf{x}+\mathbf{y}\| \leqq \|\mathbf{x}\| + \|\mathbf{y}\| \quad \text{(Triangle inequality)} \end{aligned} \tag{1.82}$$

If we use several norms, we label them by a subscript. Most important in connection with computations is the ***p-norm*** defined by

$$\|\mathbf{x}\|_p = (\|x_1\|^p + \|x_2\|^p + \cdots + \|x_n\|^p)^{1/p} \tag{1.83}$$

where p is a fixed number and $p \geqq 1$. In practice, one usually takes $p = 1$ or 2 and, as a third norm, $\|\mathbf{x}\|_\infty$ (the latter as defined below), that is,

$$\|\mathbf{x}\|_1 = |x_1| + \cdots + |x_n| \quad (\text{“}l_1\text{-norm”}) \tag{1.84}$$

$$\|\mathbf{x}\|_2 = \sqrt{x_1^2 + \cdots + x_n^2} \qquad \text{("Euclidean" or "}l_2\text{-norm")} \tag{1.85}$$

$$\|\mathbf{x}\|_\infty = \max_j |x_j| \qquad \text{("}l_\infty\text{-norm")} \tag{1.86}$$

For $n = 3$ the l_2-norm is the usual length of a vector in three-dimensional space. The l_1-norm and l_∞-norm are generally more convenient in computation. But all three norms are in common use.

Example 48 Vector Norms

If

$$\mathbf{x}^T = [2 \quad -3 \quad 0 \quad 1 \quad -4]$$

then

$$\|\mathbf{x}\|_1 = 10, \quad \|\mathbf{x}\|_2 = \sqrt{30}, \quad \|\mathbf{x}\|_\infty = 4$$

In three-dimensional space, two points with position vectors $\mathbf{x}$ and $\tilde{\mathbf{x}}$ have distance $\|\mathbf{x} - \tilde{\mathbf{x}}\|$ from each other. For a linear system $\mathbf{Ax} = \mathbf{b}$, this suggests that we take $\|\mathbf{x} - \tilde{\mathbf{x}}\|$ as a measure of inaccuraty and call it the **distance** between an exact and an approximate solution, or the **error** of $\tilde{\mathbf{x}}$.

Matrix Norm

If $\mathbf{A}$ is an $n \times n$ matrix and $\mathbf{x}$ any vector with n components, then $\mathbf{Ax}$ is a vector with n components. We now take a vector norm and consider $\|\mathbf{x}\|$ and $\|\mathbf{Ax}\|$. One can prove that there is a number c (depending on $\mathbf{A}$) such that

$$\|\mathbf{Ax}\| \leqq c\|\mathbf{x}\| \qquad \text{for all } \mathbf{x} \tag{1.87}$$

Let $\mathbf{x} \neq 0$. Then $\|\mathbf{x}\| > 0$ by (1.82b) and division gives $\|\mathbf{Ax}\|/\|\mathbf{x}\| \leqq c$. We obtain the smallest possible c valid for *all* $\mathbf{x}$ ($\neq 0$) by taking the maximum on the left. This smallest c is called the **matrix norm of A** *corresponding to the vector norm we picked* and is denoted by $\|\mathbf{A}\|$. Thus

$$\|\mathbf{A}\| = \max \frac{\|\mathbf{Ax}\|}{\|\mathbf{x}\|} \qquad (\mathbf{x} \neq 0) \tag{1.88}$$

the maximum being taken over all $\mathbf{x} \neq 0$. Alternatively,

$$\|\mathbf{A}\| = \max_{\|\mathbf{x}\|=1} \|\mathbf{Ax}\| \tag{1.89}$$

The maximum in (1.89) and thus also in (1.88) exists. And the name "matrix ***norm***" is justified because $\|\mathbf{A}\|$ satisfies (1.82) with $\mathbf{x}$ and $\mathbf{y}$ replaced by $\mathbf{A}$ and $\mathbf{B}$. Note carefully that $\|\mathbf{A}\|$ depends on the vector norm that we selected. In particular, one can show that

for the l_1-norm (1.84) one gets the column "sum" norm (1.89)

for the l_∞-norm (1.86) one gets the row "sum" norm (1.90)

By taking our best possible (our smallest) $c = \|\mathbf{A}\|$ we have from (1.87)

$$\|\mathbf{Ax}\| \leqq \|\mathbf{A}\| \, \|\mathbf{x}\| \tag{1.90}$$

This is the formula we shall need. Formula (1.88) also implies for two $n \times n$ matrices

$$\|AB\| \leqq \|A\|\,\|B\|$$

thus

$$\|A^n\| \leqq \|A\|^n \tag{1.91}$$

Before we go on, let us do a simple illustrative computation.

Example 49 **Matrix Norms**

Compute the matrix norms of the coefficient matrix A in Example 47 and of its inverse A^{-1}, assuming that we use (a) the l_1-vector norm, (b) the l_∞-vector norm.

Solution: We use (1.51) for the inverse and then (1.81) and (1.82). Thus

$$A = \begin{bmatrix} 0.9999 & -1.0001 \\ 1.0000 & -1.0000 \end{bmatrix}, \qquad A^{-1} = \begin{bmatrix} -5000.0 & 5000.5 \\ -5000.0 & 4999.5 \end{bmatrix}$$

(a) The l_1-vector norm gives the column "sum" norm (1.89); from Column 2 we thus obtain $\|A\| = \|-1.0001\| + \|-1.0000\| = 2.0001$. Similarly, $\|A^{-1}\| = 10000$.

(b) The l_∞-vector norm gives the row "sum" norm (1.92); thus $\|A\| = 2$, $\|A^{-1}\| = 10000.5$ from Row 1. We notice that $\|A^{-1}\|$ is surprisingly large, which makes the product $\|A\|\,\|A^{-1}\|$ large (20001). We shall see below that this is typical of an ill-conditioned system.

Condition Number of a Matrix

We are now ready to introduce the key concept in our discussion of ill-conditioning, the **condition number** $\kappa(A)$ of a (nonsingular) square matrix **A**, defined by

$$\kappa(A) = \|A\|\,\|A^{-1}\| \tag{1.92}$$

The role of the condition number is seen from the following theorem.

THEOREM 23 (Condition Number): *A linear system of equations* **Ax** = **b** *and its matrix* **A** *whose condition number* (1.92) *is small are well-conditioned. A large condition number indicates ill-conditioning.*

PROOF: **b** = **Ax** and (1.92) give $\|b\| \leqq \|A\|\,\|x\|$. Let $b \neq 0$ and $x \neq 0$. Then division by $\|b\|\,\|x\|$ gives

$$\frac{1}{\|x\|} \leq \frac{\|A\|}{\|b\|} \tag{1.93}$$

Multiplying (1.83) $r = A(x - \tilde{x})$ by A^{-1} from the left and interchanging sides, we have $x - \tilde{x} = A^{-1}r$. Now (1.90) with A^{-1} and **r** instead of **A** and **x** yields

$$\|x - \tilde{x}\| = \|A^{-1}r\| \leqq \|A^{-1}\|\,\|r\|$$

Division by $\|x\|$ {note that $\|x\| \neq 0$ by (1.82b)} and use of (1.93) finally gives

$$\frac{\|x - \tilde{x}\|}{\|x\|} \leq \frac{1}{\|x\|}\|A^{-1}\|\,\|r\| \leqq \frac{\|A\|}{\|b\|}\|A^{-1}\|\,\|r\| = \kappa(A)\frac{\|r\|}{\|b\|}\tilde{x} \tag{1.94}$$

Hence if $\kappa(\mathbf{A})$ is small, a small $\|\mathbf{r}\|/\|\mathbf{b}\|$ implies a small relative error $\|\mathbf{x}-\tilde{\mathbf{x}}\|/\|\mathbf{x}\|$, so that the system is well-conditioned. However, this does not hold if $\kappa(\mathbf{A})$ is large; then a small $\|\mathbf{r}\|/\|\mathbf{b}\|$ does not necessarily imply a small relative error $\|\mathbf{x}-\tilde{\mathbf{x}}\|/\|\mathbf{x}\|$.

Example 50 Condition Numbers. Gauss–Seidel Iteration

$$\mathbf{A}=\begin{bmatrix}5 & 1 & 1\\ 1 & 4 & 2\\ 1 & 2 & 4\end{bmatrix} \quad \text{has the inverse} \quad \mathbf{A}^{-1}=\frac{1}{56}\begin{bmatrix}12 & -2 & -2\\ -2 & 19 & -9\\ -2 & -9 & 19\end{bmatrix}$$

Since $\mathbf{A}$ is symmetric, (1.81) and (1.82) give the same condition number

$$\kappa(\mathbf{A}) = \|\mathbf{A}\|\,\|\mathbf{A}^{-1}\| = 7\times(1/56)\times 30 = 3.75$$

We see that a linear system $\mathbf{Ax} = \mathbf{b}$ with this $\mathbf{A}$ is well-conditioned. For instance, if $\mathbf{b} = [14 \quad 0 \quad 28]^T$, the Gauss algorithm gives the solution $\mathbf{x} = [2 \quad -5 \quad 9]^T$ (confirm this). Since the main diagonal entries of $\mathbf{A}$ are relatively large, we can expect reasonably good convergence of the Gauss–Seidel iteration. Indeed, starting from, say, $\mathbf{x}_0 = [1 \quad 1 \quad 1]^T$, we obtain the first 8 steps (3D values)

x_1	x_2	x_3
1.000	1.000	1.000
2.400	−1.100	6.950
1.630	−3.882	8.534
1.870	−4.734	8.900
1.967	−4.942	8.979
1.993	−4.988	8.996
1.998	−4.997	8.999
2.000	−5.000	9.000
2.000	−5.000	9.000

Example 51 ILL-Conditioned Linear System

Example 49 gives by (1.89) or (1.90) for the matrix in Example 46 the very large condition number $\kappa(\mathbf{A}) = 2.0001\times 10000 = 2\times 10000.5 = 200001$. This confirms that the system is very ill-conditioned. Similarly in Example 47, where by (1.56), and 6D-computation,

$$\mathbf{A}^{-1}=\frac{1}{0.0002}\begin{bmatrix}1.0001 & -1.0000\\ -1.0000 & 1.0001\end{bmatrix}=\begin{bmatrix}5000.5 & -5000.0\\ -5000.0 & 5000.5\end{bmatrix}$$

so that (1.89) gives a very large $\kappa(\mathbf{A})$, explaining the surprising result in Example 47,

$$\kappa(\mathbf{A}) = (1.0001 + 1.0000)(5000.5 + 5000.0) \approx 20002$$

In practice, A^{-1} will not be known, so that in computing the condition number $\kappa(A)$, one must estimate $\|A^{-1}\|$.

Inaccurate Matrix Entries

$\kappa(A)$ can be used for estimating the effect δx of an inaccuracy δA of A (errors of measurements of the a_{jk}, for instance). Instead of $Ax = b$ we then have

$$(A + \delta A)(x + \delta x) = b$$

Multiplying out and subtracting $Ax = b$ on both sides, we obtain

$$A\delta x + \delta A(x + \delta x) = 0$$

Multiplication by A^{-1} from the left and taking the second term to the right gives

$$\delta x = -A^{-1}\delta A(x + \delta x)$$

Applying (1.90) with A^{-1} and vector $\delta A(x + \delta x)$ instead of A and x, we get

$$\|\delta x\| = \|A^{-1}\delta A(x + \delta x)\| \leqq \|A^{-1}\|\,\|\delta A(x + \delta x)\|$$

Applying (1.90) on the right, with δA and $x - \delta x$ instead of A and x, we obtain

$$\|\delta x\| \leqq \|A^{-1}\|\,\|\delta A\|\,\|x + \delta x\|$$

Now $\|A^{-1}\| = \kappa(A)/\|A\|$ by the definition of $\kappa(A)$, so that division by $\|x + \delta x\|$ shows that the relative inaccuracy of x is related to that of A via the condition number by the inequality

$$\frac{\|\delta x\|}{\|x\|} \approx \frac{\|\delta x\|}{\|x+\delta x\|} \leqq \|A^{-1}\|\|\delta A\| = \kappa(A)\frac{\|\delta A\|}{\|A\|} \tag{1.95}$$

Conclusion. If the system is well-conditioned, small inaccuracies $\|\delta A\|/\|A\|$ can have only a small effect on the solution. However, in the case of ill-conditioning, if $\|\delta A\|/\|A\|$ is small, $\|\delta x\|/\|x\|$ ***may*** be large.

Inaccurate Right Side. You may show that, similarly, when A is accurate, an inaccuracy δb of b causes an inaccuracy δx satisfying

$$\frac{\|\delta x\|}{\|x\|} \leqq \kappa(A)\frac{\|\delta b\|}{\|b\|} \tag{1.96}$$

Hence $\|\delta x\|/\|x\|$ must remain relatively small whenever $\kappa(A)$ is small.

Example 52 **Inaccuracies. Bounds (1.95) and (1.96)**

If each of the nine entries of A in Example 50 is measured with an inaccuracy of 0.1, then $\|\delta A\| = 9 \times 0.1$ and (1.96) gives

$$\frac{\|\delta x\|}{\|x\|} \leqq 7.5\frac{3\times 0.1}{7} = 0.321$$

thus

$$\|\delta x\| \leqq 0.321\|x\| = 0.321 \times 16 = 5.14$$

By experimentation you will find that the actual inaccuracy $\|\delta \mathbf{x}\|$ is only about 30% of the bound 5.14. This is typical.

Similarly, if $\delta \mathbf{b} = [0.1 \quad 0.1 \quad 0.1]^T$, then $\|\delta \mathbf{b}\| = 0.3$ and $\|\mathbf{b}\| = 42$ in Example 50, so that (1.96) gives

$$\frac{\|\delta \mathbf{x}\|}{\|\mathbf{x}\|} \leqq 7.5 \frac{0.3}{42} = 0.0536$$

hence

$$\|\delta \mathbf{x}\| \leqq 0.0536 \times 16 = 0.857$$

but this bound is again much greater than the actual inaccuracy, which is about 0.15.

Further Comments on Condition Numbers. The following additional explainations may be helpful.

1. There is no sharp dividing line between "well-conditioned" and "ill-conditioned," but generally the situation will get worse as we go from systems with small $\kappa(\mathbf{A})$ to systems with larger $\kappa(\mathbf{A})$. Now always $\kappa(\mathbf{A}) \geqq 1$, so that values of 10 or 20 or so give no reason for concern, whereas $\kappa(\mathbf{A}) = 100$, say, calls for caution, and systems such as those in Examples 46 and 47 are extremely ill-conditioned.
2. If $\kappa(\mathbf{A})$ is large (or small) in one norm, it will be large (or small, respectively) in any other norm.
3. The literature on ill-conditioning is extensive.

This is the end of our discussion of numerics for solving linear systems. In the next section we consider curve fitting, an important area in which solutions are obtained from linear systems.

CONCEPT CHECK

1. Compute (1.84)–(1.87). Compute a corresponding **unit vector** (vector of norm 1) with respect to the l_∞-norm.

   (a) $[1 \quad -6 \quad 5]$
   (b) $[0.4 \quad -1.2 \quad 0 \quad 8.0]$
   (c) $[-4 \quad 4 \quad 3 \quad -3]$
   (d) $[0 \quad 0 \quad 1 \quad 0 \quad 0]$
   (e) $[0.3 \quad -0.1 \quad 0.5 \quad 1.0]$
   (f) $[16 \quad 21 \quad 54 \quad -119]$
   (g) $[1 \quad 1 \quad 1 \quad 1 \quad 1 \quad 1]$
   (h) $[3 \quad 0 \quad 0 \quad -3 \quad 0]$

2. Show that $\|\mathbf{x}\|_\infty \leqq \|\mathbf{x}\|_2 \leqq \|\mathbf{x}\|_1$.

3. Compute the matrix norm and the condition number corresponding to the l_1-vector norm.

   (a) $\begin{bmatrix} -3 & 4 \\ 1 & 2 \end{bmatrix}$ (b) $\begin{bmatrix} 5 & 7 \\ 7 & 10 \end{bmatrix}$

   (c) $\begin{bmatrix} \sqrt{3} & 3 \\ 0 & -\sqrt{3} \end{bmatrix}$ (d) $\begin{bmatrix} 0 & 0 & 100 \\ 0 & 100 & 0 \\ 0.01 & 0 & 0 \end{bmatrix}$

   (e) $\begin{bmatrix} 21 & 10.5 & 7 & 5.25 \\ 10.5 & 7 & 5.25 & 4.2 \\ 7 & 5.25 & 4.2 & 3.5 \\ 5.25 & 4.2 & 3.5 & 3 \end{bmatrix}$

   (f) $\begin{bmatrix} 1 & 0.1 & 0 \\ 0.1 & 1 & 0.1 \\ 0 & 0.1 & 1 \end{bmatrix}$

4. Verify (1.90) for $\mathbf{x} = [4 \quad -5 \quad 2]^T$ taken with the l_∞-norm and the matrix in Prob. 3(f).
5. Verify (1.91) for the matrices in Probs. 3(a) and 3(b).

6. Verify the calculations in Examples 50 and 51 of the text.
7. Solve $\mathbf{Ax} = \mathbf{b}_1$, $\mathbf{Ax} = \mathbf{b}_2$, compare the solutions, and comment. Compute the condition number of $\mathbf{A}$.

(a) $\mathbf{A} = \begin{bmatrix} 2 & 1.4 \\ 1.4 & 1 \end{bmatrix}, \; \mathbf{b}_1 = \begin{bmatrix} 1.4 \\ 1 \end{bmatrix}, \; \mathbf{b}_2 = \begin{bmatrix} 1.44 \\ 1 \end{bmatrix}$

(b) $\mathbf{A} = \begin{bmatrix} 5 & -7 \\ -7 & 10 \end{bmatrix}, \; \mathbf{b}_1 = \begin{bmatrix} -2 \\ 3 \end{bmatrix}, \; \mathbf{b}_2 = \begin{bmatrix} -2 \\ 3.1 \end{bmatrix}$

8. **(Residual)** For $\mathbf{Ax} = \mathbf{b}_1$ in Prob. 7(a) guess what the residual of $\tilde{\mathbf{x}} = [113 \quad -160]^T$ might be (the solution being $\mathbf{x} = [0 \quad 1]^T$). Then calculate and comment.
9. Show that $\kappa(\mathbf{A}) \geqq 1$ for the matrix norms (1.89), (1.90), and $\kappa(\mathbf{A}) \geqq \sqrt{n}$ for the Frobenius norm (1.88).

1.10 LU Decompositions

We continue our discussion of numeric methods for solving linear systems of n equations in n unknowns $x_1, \ldots, x_n$,

$$\mathbf{Ax} = \mathbf{b} \tag{1.97}$$

where $\mathbf{A} = [a_{jk}]$ is the $n \times n$ coefficient matrix and $\mathbf{x}^T = [x_1 \; \cdots \; x_n]$ and $\mathbf{b}^T = [b_1 \; \cdots \; b_n]$. We present three related methods that are modifications of the Gauss elimination, which require fewer arithmetic operations. They are named after Doolittle, Crout, and Cholesky and use the idea of the LU-factorization of **A**, which we explain first.

An **LU-factorization** of a given square matrix **A** is of the form

$$\mathbf{A} = \mathbf{LU} \tag{1.98}$$

where **L** is ***lower triangular*** and **U** is ***upper triangular***. For example,

$$\mathbf{A} = \begin{bmatrix} 2 & 3 \\ 8 & 5 \end{bmatrix} = \mathbf{LU} = \begin{bmatrix} 1 & 0 \\ 4 & 1 \end{bmatrix} \begin{bmatrix} 2 & 3 \\ 0 & -7 \end{bmatrix}$$

It can be proved that for any nonsingular matrix the rows can be reordered so that the resulting matrix **A** has an LU-factorization (1.98) in which **L** turns out to be the matrix of the ***multipliers*** m_{jk} of the Gauss elimination, with main diagonal 1, …, 1, and **U** is the matrix of the triangular system at the end of the Gauss elimination.

The ***crucial idea*** now is that **L** and **U** in (1.98) can be computed directly, without solving simultaneous equations (thus, without using the Gauss elimination). As a count shows, this needs about $n^3/3$ operations, about half as many as the Gauss elimination, which needs about $2n^3/3$. And once we have (1.98), we can use it for solving $\mathbf{Ax} = \mathbf{b}$ in two steps, involving only about n^2 operations, simply by noting that $\mathbf{Ax} = \mathbf{LUx} = \mathbf{b}$ may be written

$$\text{(a) } \mathbf{Ly} = \mathbf{b} \quad \text{and} \quad \text{(b) } \mathbf{Ux} = \mathbf{y} \tag{1.99}$$

and solving first (1.99a) for **y** and then (1.99b) for **x**. Here we can require that **L** have main diagonal 1, …, 1 as stated before; then this is called **Doolittle's method.** Both systems (1.99a) and (1.99b) are triangular, so we can solve them as in the back substitution for the Gauss elimination.

Crout's Reduction

A similar method, **Crout's method,** is obtained from (1.98) if **U** (instead of **L**) is required to have main diagonal 1, …, 1. In either case the factorization (1.98) is unique.

Example 53 Doolittle's Method

Solve the system

$$\begin{aligned} E_1&: \quad 8x_2 + 2x_3 = -7 \\ E_2&: 3x_1 + 5x_2 + 2x_3 = 8 \\ E_3&: 6x_1 + 2x_2 + 8x_3 = 26 \end{aligned}$$

by Doolittle's method.

Solution: The decomposition (1.97) is obtained from

$$A = [a_{jk}] = \begin{bmatrix} a_{11} & a_{12} & a_{13} \\ a_{21} & a_{22} & a_{23} \\ a_{31} & a_{32} & a_{33} \end{bmatrix} = \begin{bmatrix} 3 & 5 & 2 \\ 0 & 8 & 2 \\ 6 & 2 & 8 \end{bmatrix} = \begin{bmatrix} 1 & 0 & 0 \\ m_{21} & 1 & 0 \\ m_{31} & m_{32} & 1 \end{bmatrix} \begin{bmatrix} u_{11} & u_{12} & u_{13} \\ 0 & u_{22} & u_{23} \\ 0 & 0 & u_{33} \end{bmatrix}$$

by determining the m_{jk} and u_{jk}, using matrix multiplication. By going through **A** row by row we get successively

$a_{11} = 3 = 1 \times u_{11} = u_{11}$	$a_{12} = 5 = 1 \times u_{12} = u_{12}$	$a_{13} = 2 = 1 \times u_{13} = u_{13}$
$a_{21} = 0 = m_{21}u_{11}$ $m_{21} = 0$	$a_{22} = 8 = m_{21}u_{12} + u_{22}$ $u_{22} = 8$	$a_{23} = 2 = m_{23}u_{13} + u_{23}$ $u_{23} = 2$
$a_{31} = 6 = m_{31}u_{11} = m_{31} \times 3$ $m_{31} = 2$	$a_{32} = 2 = m_{31}u_{12} + m_{32}u_{22}$ $= 2 \times 5 + m_{32} \times 8$ $m_{32} = -1$	$a_{33} = 8 = m_{31}u_{13} + m_{32}u_{23} + u_{33}$ $= 2 \times 2 - 1 \times 2 + u_{33}$ $u_{33} = 6$

Thus the factorization (1.98) is

$$\begin{bmatrix} 3 & 5 & 2 \\ 0 & 8 & 2 \\ 6 & 2 & 8 \end{bmatrix} = \mathbf{LU} = \begin{bmatrix} 1 & 0 & 0 \\ 0 & 1 & 0 \\ 2 & -1 & 1 \end{bmatrix} \begin{bmatrix} 3 & 5 & 2 \\ 0 & 8 & 2 \\ 0 & 0 & 6 \end{bmatrix}$$

We first solve $\mathbf{Ly} = \mathbf{b}$, determining $y_1 = 8$, then $y_2 = -7$, then y_3 from $2y_1 - y_2 + y_3 = 16 + 7 + y_3 = 26$; thus (note the interchange in **b** because of the interchange in **A**!)

$$\begin{bmatrix} 1 & 0 & 0 \\ 0 & 1 & 0 \\ 2 & -1 & 1 \end{bmatrix} \begin{bmatrix} y_1 \\ y_2 \\ y_3 \end{bmatrix} = \begin{bmatrix} 8 \\ -7 \\ 26 \end{bmatrix}$$

Solution:

$$y = \begin{bmatrix} 8 \\ -7 \\ 3 \end{bmatrix}$$

Then we solve $\mathbf{Ux} = \mathbf{y}$, determining $x_3 = 3/6$, then x_2, then x_1, that is,

$$\begin{bmatrix} 3 & 5 & 2 \\ 0 & 8 & 2 \\ 0 & 0 & 6 \end{bmatrix} \begin{bmatrix} x_1 \\ x_2 \\ x_3 \end{bmatrix} = \begin{bmatrix} 8 \\ -7 \\ 3 \end{bmatrix}$$

Solution:

$$\mathbf{x} = \begin{bmatrix} 4 \\ -1 \\ 1/2 \end{bmatrix}$$

Our formulas in Example 53 suggest that for general n the entries of the matrices $\mathbf{L} = [m_{jk}]$ (with main diagonal 1, …, 1 and m_{jk} suggesting "multiplier") and $\mathbf{U} = [u_{jk}]$ in the **Doolittle method** are computed from

$$\begin{aligned}
u_{1k} &= a_{1k} \qquad k = 1, \ldots, n \\
m_{j1} &= a_{j1}/u_{11} \qquad j = 2, \ldots, n \\
u_{jk} &= a_{jk} - \sum_{s=1}^{j-1} m_{js} u_{sk} \qquad (k = j, \ldots, n;\ j \geqq 2) \\
m_{jk} &= \frac{1}{u_{kk}} \left(a_{jk} - \sum_{s=1}^{k-1} m_{js} u_{sk} \right) \qquad (j = k+1, \ldots, n;\ k \geqq 2)
\end{aligned} \tag{1.100}$$

Row Interchanges. Matrices, such as

$$\begin{bmatrix} 0 & 1 \\ 1 & 1 \end{bmatrix} \quad \text{or} \quad \begin{bmatrix} 0 & 1 \\ 1 & 0 \end{bmatrix}$$

have no LU-factorization (try!). This indicates that for obtaining an LU-factorization, row interchanges of **A** (and corresponding interchanges in **b**) may be necessary.

Cholesky's Method

For a ***symmetric, positive definite*** matrix **A** (thus $\mathbf{A} = \mathbf{A}^T$, $\mathbf{x}^T\mathbf{A}\mathbf{x} > 0$ for all $\mathbf{x} \neq \mathbf{0}$) we can in (1.90) even choose $\mathbf{U} = \mathbf{L}^T$, thus $u_{jk} = m_{kj}$ (but cannot impose conditions on the main diagonal entries). For example,

$$\mathbf{A} = \begin{bmatrix} 4 & 2 & 14 \\ 2 & 17 & -5 \\ 14 & -5 & -83 \end{bmatrix} = \mathbf{L}\mathbf{L}^T = \begin{bmatrix} 2 & 0 & 0 \\ 1 & 4 & 0 \\ 7 & -3 & 5 \end{bmatrix} \begin{bmatrix} 2 & 1 & 7 \\ 0 & 4 & -3 \\ 0 & 0 & 5 \end{bmatrix} \tag{1.101}$$

The popular method of solving $\mathbf{Ax} = \mathbf{b}$ based on this factorization $\mathbf{A} = \mathbf{L}\mathbf{L}^T$ is called **Cholesky's method.** In terms of the entries of $\mathbf{L} = [l_{jk}]$ the formulas for the factorization are

$$\begin{aligned}
l_{11} &= \sqrt{a_{11}} \\
l_{j1} &= \frac{a_{j1}}{l_{11}} \qquad (j = 2, \ldots, n) \\
l_{jj} &= \sqrt{a_{jj} - \sum_{s=1}^{j-1} l_{js}^2} \qquad (j = 2, \ldots, n) \\
l_{pj} &= \frac{1}{l_{jj}} \left(a_{pj} - \sum_{s=1}^{j-1} l_{js} l_{ps} \right) \qquad (p = j+1, \ldots, n;\ j \geqq 2)
\end{aligned} \tag{1.102}$$

If **A** is symmetric but not positive definite, this method could still be applied, but then leads to a *complex* matrix **L**, so that the method becomes impractical.

Example 54 **Cholesky's Method**

Solve by Cholesky's method:

$$\begin{aligned} 4x_1 + 2x_2 + 14x_3 &= 14 \\ 2x_1 + 17x_2 - 5x_3 &= -101 \\ 14x_1 - 5x_2 + 83x_3 &= 155 \end{aligned}$$

Solution: From (1.94) or from the form of the factorization

$$\begin{bmatrix} 4 & 2 & 14 \\ 2 & 17 & -5 \\ 14 & -5 & 83 \end{bmatrix} = \begin{bmatrix} l_{11} & 0 & 0 \\ l_{21} & l_{22} & 0 \\ l_{31} & l_{32} & l_{33} \end{bmatrix} \begin{bmatrix} l_{11} & l_{21} & l_{31} \\ 0 & l_{22} & l_{32} \\ 0 & 0 & l_{33} \end{bmatrix}$$

we compute, in the given order,

$$l_{11} = \sqrt{a_{11}} = 2, \qquad l_{21} = \frac{a_{21}}{l_{11}} = \frac{2}{2} = 1, \qquad l_{31} = \frac{a_{31}}{l_{11}} = \frac{14}{2} = 7$$

$$l_{22} = \sqrt{a_{22} - l_{21}^2} = \sqrt{17-1} = 4$$

$$l_{32} = \frac{1}{l_{22}}(a_{32} - l_{31}l_{21}) = \frac{1}{4}(-5 - 7 \times 1) = -3$$

$$l_{33} = \sqrt{a_{33} - l_{31}^2 - l_{32}^2} = \sqrt{83 - 7^2 - (-3)^2} = 5$$

This agrees with (1.100). We now have to solve $\mathbf{Ly} = \mathbf{b}$, that is,

$$\begin{bmatrix} 2 & 0 & 0 \\ 1 & 4 & 0 \\ 7 & -3 & 5 \end{bmatrix} \begin{bmatrix} y_1 \\ y_2 \\ y_3 \end{bmatrix} = \begin{bmatrix} 14 \\ -101 \\ 155 \end{bmatrix} \qquad \textbf{Solution:} \qquad \mathbf{y} = \begin{bmatrix} 7 \\ -27 \\ 5 \end{bmatrix}$$

As the second step, we have to solve $\mathbf{Ux} = \mathbf{L}^T\mathbf{x} = \mathbf{y}$, that is,

$$\begin{bmatrix} 2 & 1 & 7 \\ 0 & 4 & -3 \\ 0 & 0 & 5 \end{bmatrix} \begin{bmatrix} x_1 \\ x_2 \\ x_3 \end{bmatrix} = \begin{bmatrix} 7 \\ -27 \\ 5 \end{bmatrix} \qquad \textbf{Solution:} \qquad \mathbf{x} = \begin{bmatrix} 3 \\ -6 \\ 1 \end{bmatrix}$$

THEOREM 24 (Stability of the Cholesky Factorization): *The Cholesky* $\mathbf{LL}^T$*-factorization is numerically stable.*

PROOF: We have $a_{jj} = l_{j1}^2 + l_{j2}^2 + \cdots + l_{jj}^2$ by squaring the third formula in (1.101) and solving it for a_{jj}. Hence for all l_{jk} (note that $l_{jk} = 0$ for $k > j$) we obtain (the inequality being trivial)

$$l_{jk}^2 \leqq l_{j1}^2 + l_{j2}^2 + \cdots + l_{jj}^2 = a_{jj}$$

That is, l_{jk}^2 is bounded by an entry of $\mathbf{A}$, which means stability against rounding. ■

LU Decomposition from Gaussian Elimination

Another variant of the Gauss elimination is the **Gauss–Jordan elimination**, introduced by W. Jordan in 1920, in which back substitution is avoided by additional computations that reduce the matrix to diagonal form, instead of the triangular form in the Gauss elimination. But this reduction from the Gauss triangular to the diagonal form requires more operations than back substitution does, so that the method is *disadvantageous* for solving systems $\mathbf{Ax} = \mathbf{b}$. But it may be used for matrix inversion, where the situation is as follows.

The **inverse** of a nonsingular square matrix $\mathbf{A}$ may be determined in principle by solving the n systems

$$\mathbf{Ax} = \mathbf{b}_j \qquad (j = 1, \ldots, n) \tag{1.103}$$

where $\mathbf{b}_j$ is the jth column of the $n \times n$ unit matrix. However, it is preferable to produce $\mathbf{A}^{-1}$ by operating on the unit matrix $\mathbf{I}$ in the same way as the Gauss–Jordan algorithm, reducing $\mathbf{A}$ to $\mathbf{I}$.

CONCEPT CHECK

1. Show the factorization and solve by Doolittle's method.

(a) $3x_1 + 2x_2 = 15.2$
$15x_1 + 11x_2 = 77.3$

(b) $2x_1 + 9x_2 = 41$
$3x_1 - 5x_2 = -31$

(c) $4x_1 - 6x_2 = -34$
$8x_1 - 7x_2 = -53$

(d) $2x_1 + x_2 + 2x_3 = 0$
$-2x_1 + 2x_2 + x_3 = 0$
$x_1 + 2x_2 - 2x_3 = 36$

(e) $6x_1 + 4x_2 + 3x_3 = 2.0$
$4x_1 + 3x_2 + 2x_3 = 0.5$
$3x_1 + 4x_2 + 2x_3 = -2.5$

(f) $x_1 - x_2 + 2.6x_3 = -9.88$
$0.5x_1 - 3.0x_2 + 3.3x_3 = -16.54$
$-1.5x_1 - 3.5x_2 - 10.4x_3 = 21.02$

(g) $3x_1 + 9x_2 + 6x_3 = 2.3$
$18x_1 + 48x_2 + 39x_3 = 13.6$
$9x_1 - 27x_2 + 42x_3 = 4.5$

2. **TEAM PROJECT. Crout's method** factorizes $\mathbf{A} = \mathbf{LU}$, where $\mathbf{L}$ is lower triangular and $\mathbf{U}$ is upper triangular with diagonal entries $u_{jj} = 1$, $j = 1, \ldots, n$,

(a) **Formulas.** Obtain formulas for Crout's method similar to (1.100).

(b) **Examples.** Solve Probs. 1(a) and 1(g) by Crout's method.

(c) Factor the following matrix by the Doolittle, Crout, and Cholesky methods.

$$\begin{bmatrix} 1 & -4 & 2 \\ -4 & 25 & 4 \\ 2 & 4 & 24 \end{bmatrix}$$

(d) Give the formulas for factoring a tridiagonal matrix by Crout's method.

(e) When can you obtain Crout's factorization from Doolittle's by transposition?

3. Show the factorization and solve.

(a) $9x_1 + 6x_2 + 12x_3 = 87$
$6x_1 + 13x_2 + 11x_3 = 118$
$12x_1 + 11x_2 + 26x_3 = 154$

(b) $0.04x_1 + 0.12x_3 = 1.4$
$0.64x_2 + 0.32x_3 = 1.6$
$0.12x_1 + 0.32x_2 + 0.56x_3 = 5.4$

(c) $4x_1 + 6x_2 + 8x_3 = 0$
$6x_1 + 34x_2 + 52x_3 = -80$
$8x_1 + 52x_2 + 129x_3 = -226$

(d)
$$\begin{aligned} x_1 - x_2 + 3x_3 + 2x_4 &= 30 \\ -x_1 + 5x_2 - 5x_3 - 2x_4 &= -70 \\ 3x_1 - 5x_2 + 19x_3 + 3x_4 &= 188 \\ 2x_1 - 2x_2 + 3x_3 + 21x_4 &= 2 \end{aligned}$$

(e)
$$\begin{aligned} 4x_1 + 2x_2 + 4x_3 \qquad &= 20 \\ 2x_1 + 2x_2 + 3x_3 + 2x_4 &= 36 \\ 4x_1 + 3x_2 + 6x_3 + 3x_4 &= 60 \\ 2x_2 + 3x_3 + 9x_4 &= 122 \end{aligned}$$

4. **(Definiteness)** Let **A** and **B** be positive definite $n \times n$ matrices. Are $-\mathbf{A}$, $\mathbf{A}^T$, $\mathbf{A} + \mathbf{B}$, $\mathbf{A} - \mathbf{B}$ positive definite?

5. Find the inverse by the Gauss–Jordan method, showing the details.
   (a) In Prob 1(d). (b) In Prob. 1(e).
   (c) In Prob. 1(f). (d) In Prob. 1(g).

6. **(Rounding)** For the following matrix **A** find det **A**. What happens if you round off the given entries to (a) 5S, (b) 4S, (c) 3S, (d) 2S, (e) 1S? What is the practical implication of your work?

$$\mathbf{A} = \begin{bmatrix} 1/3 & 1/4 & 2 \\ -1/9 & 1 & 1/7 \\ 4/63 & -3/28 & 13/49 \end{bmatrix}$$

1.11 Solution of Tridiagonal Systems

We consider the problem of computing *all* the eigenvalues of a ***real symmetric*** matrix $\mathbf{A} = [a_{jk}]$, discussing a method widely used in practice. In the ***first stage*** we reduce the given matrix stepwise to a **tridiagonal matrix,** that is, a matrix having all its nonzero entries on the main diagonal and in the positions immediately adjacent to the main diagonal (such as $\mathbf{A}_3$ in Figure 10).

This Householder tridiagonalization will simplify the matrix without changing its eigenvalues. The latter will then be determined (approximately) by factoring the tridiagonalized matrix, as discussed later in this section.

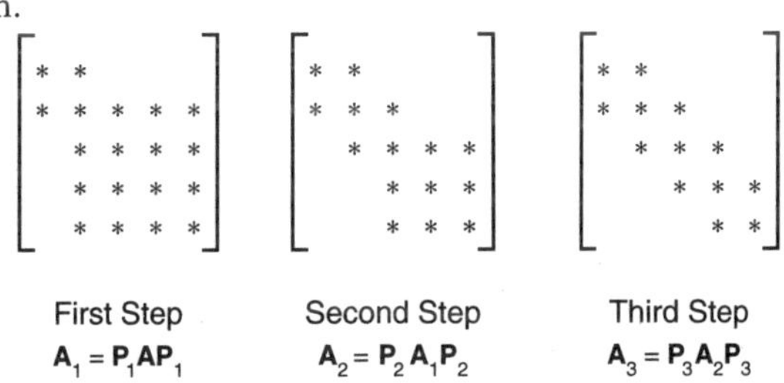

Figure 10 Householder's method for a 5×5 matrix. Positions left blank are zeros created by the method.

Householder's Tridiagonalization Method

An $n \times n$ real symmetric matrix $\mathbf{A} = [a_{jk}]$ being given, we reduce it by $n - 2$ successive similarity transformations involving matrices $\mathbf{P}_1, \ldots, \mathbf{P}_{n-2}$ to tridiagonal form. These matrices are orthogonal and symmetric. Thus $\mathbf{P}_1^{-1} = \mathbf{P}_1\mathrm{T} = \mathbf{P}_1$ and similarly for the others. These transformations produce from the given $\mathbf{A}_0 = \mathbf{A} = [a_{jk}]$ the matrices $\mathbf{A}_1 = [a_{jk}^{(1)}], \mathbf{A}_2 = [a_{jk}^{(2)}], \ldots, \mathbf{A}_{n-2} = [a_{jk}^{(n-2)}]$ in the form

$$\begin{aligned} \mathbf{A}_1 &= \mathbf{P}_1\mathbf{A}_0\mathbf{P}_1 \\ \mathbf{A}_2 &= \mathbf{P}_2\mathbf{A}_1\mathbf{P}_2 \\ &\cdots\cdots\cdots\cdots \\ \mathbf{B} = \mathbf{A}_{n-2} &= \mathbf{P}_{n-2}\,\mathbf{A}_{n-3}\,\mathbf{P}_{n-2} \end{aligned} \tag{1.104}$$

The transformations (1.104) create the necessary zeros, in the first step in Row 1 and Column 1, in the second step in Row 2 and Column 2, etc., as Figure 10 illustrates for a 5 × 5 matrix. **B** is tridiagonal. How do we determine $\mathbf{P}_1, \mathbf{P}_2, \ldots, \mathbf{P}_{n-2}$? Now, all these $\mathbf{P}_r$ are of the form

$$\mathbf{P}_r = \mathbf{I} - 2\mathbf{v}_r\mathbf{v}_r^{\mathrm{T}} \qquad (r = 1, \ldots, n-2) \tag{1.105}$$

where **I** is the $n \times n$ unit matrix and $\mathbf{v}_r = [v_{jr}]$ is a unit vector with its first r components 0; thus

$$\mathbf{v}_1 = \begin{bmatrix} 0 \\ * \\ * \\ \cdot \\ \cdot \\ \cdot \\ * \end{bmatrix}, \qquad \mathbf{v}_2 = \begin{bmatrix} 0 \\ 0 \\ * \\ \cdot \\ \cdot \\ \cdot \\ * \end{bmatrix}, \qquad \ldots, \qquad \mathbf{v}_{n-2} = \begin{bmatrix} 0 \\ 0 \\ \cdot \\ \cdot \\ \cdot \\ * \\ * \end{bmatrix} \tag{1.106}$$

where the asterisks denote the other components (which will be nonzero in general).

***Step* 1.** $\mathbf{v}_1$ has the components

$$v_{11} = 0$$

$$\text{(a)}\ v_{21} = \sqrt{\frac{1}{2}\left(1 + \frac{|a_{21}|}{S_1}\right)}$$

$$\text{(b)}\ v_{j1} = \frac{a_{j1}\,\mathrm{sgn}\,a_{21}}{2v_{21}S_1} \qquad (j = 3, 4, \ldots, n) \tag{1.107}$$

where

$$\text{(c)}\ S_1 = \sqrt{a_{21}^{\ 2} + a_{31}^{\ 2} + \cdots + a_{n1}^{\ 2}}$$

where $S_1 > 0$, and $\mathrm{sgn}\, a_{21} = +1$ if $a_{21} \geqq 0$ and $\mathrm{sgn}\, a_{21} = -1$ if $a_{21} < 0$. With this we compute $\mathbf{P}_1$ by (1.105) and then $\mathbf{A}_1$ by (1.104). This was the first step.

***Step* 2.** We compute $\mathbf{v}_2$ by (1.107) with all subscripts increased by 1 and the a_{jk} replaced by $a_{jk}^{(1)}$, the entries of $\mathbf{A}_1$ just computed. Thus [see also (1.106)]

$$v_{12} = v_{22} = 0$$

$$v_{32} = \sqrt{\frac{1}{2}\left(1 + \frac{|a_{32}^{(1)}|}{S_2}\right)} \tag{1.108}$$

$$v_{j2} = \frac{a_{j2}^{(1)}\,\mathrm{sgn}\,a_{32}^{(1)}}{2v_{32}S_2} \qquad (j = 4, 5, \ldots, n)$$

where

$$S_2 = \sqrt{(a_{32}^{(1)})^2 + (a_{42}^{(1)})^2 + \cdots + (a_{n2}^{(1)})^2}$$

With this we compute $\mathbf{P}_2$ by (1.105) and then $\mathbf{A}_2$ by (1.104).
Step **3.** We compute $\mathbf{v}_3$ by (1.108) with all subscripts increased by 1 and the $a_{jk}^{(1)}$ replaced by the entries $a_{jk}^{(2)}$ of $\mathbf{A}_2$, and so on.

Example 55 Householder Tridiagonalization

Tridiagonalize the real symmetric matrix

$$\mathbf{A} = \mathbf{A}_0 = \begin{bmatrix} 6 & 4 & 1 & 1 \\ 4 & 6 & 1 & 1 \\ 1 & 1 & 5 & 2 \\ 1 & 1 & 2 & 5 \end{bmatrix}$$

Solution:
Step **1.** We compute $S_1^2 = 4^2 + 1^2 + 1^2 = 18$ from (1.99c). Since $a_{21} = 4 > 0$, we have $\operatorname{sgn} a_{21} = +1$ in (1.107b) and get from (1.107) by straightforward computation

$$\mathbf{v}_1 = \begin{bmatrix} 0 \\ v_{32} \\ v_{31} \\ v_{41} \end{bmatrix} = \begin{bmatrix} 0 \\ 0.98559856 \\ 0.11957316 \\ 0.11957316 \end{bmatrix}$$

From this and (1.105),

$$\mathbf{P}_1 = \begin{bmatrix} 1 & 0 & 0 & 0 \\ 0 & -0.94280904 & -0.23570227 & -0.23570227 \\ 0 & -0.23570227 & 0.97140452 & -0.02859548 \\ 0 & -0.23570227 & -0.02859848 & 0.97140452 \end{bmatrix}$$

From the first line in (1.104) we now get

$$\mathbf{A}_1 = \mathbf{P}_1\mathbf{A}_0\mathbf{P}_1 = \begin{bmatrix} 6 & -\sqrt{18} & 0 & 0 \\ -\sqrt{18} & 7 & -1 & -1 \\ 0 & -1 & 9/2 & 3/2 \\ 0 & -1 & 3/2 & 9/2 \end{bmatrix}$$

Step **2.** From (1.108) we compute $S_2^2 = 2$ and

$$\mathbf{v}_2 = \begin{bmatrix} 0 \\ 0 \\ v_{32} \\ v_{32} \end{bmatrix} = \begin{bmatrix} 0 \\ 0 \\ 0.92387953 \\ 0.38268343 \end{bmatrix}$$

From this and (1.105),

$$\mathbf{P}_2 = \begin{bmatrix} 1 & 0 & 0 & 0 \\ 0 & 1 & 0 & 0 \\ 0 & 0 & -1\sqrt{2} & -1\sqrt{2} \\ 0 & 0 & -1\sqrt{2} & 1\sqrt{2} \end{bmatrix}$$

The second line in (1.104) now gives

$$\mathbf{B} = \mathbf{A}_2 = \mathbf{P}_2\mathbf{A}_1\mathbf{P}_2 \begin{bmatrix} 6 & -\sqrt{18} & 0 & 0 \\ -\sqrt{18} & 7 & \sqrt{2} & 0 \\ 0 & \sqrt{2} & 6 & 0 \\ 0 & 0 & 0 & 3 \end{bmatrix}$$

This matrix **B** is tridiagonal. Since our given matrix has order $n = 4$, we needed $n - 2 = 2$ steps to accomplish this reduction, as claimed. (Do you see that we got more zeros than we can expect in general?) **B** is similar to **A**, as we now show in general. This is essential because **B** thus has the same spectrum as **A**.

B *Similar to* A. We assert that **B** in (1.104) is similar to $\mathbf{A} = \mathbf{A}_0$. The matrix $\mathbf{P}_r$ is symmetric; indeed,

$$\mathbf{P}_r^{\mathrm{T}} = (\mathbf{I} - 2\mathbf{v}_r\mathbf{v}_r^{\mathrm{T}})^{\mathrm{T}} = \mathbf{I}^{\mathrm{T}} - 2(\mathbf{v}_r\mathbf{v}_r^{\mathrm{T}})^{\mathrm{T}} = \mathbf{I} - 2\mathbf{v}_r\mathbf{v}_r^{\mathrm{T}} = \mathbf{P}_r$$

Also, $\mathbf{P}_r$ is orthogonal because $\mathbf{v}_r$ is a unit vector, so that $\mathbf{v}_r^{\mathrm{T}}\mathbf{v}_r = 1$ and thus

$$\mathbf{P}_r\mathbf{P}_r^{\mathrm{T}} = \mathbf{P}_r^{2} = (\mathbf{I} - 2\mathbf{v}_r\mathbf{v}_r^{\mathrm{T}})^2 = \mathbf{I} - 4\mathbf{v}_r\mathbf{v}_r^{\mathrm{T}} + 4\mathbf{v}_r\mathbf{v}_r^{\mathrm{T}}\mathbf{v}_r\mathbf{v}_r^{\mathrm{T}} = \mathbf{I} - 4\mathbf{v}_r\mathbf{v}_r^{\mathrm{T}} + 4\mathbf{v}_r(\mathbf{v}_r^{\mathrm{T}}\mathbf{v}_r)\mathbf{v}_r^{\mathrm{T}} = \mathbf{I} .$$

Hence $\mathbf{P}_r^{-1} = \mathbf{P}_r^{\mathrm{T}} = \mathbf{P}_r$ and from (1.104) we now obtain

$$\begin{aligned} \mathbf{B} &= \mathbf{P}_{n-2}\mathbf{A}_{n-3}\mathbf{P}_{n-2} = \cdots \\ \cdots &= \mathbf{P}_{n-2}\mathbf{P}_{n-3} \cdots \mathbf{P}_1\mathbf{A}\mathbf{P}_1 \cdots \mathbf{P}_{n-3}\mathbf{P}_{n-2} \\ &= \mathbf{P}_{n-2}^{-1}\mathbf{P}_{n-3}^{-1} \cdots \mathbf{P}_1^{-1}\mathbf{A}\mathbf{P}_1 \cdots \mathbf{P}_{n-3}\mathbf{P}_{n-2} \\ &= \mathbf{P}^{-1}\mathbf{A}\mathbf{P} \end{aligned}$$

where $\mathbf{P} = \mathbf{P}_1\mathbf{P}_2 \cdots \mathbf{P}_{n-2}$. This proves our assertion.

CONCEPT CHECK

Tridiagonalize, showing the details:

1. $\begin{bmatrix} 3.5 & 1.0 & 1.5 \\ 1.0 & 5.0 & 3.0 \\ 1.5 & 3.0 & 3.5 \end{bmatrix}$

2. $\begin{bmatrix} 0 & 1 & 1 \\ 1 & 0 & 1 \\ 1 & 1 & 0 \end{bmatrix}$

3. $\begin{bmatrix} 0.98 & 0.04 & 0.44 \\ 0.04 & 0.56 & 0.40 \\ 0.44 & 0.40 & 0.80 \end{bmatrix}$

4. $\begin{bmatrix} 8 & 8 & 2 & 2 \\ 8 & 8 & 2 & 2 \\ 2 & 2 & 6 & 4 \\ 2 & 2 & 4 & 6 \end{bmatrix}$

1.12 QR-Factorization Method

In 1958 H. Rutishauser of Switzerland proposed the idea of using the LU-factorization in solving eigenvalue problems. An improved version of Rutishauser's method is the QR-method, independently proposed by the American J. G. F. Francis and the Russian V. N. Kublanovskaya. The QR-method uses the factorization **QR** with orthogonal **Q** and upper triangular **R**. We discuss the QR-method for a real ***symmetric*** matrix.

In this method we first transform a given real symmetric $n \times n$ matrix **A** into a tridiagonal matrix $\mathbf{B}_0 = \mathbf{B}$ by Householder's method. This creates many zeros and thus reduces the amount of further work. Then we compute $\mathbf{B}_1, \mathbf{B}_2, \ldots$ stepwise according to the following iteration method.

Step 1. Factor $\mathbf{B}_0 = \mathbf{Q}_0\mathbf{R}_0$ with orthogonal $\mathbf{R}_0$ and upper triangular $\mathbf{R}_0$. Then compute $\mathbf{B}_1 = \mathbf{R}_0\mathbf{Q}_0$.

Step 2. Factor $\mathbf{B}_1 = \mathbf{Q}_1\mathbf{R}_1$. Then compute $\mathbf{B}_2 = \mathbf{R}_1\mathbf{Q}_1$.

General Step $s + 1$.

$$\begin{aligned} &\text{(a) Factor } \mathbf{B}_s = \mathbf{Q}_s\mathbf{R}_s \\ &\text{(b) Compute } \mathbf{B}_{s+1} = \mathbf{R}_s\mathbf{Q}_s \end{aligned} \tag{1.109}$$

Here $\mathbf{Q}_s$ is orthogonal and $\mathbf{R}_s$ upper triangular. The factorization (1.109a) will be explained below.

$\mathbf{B}_{s+1}$ Similar to B. Convergence to a Diagonal Matrix. From (1.109a) we have $\mathbf{R}_s = \mathbf{Q}_s^{-1}\mathbf{B}_s$. Substitution into (1.109b) gives

$$\mathbf{B}_{s+1} = \mathbf{R}_s\mathbf{Q}_s = \mathbf{Q}_s^{-1}\mathbf{B}_s\mathbf{Q}_s \tag{1.110}$$

Thus $\mathbf{B}_{s+1}$ is similar to $\mathbf{B}_s$. Hence $\mathbf{B}_{s+1}$ is similar to $\mathbf{B}_0 = \mathbf{B}$ for all s. This implies that $\mathbf{B}_{s+1}$ has the same eigenvalues as **B**.

Also, $\mathbf{B}_{s+1}$ is symmetric. This follows by induction. Indeed, $\mathbf{B}_0 = \mathbf{B}$ is symmetric. Assuming $\mathbf{B}_s$ to be symmetric, that is, $\mathbf{B}_s^{T} = \mathbf{B}_s$, and using $\mathbf{Q}_s^{-1} = \mathbf{Q}_s^{T}$ (since $\mathbf{Q}_s$ is orthogonal), we get from (1.110) the symmetry,

$$\mathbf{B}_{s+1}^{T} = (\mathbf{Q}_s^{T}\mathbf{B}_s\mathbf{Q}_s)^{T} = \mathbf{Q}_s^{T}\mathbf{B}_s^{T}\mathbf{Q}_s = \mathbf{Q}_s^{T}\mathbf{B}_s\mathbf{Q}_s = \mathbf{B}_{s+1}$$

If the eigenvalues of **B** are different in absolute value, say, $|\lambda_1| > |\lambda_2| > \cdots > |\lambda_n|$, then

$$\underset{s\to\infty}{\text{Lt}}\ \mathbf{B}_s = \mathbf{D}$$

where **D** is diagonal, with main diagonal entries $\lambda_1, \lambda_2, \ldots, \lambda_n$.

How to Get the QR-Factorization, say, $\mathbf{B} = \mathbf{B}_0 = [b_{jk}] = \mathbf{Q}_0\mathbf{R}_0$. The tridiagonal matrix **B** has $n - 1$ generally nonzero entries below the main diagonal. These are $b_{21}, b_{32}, \ldots, b_{n,n-1}$. We multiply **B** from the left by a matrix $\mathbf{C}_2$ such that $\mathbf{C}_2\mathbf{B} = [b_{jk}^{(2)}]$ has $b_{21}^{(2)} = 0$. We multiply this by a matrix $\mathbf{C}_3$ such that $\mathbf{C}_3\mathbf{C}_2\mathbf{B} = [b_{jk}^{(3)}]$ has $b_{32}^{(3)} = 0$, etc. After $n - 1$ such multiplications we are left with an upper triangular matrix $\mathbf{R}_0$, namely,

$$\mathbf{C}_n\mathbf{C}_{n-1} \cdots \mathbf{C}_3\mathbf{C}_2\mathbf{B}_0 = \mathbf{R}_0 \tag{1.111}$$

These $n \times n$ matrices $\mathbf{C}_j$ are very simple. $\mathbf{C}_j$ has the 2×2 submatrix

$$\begin{bmatrix} \cos\theta_j & \sin\theta_j \\ -\sin\theta_j & \cos\theta_j \end{bmatrix} \quad (\theta_j \text{ suitable})$$

in Rows $j-1$ and j and Columns $j-1$ and j; everywhere else on the main diagonal the matrix $\mathbf{C}_j$ has entries 1; and all its other entries are 0. For instance, if $n=4$, writing $c_j = \cos\theta_j$, $s_j = \sin\theta_j$, we have

$$\mathbf{C}_2 = \begin{bmatrix} c_2 & s_2 & 0 & 0 \\ -s_2 & c_2 & 0 & 0 \\ 0 & 0 & 1 & 0 \\ 0 & 0 & 0 & 1 \end{bmatrix}, \quad \mathbf{C}_3 = \begin{bmatrix} 1 & 0 & 0 & 0 \\ 0 & c_3 & s_3 & 0 \\ 0 & -s_3 & c_3 & 0 \\ 0 & 0 & 0 & 1 \end{bmatrix}, \quad \mathbf{C}_4 = \begin{bmatrix} 1 & 0 & 0 & 0 \\ 0 & 1 & 0 & 0 \\ 0 & 0 & c_4 & s_4 \\ 0 & 0 & -s_4 & c_4 \end{bmatrix}$$

These $\mathbf{C}_j$ are orthogonal. Hence their product in (1.111) is orthogonal, and so is the inverse of this product. We call this inverse $\mathbf{Q}_0$. Then from (1.112),

$$\mathbf{B}_0 = \mathbf{Q}_0\mathbf{R}_0 \tag{1.112}$$

where, with $\mathbf{C}_j^{-1} = \mathbf{C}_j^{\mathrm{T}}$,

$$\mathbf{Q}_0 = (\mathbf{C}_n\mathbf{C}_{n-1}\cdots\mathbf{C}_3\mathbf{C}_2)^{-1} = \mathbf{C}_2^{\mathrm{T}}\mathbf{C}_3^{\mathrm{T}}\cdots\mathbf{C}_{n-1}^{\mathrm{T}}\mathbf{C}_n^{\mathrm{T}} \tag{1.113}$$

This is our QR-factorization of $\mathbf{B}_0$. From it we have by (1.109b) with $s=0$

$$\mathbf{B}_1 = \mathbf{R}_0\mathbf{Q}_0 = \mathbf{R}_0\mathbf{C}_2^{\mathrm{T}}\mathbf{C}_3^{\mathrm{T}}\cdots\mathbf{C}_{n-1}^{\mathrm{T}}\mathbf{C}_n^{\mathrm{T}} \tag{1.114}$$

We do not need $\mathbf{Q}_0$ explicitly, but to get $\mathbf{B}_1$ from (1.106), we first compute $\mathbf{R}_0\mathbf{C}_2^{\mathrm{T}}$, then $(\mathbf{R}_0\mathbf{C}_2^{\mathrm{T}})\mathbf{C}_3^{\mathrm{T}}$, etc. Similarly in the further steps that produce $\mathbf{B}_2$, $\mathbf{B}_3$, ….

Determination of $\cos\theta_j$ and $\sin\theta_j$. We finally show how to find the angles of rotation. $\cos\theta_2$ and $\sin\theta_2$ in $\mathbf{C}_2$ must be such that $b_{21}^{(2)} = 0$ in the product

$$\mathbf{C}_2\mathbf{B} = \begin{bmatrix} c_2 & s_2 & 0 & \cdots \\ -s_2 & c_2 & 0 & \cdots \\ \cdot & \cdot & \cdot & \cdots \\ \cdot & \cdot & \cdot & \cdots \end{bmatrix} \begin{bmatrix} b_{11} & b_{12} & b_{13} & \cdots \\ b_{21} & b_{22} & b_{23} & \cdots \\ \cdot & \cdot & \cdot & \cdots \\ \cdot & \cdot & \cdot & \cdots \end{bmatrix}$$

Now $b_{21}^{(2)}$ is obtained by multiplying the second row of $\mathbf{C}_2$ by the first column of $\mathbf{B}$,

$$b_{21}^{(2)} = -s_2b_{11} + c_2b_{21} = -(\sin\theta_2)b_{11} + (\cos\theta_2)b_{21} = 0$$

Hence $\tan\theta_2 = s_2/c_2 = b_{21}/b_{11}$, and

$$\begin{aligned} \cos\theta_2 &= \frac{1}{\sqrt{1+\tan^2\theta_2}} = \frac{1}{\sqrt{1+(b_{21}/b_{11})^2}} \\ \sin\theta_2 &= \frac{\tan\theta_2}{\sqrt{1+\tan^2\theta_2}} = \frac{b_{21}/b_{11}}{\sqrt{1+(b_{21}/b_{11})^2}} \end{aligned} \tag{1.115}$$

Similarly for θ_3, θ_4, …. The next example illustrates all this.

Example 56 QR-Factorization Method

Compute all the eigenvalues of the matrix

$$A = \begin{bmatrix} 6 & 4 & 1 & 1 \\ 4 & 6 & 1 & 1 \\ 1 & 1 & 5 & 2 \\ 1 & 1 & 2 & 5 \end{bmatrix}$$

Solution: We first reduce **A** to tridiagonal form. Applying Householder's method (explained in next section), we obtain

$$A_2 = \begin{bmatrix} 6 & -\sqrt{18} & 0 & 0 \\ -\sqrt{18} & 7 & \sqrt{2} & 0 \\ 0 & \sqrt{2} & 6 & 0 \\ 0 & 0 & 0 & 3 \end{bmatrix}$$

From the characteristic determinant we see that A_2, hence **A**, has the eigenvalue 3. (Can you see this directly from A_2?) Hence it suffices to apply the QR-method to the tridiagonal 3×3 matrix

$$B_0 = B = \begin{bmatrix} 6 & -\sqrt{18} & 0 \\ -\sqrt{18} & 7 & \sqrt{2} \\ 0 & \sqrt{2} & 6 \end{bmatrix}$$

Step 1. We multiply **B** from the left by

$$C_2 = \begin{bmatrix} \cos\theta_2 & \sin\theta_2 & 0 \\ -\sin\theta_2 & \cos\theta_2 & 0 \\ 0 & 0 & 1 \end{bmatrix}$$

and the C_2B by

$$C_3 = \begin{bmatrix} 1 & 0 & 0 \\ 0 & \cos\theta_3 & \sin\theta_3 \\ 0 & -\sin\theta_3 & \cos\theta_3 \end{bmatrix}$$

Here $(-\sin\theta_2) \times 6 + (\cos\theta_2)(-\sqrt{18}) = 0$ gives $\cos\theta_2 = 0.81649658$ and $\sin\theta_2 = -0.57735027$. With these values we compute

$$C_2B = \begin{bmatrix} 7.34846923 & -7.50555350 & -0.81649658 \\ 0 & 3.26598632 & 1.15470054 \\ 0 & 1.41421356 & 6.00000000 \end{bmatrix}$$

In C_3 we get from $(-\sin\theta_3) \times 3.26598632 + (\cos\theta_3) \times 1.41421356 = 0$ the values $\cos\theta_3 = 0.91766294$ and $\sin\theta_3 = 0.39735971$. This gives

$$\mathbf{R}_0 = \mathbf{C}_3\mathbf{C}_2\mathbf{B} = \begin{bmatrix} 7.34846923 & -7.50555350 & -0.81649658 \\ 0 & 3.55902608 & 3.44378413 \\ 0 & 0 & 5.04714615 \end{bmatrix}$$

From this we compute

$$\mathbf{B}_1 = \mathbf{R}_0\mathbf{C}_2^{\mathrm{T}}\mathbf{C}_3^{\mathrm{T}} = \begin{bmatrix} 10.33333333 & -2.05480467 & 0 \\ -2.05480467 & 4.03508772 & 2.00553251 \\ 0 & 2.00553251 & 4.63157895 \end{bmatrix}$$

which is symmetric and tridiagonal. The off-diagonal entries in $\mathbf{B}_1$ are still large in absolute value. Hence we have to go on.

***Step* 2.** We do the same computations as in the first step, with $\mathbf{B}_0 = \mathbf{B}$ replaced by $\mathbf{B}_1$ and $\mathbf{C}_2$ and $\mathbf{C}_3$ changed accordingly, the new angles being $\theta_2 = -0.196291533$ and $\theta_3 = 0.513415589$. We obtain

$$\mathbf{R}_1 = \begin{bmatrix} 10.53565375 & -2.80232241 & -0.39114588 \\ 0 & 4.08329584 & 3.98824028 \\ 0 & 0 & 3.06832668 \end{bmatrix}$$

and from this

$$\mathbf{B}_2 = \begin{bmatrix} 10.87987988 & -0.79637918 & 0 \\ -0.79637918 & 5.44738664 & 1.50702500 \\ 0 & 1.50702500 & 2.67273348 \end{bmatrix}$$

We see that the off-diagonal entries are somewhat smaller in absolute value than those of $\mathbf{B}_1$, but still much too large for the diagonal entries to be good approximations of the eigenvalues of $\mathbf{B}$.

Further Steps. We list the main diagonal entries and the absolutely largest off-diagonal entry, which is $|b_{12}^{(j)}| = |b_{21}^{(j)}|$ in all steps. You may show that the given matrix $\mathbf{A}$ has the spectrum 11, 6, 3, 2.

Step j	$b_{11}^{(j)}$	$b_{22}^{(j)}$	$b_{33}^{(j)}$	$\max_{j \neq k} \lvert b_{jk}^{(j)} \rvert$
3	10.9668929	5.94589856	2.08720851	0.58523582
5	10.9970872	6.00181541	2.00109738	0.12065334
7	10.9997421	6.00024439	2.00001355	0.03591107
9	10.9999772	6.00002267	2.00000017	0.01068477

Looking back at our discussion, we recognize that the purpose of applying Householder's tridiagonalization before the QR-factorization method is a substantial reduction of cost in each QR-factorization, in particular if $\mathbf{A}$ is large.

Convergence acceleration and thus further reduction of cost can be achieved by a **spectral shift**, that is, by taking $\mathbf{B}_s - k_s\mathbf{I}$ instead of $\mathbf{B}_s$ with a suitable k_s.

CONCEPT CHECK

Do three QR-steps to find approximations of the eigenvalues of:

1. The matrix in the answer to

$$\begin{bmatrix} 3.5 & 1.0 & 1.5 \\ 1.0 & 5.0 & 3.0 \\ 1.5 & 3.0 & 3.5 \end{bmatrix}$$

2. The matrix in the answer to

$$\begin{bmatrix} 0.98 & 0.04 & 0.44 \\ 0.04 & 0.56 & 0.40 \\ 0.44 & 0.40 & 0.80 \end{bmatrix}$$

3. $\begin{bmatrix} 18 & 2 & 0 \\ 2 & 8 & 2 \\ 0 & 0 & 2 \end{bmatrix}$

4. $\begin{bmatrix} 16.2 & -0.1 & 0 \\ -0.1 & -4.3 & 0.2 \\ 0 & 0.2 & 4.1 \end{bmatrix}$

5. $\begin{bmatrix} 7.0 & 0.1 & 0 \\ 0.1 & 4.0 & 0.1 \\ 0 & 0.1 & 1.0 \end{bmatrix}$

Solved Problems

Q 1. Show that the system of equations: $3x+3y+2z=1$, $x+2y=4$, $10y+3z=-2$, $2x-3y-z=5$ is consistent and hence solve it. [JNTU Aug./Sep. 2008, May/June 2008]

OR

Prove that the following set of equations are consistent and solve them.

$$\begin{aligned} 3x+3y+2z&=1 \\ x+2y\quad\quad&=4 \\ 10y+3z&=-2 \\ 2x-3y-z&=5 \end{aligned}$$

[JNTU April/May 2007, May/June 2006]

Answer: Given equations are,

$$\begin{aligned} 3x+3y+2z&=1 \\ x+2y\quad\quad&=4 \\ 10y+3z&=-2 \\ 2x-3y-z&=5 \end{aligned}$$

Writing the above system of equations in matrix form, **AX = B**

$$\begin{bmatrix} 3 & 3 & 2 \\ 1 & 2 & 0 \\ 0 & 10 & 3 \\ 2 & -3 & -1 \end{bmatrix}\begin{bmatrix} x \\ y \\ z \end{bmatrix}=\begin{bmatrix} 1 \\ 4 \\ -2 \\ 5 \end{bmatrix}$$

Given set of equations would be consistent if

$$\rho(A) = \rho(A \mid B)$$

Augmented matrix is,

$$[A \mid B] = \begin{bmatrix} 3 & 3 & 2 & 1 \\ 1 & 2 & 0 & 4 \\ 0 & 10 & 3 & -2 \\ 2 & -3 & -1 & 5 \end{bmatrix}$$

Reducing the augmented matrix to Echelon form by applying elementary row operation.

$$[A \mid B] = \begin{bmatrix} 3 & 3 & 2 & 1 \\ 1 & 2 & 0 & 4 \\ 0 & 10 & 3 & -2 \\ 2 & -3 & -1 & 5 \end{bmatrix} \quad \text{Row 1} \leftrightarrow \text{Row 2}$$

$$= \begin{bmatrix} 1 & 2 & 0 & 4 \\ 3 & 3 & 2 & 1 \\ 0 & 10 & 3 & -2 \\ 2 & -3 & -1 & 5 \end{bmatrix} \quad \begin{matrix} \text{Row 2} \to \text{Row 2} - 3\,\text{Row 1} \\ \text{Row 4} \to \text{Row 4} - 2\,\text{Row 1} \end{matrix}$$

$$= \begin{bmatrix} 1 & 2 & 0 & 4 \\ 0 & -3 & 2 & -11 \\ 0 & 10 & 3 & -2 \\ 0 & -7 & -1 & -3 \end{bmatrix} \quad \begin{matrix} \text{Row 3} \to 3\,\text{Row 3} + 10\,\text{Row 2} \\ \text{Row 4} \to 3\,\text{Row 4} - 7\,\text{Row 2} \end{matrix}$$

$$= \begin{bmatrix} 1 & 2 & 0 & 4 \\ 0 & -3 & 2 & -11 \\ 0 & 0 & 29 & -116 \\ 0 & 0 & -17 & 68 \end{bmatrix} \quad \begin{matrix} \text{Row 3} \to \text{Row 3}/29 \\ \text{Row 4} \to \text{Row 4}/(-17) \end{matrix}$$

$$= \begin{bmatrix} 1 & 2 & 0 & 4 \\ 0 & -3 & 2 & -11 \\ 0 & 0 & 1 & -4 \\ 0 & 0 & 1 & -4 \end{bmatrix} \quad \text{Row 4} \to \text{Row 4} - \text{Row 3}$$

$$= \begin{bmatrix} 1 & 2 & 0 & 4 \\ 0 & -3 & 2 & -11 \\ 0 & 0 & 1 & -4 \\ 0 & 0 & 0 & 0 \end{bmatrix}$$

$$\rho(A) = \rho(A \mid B) = 3$$

Therefore the equations are consistent and have unique solution. Now, writing the reduced matrix in equation form, we have

$$x + 2y = 4 \quad (1.116)$$

$$-3y + 2z = -11 \quad (1.117)$$

$$z = -4 \quad (1.118)$$

Solving we get

$$-3y + 2z = -11$$

$$-3y - 8 = -11 \text{ (putting } z = -4)$$

$$-3y = -11 + 8$$

$$-3y = -3$$

$$y = 1$$

From (1.108), we get,

$$x + 2y = 4$$

$$x + 2 = 4$$

$$x = 2 \text{ (putting } y = 1)$$

The solution set is

$$x = 2, \ y = 1, \ z = -4$$

Q 2. Show that the only real value of λ for which the following equations have nontrivial solution is 6 and solve them when $\lambda = 6$.

$$x + 2y + 3z = \lambda x$$

$$3x + y + 2z = \lambda y$$

$$2x + 3y + z = \lambda z$$

[JNTU April/May 2007, Aug./Sep. 2006, May/June 2006, May 2005]

Answer: The given set of equations are,

$$x + 2y + 3z = \lambda x$$

$$3x + y + 2z = \lambda y$$

$$2x + 3y + z = \lambda z$$

The above set of equations can be written as

$$(1 - \lambda)x + 2y + 3z = 0$$

$$3x + (1 - \lambda)y + 2z = 0$$

$$2x + 3y + (1 - \lambda)z = 0$$

Writing the above system of equation in matrix form we get

$$\begin{bmatrix} 1-\lambda & 2 & 3 \\ 3 & 1-\lambda & 2 \\ 2 & 3 & 1-\lambda \end{bmatrix} \begin{bmatrix} x \\ y \\ z \end{bmatrix} = \begin{bmatrix} 0 \\ 0 \\ 0 \end{bmatrix} \quad (1.119)$$

that is $\mathbf{AX} = 0$ where,

$$\mathbf{A} = \begin{bmatrix} 1-\lambda & 2 & 3 \\ 3 & 1-\lambda & 2 \\ 2 & 3 & 1-\lambda \end{bmatrix} \quad \text{and} \quad \mathbf{X} = \begin{bmatrix} x \\ y \\ z \end{bmatrix}$$

When the rank of the coefficient matrix is less than the number of unknowns, then the system is said to possess a nonzero solution or a nontrivial solution. If the rank of the coefficient matrix is less than 3 then $|\mathbf{A}| = 0$. Therefore number of unknowns = 3.

$$|\mathbf{A}| = 0$$

$$\begin{vmatrix} 1-\lambda & 2 & 3 \\ 3 & 1-\lambda & 2 \\ 2 & 3 & 1-\lambda \end{vmatrix} = 0 \quad \text{Row } 1 \to \text{Row } 1 + \text{Row } 2 + \text{Row } 3$$

$$\begin{vmatrix} 6-\lambda & 6-\lambda & 6-\lambda \\ 3 & 1-\lambda & 2 \\ 2 & 3 & 1-\lambda \end{vmatrix} = 0$$

Taking $(6 - \lambda)$ common from Row 1 we have

$$(6-\lambda)\begin{vmatrix} 1 & 1 & 1 \\ 3 & 1-\lambda & 2 \\ 2 & 3 & 1-\lambda \end{vmatrix} = 0$$

$$(6-\lambda)\begin{vmatrix} 1 & 0 & 0 \\ 3 & -\lambda-2 & -1 \\ 2 & 1 & -\lambda-1 \end{vmatrix} = 0 \quad \begin{matrix} \text{Column } 2 \to \text{Column } 2 - \text{Column } 1 \\ \text{Column } 3 \to \text{Column } 3 - \text{Column } 1 \end{matrix}$$

Solving the matrix we have

$$(6 - \lambda)\,[(\lambda + 2)(\lambda + 1) + 1] = 0$$

$$(6 - \lambda)(\lambda^2 + 3\lambda + 3) = 0$$

$$(6 - \lambda) = 0 \quad \text{and} \quad (\lambda^2 + 3\lambda + 3) = 0$$

$$\lambda = 6, \qquad \lambda = \frac{-3 \pm \sqrt{9-12}}{2}$$

Therefore

$$\lambda = 6, \qquad \lambda = \frac{-3 \pm \sqrt{-3}}{2}$$

Now $\lambda = 6$ is the only real value for which the given system has nontrivial solution or nonzero solution. Substituting the value $\lambda = 6$ in (1.119), we have,

$$\begin{bmatrix} -5 & 2 & 3 \\ 3 & -5 & 2 \\ 2 & 3 & -5 \end{bmatrix}\begin{bmatrix} x \\ y \\ z \end{bmatrix} = \begin{bmatrix} 0 \\ 0 \\ 0 \end{bmatrix}$$

$$\begin{bmatrix} -5 & 2 & 3 \\ 0 & -19 & 19 \\ 0 & 19 & -19 \end{bmatrix}\begin{bmatrix} x \\ y \\ z \end{bmatrix} = \begin{bmatrix} 0 \\ 0 \\ 0 \end{bmatrix} \quad \begin{matrix} \text{Row } 2 \to 5 \text{ Row } 2 + 3 \text{ Row } 1 \\ \text{Row } 3 \to 5 \text{ Row } 3 + 2 \text{ Row } 1 \end{matrix}$$

$$\begin{bmatrix} -5 & 2 & 3 \\ 0 & -19 & 19 \\ 0 & 0 & 0 \end{bmatrix}\begin{bmatrix} x \\ y \\ z \end{bmatrix} = \begin{bmatrix} 0 \\ 0 \\ 0 \end{bmatrix} \quad \text{Row } 3 \to \text{Row } 3 + \text{Row } 2 \qquad (1.120)$$

This is in Echelon form $\Rightarrow r = \rho(A) = 2$

Number of unknowns, $n = 3$ (since $n > r$)

Number of independent solutions $= n - r = 3 - 2 = 1$

From (1.120), we have

$$-5x + 2y + 3z = 0$$

$$-19y + 19z = 0 \Rightarrow y = z$$

Putting $y = z$ we get

$$-5x + 2y + 3y = 0$$

$$-5x + 5y = 0 \Rightarrow x = y$$

Let $x = k$ then

$$y = k \quad \text{(since } x = y)$$

$$z = k \quad \text{(since } y = z)$$

Therefore the solution is

$$\begin{bmatrix} x \\ y \\ z \end{bmatrix} = \begin{bmatrix} k \\ k \\ k \end{bmatrix} = k\begin{bmatrix} 1 \\ 1 \\ 1 \end{bmatrix}$$

Q 3. Solve by matrix method the equations $3x + y + 2z = 3;\ 2x - 3y - z = -3;\ x + 2y + z = 4.$

[JNTU Aug./Sep. 2008]

OR

Solve the system of equations $3x + y + 2z = 3,\ 2x - 3y - z = -3,\ x + 2y + z = 4.$

[JNTU May/June 2008]

Answer: Given equations can be expressed in the matrix form as **AX = B**. We frame an augmented matrix **C** = [**A** | **B**] and proceed with the solution,

$$\mathbf{C} = \begin{bmatrix} 3 & 1 & 2 & 3 \\ 2 & -3 & -1 & -3 \\ 1 & 2 & 1 & 4 \end{bmatrix} \quad \text{Row } 1 \to \text{Row } 1 - \text{Row } 2$$

$$= \begin{bmatrix} 1 & 4 & 3 & 6 \\ 2 & -3 & -1 & -3 \\ 1 & 2 & 1 & 4 \end{bmatrix} \begin{matrix} \text{Row } 2 \to \text{Row } 2 - 2 \text{ Row } 1 \\ \text{Row } 3 \to \text{Row } 3 - \text{Row } 1 \end{matrix}$$

$$= \begin{bmatrix} 1 & 4 & 3 & 6 \\ 0 & -11 & -7 & -15 \\ 0 & -2 & -2 & -2 \end{bmatrix} \text{Row } 3 \to \text{Row } 3/(-2)$$

$$= \begin{bmatrix} 1 & 4 & 3 & 6 \\ 0 & -11 & -7 & -15 \\ 0 & 1 & 1 & 1 \end{bmatrix} \text{Row } 3 \to \text{Row } 3 \times 11 + \text{Row } 2$$

$$= \begin{bmatrix} 1 & 4 & 3 & 6 \\ 0 & -11 & -7 & -15 \\ 0 & 0 & 4 & -4 \end{bmatrix}$$

$$x + 4y + 3z = 6 \tag{1.121}$$

$$-11y - 7z = -15 \Rightarrow 4z = -4 \Rightarrow z = -1 \tag{1.122}$$

Substituting the value of z in (1.122) we get

$$-11y - 7(-1) = -15$$

$$-11y = -22 \Rightarrow y = 2$$

Substituting $y = 2,\ z = -1$ in (1.121) we have

$$x + 4(2) + 3(-1) = 6$$

$$x = 6 - 5 = 1$$

Therefore $x = 1,\ y = 2,\ z = -1.$

Q 4. Find a LU decomposition of the matrix **A** and solve the linear system **AX** = **B**.

$$\begin{bmatrix} -3 & 12 & -6 \\ 1 & -2 & 2 \\ 0 & 1 & 1 \end{bmatrix} \begin{bmatrix} x \\ y \\ z \end{bmatrix} = \begin{bmatrix} -33 \\ 7 \\ -1 \end{bmatrix}$$

[JNTU April/May 2007, May/June 2006]

Answer: The given system of equation, **AX** = **B**, is

$$\begin{bmatrix} -3 & 12 & -6 \\ 1 & -2 & 2 \\ 0 & 1 & 1 \end{bmatrix} \begin{bmatrix} x \\ y \\ z \end{bmatrix} = \begin{bmatrix} -33 \\ 7 \\ -1 \end{bmatrix}$$

Let,

$$A = \begin{bmatrix} -3 & 12 & -6 \\ 1 & -2 & 2 \\ 0 & 1 & 1 \end{bmatrix} = LU = \begin{bmatrix} 1 & 0 & 0 \\ l_{21} & 1 & 0 \\ l_{31} & l_{32} & 1 \end{bmatrix} \begin{bmatrix} u_{11} & u_{12} & u_{13} \\ 0 & u_{22} & u_{23} \\ 0 & 0 & u_{33} \end{bmatrix}$$

$$\begin{bmatrix} u_{11} & u_{12} & u_{13} \\ l_2 u_{11} & l_{21} u_{12} + u_{22} & l_{21} u_{13} + u_{23} \\ l_{31} u_{11} & l_{31} u_{12} + l_{32} u_{22} & l_{31} u_{13} + l_{32} u_{23} + u_{33} \end{bmatrix} = \begin{bmatrix} -3 & 12 & -6 \\ 1 & -2 & 2 \\ 0 & 1 & 1 \end{bmatrix}$$

Now equating both sides we get $u_{11} = -3$, $u_{12} = 12$, $u_{13} = -6$. Also

$$l_{21}\, u_{11} = 1 \Rightarrow l_{21}\,(-3) = 1 \Rightarrow l_{21} = -1/3$$

$$l_{21}\, u_{12} + u_{22} = -2 \Rightarrow [(-1)/(-3)]\,(12) + u_{22} = -2 \Rightarrow -4 + u_{22} = -2 \Rightarrow u_{22} = 2$$

$$l_{21}\, u_{13} + u_{23} = 2 \Rightarrow (-1/3)\,(-6) + u_{23} = 2 \Rightarrow 2 + u_{23} = 2 \Rightarrow u_{23} = 0$$

$$l_{31}\, u_{11} = 0 \Rightarrow l_{31}(-3) = 0 \Rightarrow l_{31} = 0$$

$$l_{31}\, u_{12} + l_{32}\, u_{22} = 1 \Rightarrow 0 + l_{32}(2) = 1 \Rightarrow l_{32} = 1/2$$

$$l_{31}\, u_{13} + l_{32}\, u_{23} + u_{33} = 1 \Rightarrow 0 + (1/2) \times 0 + u_{33} = 1 \Rightarrow u_{33} = 1$$

Therefore

$$LU = \begin{bmatrix} 1 & 0 & 0 \\ -1/3 & 1 & 0 \\ 0 & 1/2 & 1 \end{bmatrix} \begin{bmatrix} -3 & 12 & -6 \\ 0 & 2 & 0 \\ 0 & 0 & 1 \end{bmatrix}$$

Let **LY** = **B** where **Y** = **UX**. Then

$$\begin{bmatrix} 1 & 0 & 0 \\ l_{21} & 1 & 0 \\ l_{31} & l_{32} & 1 \end{bmatrix} \begin{bmatrix} y_1 \\ y_2 \\ y_3 \end{bmatrix} = \begin{bmatrix} -33 \\ 7 \\ -1 \end{bmatrix}$$

$$\begin{bmatrix} 1 & 0 & 0 \\ -1/3 & 1 & 0 \\ 0 & 1/2 & 1 \end{bmatrix} \begin{bmatrix} y_1 \\ y_2 \\ y_3 \end{bmatrix} = \begin{bmatrix} -33 \\ 7 \\ -1 \end{bmatrix}$$

Writing in equation form we get

$$y_1 = -33$$

$$-\frac{1}{3} y_1 + y_2 = 7$$

$$0 + \frac{1}{2} y_2 + y_3 = -1$$

Using $y_1 = -33$ we solve the set of equations above to get

$$-\frac{1}{3}(-33)+y_2=7$$

$$11+y_2=7$$

$$y_2=-4$$

Using $y_2=-4$ we get

$$\frac{1}{2}(-4)+y_3=-1$$

$$-2+y_3=-1$$

$$y_3=1$$

To find **X**. We use **UX** = **Y**. Therefore

$$\begin{bmatrix} u_{11} & u_{12} & u_{13} \\ 0 & u_{22} & u_{23} \\ 0 & 0 & u_{33} \end{bmatrix}\begin{bmatrix} x_1 \\ x_2 \\ x_3 \end{bmatrix}=\begin{bmatrix} y_1 \\ y_2 \\ y_3 \end{bmatrix}$$

$$\begin{bmatrix} -3 & 12 & -6 \\ 0 & 2 & 0 \\ 0 & 0 & 1 \end{bmatrix}\begin{bmatrix} x_1 \\ x_2 \\ x_3 \end{bmatrix}=\begin{bmatrix} -33 \\ -4 \\ 1 \end{bmatrix}$$

In equation form we get

$$-3x_1+12x_2-6x_3=-33$$

$$2x_2=-4 \Rightarrow x_2=-2$$

$$x_3=1$$

Substituting $x_2=-2$ and $x_3=1$ we get

$$-3x_1+12(-2)-6(1)=-33$$

$$-3x_1-24-6=-33$$

$$-3x_1=-3$$

$$x_1=1$$

Therefore $x_1=1, x_2=-2, x_3=1$.

Q 5. Express the following system in matrix form and solve by Gauss elimination method:

$$2x_1+x_2+2x_3+x_4=6$$

$$6x_1-6x_2+6x_3+12x_4=36$$

$$4x_1+3x_2+3x_3-3x_4=-1$$

$$2x_1+2x_2-x_3+x_4=10$$

[JNTU Aug./Sep. 2008, May/June 2008]

Answer: The given set of equations are

$$\begin{aligned}2x_1 + x_2 + 2x_3 + x_4 &= 6\\ 6x_1 - 6x_2 + 6x_3 + 12x_4 &= 36\\ 4x_1 + 3x_2 + 3x_3 - 3x_4 &= -1\\ 2x_1 + 2x_2 - x_3 + x_4 &= 10\end{aligned}$$

Assembling it into a matrix form, we get

$$\begin{bmatrix} 2 & 1 & 2 & 1 \\ 6 & -6 & 6 & 12 \\ 4 & 3 & 3 & -3 \\ 2 & 2 & -1 & 1 \end{bmatrix} \begin{bmatrix} x_1 \\ x_2 \\ x_3 \\ x_4 \end{bmatrix} = \begin{bmatrix} 6 \\ 36 \\ -1 \\ 10 \end{bmatrix}$$

The augmented matrix is

$$[A \mid B] = \begin{bmatrix} 2 & 1 & 2 & 1 & 6 \\ 6 & -6 & 6 & 12 & 36 \\ 4 & 3 & 3 & -3 & -1 \\ 2 & 2 & -1 & 1 & 10 \end{bmatrix}$$

Applying row operation to this matrix we get

$$[A \mid B] = \begin{bmatrix} 2 & 1 & 2 & 1 & 6 \\ 1 & -1 & 1 & 2 & 6 \\ 4 & 3 & 3 & -3 & -1 \\ 2 & 2 & -1 & 1 & 10 \end{bmatrix} \quad \text{Row } 2 \rightarrow \text{Row } 2/6$$

$$= \begin{bmatrix} 1 & 2 & 1 & -1 & 0 \\ 1 & -1 & 1 & 2 & 6 \\ 4 & 3 & 3 & -3 & -1 \\ 2 & 2 & -1 & 1 & 10 \end{bmatrix} \quad \text{Row } 1 \rightarrow \text{Row } 1 - \text{Row } 2$$

$$= \begin{bmatrix} 1 & 2 & 1 & -1 & 0 \\ 0 & -3 & 0 & 3 & 6 \\ 0 & -5 & -1 & 1 & -1 \\ 0 & -2 & -3 & 3 & 10 \end{bmatrix} \quad \begin{array}{l} \text{Row } 2 \rightarrow \text{Row } 2 - \text{Row } 1 \\ \text{Row } 3 \rightarrow \text{Row } 3 - 4 \text{ Row } 1 \\ \text{Row } 4 \rightarrow \text{Row } 4 - 2 \text{ Row } 1 \end{array}$$

$$= \begin{bmatrix} 1 & 2 & 1 & -1 & 0 \\ 0 & -1 & 0 & 1 & 2 \\ 0 & -5 & -1 & 1 & -1 \\ 0 & -2 & -3 & 3 & 10 \end{bmatrix} \quad \text{Row } 2 \rightarrow \text{Row } 2/3$$

$$= \begin{bmatrix} 1 & 1 & 1 & -1 & 0 \\ 0 & -1 & 0 & 1 & 2 \\ 0 & 0 & -1 & -4 & -11 \\ 0 & 0 & -3 & 1 & 6 \end{bmatrix} \quad \begin{matrix} \text{Row } 3 \to \text{Row } 3 - 5 \text{ Row } 2 \\ \text{Row } 4 \to \text{Row } 4 - 2 \text{ Row } 2 \end{matrix}$$

$$= \begin{bmatrix} 1 & 2 & 1 & -1 & 0 \\ 0 & -1 & 0 & 1 & 2 \\ 0 & 0 & -1 & -4 & -11 \\ 0 & 0 & 0 & 13 & 39 \end{bmatrix} \quad \text{Row } 4 \to \text{Row } 4 - 3 \text{ Row } 3$$

$$= \begin{bmatrix} 1 & 2 & 1 & -1 & 0 \\ 0 & -1 & 0 & 1 & 2 \\ 0 & 0 & -1 & -4 & -11 \\ 0 & 0 & 0 & 1 & 3 \end{bmatrix} \quad \text{Row } 4 \to \text{Row } 4/13$$

In equation form we get

$$x_1 + 2x_2 + x_3 - x_4 = 0 \tag{1.123}$$

$$-x_2 + x_4 = 2 \tag{1.124}$$

$$-x_3 - 4x_4 = -11 \tag{1.125}$$

$$x_4 = 3 \tag{1.126}$$

Substituting the value of x_4 in (1.125), we get

$$-x_3 - 12 = -11$$

$$x_3 = -1$$

Substituting the value of x_4 in (1.124), we get

$$-x_2 = 2 - 3$$

$$x_2 = 1$$

Finally substituting the values of x_2, x_3, x_4 in (1.123), we get

$$x_1 + 2 - 1 - 3 = 0$$

$$x_1 = 2$$

Therefore the required solution set is $x_1 = 2$, $x_2 = 1$, $x_3 = -1$, $x_4 = 3$.

Summary

An $m \times n$ **matrix** $\mathbf{A} = [a_{jk}]$ is a rectangular array of numbers or functions ("entries," "elements") arranged in m horizontal **rows** and n vertical **columns.** If $m = n$, the matrix is called **square.** A $1 \times n$ matrix is called a **row vector** and an $m \times 1$ matrix a **column vector.**

The **sum** $\mathbf{A} + \mathbf{B}$ of matrices of the same **size** (i.e., both $m \times n$) is obtained by adding corresponding entries. The **product** of $\mathbf{A}$ by a scalar c is obtained by multiplying each a_{jk} by c.

The **product** $\mathbf{C} = \mathbf{AB}$ of an $m \times n$ matrix $\mathbf{A}$ by an $r \times p$ matrix $\mathbf{B} = [b_{jk}]$ is defined only when $r = n$, and is the $m \times p$ matrix $\mathbf{C} = [c_{jk}]$ with entries

$$c_{jk} = a_{j1}b_{1k} + a_{j2}b_{2k} + \cdots + a_{jn}b_{nk}$$
$$\text{(row } j \text{ of } \mathbf{A} \text{ times column } k \text{ of } \mathbf{B}) \qquad (1.127)$$

This multiplication is motivated by the composition of **linear transformations**. It is associative, but is ***not commutative:*** if $\mathbf{AB}$ is defined, $\mathbf{BA}$ may not be defined, but even if $\mathbf{BA}$ is defined, $\mathbf{AB} \neq \mathbf{BA}$ in general. Also $\mathbf{AB} = \mathbf{0}$ may not imply $\mathbf{A} = \mathbf{0}$ or $\mathbf{B} = \mathbf{0}$ or $\mathbf{BA} = \mathbf{0}$. Illustrations:

$$\begin{bmatrix} 1 & 1 \\ 2 & 2 \end{bmatrix}\begin{bmatrix} -1 & 1 \\ 1 & -1 \end{bmatrix} = \begin{bmatrix} 0 & 0 \\ 0 & 0 \end{bmatrix}$$

$$\begin{bmatrix} -1 & 1 \\ 1 & -1 \end{bmatrix}\begin{bmatrix} 1 & 1 \\ 2 & 2 \end{bmatrix} = \begin{bmatrix} 1 & 1 \\ -1 & -1 \end{bmatrix}$$

$$[1 \quad 2]\begin{bmatrix} 3 \\ 4 \end{bmatrix} = [11], \quad \begin{bmatrix} 3 \\ 4 \end{bmatrix}[1 \quad 2] = \begin{bmatrix} 3 & 6 \\ 4 & 8 \end{bmatrix}$$

The **transpose** $\mathbf{A}^\mathsf{T}$ of a matrix $\mathbf{A} = [a_{jk}]$ is $\mathbf{A}^\mathsf{T} = [a_{kj}]$; rows become columns and conversely. Here, $\mathbf{A}$ need not be square. If it is and $\mathbf{A} = \mathbf{A}^\mathsf{T}$, then $\mathbf{A}$ is called **symmetric;** if $\mathbf{A} = -\mathbf{A}^\mathsf{T}$, it is called **skew-symmetric.** For a product, $(\mathbf{AB})^\mathsf{T} = \mathbf{B}^\mathsf{T}\mathbf{A}^\mathsf{T}$.

A main application of matrices concerns **linear systems of equations**

$$\mathbf{Ax} = \mathbf{b} \qquad (1.128)$$

(m equations in n unknowns $x_1, \ldots, x_n$; $\mathbf{A}$ and $\mathbf{b}$ given). The most important method of solution is the **Gauss elimination** which reduces the system to "triangular" form by *elementary row operations,* which leave the set of solutions unchanged.

Cramer's rule represents the unknowns in a system (1.120) of n equations in n unknowns as quotients of determinants; for numeric work it is impractical. **Determinants** have decreased in importance, but will retain their place in eigenvalue problems, elementary geometry, etc.

The **inverse** $\mathbf{A}^{-1}$ of a square matrix satisfies $\mathbf{AA}^{-1} = \mathbf{A}^{-1}\mathbf{A} = \mathbf{I}$. It exists if and only if $\det \mathbf{A} \neq 0$. It can be computed by the *Gauss–Jordan elimination.*

The **rank** r of a matrix $\mathbf{A}$ is the maximum number of linearly independent rows or columns of $\mathbf{A}$ or, equivalently, the number of rows of the largest square submatrix of $\mathbf{A}$ with nonzero determinant.

The system (1.128) has solutions if and only if $\text{rank } \mathbf{A} = \text{rank } [\mathbf{A} \quad \mathbf{b}]$, where $[\mathbf{A} \quad \mathbf{b}]$ is the **augmented matrix** (Fundamental Theorem).

The **homogeneous system**

$$\mathbf{Ax} = \mathbf{0} \qquad (1.129)$$

has unique solution or trivial solution if $\det \mathbf{A} \neq 0$ ("nontrivial solutions") if and only if $\text{rank } \mathbf{A} < n$, in the case $m = n$ equivalently if and only if $\det \mathbf{A} = 0$.

Vector spaces, inner product spaces, and linear transformations are also discussed.

Main tasks are the numeric solution of linear systems, curve fitting, and eigenvalue problems.

Linear systems $\mathbf{Ax} = \mathbf{b}$ with $\mathbf{A} = [a_{jk}]$, written out

$$\begin{array}{ll} E_1: & a_{11}x_1 + \cdots + a_{1n}x_n = b_1 \\ E_2: & a_{21}x_1 + \cdots + a_{2n}x_n = b_2 \\ \multicolumn{2}{c}{\cdots\cdots\cdots\cdots\cdots\cdots\cdots\cdots} \\ E_n: & a_{n1}x_1 + \cdots + a_{nn}x_n = b_n \end{array} \qquad (1.130)$$

can be solved by a **direct method** (one in which the number of numeric operations can be specified in advance, e.g., Gauss's elimination) or by an **indirect** or **iterative method** (in which an initial approximation is improved stepwise).

The **Gauss elimination** is direct, namely, a systematic elimination process that reduces (1.130) stepwise to triangular form. In Step 1 we eliminate x_1 from equations E_2 to E_n by subtracting (a_{21}/a_{11}) E_1 from E_2, then (a_{31}/a_{11}) E_1 from E_3, etc. Equation E_1 is called the **pivot equation** in this step and a_{11} the **pivot.** In Step 2 we take the new second equation as pivot equation and eliminate x_2, etc. If the triangular form is reached, we get x_n from the last equation, then x_{n-1} from the second last, etc. **Partial pivoting** (= interchange of equations) is *necessary* if candidates for pivots are zero, and *advisable* if they are small in absolute value.

Doolittle's, Crout's, and **Cholesky's methods** are variants of the Gauss elimination. They factor

$\mathbf{A} = \mathbf{LU}$ (L lower triangular, U upper triangular) and solve $\mathbf{Ax} = \mathbf{LUx} = \mathbf{b}$ by solving $\mathbf{Ly} = \mathbf{b}$ for $\mathbf{y}$ and then $\mathbf{Ux} = \mathbf{y}$ for $\mathbf{x}$.

In the Gauss–Seidel iteration we make $a_{11} = a_{22} = \dots = a_{nn} = 1$ (by division) and write $\mathbf{Ax} = (\mathbf{I} + \mathbf{L} + \mathbf{U})\mathbf{x} = \mathbf{b}$; thus $\mathbf{x} = \mathbf{b} - (\mathbf{L} + \mathbf{U})\mathbf{x}$, which suggests the iteration formula

$$\mathbf{x}^{(m+1)} = \mathbf{b} - \mathbf{L}\mathbf{x}^{(m+1)} - \mathbf{U}\mathbf{x}^{(m)} \qquad (1.131)$$

in which we always take the most recent approximate x_j's on the right. If $\|\mathbf{C}\| < 1$, where $\mathbf{C} = -(\mathbf{I} + \mathbf{L})^{-1}\mathbf{U}$, then this process converges. Here, $\|\mathbf{C}\|$ denotes any matrix norm.

Review Questions and Problems

1. What properties of matrix multiplication differ from those of the multiplication of numbers? What about division of matrices?
2. Let $\mathbf{A}$ be a 50×50 matrix and $\mathbf{B}$ a 50×20 matrix. Are the following expressions defined or not? $\mathbf{A} + \mathbf{B}$, $\mathbf{A}^2$, $\mathbf{B}^2$, $\mathbf{AB}$, $\mathbf{BA}$, $\mathbf{AA}^T$, $\mathbf{B}^T\mathbf{A}$, $\mathbf{B}^T\mathbf{B}$, $\mathbf{BB}^T$, $\mathbf{B}^T\mathbf{AB}$. (Give reasons.)
3. How is matrix multiplication motivated?
4. Are there any linear systems without solutions? With one solution? With more than one solution? Give simple examples.
5. How can you give the rank of a matrix in terms of row vectors? Of column vectors? Of determinants?
6. What is the role of rank in connection with solving linear systems?
7. What is the row space of a matrix? The column space? The null space?
8. What is the idea of Gauss elimination and back substitution?
9. What is the inverse of a matrix? When does it exist? How would you determine it?
10. What is Cramer's rule? When would you apply it?
11. Find all solutions or indicate that no solution exists. (Show the details of your work.)

a. $9x - 3y = 15$
$5x + 4y = 48$

b. $-2x - 4y + 7z = -6$
$x + 2y + 16z = 3$

c. $3x + 5y - 8z = 18$
$x + 2y - 3z = 6$

d. $5x - 10y = 2$
$3x + y = 13$
$-x + 6y = 6$

e. $-8x + 2z = 1$
$6y + 4z = 3$
$12x + 2y = 2$

f. $2y + z = -1$
$2x + 3y - z = -12$
$5x - 4y + 3z = 32$

g. $3x + 7y = 0$
$5x - 4y = 47$
$6x + 9y = 15$

h. $-x + 4y - 2z = 1$
$3x + 4y + 6z = 1$
$x - 2y + 2z = -1/2$

i. $7x + 9y - 14z = 36$
$-x - 3y + 2z = -12$
$2x + y - 4z = 4$

12. Calculate the following expressions (showing the details of your work) or indicate why they do not exist, when

$$\mathbf{A} = \begin{bmatrix} 9 & 2 & 8 \\ 2 & 18 & 10 \\ 8 & 10 & 15 \end{bmatrix}, \quad \mathbf{B} = \begin{bmatrix} 0 & 2 & 6 \\ -2 & 0 & -3 \\ -6 & 3 & 0 \end{bmatrix}$$

$$\mathbf{a} = \begin{bmatrix} 3 \\ 7 \\ 1 \end{bmatrix}, \qquad \mathbf{b} = \begin{bmatrix} 4 \\ 0 \\ 2 \end{bmatrix}$$

a. $\mathbf{AB}$, $\mathbf{BA}$
b. $\mathbf{A} - \mathbf{A}^T$
c. $\mathbf{A}^2 + \mathbf{B}^2$
d. det $\mathbf{A}$, det $\mathbf{B}$, det $\mathbf{AB}$
e. $\mathbf{AA}^T$, $\mathbf{A}^T\mathbf{A}$
f. $0.2\mathbf{BB}^T$
g. $\mathbf{Aa}$, $\mathbf{a}^T\mathbf{A}$, $\mathbf{a}^T\mathbf{Aa}$
h. $\mathbf{a}^T\mathbf{b}$, $\mathbf{b}^T\mathbf{a}$, $\mathbf{ab}^T$
i. $\mathbf{b}^T\mathbf{Bb}$
j. $\mathbf{a}^T\mathbf{B}$, $\mathbf{B}^T\mathbf{a}$
k. $0.1(\mathbf{A} + \mathbf{A}^T)(\mathbf{B} - \mathbf{B}^T)$

13. Determine the ranks of the coefficient matrix and the augmented matrix and state how many solutions the linear system will have.
    a. In Prob. 11(c) b. In Prob. 11(b)
    c. In Prob. 11(g) d. In Prob. 11(d)
    e. In Prob. 11(i) f. In Prob. 11(h)
14. Find the inverse or state why it does not exist. (Show details.)
    a. Of the coefficient matrix in Prob. 11(a)
    b. Of the coefficient matrix in Prob. 11(e)
    c. Of the coefficient matrix in Prob. 11(f)
    d. Of the coefficient matrix in Prob. 11(h)
    e. Of the augmented matrix in Prob. 11(d)
    f. Of the diagonal matrix with entries 3, −1, 5
15. Find the currents in the following networks.

a.

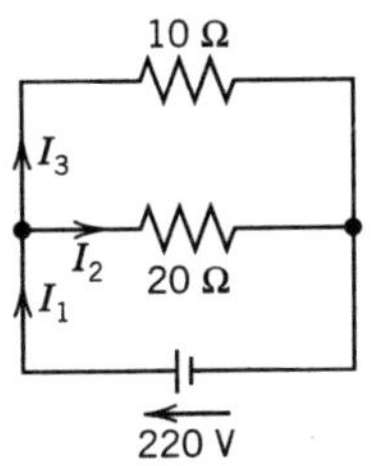

b.

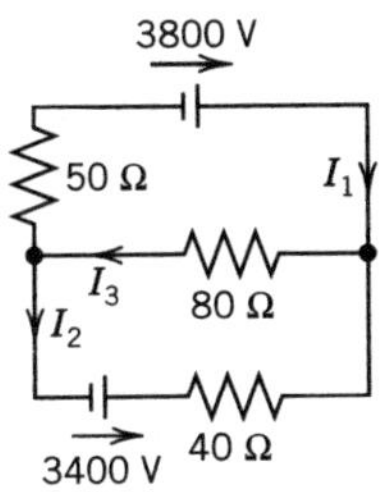

c.

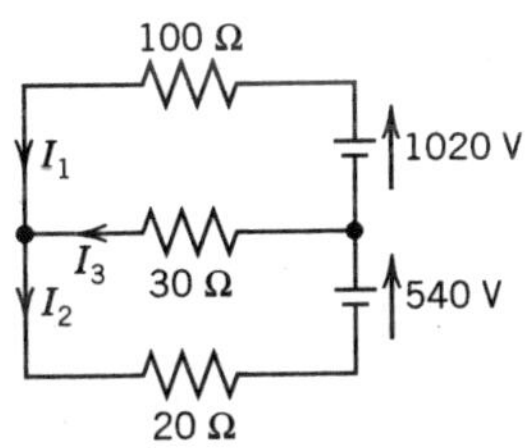

d. What are the main problem areas in numeric linear algebra?
e. What is pivoting? When and how would you apply it?
f. What happens if you apply Gauss elimination to a system that has no solutions?
g. What is Doolittle's method? Its connection to Gauss elimination?
h. What is Cholesky's method? When would you apply it?
i. What do you know about the convergence of the Gauss–Seidel method?
j. What is ill-conditioning? What is the condition number and its significance?
k. What is tridiagonalization? When would you apply it?
l. What is the idea of the QR-method? When would you apply the method?

16. Solve:

a.
$$\begin{aligned} 4x_2 - 3x_3 &= 11.8 \\ 5x_1 + 3x_2 + x_3 &= 34.2 \\ 6x_1 - 7x_2 + 2x_3 &= -3.1 \end{aligned}$$

b.
$$\begin{aligned} x_1 + x_2 + x_3 &= 5 \\ x_1 + 2x_2 + 2x_3 &= 6 \\ x_1 + 2x_2 + 3x_3 &= 8 \end{aligned}$$

c.
$$\begin{aligned} 5x_1 + x_2 - 3x_3 &= 17 \\ -5x_2 + 15x_3 &= -10 \\ 2x_1 - 3x_2 + 9x_3 &= 0 \end{aligned}$$

d.
$$\begin{aligned} 2x_1 + 3x_3 &= 15 \\ 4x_2 - x_3 &= -13 \\ 3x_1 - x_2 + 5x_3 &= 26 \end{aligned}$$

e. Solve Prob. 11(g) by Doolittle's method.
f. Solve Prob. 11(g) by Cholesky's method.

17. Compute the inverse of:

a. $\begin{bmatrix} 1.0 & 2.0 & 0.5 \\ 0.5 & 1.0 & 0.5 \\ 1.5 & 2.0 & 1.0 \end{bmatrix}$

b. $\begin{bmatrix} 1.5 & 2.0 & 1.0 \\ 2.0 & 3.5 & 1.5 \\ 1.0 & 1.5 & 9.0 \end{bmatrix}$ c. $\begin{bmatrix} 5 & 1 & 1 \\ 1 & 6 & 0 \\ 1 & 0 & 8 \end{bmatrix}$

18. Do 3 steps without scaling, starting from $[1 \quad 1 \quad 1]^T$:

a.
$$\begin{aligned} x_1 + 15x_2 - x_3 &= 11 \\ 10x_1 + 3x_2 &= -17 \\ 2x_1 - x_2 + 5x_3 &= 5 \end{aligned}$$

b.
$$\begin{aligned} 4x_1 - x_2 &= 5.5 \\ 4x_2 - x_3 &= 0.4 \\ -x_1 + 4x_3 &= 11.2 \end{aligned}$$

2 Eigenvalues and Eigenvectors

LEARNING OBJECTIVES

After reading this chapter, you will be able to understand the following:

- Eigenvalues and eigenvectors.
- Properties of eigenvalues and eigenvectors.
- Cayley–Hamilton theorem.
- Diagonalization.
- Power and inverse of matrix.
- Modal and spectral matrices.

Matrix eigenvalue problems concern the solutions of vector equations

$$\mathbf{Ax} = \lambda\mathbf{x} \tag{2.1}$$

where **A** is a given ***square*** matrix and vector **x** and scalar λ are unknown. Clearly, $\mathbf{x} = \mathbf{0}$ is a solution of (2.1), giving $\mathbf{0} = \mathbf{0}$. But this of no interest, and we want to find solution vectors $\mathbf{x} \neq \mathbf{0}$ of (2.1), called **eigenvectors** of **A**. We shall see that eigenvectors can be found only for certain values of the scalar λ; these values λ for which an eigenvector exists are called the **eigenvalues** of **A**. Geometrically, solving (2.1) in this way means that we are looking for vectors **x** for which the multiplication of **x** by the matrix **A** has the same effect as the multiplication of **x** by a scalar λ, giving a vector $\lambda\mathbf{x}$ with components proportional to those of **x**, and λ as the factor of proportionality.

Eigenvalue problems are of greatest practical interest to the engineer, physicist, and mathematician, and we shall see that their theory makes up a beautiful chapter in linear algebra that has found numerous applications.

2.1 An Introduction to Eigenvalues and Eigenvectors

From the viewpoint of engineering applications, eigenvalue problems are among the most important problems in connection with matrices, and the student should follow the present discussion with particular attention. We begin by defining the basic concepts and show how to solve these problems, by examples as well as in general. Then we shall turn to applications.

Let $\mathbf{A} = [a_{jk}]$ be a given $n \times n$ matrix and consider the vector equation

$$\mathbf{Ax} = \lambda\mathbf{x} \tag{2.2}$$

Here **x** is an unknown vector and λ an unknown scalar. Our task is to determine **x**'s and λ's that satisfy (2.2). Geometrically, we are looking for vectors **x** for which the multiplication by **A** has the same effect as the multiplication by a scalar λ; in other words, **Ax** should be proportional to **x**.

Clearly, the zero vector $\mathbf{x} = \mathbf{0}$ is a solution of (2.2) for any value of λ, because $\mathbf{A0} = \mathbf{0}$. This is of no interest. A value of λ for which (2.2) has a solution $\mathbf{x} \neq \mathbf{0}$ is called an **eigenvalue** or *characteristic value* (or *latent root*) of the matrix **A**. ("Eigen" is a German word and means "proper" or "characteristic.") The corresponding solutions $\mathbf{x} \neq \mathbf{0}$ of (2.2) are called the **eigenvectors** or *characteristic vectors* of **A** corresponding to that eigenvalue λ. The set of all the eigenvalues of **A** is called the **spectrum** of **A**. We shall see that the spectrum consists of at least one eigenvalue and at most of n numerically different eigenvalues. The largest of the absolute values of the eigenvalues of **A** is called the *spectral radius* of **A**, a name to be motivated later.

How to Find Eigenvalues and Eigenvectors

The problem of determining the eigenvalues and eigenvectors of a matrix is called an **eigenvalue problem.** (More precisely: an *algebraic* eigenvalue problem, as opposed to an eigenvalue problem involving an ODE, PDE). Such problems occur in physical, technical, geometric, and other applications, as we shall see. We show how to solve them, first by an example and then in general. Some typical applications will follow afterwards.

Example 1 Determination of Eigenvalues and Eigenvectors

We illustrate all the steps in terms of the matrix

$$\mathbf{A} = \begin{bmatrix} -5 & 2 \\ 2 & -2 \end{bmatrix}$$

Solution:

(a) ***Eigenvalues.*** These must be determined ***first.*** Equation (2.2) is

$$\mathbf{Ax} = \begin{bmatrix} -5 & 2 \\ 2 & -2 \end{bmatrix}\begin{bmatrix} x_1 \\ x_2 \end{bmatrix} = \lambda \begin{bmatrix} x_1 \\ x_2 \end{bmatrix}$$

In components,

$$\begin{aligned} -5x_1 + 2x_2 &= \lambda x_1 \\ 2x_1 - 2x_2 &= \lambda x_2 \end{aligned}$$

Transferring the terms on the right to the left, we get

$$\begin{aligned} (-5-\lambda)x_1 + 2x_2 &= 0 \\ 2x_1 + (-2-\lambda)x_2 &= 0 \end{aligned} \tag{2.3}$$

This can be written in matrix notation as

$$(\mathbf{A} - \lambda\mathbf{I})\mathbf{x} = \mathbf{0} \tag{2.4}$$

because (2.2) is $\mathbf{Ax} - \lambda\mathbf{x} = \mathbf{Ax} - \lambda\mathbf{Ix} = (\mathbf{A} - \lambda\mathbf{I})\mathbf{x} = \mathbf{0}$, which gives (2.4). We see that this is a ***homogeneous*** linear system. By Cramer's theorem it has a nontrivial solution $\mathbf{x} \neq \mathbf{0}$ (an eigenvector of **A** we are looking for) if and only if its coefficient determinant is zero, that is,

$$D(\lambda) = \det(\mathbf{A} - \lambda\mathbf{I}) = \begin{vmatrix} -5-\lambda & 2 \\ 2 & -2-\lambda \end{vmatrix} = (-5-\lambda)(-2-\lambda) - 4 = \lambda^2 + 7\lambda + 6 = 0 \tag{2.5}$$

We call $D(\lambda)$ the **characteristic determinant** or, if expanded, the **characteristic polynomial,** and $D(\lambda) = 0$ the **characteristic equation** of **A**. The solutions of this quadratic equation are $\lambda_1 = -1$ and $\lambda_2 = -6$. These are the eigenvalues of **A**.

(b) ***Eigenvector of* A *corresponding to* λ_1.** This vector is obtained from (2.3) with $\lambda = \lambda_1 = -1$, that is,

$$\begin{aligned} -4x_1 + 2x_2 &= 0 \\ 2x_1 - x_2 &= 0 \end{aligned}$$

A solution is $x_2 = 2x_1$, as we see from either of the two equations, so that we need only one of them. This determines an eigenvector corresponding to $\lambda_1 = -1$ up to a scalar multiple. If we choose $x_1 = 1$, we obtain the eigenvector

$$\mathbf{x}_1 = \begin{bmatrix} 1 \\ 2 \end{bmatrix}$$

Check:

$$\mathbf{A}\mathbf{x}_1 = \begin{bmatrix} -5 & 2 \\ 2 & -2 \end{bmatrix} \begin{bmatrix} 1 \\ 2 \end{bmatrix} = \begin{bmatrix} -1 \\ -2 \end{bmatrix} = (-1)\mathbf{x}_1 = \lambda_1 \mathbf{x}_1$$

(c) ***Eigenvector of* A *corresponding to* λ_2.** For $\lambda = \lambda_2 = -6$, (2.3) becomes

$$\begin{aligned} x_1 + 2x_2 &= 0 \\ 2x_1 - 4x_2 &= 0 \end{aligned}$$

A solution is $x_2 = -x_1/2$ with arbitrary x_1. If we choose $x_1 = 2$, we get $x_2 = -1$. Thus an eigenvector of A corresponding to $\lambda_2 = -6$ is

$$\mathbf{x}_2 = \begin{bmatrix} 2 \\ -1 \end{bmatrix}$$

Check:

$$\mathbf{A}\mathbf{x}_2 = \begin{bmatrix} -5 & 2 \\ 2 & -2 \end{bmatrix} \begin{bmatrix} 2 \\ -1 \end{bmatrix} = \begin{bmatrix} -12 \\ 6 \end{bmatrix} = (-6)\mathbf{x}_2 = \lambda_2 \mathbf{x}_2$$

This example illustrates the general case as follows. Equation (2.2) written in components is

$$\begin{aligned} a_{11}x_1 + \cdots + a_{1n}x_n &= \lambda x_1 \\ a_{21}x_1 + \cdots + a_{2n}x_n &= \lambda x_2 \\ \cdots\cdots\cdots\cdots\cdots\cdots & \\ a_{n1}x_1 + \cdots + a_{nn}x_n &= \lambda x_n \end{aligned}$$

Transferring the terms on the right side to the left side, we have

$$\begin{aligned} (a_{11} - \lambda)x_1 + a_{12}x_2 + \cdots + a_{1n}x_n &= 0 \\ a_{21}x_1 + (a_{22} - \lambda)x_2 + \cdots + a_{2n}x_n &= 0 \\ \cdots\cdots\cdots\cdots\cdots\cdots\cdots\cdots & \\ an_1x_1 + an_2x_2 + \cdots + (a_{nn} - \lambda)x_n &= 0 \end{aligned} \tag{2.6}$$

In matrix notation,

$$(\mathbf{A} - \lambda\mathbf{I})\mathbf{x} = 0 \tag{2.7}$$

By Cramer's theorem, this homogeneous linear system of equations has a nontrivial solution if and only if the corresponding determinant of the coefficients is zero:

$$D(\lambda) = \det(\mathbf{A} - \lambda\mathbf{I}) = \begin{vmatrix} a_{11} - \lambda & a_{12} & \cdots & a_{1n} \\ a_{21} & a_{22} - \lambda & \cdots & a_{2n} \\ \cdot & \cdot & \cdots & \cdot \\ a_{n1} & a_{n2} & \cdots & a_{nn} - \lambda \end{vmatrix} = 0 \qquad (2.8)$$

$\mathbf{A} - \lambda\mathbf{I}$ is called the **characteristic matrix** and $D(\lambda)$ the **characteristic determinant** of **A**. Equation (2.8) is called the **characteristic equation** of **A**. By developing $D(\lambda)$ we obtain a polynomial of nth degree in λ. This is called the **characteristic polynomial** of **A**. This proves the following important theorem.

THEOREM 1 (Eigenvalues): *The eigenvalues of a square matrix* **A** *are the roots of the characteristic equation* (2.8) *of* **A**. *Hence an* $n \times n$ *matrix has at least one eigenvalue and at most n numerically different eigenvalues.*

For larger n, the actual computation of eigenvalues will in general require the use of Newton's method or another numeric approximation method discussed in later chapters.

The eigenvalues must be determined first. Once these are known, corresponding eigen*vectors* are obtained from the system (2.6), for instance, by the Gauss elimination, where λ is the eigenvalue for which an eigenvector is wanted. This is what we did in Example 1 and shall do again in the examples below.

Eigenvectors have the following properties.

THEOREM 2 (Eigenvectors, Eigenspace): *If* **w** *and* **x** *are eigenvectors of a matrix* **A** *corresponding to* ***the same*** *eigenvalue* λ, *so are* $\mathbf{w} + \mathbf{x}$ (provided $\mathbf{x} \neq -\mathbf{w}$) *and* $k\mathbf{x}$ *for any* $k \neq 0$. *Hence the eigenvectors corresponding to one and the same eigenvalue* λ *of* **A**, *together with* **0**, *form a vector space, called the* **eigenspace** *of* **A** *corresponding to that* λ.

PROOF: $\mathbf{Aw} = \lambda\mathbf{w}$ and $\mathbf{Ax} = \lambda\mathbf{x}$ imply $\mathbf{A}(\mathbf{w} + \mathbf{x}) = \mathbf{Aw} + \mathbf{Ax} = \lambda\mathbf{w} + \lambda\mathbf{x} = \lambda(\mathbf{w} + \mathbf{x})$ and $\mathbf{A}(k\mathbf{w}) = k(\mathbf{Aw}) = k(\lambda\mathbf{w}) = \lambda(k\mathbf{w})$; hence $\mathbf{A}(k\mathbf{w} + \ell\mathbf{x}) = \lambda(k\mathbf{w} + \ell\mathbf{x})$. ■

In particular, *an eigenvector* **x** *is determined only up to a constant factor.* Hence we can **normalize x**, that is, multiply it by a scalar to get a unit vector. For instance, $\mathbf{x}_1 = [1 \quad 2]^T$ in Example 1 has the length $\|\mathbf{x}_1\| = \sqrt{1^2 + 2^2} = \sqrt{5}$; hence $[1/\sqrt{5} \quad 2/\sqrt{5}]^T$ is a normalized eigenvector (a unit eigenvector).

Examples 2 and 3 will illustrate that an $n \times n$ matrix may have n linearly independent eigenvectors, or it may have fewer than n. In Example 4 we shall see that a *real* matrix may have *complex* eigenvalues and eigenvectors.

Example 2 Multiple Eigenvalues

Find the eigenvalues and eigenvectors of

$$\mathbf{A} = \begin{bmatrix} -2 & 2 & -3 \\ 2 & 1 & -6 \\ -1 & -2 & 0 \end{bmatrix}$$

Solution: For our matrix, the characteristic determinant gives the characteristic equation

$$-\lambda^3 - \lambda^2 + 21\lambda + 45 = 0$$

The roots (eigenvalues of **A**) are $\lambda_1 = 5$, $\lambda_2 = \lambda_3 = -3$. To find eigenvectors, we apply the Gauss elimination to the system $(\mathbf{A} - \lambda\mathbf{I})\mathbf{x} = \mathbf{0}$, first with $\lambda = 5$ and then with $\lambda = -3$. For $\lambda = 5$ the characteristic matrix is

$$\mathbf{A} - \lambda\mathbf{I} = \mathbf{A} - 5\mathbf{I} = \begin{bmatrix} -7 & 2 & -3 \\ 2 & -4 & -6 \\ -1 & -2 & -5 \end{bmatrix}$$

It row-reduces to

$$\begin{bmatrix} -7 & 2 & -3 \\ 0 & -24/7 & -48/7 \\ 0 & 0 & 0 \end{bmatrix}$$

Hence it has rank 2. Choosing $x_3 = -1$ we have $x_2 = 2$ from $-(24/7)x_2 - (48/7)x_3 = 0$ and then $x_1 = 1$ from $-7x_1 + 2x_2 - 3x_3 = 0$. Hence an eigenvector of **A** coresponding to $\lambda = 5$ is $\mathbf{x}_1 = [1 \quad 2 \quad -1]^T$. For $\lambda = -3$ the characteristic matrix

$$\mathbf{A} - \lambda\mathbf{I} = \mathbf{A} + 3\mathbf{I} = \begin{bmatrix} 1 & 2 & -3 \\ 2 & 4 & -6 \\ -1 & -2 & 3 \end{bmatrix}$$

row-reduces to

$$\begin{bmatrix} 1 & 2 & -3 \\ 0 & 0 & 0 \\ 0 & 0 & 0 \end{bmatrix}$$

Hence it has rank 1. From $x_1 + 2x_2 - 3x_3 = 0$ we have $x_1 = -2x_2 + 3x_3$. Choosing $x_2 = 1, x_3 = 0$ and $x_2 = 0, x_3 = 1$, we obtain two linearly independent eigenvectors of **A** corresponding to $\lambda = -3$ [as they must exist with rank = 1 and $n = 3$],

$$\mathbf{x}_2 = \begin{bmatrix} -2 \\ 1 \\ 0 \end{bmatrix} \quad \text{and} \quad \mathbf{x}_3 = \begin{bmatrix} 3 \\ 0 \\ 1 \end{bmatrix}$$

The order M_λ of an eigenvalue λ as a root of the characteristic polynomial is called the **algebraic multiplicity** of λ. The number m_λ of linearly independent eigenvectors corresponding to λ is called the **geometric multiplicity** of λ. Thus m_λ is the dimension of the eigenspace corresponding to this λ. Since the characteristic polynomial has degree n, the sum of all the algebraic multiplicities must equal n. In Example 2 for $\lambda = -3$ we have $m_\lambda = M_\lambda = 2$. In general, $m_\lambda \leqq M_\lambda$, as can be shown. The difference $\Delta_\lambda = M_\lambda - m_\lambda$ is called the **defect** of λ. Thus $\Delta_{-3} = 0$ in Example 2, but positive defects Δ_λ can easily occur.

Example 3 Algebraic Multiplicity, Geometric Multiplicity. Positive Defect

The characteristic equation of the matrix

$$\mathbf{A} = \begin{bmatrix} 0 & 1 \\ 0 & 0 \end{bmatrix} \quad \text{is} \quad \det(\mathbf{A} - \lambda\mathbf{I}) = \begin{vmatrix} -\lambda & 1 \\ 0 & -\lambda \end{vmatrix} = \lambda^2 = 0$$

Hence $\lambda = 0$ is an eigenvalue of algebraic multiplicity $M_0 = 2$. But its geometric multiplicity is only $m_0 = 1$, since eigenvectors result from $-0x_1 + x_2 = 0$, hence $x_2 = 0$, in the form $[x_1 \quad 0]^T$. Hence for $\lambda = 0$ the defect is $\Delta_0 = 1$.
Similarly, the characteristic equation of the matrix

$$\mathbf{A} = \begin{bmatrix} 3 & 2 \\ 0 & 3 \end{bmatrix} \quad \text{is} \quad \det(\mathbf{A} - \lambda\mathbf{I}) = \begin{vmatrix} 3-\lambda & 2 \\ 0 & 3-\lambda \end{vmatrix} = (3-\lambda)^2 = 0$$

Hence $\lambda = 3$ is an eigenvalue of algebraic multiplicity $M_3 = 2$, but its geometric multiplicity is only $m_3 = 1$, since eigenvectors result from $0x_1 + 2x_2 = 0$ in the form $[x_1 \quad 0]^T$.

Example 4 Real Matrices with Complex Eigenvalues and Eigenvectors

Since real polynomials may have complex roots (which then occur in conjugate pairs), a real matrix may have complex eigenvalues and eigenvectors. For instance, the characteristic equation of the skew-symmetric matrix

$$\mathbf{A} = \begin{bmatrix} 0 & 1 \\ -1 & 0 \end{bmatrix} \quad \text{is} \quad \det(\mathbf{A} - \lambda\mathbf{I}) = \begin{vmatrix} -\lambda & 1 \\ -1 & -\lambda \end{vmatrix} = \lambda^2 + 1 = 0$$

It gives the eigenvalues $\lambda_1 = i \ (= \sqrt{-1})$, $\lambda_2 = -i$. Eigenvectors are obtained from $-ix_1 + x_2 = 0$ and $ix_1 + x_2 = 0$, respectively, and we can choose $x_1 = 1$ to get

$$\begin{bmatrix} 1 \\ i \end{bmatrix} \quad \text{and} \quad \begin{bmatrix} 1 \\ -i \end{bmatrix}$$

In the next section we shall need the following simple theorem.

THEOREM 3 (Eigenvalues of the Transpose): *The transpose* $\mathbf{A}^T$ *of a square matrix* $\mathbf{A}$ *has the same eigenvalues as* $\mathbf{A}$.

PROOF: Transposition does not change the value of the characteristic determinant, as follows from Theorem 2d in Chapter 1. ■

Having gained a first impression of matrix eigenvalue problems, in the next section we will illustrate their importance with some typical applications.

CONCEPT CHECK

1. Find the eigenvalues and eigenvectors of the following matrices. (Use the given λ or factors.)

(a) $\begin{bmatrix} -2 & 0 \\ 0 & 0.4 \end{bmatrix}$ (b) $\begin{bmatrix} a & b \\ 0 & c \end{bmatrix}$

(c) $\begin{bmatrix} 4 & 0 \\ 2 & -4 \end{bmatrix}$ (d) $\begin{bmatrix} 0 & 0 \\ 0 & 0 \end{bmatrix}$

(e) $\begin{bmatrix} 5 & -2 \\ 9 & -6 \end{bmatrix}$ (f) $\begin{bmatrix} a & b \\ -b & a \end{bmatrix}$

(g) $\begin{bmatrix} 0.8 & -0.6 \\ 0.6 & 0.8 \end{bmatrix}$ (h) $\begin{bmatrix} 1 & 0 \\ 0 & 1 \end{bmatrix}$

(i) $\begin{bmatrix} 1 & 2 \\ 2 & 4 \end{bmatrix}$

(j) $\begin{bmatrix} \cos\theta & -\sin\theta \\ \sin\theta & \cos\theta \end{bmatrix}$

(k) $\begin{bmatrix} 4 & 0 & 0 \\ 0 & 0 & 0 \\ 0 & 0 & -1 \end{bmatrix}$

(l) $\begin{bmatrix} 85 & -28 & -28 \\ -10 & -11 & -11 \\ -46 & -2 & -2 \end{bmatrix}$

(m) $\begin{bmatrix} 6 & 2 & -2 \\ 2 & 5 & 0 \\ -2 & 0 & 7 \end{bmatrix}, \lambda = 3$

(n) $\begin{bmatrix} 2 & 0 & -2 \\ 0 & 0 & -2 \\ -2 & -2 & 1 \end{bmatrix}, \lambda = 1$

(o) $\begin{bmatrix} 1 & 0 & 0 \\ 2 & 4 & 0 \\ 6 & 4 & 2 \end{bmatrix}$

(p) $\begin{bmatrix} 0.5 & 0.2 & 0.1 \\ 0 & 1.0 & 1.5 \\ 0 & 0 & 3.5 \end{bmatrix}$

(q) $\begin{bmatrix} 4 & -2 & 3 \\ -2 & 1 & 6 \\ 1 & 2 & 2 \end{bmatrix}, \lambda = -3$

(r) $\begin{bmatrix} 3 & 0 & 12 \\ -6 & 3 & 0 \\ 9 & 6 & 3 \end{bmatrix}, \lambda = 9$

(s) $\begin{bmatrix} 13 & 5 & 2 \\ 2 & 7 & -8 \\ 5 & 4 & 7 \end{bmatrix}$

(t) $\begin{bmatrix} 0 & 0 & -5 & 7 \\ 0 & 0 & 7 & -5 \\ 0 & 0 & 19 & -1 \\ 0 & 0 & -1 & 19 \end{bmatrix}$

(u) $\begin{bmatrix} 0 & -2 & 2 & 0 \\ -4 & 2 & -2 & 4 \\ 0 & 2 & 2 & -4 \\ 0 & 2 & -6 & 4 \end{bmatrix}, \lambda = 4$

(v) $\begin{bmatrix} -3 & 0 & 4 & 2 \\ 0 & 1 & -2 & 4 \\ 2 & 4 & -1 & -2 \\ 0 & 2 & -2 & 3 \end{bmatrix}, (\lambda - 3)^2$

(w) $\begin{bmatrix} 2 & 0 & 0 & 0 \\ 1 & 1 & 0 & 0 \\ 2 & 0 & 3 & 0 \\ 1 & 4 & 2 & -6 \end{bmatrix}$

(x) $\begin{bmatrix} -3 & 0 & -2 & 8 \\ 0 & 1 & 4 & -2 \\ -4 & 10 & -1 & -2 \\ 6 & -4 & -2 & 3 \end{bmatrix}, \begin{matrix} \lambda = 3 \\ \lambda = -5 \end{matrix}$

(y) $\begin{bmatrix} -1 & 0 & 12 & 0 \\ 0 & -1 & 0 & 12 \\ 0 & 0 & -1 & -4 \\ 0 & 0 & -4 & -1 \end{bmatrix}, (\lambda + 1)^2$

2. **(Multiple eigenvalues)** Find further 2×2 and 3×3 matrices with multiple eigenvalues. (See Example 2.)

3. **(Nonzero defect)** Find further 2×2 and 3×3 matrices with positive defect. (See Example 3.)

4. **(Transpose)** Illustrate Theorem 3 with examples of your own.

5. **(Complex eigenvalues)** Show that the eigenvalues of a real matrix are real or complex conjugate in pairs.

6. **(Inverse)** Show that the inverse $\mathbf{A}^{-1}$ exists if and only if none of the eigenvalues $\lambda_1, \ldots, \lambda_n$ of $\mathbf{A}$ is zero, and then $\mathbf{A}^{-1}$ has the eigenvalues $1/\lambda_1, \ldots, 1/\lambda_n$.

2.2 Properties of Eigenvalues and Eigenvectors

In this section we discuss a few typical examples from the range of applications of matrix eigenvalue problems, which is incredibly large.

Example 5 Stretching of an Elastic Membrane

An elastic membrane in the x_1x_2-plane with boundary circle $x_1^2 + x_2^2 = 1$ (Figure 1) is stretched so that a point P: (x_1, x_2) goes over into the point Q: (y_1, y_2) given by

$$\mathbf{y} = \begin{bmatrix} y_1 \\ y_2 \end{bmatrix} = \mathbf{Ax} = \begin{bmatrix} 5 & 3 \\ 3 & 5 \end{bmatrix} \begin{bmatrix} x_1 \\ x_2 \end{bmatrix}$$

In equation form this is

$$\begin{aligned} y_1 &= 5x_1 + 3x_2 \\ y_2 &= 3x_1 + 5x_2 \end{aligned} \tag{2.9}$$

Find the **principal directions**, that is, the directions of the position vector $\mathbf{x}$ of P for which the direction of the position vector $\mathbf{y}$ of Q is the same or exactly opposite. What shape does the boundary circle take under this deformation?

Solution: We are looking for vectors $\mathbf{x}$ such that $\mathbf{y} = \lambda\mathbf{x}$. Since $\mathbf{y} = \mathbf{Ax}$, this gives $\mathbf{Ax} = \lambda\mathbf{x}$, the equation of an eigenvalue problem. In components, $\mathbf{Ax} = \lambda\mathbf{x}$ is

$$\begin{aligned} 5x_1 + 3x_2 &= \lambda x_1 \\ 3x_1 + 5x_2 &= \lambda x_2 \end{aligned} \quad \text{or} \quad \begin{aligned} (5-\lambda)\, x_1 + 3x_2 &= 0 \\ 3x_1 + (5-\lambda)\, x_2 &= 0 \end{aligned} \tag{2.10}$$

The characteristic equation is

$$\begin{vmatrix} 5-\lambda & 3 \\ 3 & 5-\lambda \end{vmatrix} = (5-\lambda)^2 - 9 = 0 \tag{2.11}$$

Its solutions are $\lambda_1 = 8$ and $\lambda_2 = 2$. These are the eigenvalues of our problem. For $\lambda = \lambda_1 = 8$, our system (2.10) becomes

$$\begin{aligned} -3x_1 + 3x_2 &= 0 \\ 3x_1 - 3x_2 &= 0 \end{aligned}$$

The solution is $x_2 = x_1$, where x_1 is arbitrary; for instance, $x_1 = x_2 = 1$. For $\lambda_2 = 2$, our system (2.10) becomes

$$\begin{aligned} 3x_1 + 3x_2 &= 0 \\ 3x_1 + 3x_2 &= 0 \end{aligned}$$

The solution is $x_2 = -x_1$, where x_1 is arbitrary; for instance, $x_1 = 1$, $x_2 = -1$. We thus obtain as eigenvectors of $\mathbf{A}$, for instance, $[1\ \ 1]^T$ corresponding to λ_1 and $[1\ \ -1]^T$ corresponding to λ_2 (or a nonzero scalar multiple of these). These vectors make 45° and 135° angles with the positive x_1-direction. They give the principal directions, the answer to our problem. The eigenvalues show that in the principal directions the membrane is stretched by factors 8 and 2, respectively; see Figure 1. Accordingly, if we choose the principal directions as directions of a new Cartesian u_1u_2-coordinate system, say, with the positive u_1-semi-axis in the first quadrant and the positive u_2-semi-axis in the second quadrant of the x_1x_2-system, and if we set $u_1 = r\cos\phi$, $u_2 = r\sin\phi$, then a boundary point of the unstretched circular membrane has coordinates $\cos\phi$, $\sin\phi$. Hence, after the stretch we have

$$z_1 = 8\cos\phi, \qquad z_2 = 2\sin\phi$$

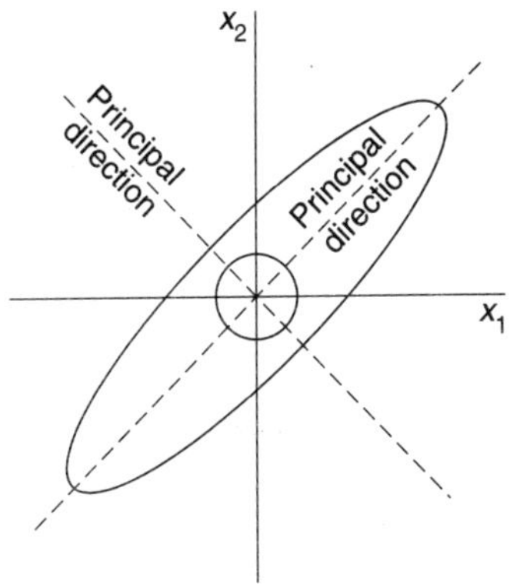

Figure 1 Undeformed and deformed membrane in Example 5.

Since $\cos^2 \phi + \sin^2 \phi = 1$, this shows that the deformed boundary is an ellipse (Figure 1)

$$\frac{z_1^2}{8^2} + \frac{z_2^2}{2^2} = 1 \tag{2.12}$$

Example 6 Eigenvalue Problems Arising from Markov Processes

Markov processes as considered in Example 19, Chapter 1 lead to eigenvalue problems if we ask for the limit state of the process in which the state vector **x** is reproduced under the multiplication by the stochastic matrix **A** governing the process, that is, $\mathbf{Ax} = \mathbf{x}$. Hence **A** should have the eigenvalue 1, and **x** should be a corresponding eigenvector. This is of practical interest because it shows the long-term tendency of the development modeled by the process. In that example,

$$\mathbf{A} = \begin{bmatrix} 0.7 & 0.1 & 0 \\ 0.2 & 0.9 & 0.2 \\ 0.1 & 0 & 0.8 \end{bmatrix}$$

For the transpose,

$$\begin{bmatrix} 0.7 & 0.2 & 0.1 \\ 0.1 & 0.9 & 0 \\ 0 & 0.2 & 0.8 \end{bmatrix} \begin{bmatrix} 1 \\ 1 \\ 1 \end{bmatrix} = \begin{bmatrix} 1 \\ 1 \\ 1 \end{bmatrix}$$

Hence $\mathbf{A}^{\mathrm{T}}$ has the eigenvalue 1, and the same is true for **A** by Theorem 3 in Sec. 2.1. An eigenvector **x** of **A** for $\lambda = 1$ is obtained from

$$\mathbf{A} - \mathbf{I} = \begin{bmatrix} -0.3 & 0.1 & 0 \\ 0.2 & -0.1 & 0.2 \\ 0.1 & 0 & -0.2 \end{bmatrix}$$

row-reduced to

$$\begin{bmatrix} -3/10 & 1/10 & 0 \\ 0 & -1/30 & 1/5 \\ 0 & 0 & 0 \end{bmatrix}$$

Taking $x_3 = 1$, we get $x_2 = 6$ from $(-x_2/30) + (x_3/5) = 0$ and then $x_1 = 2$ from $(-3x_1/10) + (x_2/10) = 0$. This gives $\mathbf{x} = [2 \quad 6 \quad 1]^T$. It means that in the long run, the ratio Commercial:Industrial:Residential will approach 2:6:1, provided that the probabilities given by **A** remain (about) the same. (We switched to ordinary fractions to avoid rounding errors.)

Example 7 Eigenvalue Problems Arising from Population Models. Leslie Model

The Leslie model describes age-specified population growth, as follows. Let the oldest age attained by the females in some animal population be 9 years. Divide the population into three age classes of 3 years each. Let the "*Leslie matrix*" be

$$\mathbf{L} = [l_{jk}] = \begin{bmatrix} 0 & 2.3 & 0.4 \\ 0.6 & 0 & 0 \\ 0 & 0.3 & 0 \end{bmatrix} \qquad (2.13)$$

where l_{1k} is the average number of daughters born to a single female during the time she is in age class k, and $l_{j,\,j-1}$ $(j = 2, 3)$ is the fraction of females in age class $j - 1$ that will survive and pass into class j. (a) What is the number of females in each class after 3, 6, 9 years if each class initially consists of 400 females? (b) For what initial distribution will the number of females in each class change by the same proportion? What is this rate of change?

Solution: (a) Initially, $\mathbf{x}_{(0)}{}^T = [400 \quad 400 \quad 400]$. After 3 years,

$$\mathbf{x}_{(3)} = \mathbf{L}\mathbf{x}_{(0)} = \begin{bmatrix} 0 & 2.3 & 0.4 \\ 0.6 & 0 & 0 \\ 0 & 0.3 & 0 \end{bmatrix} \begin{bmatrix} 400 \\ 400 \\ 400 \end{bmatrix} = \begin{bmatrix} 1080 \\ 240 \\ 120 \end{bmatrix}$$

Similarly, after 6 years the number of females in each class is given by $\mathbf{x}_{(9)}{}^T = (\mathbf{L}\mathbf{x}_{(3)})^T = [600 \quad 648 \quad 72]$, and after 9 years we have $\mathbf{x}_{(9)}{}^T = (\mathbf{L}\mathbf{x}_{(6)})^T = [1519.2 \quad 360 \quad 194.4]$.
(b) Proportional change means that we are looking for a distribution vector **x** such that $\mathbf{Lx} = \lambda\mathbf{x}$, where λ is the rate of change (growth if $\lambda > 1$, decrease if $\lambda < 1$). The characteristic equation is (develop the characteristic determinant by the first column)

$$\det(\mathbf{L} - \lambda\mathbf{I}) = -\lambda^3 - 0.6(-2.3\lambda - 0.3 \times 0.4) = -\lambda^3 + 1.38\lambda + 0.072 = 0$$

A positive root is found to be $\lambda = 1.2$. A corresponding eigenvector **x** can be determined from the characteristic matrix

$$\mathbf{A}-1.2\mathbf{I}=\begin{bmatrix}-1.2 & 2.3 & 0.4\\ 0.6 & -1.2 & 0\\ 0 & 0.3 & -1.2\end{bmatrix} \qquad \text{say,} \qquad \mathbf{x}=\begin{bmatrix}1\\ 0.5\\ 0.125\end{bmatrix}$$

where $x_3 = 0.125$ is chosen, $x_2 = 0.5$ then follows from $0.3x_2 - 1.2x_3 = 0$, and $x_1 = 1$ from $-1.2x_1 + 2.3x_2 + 0.4x_3 = 0$. To get an initial population of 1200 as before, we multiply **x** by $1200/(1 + 0.5 + 0.125) = 738$.

Answer: Proportional growth of the numbers of females in the three classes will occur if the initial values are 738, 369, 92 in classes 1, 2, 3, respectively. The growth rate will be 1.2 per 3 years.

CONCEPT CHECK

1. Find the matrix **A** in the indicated linear transformation $\mathbf{y} = \mathbf{Ax}$. Explain the geometric significance of the eigenvalues and eigenvectors of **A**. Show the details.

   (a) Reflection about the y axis in R^2
   (b) Reflection about the xy plane in R^3
   (c) Orthogonal projection (perpendicular projection) of R^2 onto the x axis
   (d) Orthogonal projection of R^3 onto the plane $y = x$
   (e) Dilatation (uniform stretching) in R^2 by a factor 5
   (f) Counterclockwise rotation through the angle $\pi/2$ about the origin in R^2

2. Given **A** in a deformation $\mathbf{y} = \mathbf{Ax}$, find the principal directions and corresponding factors of extension or contraction. Show the details.

   (a) $\begin{bmatrix}3 & 5\\ 5 & 3\end{bmatrix}$ (b) $\begin{bmatrix}0.4 & 0.8\\ 0.8 & 0.4\end{bmatrix}$

   (c) $\begin{bmatrix}2.5 & 1.5\\ 1.5 & 6.5\end{bmatrix}$ (d) $\begin{bmatrix}5 & 4\\ 4 & 11\end{bmatrix}$

   (e) $\begin{bmatrix}7 & \sqrt{6}\\ \sqrt{6} & 2\end{bmatrix}$ (f) $\begin{bmatrix}5 & 2\\ 2 & 13\end{bmatrix}$

   (g) $\begin{bmatrix}-2 & 3\\ 3 & -2\end{bmatrix}$ (h) $\begin{bmatrix}10.5 & 1/\sqrt{2}\\ 1/\sqrt{2} & 10.0\end{bmatrix}$

3. **(Leontief input–output model)** Suppose that three industries are interrelated so that their outputs are used as inputs by themselves, according to the 3×3 **consumption matrix**

$$\mathbf{A}=[a_{jk}]=\begin{bmatrix}0.2 & 0.5 & 0\\ 0.6 & 0 & 0.3\\ 0.2 & 0.5 & 0.7\end{bmatrix}$$

   where a_{jk} is the fraction of the output of industry k consumed (purchased) by industry j. Let p_j be the price charged by industry j for its total output. A problem is to find prices so that for each industry, total expenditures equal total income. Show that this leads to $\mathbf{Ap} = \mathbf{p}$, where $\mathbf{p} = [p_1 \;\; p_2 \;\; p_3]^T$, and find a solution **p** with nonnegative p_1, p_2, p_3.

4. Show that a consumption matrix as considered in Prob. 3 must have column sums 1 and always has the eigenvalue 1.

5. **(Open Leontief input–output model)** If not the whole output but only a portion of it is consumed by the industries themselves, then instead of $\mathbf{Ax} = \mathbf{x}$ (as in Prob. 3), we have $\mathbf{x} - \mathbf{Ax} = \mathbf{y}$, where $\mathbf{x} = [x_1 \;\; x_2 \;\; x_3]^T$ is produced, $\mathbf{Ax}$ is consumed by the industries, and, thus, **y** is the net production available for other consumers. Find for what production **x** a given **demand vector** $\mathbf{y} = [0.136 \;\; 0.272 \;\; 0.136]^T$ can be achieved if the consumption matrix is

$$A = \begin{bmatrix} 0.2 & 0.4 & 0.2 \\ 0.3 & 0 & 0.1 \\ 0.2 & 0.4 & 0.5 \end{bmatrix}$$

6. Find limit states of the Markov processes modeled by the following matrices. (Show the details.)

(a) $\begin{bmatrix} 0.1 & 0.4 \\ 0.9 & 0.6 \end{bmatrix}$ (b) $\begin{bmatrix} 0.5 & 0.3 & 0.2 \\ 0.3 & 0.5 & 0.2 \\ 0.2 & 0.2 & 0.6 \end{bmatrix}$

(c) $\begin{bmatrix} 0.6 & 0.1 & 0.2 \\ 0.4 & 0.1 & 0.4 \\ 0 & 0.8 & 0.4 \end{bmatrix}$

7. Find the growth rate in the Leslie model (see Example 3) with the matrix as given. (Show details.)

(a) $\begin{bmatrix} 0 & 3.45 & 0.60 \\ 0.90 & 0 & 0 \\ 0 & 0.45 & 0 \end{bmatrix}$

(b) $\begin{bmatrix} 0 & 12.0 & 0 \\ 0.75 & 0 & 0 \\ 0 & 0.30 & 0 \end{bmatrix}$

(c) $\begin{bmatrix} 0 & 7.280 & 2.975 \\ 0.560 & 0 & 0 \\ 0 & 0.420 & 0 \end{bmatrix}$

8. **TEAM PROJECT. General Properties of Eigenvalues and Eigenvectors.** Prove the following statements and illustrate them with examples of your own choice. Here, $\lambda_1, \ldots, \lambda_n$ are the (not necessarily distinct) eigenvalues of a given $n \times n$ matrix $\mathbf{A} = [a_{jk}]$.

(a) **Trace.** The sum of the main diagonal entries is called the **trace** of **A**. It equals the sum of the eigenvalues.

(b) **Spectral shift.** $\mathbf{A} - k\mathbf{I}$ has the eigenvalues $\lambda_1 - k, \ldots, \lambda_n - k$ and the same eigenvectors as **A**.

(c) **Scalar multiples, powers.** $k\mathbf{A}$ has the eigenvalues $k\lambda_1, \ldots, k\lambda_n$. $\mathbf{A}^m$ $(m = 1, 2, \ldots)$ has the eigenvalues $\lambda_1^m, \ldots, \lambda_n^m$. The eigenvectors are those of **A**.

(d) **Spectral mapping theorem.** The "polynomial matrix"

$$p(\mathbf{A}) = k_m\mathbf{A}^m + k_{m-1}\mathbf{A}^{m-1} + \cdots + k_1\mathbf{A} + k_0\mathbf{I}$$

has the eigenvalues

$$p(\lambda_j) = k_m\lambda_j^m + k_{m-1}\lambda_j^{m-1} + \cdots + k_1\lambda_j + k_0$$

where $j = 1, \ldots, n$, and the same eigenvectors as **A**.

(e) **Perron's theorem.** Show that a Leslie matrix **L** with positive $l_{12}, l_{13}, l_{21}, l_{32}$ has a positive eigenvalue.

2.3 Cayley–Hamilton Theorem

THEOREM 4: *Every square matrix satisfies its own characteristic equation. That is if*

$$|\mathbf{A} - \lambda\mathbf{I}| = (-1)^n(\lambda^n + a_1\lambda^{n-1} + a_2\lambda^{n-2} + \cdots + a_n)$$

is the characteristic polynomial of a matrix **A** *of order* $n \times n$*, then the matrix equation*

$$\mathbf{X}^n + a_1\mathbf{X}^{n-1} + a_2\mathbf{X}^{n-2} + \cdots + a_n\mathbf{I} = 0$$

is satisfied by $\mathbf{X} = \mathbf{A}$*. Mathematically*

$$\mathbf{A}^n + a_1\mathbf{A}^{n-1} + a_2\mathbf{A}^{n-2} + \cdots + a_n\mathbf{I} = 0$$

PROOF: Let

$$\mathbf{A} = \begin{bmatrix} a_{11} & a_{12} & . & . & . & a_{1n} \\ a_{21} & a_{22} & . & . & . & a_{2n} \\ a_{31} & a_{32} & . & . & . & a_{3n} \\ . & . & . & . & . & . \\ . & . & . & . & . & . \\ a_{n1} & a_{n2} & . & . & . & a_{nn} \end{bmatrix}$$

be a square matrix of order n and let

$$|\mathbf{A} - \lambda\mathbf{I}| = (-1)^n(\lambda^n + a_1\lambda^{n-1} + a_2\lambda^{n-2} + \cdots + a_n)$$

Since the elements of $\mathbf{A} - \lambda\mathbf{I}$ are at most of the first degree in λ, the elements of adj $(\mathbf{A} - \lambda\mathbf{I})$ are at most of degree $(n - 1)$ in λ. Thus adj $(\mathbf{A} - \lambda\mathbf{I})$ can be expressed as a matrix polynomial in λ as

$$\lambda \,\text{adj}\,(\mathbf{A} - \lambda\mathbf{I}) = \mathbf{B}_0\lambda^n + \mathbf{B}_1\lambda^{n-1} + \mathbf{B}_2\lambda^{n-2} + \cdots + \mathbf{B}_{n-1}$$

where $\mathbf{B}_0, \mathbf{B}_1, \mathbf{B}_2, \ldots, \mathbf{B}_{n-1}$ are $n \times n$ matrices, their elements being polynomials in λ. We show that

$$(\mathbf{A} - \lambda\mathbf{I})\,\text{adj}\,(\mathbf{A} - \lambda\mathbf{I}) = (\mathbf{A} - \lambda\mathbf{I})\mathbf{I}$$

$$(\mathbf{A} - \lambda\mathbf{I})(\mathbf{B}_0\lambda^n + \mathbf{B}_1\lambda^{n-1} + \mathbf{B}_2\lambda^{n-2} + \cdots + \mathbf{B}_{n-1}) = (-1)^n(\lambda^n + a_1\lambda^{n-1} + a_2\lambda^{n-2} + \cdots + a_n)\mathbf{I}$$

Equating the coefficients of equal powers of λ on both sides we get

$$-\mathbf{B}_0 = (-1)^n\mathbf{I}$$

$$\mathbf{A}\mathbf{B}_0 - \mathbf{B}_1 = (-1)^n a_1\mathbf{I}$$

$$\mathbf{A}\mathbf{B}_1 - \mathbf{B}_2 = (-1)^n a_2\mathbf{I}$$

....

$$\mathbf{A}\mathbf{B}^{n-1} = (-1)^n a_n\mathbf{I}$$

Multiplying both sides of the equation by $\mathbf{A}^n, \mathbf{A}^{n-1}, \ldots, \mathbf{I}$ respectively and adding we obtain

$$0 = (-1)^n(\mathbf{A} + a_1\mathbf{A}^{n-1} + a_2\mathbf{A}^{n-2} + \cdots + a_n\mathbf{I})$$

Hence

$$\mathbf{A}^n + a_1\mathbf{A}^{n-1} + a_2\mathbf{A}^{n-2} + \cdots + a_n\mathbf{I} = 0$$ ■

Students should refer to "Solved Problems" section for worked out problems.

CONCEPT CHECK

1. Verify Cayley–Hamilton theorem for the matrix

$$\mathbf{A} = \begin{bmatrix} 1 & 0 & 2 \\ 0 & 2 & 1 \\ 2 & 0 & 3 \end{bmatrix}$$

2.4 Inverse of a Matrix by Cayley–Hamilton Theorem

If $|\mathbf{A} - \lambda \mathbf{I}| = a_0\lambda^n + a_1\lambda^{n-1} + a_2\lambda^{n-2} + \cdots + a_n$ is the characteristic polynomial of **A**, we have

$$\mathbf{A}^n + a_1\mathbf{A}^{n-1} + a_2\mathbf{A}^{n-2} + \cdots + a_n\mathbf{I} = 0 \tag{2.14}$$

Now pre-multiplying (2.14) by $\mathbf{A}^{-1}$, we have

$$a_0\mathbf{A}^{n-1} + a_1\mathbf{A}^{n-2} + a_2\mathbf{A}^{n-3} + \cdots + a_{n-1}\mathbf{I} + a_n\mathbf{A}^{-1} = 0$$

$$-a_n\mathbf{A}^{-1} = a_0\mathbf{A}^{n-1} + a_1\mathbf{A}^{n-2} + a_2\mathbf{A}^{n-3} + \cdots + a_{n-1}\mathbf{I}$$

$$\mathbf{A}^{-1} = -\frac{1}{a_n}\,[a_0\mathbf{A}^{n-1} + a_1\mathbf{A}^{n-2} + a_2\mathbf{A}^{n-3} + \cdots + a_{n-1}\mathbf{I}]$$

CONCEPT CHECK

1. Find $\mathbf{A}^{-1}$ where

$$\mathbf{A} = \begin{bmatrix} 2 & 0 & 2 \\ 0 & 3 & 1 \\ 2 & 0 & 3 \end{bmatrix}$$

using Cayley–Hamilton theorem.

2.5 Diagonalization and Powers of a Matrix

So far we have emphasized properties of *eigenvalues.* We now turn to general properties of eigenvectors. Eigenvectors of an $n \times n$ matrix A may (or may not!) form a basis for R^n. If we are interested in a transformation $\mathbf{y} = \mathbf{A}\mathbf{x}$, such an "**eigenbasis**" (basis of eigenvectors) – if it exists – is of great advantage because then we can represent any **x** in R^n uniquely as a linear combination of the eigenvectors $\mathbf{x}_1, \ldots, \mathbf{x}_n$, say,

$$\mathbf{x} = c_1\mathbf{x}_1 + c_2\mathbf{x}_2 + \cdots + c_n\mathbf{x}_n$$

Denoting the corresponding (not necessarily distinct) eigenvalues of the matrix **A** by $\lambda_1, \ldots, \lambda_n$, we have $\mathbf{A}\mathbf{x}_j = \lambda_j\mathbf{x}_j$, so that we simply obtain

$$\begin{aligned} \mathbf{y} = \mathbf{Ax} &= \mathbf{A}(c_1\mathbf{x}_1 + \cdots + c_n\mathbf{x}_n) \\ &= c_1\mathbf{Ax}_1 + \cdots + c_n\mathbf{Ax}_n \\ &= c_1\lambda_1\mathbf{x}_1 + \cdots + c_n\lambda_n\mathbf{x}_n \end{aligned} \tag{2.15}$$

This shows that we have decomposed the complicated action of **A** on an arbitrary vector **x** into a sum of simple actions (multiplication by scalars) on the eigenvectors of **A**. This is the point of an eigenbasis.

Now if the n eigenvalues are all different, we do obtain a basis.

THEOREM 5 (Basis of Eigenvectors): *If an $n \times n$ matrix* **A** *has n* ***distinct*** *eigenvalues, then* **A** *has a basis of eigenvectors* $\mathbf{x}_1, \ldots, \mathbf{x}_n$ *for* R^n.

PROOF: All we have to show is that $\mathbf{x}_1, \ldots, \mathbf{x}_n$ are linearly independent. Suppose they are not. Let r be the largest integer such that $\{\mathbf{x}_1, \ldots, \mathbf{x}_r\}$ is a linearly independent set. Then $r < n$ and the set $\{\mathbf{x}_1, \ldots, \mathbf{x}_r, \mathbf{x}_{r+1}\}$ is linearly dependent. Thus there are scalars $c_1, \ldots, c_{r+1}$, not all zero, such that

$$c_1\mathbf{x}_1 + \cdots + c_{r+1}\mathbf{x}_{r+1} = \mathbf{0} \tag{2.16}$$

Multiplying both sides by **A** and using $\mathbf{Ax}_j = \lambda_j\mathbf{x}_j$, we obtain

$$c_1\lambda_1\mathbf{x}_1 + \cdots + c_{r+1}\lambda_{r+1}\mathbf{x}_{r+1} = \mathbf{0} \tag{2.17}$$

To get rid of the last term, we subtract λ_{r+1} times (2.16) from this, obtaining

$$c_1(\lambda_1 - \lambda_{r+1})\mathbf{x}_1 + \cdots + c_r(\lambda_r - \lambda_{r+1})\mathbf{x}_r = \mathbf{0}$$

Here $c_1(\lambda_1 - \lambda_{r+1}) = 0, \ldots, c_r(\lambda_r - \lambda_{r+1}) = 0$ since $\{x_1, \ldots, x_r\}$ is linearly independent. Hence $c_1 = \cdots = c_r = 0$, since all the eigenvalues are distinct. But with this, (2.16) reduces to $c_{r+1}\mathbf{x}_{r+1} = \mathbf{0}$, hence $c_{r+1} = 0$, since $\mathbf{x}_{r+1} \neq \mathbf{0}$ (an eigenvector!). This contradicts the fact that not all scalars in (2.16) are zero. Hence the conclusion of the theorem must hold. ■

Example 8 Eigenbasis. Nondistinct Eigenvalues. Nonexistence

The matrix

$$\mathbf{A} = \begin{bmatrix} 5 & 3 \\ 3 & 5 \end{bmatrix}$$

has a basis of eigenvectors

$$\begin{bmatrix} 1 \\ 1 \end{bmatrix}, \quad \begin{bmatrix} 1 \\ -1 \end{bmatrix}$$

corresponding to the eigenvalues $\lambda_1 = 8$, $\lambda_2 = 2$. Even if not all n eigenvalues are different, a matrix **A** may still provide an eigenbasis for R^n.

On the other hand, **A** ***may not have enough linearly independent eigenvectors to make up a basis.***

$$\mathbf{A} = \begin{bmatrix} 0 & 1 \\ 0 & 0 \end{bmatrix}$$

and has only one eigenvector

$$\begin{bmatrix} k \\ 0 \end{bmatrix} \qquad (k \neq 0, \text{ arbitrary})$$

Actually, eigenbases exist under much more general conditions than those in Theorem 5. An important case is the following.

THEOREM 6 (Symmetric Matrices): *A symmetric matrix has an orthonormal basis of eigenvectors for R^n.* The proof is out of scope of this book.

Example 9 Orthonormal Basis of Eigenvectors

The first matrix in Example 8 is symmetric, and an orthonormal basis of eigenvectors is $[1/\sqrt{2} \quad 1/\sqrt{2}]^T, [1/\sqrt{2} \quad -1/\sqrt{2}]^T$.

Diagonalization of Matrices

Eigenbases also play a role in reducing a matrix **A** to a diagonal matrix whose entries are the eigenvalues of **A**. This is done by a "similarity transformation," which is defined as follows.

DEFINITION **(Similar Matrices. Similarity Transformation)** An $n \times n$ matrix $\hat{\mathbf{A}}$ is called **similar** to an $n \times n$ matrix **A** if

$$\hat{\mathbf{A}} = \mathbf{P}^{-1}\mathbf{A}\mathbf{P} \tag{2.18}$$

for some (nonsingular!) $n \times n$ matrix **P**. This transformation, which gives $\hat{\mathbf{A}}$ from **A**, is called a **similarity transformation.**

The key property of this transformation is that it preserves the eigenvalues of **A**.

THEOREM 7 (Eigenvalues and Eigenvectors of Similar Matrices): *If $\hat{\mathbf{A}}$ is similar to* **A**, *then $\hat{\mathbf{A}}$ has the same eigenvalues as* **A**. *Furthermore, if* **x** *is an eigenvector of* **A**, then $\mathbf{y} = \mathbf{P}^{-1}\mathbf{x}$ *is an eigenvector of $\hat{\mathbf{A}}$ corresponding to the same eigenvalue.*

PROOF: From $\mathbf{Ax} = \lambda\mathbf{x}$ (λ is an eigenvalue, $\mathbf{x} \neq 0$) we get $\mathbf{P}^{-1}\mathbf{Ax} = \lambda\mathbf{P}^{-1}\mathbf{x}$. Now $\mathbf{I} = \mathbf{PP}^{-1}$. By this ***"identity trick"*** the previous equation gives

$$\mathbf{P}^{-1}\mathbf{Ax} = \mathbf{P}^{-1}\mathbf{AIx} = \mathbf{P}^{-1}\mathbf{APP}^{-1}\mathbf{x} = \hat{\mathbf{A}}(\mathbf{P}^{-1}\mathbf{x}) = \lambda\mathbf{P}^{-1}\mathbf{x}$$

Hence λ is an eigenvalue of $\hat{\mathbf{A}}$ and $\mathbf{P}^{-1}\mathbf{x}$ a corresponding eigenvector. Indeed, $\mathbf{P}^{-1}\mathbf{x} = 0$ would give $\mathbf{x} = \mathbf{Ix} = \mathbf{PP}^{-1}\mathbf{x} = \mathbf{P} \times 0 = 0$, contradicting $\mathbf{x} \neq 0$. ■

Example 10 Eigenvalues and Vectors of Similar Matrices

Let

$$\mathbf{A} = \begin{bmatrix} 6 & -3 \\ 4 & -1 \end{bmatrix} \quad \text{and} \quad \mathbf{P} = \begin{bmatrix} 1 & 3 \\ 1 & 4 \end{bmatrix}$$

Then

$$\hat{\mathbf{A}} = \begin{bmatrix} 4 & -3 \\ -1 & 1 \end{bmatrix} \begin{bmatrix} 6 & -3 \\ 4 & -1 \end{bmatrix} \begin{bmatrix} 1 & 3 \\ 1 & 4 \end{bmatrix} = \begin{bmatrix} 3 & 0 \\ 0 & 2 \end{bmatrix}$$

Here $\mathbf{P}^{-1}$ was obtained from (1.51) in Chapter 1 with det $\mathbf{P} = 1$. We see that $\hat{\mathbf{A}}$ has the eigenvalues $\lambda_1 = 3$, $\lambda_2 = 2$. The characteristic equation of $\mathbf{A}$ is $(6 - \lambda)(-1 - \lambda) + 12 = \lambda^2 - 5\lambda + 6 = 0$. It has the roots (the eigenvalues of A) $\lambda_1 = 3$, $\lambda_2 = 2$, confirming the first part of Theorem 7.

We confirm the second part. From the first component of $(\mathbf{A} - \lambda\mathbf{I})\mathbf{x} = \mathbf{0}$ we have $(6 - \lambda)x_1 - 3x_2 = 0$. For $\lambda = 3$ this gives $3x_1 - 3x_2 = 0$, say, $\mathbf{x}_1 = [1 \quad 1]^T$. For $\lambda = 2$ it gives $4x_1 - 3x_2 = 0$, say, $\mathbf{x}_2 = [3 \quad 4]^T$. In Theorem 3 we thus have

$$\mathbf{y}_1 = \mathbf{P}^{-1}\mathbf{x}_1 = \begin{bmatrix} 4 & -3 \\ -1 & 1 \end{bmatrix} \begin{bmatrix} 1 \\ 1 \end{bmatrix} = \begin{bmatrix} 1 \\ 0 \end{bmatrix}, \quad \mathbf{y}_2 = \mathbf{P}^{-1}\mathbf{x}_2 = \begin{bmatrix} 4 & -3 \\ -1 & 1 \end{bmatrix} \begin{bmatrix} 3 \\ 4 \end{bmatrix} = \begin{bmatrix} 0 \\ 1 \end{bmatrix}$$

Indeed, these are eigenvectors of the diagonal matrix $\hat{\mathbf{A}}$.

Perhaps we see that $\mathbf{x}_1$ and $\mathbf{x}_2$ are the columns of $\mathbf{P}$. This suggests the general method of transforming a matrix $\mathbf{A}$ to diagonal form $\mathbf{D}$ by using $\mathbf{P} = \mathbf{X}$, the matrix with eigenvectors as columns.

THEOREM 8 (Diagonalization of a Matrix): *If an $n \times n$ matrix* $\mathbf{A}$ *has a basis of eigenvectors, then*

$$\mathbf{D} = \mathbf{X}^{-1}\mathbf{A}\mathbf{X} \tag{2.19}$$

is diagonal, with the eigenvalues of $\mathbf{A}$ *as the entries on the main diagonal. Here* $\mathbf{X}$ *is the matrix with these eigenvectors as column vectors. Also,*

$$\mathbf{D}^m = \mathbf{X}^{-1}\mathbf{A}^m\mathbf{X} \qquad (m = 2, 3, \ldots) \tag{2.20}$$

PROOF: Let $\mathbf{x}_1, \ldots, \mathbf{x}_n$ constitute a basis of eigenvectors of $\mathbf{A}$ for R^n. Let the corresponding eigenvalues of $\mathbf{A}$ be $\lambda_1, \ldots, \lambda_n$, respectively, so that $\mathbf{A}\mathbf{x}_1 = \lambda_1\mathbf{x}_1, \ldots, \mathbf{A}\mathbf{x}_n = \lambda_n\mathbf{x}_n$. Then $\mathbf{X} = [\mathbf{x}_1 \quad \cdots \quad \mathbf{x}_n]$ has rank n. Hence $\mathbf{X}^{-1}$ exists by Theorem 17 in Chapter 1. We claim that

$$\mathbf{A}\mathbf{X} = \mathbf{A}[\mathbf{x}_1 \quad \cdots \quad \mathbf{x}_n] = [\mathbf{A}\mathbf{x}_1 \quad \cdots \quad \mathbf{A}\mathbf{x}_n] = [\lambda_1\mathbf{x}_1 \quad \cdots \quad \lambda_n\mathbf{x}_n] = \mathbf{X}\mathbf{D} \tag{2.21}$$

where $\mathbf{D}$ is the diagonal matrix as in (2.20). The fourth equality in (2.21) follows by direct calculation. (Try it for $n = 2$ and then for general n.) The third equality uses $\mathbf{A}\mathbf{x}_k = \lambda_k\mathbf{x}_k$. The second equality results if we note that the first column of $\mathbf{AX}$ is $\mathbf{A}$ times the first column of $\mathbf{X}$, and so on. For instance, when $n = 2$ and we write $\mathbf{x}_1 = [x_{11} \quad x_{21}]^T$, $\mathbf{x}_2 = [x_{12} \quad x_{22}]^T$, we have

$$\mathbf{A}\mathbf{X} = \mathbf{A}[\mathbf{x}_1\ \mathbf{x}_2] = \begin{bmatrix} a_{11} & a_{12} \\ a_{21} & a_{22} \end{bmatrix} \begin{bmatrix} x_{11} & x_{12} \\ x_{21} & x_{22} \end{bmatrix}$$

$$= \begin{bmatrix} a_{11}x_{11} + a_{12}x_{21} & a_{11}x_{12} + a_{12}x_{22} \\ a_{21}x_{11} + a_{22}x_{21} & a_{21}x_{12} + a_{22}x_{22} \end{bmatrix} = [\mathbf{A}\mathbf{x}_1\ \mathbf{A}\mathbf{x}_2]$$

Column 1 Column 2

If we multiply (2.21) by $\mathbf{X}^{-1}$ from the left, we obtain (2.19). Since (2.19) is a similarity transformation, Theorem 7 implies that $\mathbf{D}$ has the same eigenvalues as $\mathbf{A}$. Equation (2.20) follows if we note that

$$\mathbf{D}^2 = \mathbf{D}\mathbf{D} = \mathbf{X}^{-1}\mathbf{A}\mathbf{X}\mathbf{X}^{-1}\mathbf{A}\mathbf{X} = \mathbf{X}^{-1}\mathbf{A}\mathbf{A}\mathbf{X} = \mathbf{X}^{-1}\mathbf{A}^2\mathbf{X}, \qquad \text{etc.}$$

■

Example 11 Diagonalization

Diagonalize

$$\mathbf{A} = \begin{bmatrix} 7.3 & 0.2 & -3.7 \\ -11.5 & 1.0 & 5.5 \\ 17.7 & 1.8 & -9.3 \end{bmatrix}$$

Solution: The characteristic determinant gives the characteristic equation $-\lambda^3 - \lambda^2 + 12\lambda = 0$. The roots (eigenvalues of **A**) are $\lambda_1 = 3$, $\lambda_2 = -4$, $\lambda_3 = 0$. By the Gauss elimination applied to $(\mathbf{A} - \lambda\mathbf{I})\mathbf{x} = \mathbf{0}$ with $\lambda = \lambda_1, \lambda_2, \lambda_3$ we find eigenvectors and then $\mathbf{X}^{-1}$ by the Gauss–Jordan elimination. The results are

$$\begin{bmatrix} -1 \\ 3 \\ -1 \end{bmatrix}, \begin{bmatrix} 1 \\ -1 \\ 3 \end{bmatrix}, \begin{bmatrix} 2 \\ 1 \\ 4 \end{bmatrix}, \quad \mathbf{X} = \begin{bmatrix} -1 & 1 & 2 \\ 3 & -1 & 1 \\ -1 & 3 & 4 \end{bmatrix}, \quad \mathbf{X}^{-1} = \begin{bmatrix} -0.7 & 0.2 & 0.3 \\ -1.3 & -0.2 & 0.7 \\ 0.8 & 0.2 & -0.2 \end{bmatrix}$$

Calculating **AX** and multiplying by $\mathbf{X}^{-1}$ from the left, we thus obtain

$$\mathbf{D} = \mathbf{X}^{-1}\mathbf{A}\mathbf{X} = \begin{bmatrix} -0.7 & 0.2 & 0.3 \\ -1.3 & -0.2 & 0.7 \\ 0.8 & 0.2 & -0.2 \end{bmatrix} \begin{bmatrix} -3 & -4 & 0 \\ 9 & 4 & 0 \\ -3 & -12 & 0 \end{bmatrix} = \begin{bmatrix} 3 & 0 & 0 \\ 0 & -4 & 0 \\ 0 & 0 & 0 \end{bmatrix}$$

CONCEPT CHECK

1. Find an eigenbasis (a basis of eigenvectors) and diagonalize. (Show the details.)

(a) $\begin{bmatrix} 3 & 2 \\ 2 & 6 \end{bmatrix}$ (b) $\begin{bmatrix} 0 & 16 \\ 4 & 0 \end{bmatrix}$

(c) $\begin{bmatrix} 5 & 1 \\ 1 & 5 \end{bmatrix}$ (d) $\begin{bmatrix} 3 & 2 \\ -5 & -4 \end{bmatrix}$

(e) $\begin{bmatrix} 1.0 & 6.0 \\ 1.5 & 1.0 \end{bmatrix}$ (f) $\begin{bmatrix} 2 & 7 \\ 6 & -9 \end{bmatrix}$

(g) $\begin{bmatrix} 1 & 0 & 1 \\ 0 & 3 & 2 \\ 0 & 0 & 2 \end{bmatrix}$ (h) $\begin{bmatrix} -6 & -6 & 10 \\ -5 & -5 & 5 \\ -9 & -9 & 13 \end{bmatrix}$

(i) $\begin{bmatrix} 3 & 10 & -15 \\ -18 & 39 & 9 \\ -24 & 40 & -15 \end{bmatrix}$

2. **(Orthonormal basis)** Illustrate Theorem 6 with further examples.

3. **(No basis)** Find further 2×2 and 3×3 matrices without eigenbases.

4. **PROJECT. Similarity of Matrices.** Similarity is basic, for instance in designing numeric methods.

(a) **Trace.** By definition, the **trace** of an $n \times n$ *matrix* $\mathbf{A} = [a_{jk}]$ is the sum of the diagonal entries,

$$\text{trace } \mathbf{A} = a_{11} + a_{22} + \cdots + a_{nn}$$

Show that the trace equals the sum of the eigenvalues, each counted as often as its algebraic multiplicity indicates. Illustrate this with the matrices in Probs. 1(a), 1(c), 1(e), 1(g), 1(i).

(b) **Trace of product.** Let $\mathbf{B} = [b_{jk}]$ be $n \times n$. Show that similar matrices have equal traces, by first proving

$$\text{trace } \mathbf{AB} = \sum_{i=1}^{n}\sum_{l=1}^{n} a_{il}b_{il} = \text{trace } \mathbf{BA}$$

(c) Find a relationship between $\hat{\mathbf{A}}$ in (2.18) and $\tilde{\mathbf{A}} = \mathbf{PAP}^{-1}$.

(d) **Diagonalization.** What can you do in (2.18) if you want to change the order of the eigenvalues in **D**, for instance, interchange $d_{11} = \lambda_1$ and $d_{22} = \lambda_2$?

2.6 Eigenvalue Problem

An **eigenvalue** or **characteristic value** (or *latent root*) of a given $n \times n$ matrix $\mathbf{A} = [a_{jk}]$ is a real or complex number λ such that the vector equation

$$\mathbf{Ax} = \lambda\mathbf{x} \tag{2.22}$$

has a nontrivial solution, that is, a solution $\mathbf{x} \neq \mathbf{0}$, which is then called an **eigenvector** or **characteristic vector** of **A** corresponding to that eigenvalue λ. The set of all eigenvalues of **A** is called the **spectrum** of **A**. Equation (2.22) can be written

$$(\mathbf{A} - \lambda\mathbf{I})\mathbf{x} = \mathbf{0} \tag{2.23}$$

where **I** is the $n \times n$ unit matrix. This homogeneous system has a nontrivial solution if and only if the **characteristic determinant** det $(\mathbf{A} - \lambda\mathbf{I})$ is 0.

THEOREM 9 (Eigenvalues): *The eigenvalues of* **A** *are the solutions* λ *of the* ***characteristic equation***

$$\det(\mathbf{A} - \lambda\mathbf{I}) = \begin{vmatrix} a_{11} - \lambda & a_{12} & \cdots & a_{1n} \\ a_{21} & a_{22} - \lambda & \cdots & a_{2n} \\ \cdot & \cdot & \cdots & \cdot \\ a_{n1} & a_{n2} & \cdots & a_{nn} - \lambda \end{vmatrix} = 0 \tag{2.24}$$

Developing the characteristic determinant, we obtain the **characteristic polynomial** of **A**, which is of degree n in λ. Hence **A** has at least one and at most n numerically different eigenvalues. If **A** is real, so are the coefficients of the characteristic polynomial. By familiar algebra it follows that then the roots (the eigenvalues of **A**) are ***real*** or ***complex conjugates*** in pairs.

We shall usually denote the eigenvalues of **A** by

$$\lambda_1, \lambda_2, \ldots, \lambda_n$$

with the understanding that some (or all) of them may be equal. The sum of these n eigenvalues equals the sum of the entries on the main diagonal of **A**, called the **trace of A**; thus

$$\text{trace } \mathbf{A} = \sum_{j=1}^{n} a_{jj} = \sum_{k=1}^{n} \lambda_k \tag{2.25}$$

Also, the product of the eigenvalues equals the determinant of **A**,

$$\det \mathbf{A} = \lambda_1\lambda_2 \cdots \lambda_n \tag{2.26}$$

Both formulas follow from the product representation of the characteristic polynomial, which we denote by $f(\lambda)$,

$$f(\lambda) = (-1)^n(\lambda - \lambda_1)(\lambda - \lambda_2) \cdots (\lambda - \lambda_n)$$

If we take equal factors together and denote the *numerically distinct* eigenvalues of **A** by $\lambda_1, \ldots, \lambda_r$ ($r \leqq n$), then the product becomes

$$f(\lambda) = (-1)^n(\lambda - \lambda_1)^{m_1}(\lambda - \lambda_2)^{m_2} \cdots (\lambda - \lambda_r)^{m_r} \tag{2.27}$$

The exponent m_j is called the **algebraic multiplicity** of λ_j. The maximum number of linearly independent eigenvectors corresponding to λ_j is called the **geometric multiplicity** of λ_j. It is equal to or smaller than m_j.

DEFINITION A subspace S of R^n or C^n (if **A** is complex) is called an **invariant subspace of A** if for every **v** in S the vector **Av** is also in S. **Eigenspaces** of **A** are important invariant subspaces of **A**.

DEFINITION An $n \times n$ matrix **B** is called **similar to A** if there is a nonsingular $n \times n$ matrix **T** such that

$$\mathbf{B} = \mathbf{T}^{-1}\mathbf{A}\mathbf{T} \tag{2.28}$$

Similarity is important for the following reason.

THEOREM 10 (Similar Matrices): *Similar matrices have the same eigenvalues. If* **x** *is an eigenvector of* **A**, *then* $\mathbf{y} = \mathbf{T}^{-1}\mathbf{x}$ *is an eigenvector of* **B** *in* (2.28) *corresponding to the same eigenvalue.*

Another theorem that has various applications in numerics is as follows.

THEOREM 11 (Spectral Shift): *If* **A** *has the eigenvalues* $\lambda_1, \ldots, \lambda_n$, *then* $\mathbf{A} - k\mathbf{I}$ *with arbitrary* k *has the eigenvalues* $\lambda_1 - k, \ldots, \lambda_n - k$.

This theorem is a special case of the following **spectral mapping theorem**.

THEOREM 12 (Polynomial Matrices): *If* λ *is an eigenvalue of* **A**, *then*

$$q(\lambda) = \alpha_s\lambda^s + \alpha_{s-1}\lambda^{s-1} + \cdots + \alpha_1\lambda + \alpha_0$$

is an eigenvalue of the **polynomial matrix**

$$q(\mathbf{A}) = \alpha_s\mathbf{A}^s + \alpha_{s-1}\mathbf{A}^{s-1} + \cdots + \alpha_1\mathbf{A} + \alpha_0\mathbf{I}$$

PROOF: $\mathbf{A}\mathbf{x} = \lambda\mathbf{x}$ implies $\mathbf{A}^2\mathbf{x} = \mathbf{A}\lambda x = \lambda\mathbf{A}\mathbf{x} = \lambda^2\mathbf{x}$, $\mathbf{A}^3\mathbf{x} = \lambda^3\mathbf{x}$, etc. Thus

$$\begin{aligned} q(\mathbf{A})\mathbf{x} &= (\alpha_s\mathbf{A}^s + \alpha_{s-1}\mathbf{A}^{s-1} + \cdots)\mathbf{x} \\ &= \alpha_s\mathbf{A}^s\mathbf{x} + \alpha_{s-1}\mathbf{A}^{s-1}\mathbf{x} + \cdots \\ &= \alpha_s\lambda^s\mathbf{x} + \alpha_{s-1}\lambda^{s-1}\mathbf{x} + \cdots = q(\lambda)\mathbf{x} \end{aligned}$$

■

The eigenvalues of important special matrices can be characterized as follows.

THEOREM 13 (Special Matrices): *The eigenvalues of Hermitian matrices (i.e.,* $\bar{\mathbf{A}}^{\mathrm{T}} = \mathbf{A}$*), hence of real symmetric matrices (i.e.,* $\mathbf{A}^{\mathrm{T}} = \mathbf{A}$*) are real. The eigenvalues of skew-Hermitian matrices (i.e.,* $\bar{\mathbf{A}}^{\mathrm{T}} = -\mathbf{A}$*), hence of real skew-symmetric matrices (i.e.,* $\mathbf{A}^{\mathrm{T}} = -\mathbf{A}$*) are pure imaginary or* 0. *The eigenvalues of unitary matrices (i.e.,* $\bar{\mathbf{A}}^{\mathrm{T}} = \mathbf{A}^{-1}$*), hence of orthogonal matrices (i.e.,* $\mathbf{A}^{\mathrm{T}} = \mathbf{A}^{-1}$*), have absolute value* 1.

The **choice of a numeric method** for matrix eigenvalue problems depends essentially on two circumstances, on the kind of matrix (real symmetric, real general, complex, sparse, or full) and on the kind of information to be obtained, that is, whether one wants to know all eigenvalues or merely specific ones, for instance, the largest eigenvalue, whether eigenvalues and eigenvectors are wanted, and so on. It is clear that we cannot enter into a systematic discussion of all these and further possibilities that arise in practice, but we shall concentrate on some basic aspects and methods that will give us a general understanding of this fascinating field.

Inclusion of Matrix Eigenvalues

The whole numerics for matrix eigenvalues is motivated by the fact that except for a few trivial cases we cannot determine eigenvalues *exactly* by a finite process because these values are the roots of a polynomial of nth degree. Hence we must mainly use iteration.

In this section we state a few general theorems that give approximations and error bounds for eigenvalues. Our matrices will continue to be real, but since (nonsymmetric) matrices may have complex eigenvalues, complex numbers will play a (very modest) role in this section.

The important theorem by Gerschgorin gives a region consisting of closed circular disks in the complex plane and including all the eigenvalues of a given matrix. Indeed, for each $j = 1, \ldots, n$ the inequality (2.29) in the theorem determines a closed circular disk in the complex λ-plane with center a_{jj} and radius given by the right side of (2.29); and Theorem 1 states that each of the eigenvalues of $\mathbf{A}$ lies in one of these n disks.

THEOREM 14 (Gerschgorin's Theorem): *Let λ be an eigenvalue of an arbitrary $n \times n$ matrix $\mathbf{A} = [a_{jk}]$. Then for some integer j $(1 \leqq j \leqq n)$ we have*

$$|a_{jj} - \lambda| \leqq |a_{j1}| + |a_{j2}| + \cdots + |a_{j,j-1}| + |a_{j,j+1}| + \cdots + |a_{jn}| \tag{2.29}$$

PROOF: Let $\mathbf{x}$ be an eigenvector corresponding to an eigenvalue λ of $\mathbf{A}$. Then

$$\mathbf{Ax} = \lambda\mathbf{x} \qquad \text{or} \qquad (\mathbf{A} - \lambda\mathbf{I})\mathbf{x} = \mathbf{0} \tag{2.30}$$

Let x_j be a component of $\mathbf{x}$ that is largest in absolute value. Then we have $|x_m/x_j| \leqq 1$ for $m = 1, \ldots, n$. The vector equation (2.30) is equivalent to a system of n equations for the n components of the vectors on both sides. The jth of these n equations with j as just indicated is

$$a_{j1}x_1 + \cdots + a_{j,j-1}x_{j-1} + (a_{jj} - \lambda)x_j + a_{j,j+1}x_{j+1} + \cdots + a_{jn}x_n = 0$$

Division by x_j (which cannot be zero; why?) and reshuffling terms gives

$$a_{jj} - \lambda = -a_{j1}\frac{x_1}{x_j} - \cdots - a_{j,j-1}\frac{x_{j-1}}{x_j} - a_{j,j+1}\frac{x_{j+1}}{x_j} - \cdots - a_{jn}\frac{x_n}{x_j}$$

By taking absolute values on both sides of this equation, applying the triangle inequality $|a + b| \leqq |a| + |b|$ (where a and b are any complex numbers), and observing that because of the choice of j (which is crucial!), $|x_1/x_j| \leqq 1, \ldots, |x_n/x_j| \leqq 1$, we obtain (2.29), and the theorem is proved. ■

Example 12 Gerschgorin's Theorem

For the eigenvalues of the matrix

$$\mathbf{A} = \begin{bmatrix} 0 & 1/2 & 1/2 \\ 1/2 & 5 & 1 \\ 1/2 & 1 & 1 \end{bmatrix}$$

we get the Gerschgorin disks (Figure 2)

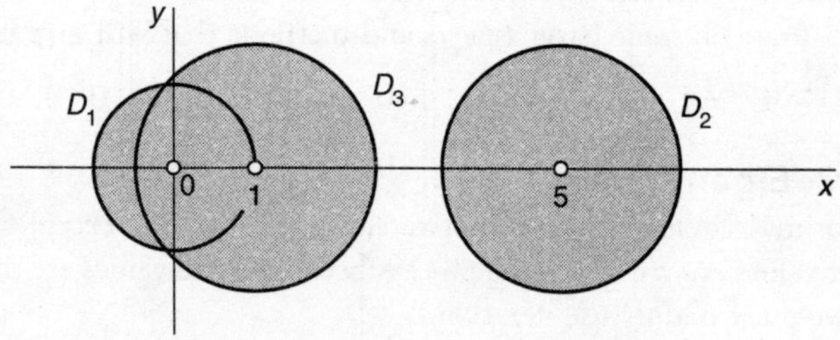

Figure 2 Gerschgorin disks in Example 12.

D_1: Center 0, radius 1; D_2: Center 5, radius 1.5; D_3: Center 1, radius 1.5

The centers are the main diagonal entries of **A**. These would be the eigenvalues of **A** if **A** were diagonal. We can take these values as crude approximations of the unknown eigenvalues (3D values) $\lambda_1 = -0.209$, $\lambda_2 = 5.305$, $\lambda_3 = 0.904$ (verify this); then the radii of the disks are corresponding error bounds. Since **A** is symmetric, it follows that the spectrum of **A** must actually lie in the intervals [−1, 2.5] and [3.5, 6.5].

It is interesting that here the Gerschgorin disks form two disjoint sets, namely, $D_1 \cup D_3$, which contains two eigenvalues, and D_2, which contains one eigenvalue. This is typical, as the following theorem shows.

THEOREM 15 (Extension of Gerschgorin's Theorem): *If p Gerschgorin disks form a set S that is disjoint from the n − p other disks of a given matrix* **A**, *then S contains precisely p eigenvalues of* **A** (*each counted with its algebraic multiplicity*).

Idea of the Proof: Set $\mathbf{A} = \mathbf{B} + \mathbf{C}$, where **B** is the diagonal matrix with entries a_{jj}, and apply Theorem 14 to $\mathbf{A}_t = \mathbf{B} + t\mathbf{C}$ with real t growing from 0 to 1. ■

Example 13 Another Application of Gerschgorin's Theorem. Similarity

Suppose that we have diagonalized a matrix by some numeric method that left us with some off-diagonal entries of size 10^{-5}, say,

$$\mathbf{A} = \begin{bmatrix} 2 & 10^{-5} & 10^{-5} \\ 10^{-5} & 2 & 10^{-5} \\ 10^{-5} & 10^{-5} & 4 \end{bmatrix}$$

What can we conclude about deviations of the eigenvalues from the main diagonal entries?

Solution: By Theorem 15, one eigenvalue must lie in the disk of radius 2×10^{-5} centered at 4 and two eigenvalues (or an eigenvalue of algebraic multiplicity 2) in the disk of radius 2×10^{-5} centered at 2. Actually, since the matrix is symmetric, these eigenvalues must lie in the intersections of these disks and the real axis.

We show how an isolated disk can always be reduced in size by a similarity transformation. The matrix

$$\mathbf{B} = \mathbf{T}^{-1}\mathbf{A}\mathbf{T} = \begin{bmatrix} 1 & 0 & 0 \\ 0 & 1 & 0 \\ 0 & 0 & 10^{-5} \end{bmatrix} \begin{bmatrix} 2 & 10^{-5} & 10^{-5} \\ 10^{-5} & 2 & 10^{-5} \\ 10^{-5} & 10^{-5} & 4 \end{bmatrix} \begin{bmatrix} 1 & 0 & 0 \\ 0 & 1 & 0 \\ 0 & 0 & 10^{5} \end{bmatrix}$$

$$= \begin{bmatrix} 2 & 10^{-5} & 1 \\ 10^{-5} & 2 & 1 \\ 10^{-10} & 10^{-10} & 4 \end{bmatrix}$$

is similar to **A**. Hence by Theorem 10 it has the same eigenvalues as **A**. From Row 3 we get the smaller disk of radius 2×10^{-10}. Note that the other disks got bigger, approximately by a factor of 10^5. And in choosing **T** we have to watch that the new disks do not overlap with the disk whose size we want to decrease.

By definition, a **diagonally dominant** matrix $\mathbf{A} = [a_{jk}]$ is an $n \times n$ matrix such that

$$|a_{jj}| \geqq \sum_{k \neq j} |a_{jk}| \quad j = 1, \ldots, n \tag{2.31}$$

where we sum over all off-diagonal entries in Row j. The matrix is said to be **strictly diagonally dominant** if $>$ in (2.31) for all j. Use Theorem 14 to prove the following basic property.

THEOREM 16 (Strict Diagonal Dominance): *Strictly diagonally dominant matrices are nonsingular.*

Further Inclusion Theorems

An **inclusion theorem** is a theorem that specifies a set which contains at least one eigenvalue of a given matrix. Thus, Theorems 14 and 15 are inclusion theorems; they even include the whole spectrum. We now discuss some famous theorems that yield further inclusions of eigenvalues. We state the first two of these without proofs (which would exceed the level of this book).

THEOREM 17 (Schur's Theorem): *Let* $\mathbf{A} = [a_{jk}]$ *be an* $n \times n$ *matrix. Then for each of its eigenvalues* $\lambda_1, \ldots, \lambda_n$,

$$|\lambda_m|^2 \leqq \sum_{i=1}^{n} |\lambda_i|^2 \leqq \sum_{j=1}^{n}\sum_{k=1}^{n} |a_{jk}|^2 \qquad \text{(Schur's inequality)} \tag{2.32}$$

In (2.32) *the second equality sign holds if and only if* **A** *is such that*

$$\bar{\mathbf{A}}^{\mathrm{T}}\mathbf{A} = \mathbf{A}\bar{\mathbf{A}}^{\mathrm{T}} \tag{2.33}$$

Matrices that satisfy (2.33) are called **normal matrices**. It is not difficult to see that Hermitian, skew-Hermitian, and unitary matrices are normal, and so are real symmetric, skew-symmetric, and orthogonal matrices. These matrices will be discussed in the next chapter.

Example 14 Bounds for Eigenvalues Obtained from Schur's Inequality

For the matrix

$$A = \begin{bmatrix} 26 & -2 & 2 \\ 2 & 21 & 4 \\ 4 & 2 & 28 \end{bmatrix}$$

we obtain from Schur's inequality $|\lambda| \leqq \sqrt{1949} = 44.1475$. You may verify that the eigenvalues are 30, 25, and 20. Thus $30^2 + 25^2 + 20^2 = 1925 < 1949$; in fact, **A** is not normal.

The preceding theorems are valid for every real or complex square matrix. Other theorems hold for special classes of matrices only. Famous is the following.

THEOREM 18 (Perron's Theorem): *Let* **A** *be a real* $n \times n$ *matrix whose entries are all positive. Then* **A** *has a positive real eigenvalue* $\lambda = \rho$ *of multiplicity* 1. *The corresponding eigenvector can be chosen with all components positive. (The other eigenvalues are less than* ρ *in absolute value.)*

The theorem also holds for matrices with *nonnegative* real entries (**"Perron–Frobenius Theorem"**) provided **A** is **irreducible**, that is, it cannot be brought to the following form by interchanging rows and columns; here **B** and **F** are square and **0** is a zero matrix.

$$\begin{bmatrix} \mathbf{B} & \mathbf{C} \\ \mathbf{0} & \mathbf{F} \end{bmatrix}$$

Perron's theorem has various applications, for instance, in economics. It is interesting that one can obtain from it a theorem that gives a numeric algorithm:

THEOREM 19 (Collatz Inclusion Theorem): *Let* $\mathbf{A} = [a_{jk}]$ *be a real* $n \times n$ *matrix whose elements are all positive. Let* **x** *be any real vector whose components* $x_1, \ldots, x_n$ *are positive, and let* $y_1, \ldots, y_n$ *be the components of the vector* $\mathbf{y} = \mathbf{Ax}$. *Then the closed interval on the real axis bounded by the smallest and the largest of the n quotients* $q_j = y_j/x_j$ *contains at least one eigenvalue of* **A**.

PROOF: We have

$$\mathbf{Ax} = \mathbf{y} \quad \text{or} \quad \mathbf{y} - \mathbf{Ax} = \mathbf{0} \tag{2.34}$$

The transpose $\mathbf{A}^T$ satisfies the conditions of Theorem 18. Hence $\mathbf{A}^T$ has a positive eigenvalue λ and, corresponding to this eigenvalue, an eigenvector **u** whose components u_j are all positive. Thus $\mathbf{A}^T\mathbf{u} = \lambda\mathbf{u}$, and by taking the transpose we obtain $\mathbf{u}^T\mathbf{A} = \lambda\mathbf{u}^T$. From this and (2.34) we have

$$\mathbf{u}^T(\mathbf{y} - \mathbf{Ax}) = \mathbf{u}^T\mathbf{y} - \mathbf{u}^T\mathbf{Ax} = \mathbf{u}^T\mathbf{y} - \lambda\mathbf{u}^T\mathbf{x} = \mathbf{u}^T(\mathbf{y} - \lambda\mathbf{x}) = 0$$

or written out

$$\sum_{j=1}^{n} u_j(y_j - \lambda x_j) = 0$$

Since all the components u_j are positive, it follows that

$$\begin{aligned} &y_j - \lambda x_j \geqq 0, && \text{that is,} && q_j \geqq \lambda && \text{for at least one } j \\ &y_j - \lambda x_j \leqq 0, && \text{that is,} && q_j \leqq \lambda && \text{for at least one } j \end{aligned} \tag{2.35}$$

Since **A** and $\mathbf{A}^T$ have the same eigenvalues, λ is an eigenvalue of **A**, and from (2.35) the statement of the theorem follows. ■

Example 15 Bounds for Eigenvalues from Collatz's Theorem. Iteration

For a given matrix $\mathbf{A}$ with positive entries we choose an $\mathbf{x} = \mathbf{x}_0$ and **iterate**, that is, we compute $\mathbf{x}_1 = \mathbf{A}\mathbf{x}_0$. $\mathbf{x}_2 = \mathbf{A}\mathbf{x}_1, \ldots, \mathbf{x}_{20} = \mathbf{A}\mathbf{x}_{19}$. In each step, taking $\mathbf{x} = \mathbf{x}_j$ and $\mathbf{y} = \mathbf{A}\mathbf{x}_j = \mathbf{x}_{j+1}$ we compute an inclusion interval by Collatz's theorem. This gives (6S)

$$\mathbf{A} = \begin{bmatrix} 0.49 & 0.02 & 0.22 \\ 0.02 & 0.28 & 0.20 \\ 0.22 & 0.20 & 0.40 \end{bmatrix}, \quad \mathbf{x}_0 = \begin{bmatrix} 1 \\ 1 \\ 1 \end{bmatrix}, \quad \mathbf{x}_1 = \begin{bmatrix} 0.73 \\ 0.50 \\ 0.82 \end{bmatrix}, \quad \mathbf{x}_2 = \begin{bmatrix} 0.5481 \\ 0.3186 \\ 0.5886 \end{bmatrix}, \ldots,$$

$$\mathbf{x}_{19} = \begin{bmatrix} 0.00216309 \\ 0.00108155 \\ 0.00216309 \end{bmatrix}, \quad \mathbf{x}_{20} = \begin{bmatrix} 0.00155743 \\ 0.000778713 \\ 0.00155743 \end{bmatrix}$$

and the intervals $0.5 \leqq \lambda \leqq 0.82$, $0.3186/0.50 = 0.6372 \leqq \lambda \leqq 0.5481/0.73 = 0.750822$, etc. These intervals have length

j	1	2	3	10	15	20
Length	0.32	0.113622	0.0539835	0.0004217	0.0000132	0.0000004

Using the characteristic polynomial, you may verify that the eigenvalues of $\mathbf{A}$ are 0.72, 0.36, 0.09, so that those intervals include the largest eigenvalue, 0.72. Their lengths decreased with j, so that the iteration was worthwhile. The reason will appear in the next section, where we discuss an iteration method for eigenvalues.

CONCEPT CHECK

1. Find and sketch disks or intervals that contain the eigenvalues.

(a) $\begin{bmatrix} 6 & -1 & 0 \\ -1 & 7 & 1 \\ 0 & 1 & 8 \end{bmatrix}$

(b) $\begin{bmatrix} 5 & 10^{-2} & 10^{-2} \\ 10^{-2} & 8 & 10^{-2} \\ 10^{-2} & 10^{-2} & 9 \end{bmatrix}$

(c) $\begin{bmatrix} 5 & -2 & 2 \\ 2 & 0 & 4 \\ 4 & 2 & 7 \end{bmatrix}$

(d) $\begin{bmatrix} 1+i & 0.3 & 0.5i \\ 0.3i & -3+2i & 0.1 \\ 0.1i & 0.2 & 4-i \end{bmatrix}$

(e) $\begin{bmatrix} 4i & 0.1i & 1+i \\ 0.1i & 0 & 0 \\ -1+i & 0 & 9i \end{bmatrix}$

(f) $\begin{bmatrix} 10 & 0.1 & -0.2 \\ 0.1 & 6 & 0 \\ -0.2 & 0 & 3 \end{bmatrix}$

2. **(Similarity)** Find $\mathbf{T}^{-1}\mathbf{A}\mathbf{T}$ such that in Prob. 1(b) the radius of the Gerschgorin circle with center 5 is reduced by a factor 1/100.
3. By what integer factor can you at most reduce the Gerschgorin circle with center 3 in Prob. 1(f)?
4. If a symmetric $n \times n$ matrix $\mathbf{A} = [a_{jk}]$ has been diagonalized except for small off-diagonal entries of size 10^{-6}, what can you say about the eigenvalues?

5. **(Extended Gerschgorin theorem)** Prove Theorem 15.
6. Prove Theorem 16.
7. **(Normal matrices)** Show that Hermitian, skew-Hermitian, and unitary matrices (hence real symmetric, skew-symmetric, and orthogonal matrices) are normal. Why is this of practical interest?
8. **[Spectral radius $\rho(\mathbf{A})$]** Show that $\rho(\mathbf{A})$ cannot be greater than the row sum norm of **A**.
9. **(Eigenvalues on the circle)** Illustrate with a 2×2 matrix that an eigenvalue may very well lie on a Gerschgorin circle (so that Gerschgorin disks can generally not be replaced with smaller disks without losing the inclusion property).
10. Use (2.32) to obtain an upper bound for the spectral radius:

    (a) In Prob. 1(a)

    (b) In Prob. 1(f)

    (c) In Prob. 1(c)
11. Apply Theorem 19, choosing the given vectors as vectors **x**.

    (a) $\begin{bmatrix} 10 & 1 & 1 \\ 1 & 9 & 3 \\ 1 & 3 & 11 \end{bmatrix}, \begin{bmatrix} 1 \\ 1 \\ 1 \end{bmatrix}, \begin{bmatrix} 1 \\ 2 \\ 1 \end{bmatrix}, \begin{bmatrix} 2 \\ 3 \\ 2 \end{bmatrix}$

    (b) $\begin{bmatrix} 4 & 2 & 2 \\ 2 & 4 & 2 \\ 2 & 2 & 4 \end{bmatrix}, \begin{bmatrix} 1 \\ 2 \\ 1 \end{bmatrix}, \begin{bmatrix} 1 \\ 1 \\ 1 \end{bmatrix}$

2.7 Power Method for Eigenvalues

A simple standard procedure for computing approximate values of the eigenvalues of an $n \times n$ matrix $\mathbf{A} = [a_{jk}]$ is the **power method**. In this method we start from any vector $\mathbf{x}_0$ ($\neq \mathbf{0}$) with n components and compute successively

$$\mathbf{x}_1 = \mathbf{A}\mathbf{x}_0, \qquad \mathbf{x}_2 = \mathbf{A}\mathbf{x}_1, \qquad \ldots, \qquad \mathbf{x}_s = \mathbf{A}\mathbf{x}_{s-1}$$

For simplifying notation, we denote $\mathbf{x}_{s-1}$ by $\mathbf{x}$ and $\mathbf{x}_s$ by $\mathbf{y}$, so that $\mathbf{y} = \mathbf{A}\mathbf{x}$. The method applies to any $n \times n$ matrix **A** that has **a dominant eigenvalue** (a λ such that $|\lambda|$ is greater than the absolute values of the other eigenvalues). If **A** is *symmetric*, it also gives the error bound (2.36), in addition to the approximation (2.35).

THEOREM 20 (Power Method, Error Bounds): *Let* **A** *be an $n \times n$ real symmetric matrix. Let* **x** ($\neq \mathbf{0}$) *be any real vector with n components. Furthermore, let*

$$\mathbf{y} = \mathbf{A}\mathbf{x}, \qquad m_0 = \mathbf{x}^{\mathrm{T}}\mathbf{x}, \qquad m_1 = \mathbf{x}^{\mathrm{T}}\mathbf{y}, \qquad m_2 = \mathbf{y}^{\mathrm{T}}\mathbf{y}$$

Then the quotient

$$q = \frac{m_1}{m_0} \quad \textit{(Rayleigh quotient)} \tag{2.36}$$

is an approximation for an eigenvalue λ of **A** *(usually that which is greatest in absolute value, but no general statements are possible). Furthermore, if we set $q = \lambda - \epsilon$, so that ϵ is the error of q, then*

$$|\epsilon| \leq \delta = \sqrt{\frac{m_2}{m_0} - q^2} \tag{2.37}$$

PROOF: δ^2 denotes the radicand in (2.37). Since $m_1 = qm_0$ by (2.36), we have

$$(\mathbf{y} - q\mathbf{x})^{\mathrm{T}}(\mathbf{y} - q\mathbf{x}) = m_2 - 2qm_1 + q^2 m_0 = m_2 - q^2 m_0 = \delta^2 m_0 \tag{2.38}$$

Since **A** is real symmetric, it has an orthogonal set of n real unit eigenvectors $\mathbf{z}_1, \ldots, \mathbf{z}_n$ corresponding to the eigenvalues $\lambda_1, \ldots, \lambda_n$, respectively (some of which may be equal). Then **x** has a representation of the form

$$\mathbf{x} = a_1\mathbf{z}_1 + \cdots + a_n\mathbf{z}_n$$

Now $\mathbf{A}\mathbf{z}_1 = \lambda_1\mathbf{z}_1$, etc., and we obtain

$$\mathbf{y} = \mathbf{A}\mathbf{x} = a_1\lambda_1\mathbf{z}_1 + \cdots + a_n\lambda_n\mathbf{z}_n$$

and, since the $\mathbf{z}_j$ are orthogonal unit vectors,

$$m_0 = \mathbf{x}^{\mathrm{T}}\mathbf{x} = a_1^{\,2} + \cdots + a_n^{\,2} \tag{2.39}$$

It follows that in (2.38),

$$\mathbf{y} - q\mathbf{x} = a_1(\lambda_1 - q)\mathbf{z}_1 + \cdots + a_n(\lambda_n - q)\mathbf{z}_n$$

Since the $\mathbf{z}_j$ are orthogonal unit vectors, we thus obtain from (2.38)

$$\delta^2 m_0 = (\mathbf{y} - q\mathbf{x})^{\mathrm{T}}(\mathbf{y} - q\mathbf{x}) = a_1^{\,2}(\lambda_1 - q)^2 + \cdots + a_n^{\,2}(\lambda_n - q)^2 \tag{2.40}$$

Now let λ_c be an eigenvalue of **A** to which q is closest, where c suggests "closest." Then $(\lambda_c - q)^2 \leqq (\lambda_j - q)^2$ for $j = 1, \ldots, n$. From this and (2.40) we obtain the inequality

$$\delta^2 m_0 \geqq (\lambda_c - q)^2(a_1^{\,2} + \cdots + a_n^{\,2}) = (\lambda_c - q)^2 m_0$$

Dividing by m_0, taking square roots, and recalling the meaning of δ^2 gives

$$\delta = \sqrt{\frac{m_2}{m_0} - q^2} \geqq |\lambda_c - q|$$

This shows that δ is a bound for the error ϵ of the approximation q of an eigenvalue of **A** and completes the proof. ■

The main advantage of the method is its simplicity. And it can handle *sparse matrices* too large to store as a full square array. Its disadvantage is its possibly slow convergence. From the proof of Theorem 20 we see that the speed of convergence depends on the ratio of the dominant eigenvalue to the next in absolute value (2:1 in Example 16).

If we want a convergent sequence of **eigenvectors**, then at the beginning of each step we **scale** the vector, say, by dividing its components by an absolutely largest one, as in Example 16, as follows.

Example 16 Application of Theorem 20. Scaling

For the symmetric matrix **A** in Example 15 and $\mathbf{x}_0 = [1 \quad 1 \quad 1]^{\mathrm{T}}$ we obtain from (2.36) and (2.37) and the indicated scaling

$$\mathbf{A} = \begin{bmatrix} 0.49 & 0.02 & 0.22 \\ 0.02 & 0.28 & 0.20 \\ 0.22 & 0.20 & 0.40 \end{bmatrix}, \quad \mathbf{x}_0 = \begin{bmatrix} 1 \\ 1 \\ 1 \end{bmatrix}, \quad \mathbf{x}_1 = \begin{bmatrix} 0.890244 \\ 0.609756 \\ 1 \end{bmatrix}, \quad \mathbf{x}_2 = \begin{bmatrix} 0.931193 \\ 0.541284 \\ 1 \end{bmatrix}$$

$$\mathbf{x}_5 = \begin{bmatrix} 0.990663 \\ 0.504682 \\ 1 \end{bmatrix}, \quad \mathbf{x}_{10} = \begin{bmatrix} 0.999707 \\ 0.500146 \\ 1 \end{bmatrix}, \quad \mathbf{x}_{15} = \begin{bmatrix} 0.999991 \\ 0.500005 \\ 1 \end{bmatrix}$$

Here $\mathbf{Ax}_0 = [0.73 \quad 0.5 \quad 0.82]^T$, scaled to $\mathbf{x}_1 = [0.73/0.82 \quad 0.5/0.82 \quad 1]^T$, etc. The dominant eigenvalue is 0.72, an eigenvector $[1 \quad 0.5 \quad 1]^T$. The corresponding q and δ are computed each time before the next scaling. Thus in the first step,

$$q = \frac{m_1}{m_0} = \frac{\mathbf{x}_0^T \mathbf{Ax}_0}{\mathbf{x}_0^T \mathbf{x}_0} = \frac{2.05}{3} = 0.683333$$

$$\delta = \left(\frac{m_2}{m_0} - q^2\right)^{1/2} = \left(\frac{(\mathbf{Ax}_0)^T \mathbf{Ax}_0}{\mathbf{x}_0^T \mathbf{x}_0} - q^2\right)^{1/2} = \left(\frac{1.4553}{3} - q^2\right)^{1/2} = 0.134743$$

This gives the following values of q, δ, and the error $\epsilon = 0.72 - q$ (calculations with 10**D**, rounded to 6**D**):

j	1	2	5	10
q	0.683333	0.716048	0.719944	0.720000
δ	0.134743	0.038887	0.004499	0.000141
ϵ	0.036667	0.003952	0.000056	5×10^{-8}

The error bounds are much larger than the actual errors. This is typical, although the bounds cannot be improved; that is, for special symmetric matrices they agree with the errors.

Spectral shift, the transition from **A** to $\mathbf{A} - k\mathbf{I}$, shifts every eigenvalue by $-k$. Although finding a good k can hardly be made automatic, it may be helped by some other method or small preliminary computational experiments. In Example 16, Gerschgorin's theorem gives $-0.02 \leqq \lambda \leqq 0.82$ for the whole spectrum (verify!). Shifting by -0.4 might be too much (then $-0.42 \leqq \lambda \leqq 0.42$), so let us try -0.2.

Example 17 Power Method with Spectral Shift

For $\mathbf{A} - 0.2\mathbf{I}$ with **A** as in Example 16 we obtain the following substantial improvements (where the index 1 refers to Example 16 and the index 2 to the present example).

j	1	2	5	10
δ_1	0.134743	0.038887	0.004499	0.000141
δ_2	0.134743	0.034474	0.000693	1.8×10^{-6}
ϵ_1	0.036667	0.003952	0.000056	5×10^{-8}
ϵ_2	0.036667	0.002477	1.3×10^{-6}	9×10^{-12}

CONCEPT CHECK

1. Apply the power method (3 steps) with scaling, using $\mathbf{x}_0 = [1 \quad 1]^T$ or $[1 \quad 1 \quad 1]^T$ or $[1 \quad 1 \quad 1 \quad 1]^T$, as applicable. Give Rayleigh quotients and error bounds. Show the details of your work.

(a) $\begin{bmatrix} 3.5 & 2.0 \\ 2.0 & 0.5 \end{bmatrix}$ (b) $\begin{bmatrix} 0.6 & 0.8 \\ 0.8 & -0.6 \end{bmatrix}$

(c) $\begin{bmatrix} 5 & 2 \\ 2 & 7 \end{bmatrix}$ (d) $\begin{bmatrix} -2 & 2 & 3 \\ 2 & 1 & 6 \\ 3 & 6 & -2 \end{bmatrix}$

(e) $\begin{bmatrix} 2 & -1 & 1 \\ -1 & 3 & 2 \\ 1 & 2 & 3 \end{bmatrix}$ (f) $\begin{bmatrix} 0 & 4 & 0 & 1 \\ 4 & -1 & 2 & 8 \\ 0 & 2 & 3 & 2 \\ 1 & 8 & 2 & -2 \end{bmatrix}$

(g) $\begin{bmatrix} 5 & 1 & 0 & 0 \\ 1 & 3 & 1 & 0 \\ 0 & 1 & 3 & 1 \\ 0 & 0 & 1 & 5 \end{bmatrix}$

2. **(Optimality of δ)** In Prob. 1(b) choose $\mathbf{x}_0 = [3 \quad -1]^T$ and show that $q = 0$ and $\delta = 1$ for all steps and that the eigenvalues are ± 1, so that the interval $[q - \delta, q + \delta]$ cannot be shortened in general! Experiment with other $\mathbf{x}_0$.
3. Prove that if $\mathbf{x}$ is an eigenvector, then $\delta = 0$ in (2.37). Give two examples.
4. **(Rayleigh quotient)** Why does q generally approximate the eigenvalue of greatest absolute value? When will q be a good approximation?
5. **(Spectral shift, smallest eigenvalue)** In Prob. 1(e) set $\mathbf{B} = \mathbf{A} - 3\mathbf{I}$ (as perhaps suggested by the diagonal entries) and try whether you may get a sequence of q's converging to an eigenvalue of $\mathbf{A}$ that *is smallest* (not largest) in absolute value. Use $\mathbf{x}_0 = [1 \quad 1 \quad 1]^T$. Do 8 steps. Verify that $\mathbf{A}$ has the spectrum $\{0, 3, 5\}$.

2.8 Modal and Spectral Matrices

Modal Matrices

DEFINITION A square matrix obtained by using as its columns any set of linearly independent eigenvectors of a matrix **A** is called a modal matrix of **A**, and may be denoted by **M**.

Notes: (i) There are infinitely many modal matrices for a given matrix A, since any multiple of an eigenvector is also an eigenvector. (ii) A set of normalized eigenvectors is sometimes convenient to use.

When using normalized eigenvectors, the modal matrix may be denoted by **N**. For an $n \times n$ matrix **A**, there are 2^n possibilities for **N** since each of the n columns has two possibilities.

Spectral Matrices

DEFINITION If $\lambda_1, \lambda_2, \lambda_3, \ldots, \lambda_n$ are the eigenvalues of an $n \times n$ matrix **A**, then the diagonal matrix

$$\begin{bmatrix} \lambda_1 & 0 & 0 & \cdot & \cdot & 0 \\ 0 & \lambda_2 & 0 & \cdot & \cdot & 0 \\ 0 & 0 & \lambda_3 & \cdot & \cdot & 0 \\ \cdot & \cdot & \cdot & \cdot & \cdot & \cdot \\ 0 & 0 & 0 & \cdot & \cdot & \lambda_n \end{bmatrix}$$

is called the spectral matrix of **A** and is denoted by **S**.

CONCEPT CHECK

1. Define spectral matrices. [JNTU Aug./Sep. 2006, May 2005]
2. Define modal matrices.

Solved Problems

Q 1. Find the eigenvalues and eigenvectors of

$$\begin{bmatrix} 1 & -6 & -4 \\ 0 & 4 & 2 \\ 0 & -6 & -3 \end{bmatrix}$$

[JNTU Aug./Sep. 2008, May/June 2008]

Answer: Consider, $|\mathbf{A} - \lambda\mathbf{I}| = 0$. This implies

$$\begin{vmatrix} 1 & -6 & -4 \\ 0 & 4 & 2 \\ 0 & -6 & -3 \end{vmatrix} - \lambda \begin{vmatrix} 1 & 0 & 0 \\ 0 & 1 & 0 \\ 0 & 0 & 1 \end{vmatrix} = 0$$

$$\begin{vmatrix} 1-\lambda & -6 & -4 \\ 0 & 4-\lambda & 2 \\ 0 & -6 & -3-\lambda \end{vmatrix} = 0$$

(a)

$$(1-\lambda)\,[(4-\lambda)(-3-\lambda) + 12] = 0$$

$$(1-\lambda) = 0$$

$$1-\lambda = 0$$

$$-\lambda = -1$$

$$\lambda = 1$$

(b)

$$(4-\lambda)(-3-\lambda) + 12 = 0$$

$$-12 - 4\lambda + 3\lambda + \lambda^2 + 12 = 0$$

$$\lambda^2 - \lambda = 0$$

$$\lambda(\lambda - 1) = 0$$

$$\lambda = 0,\ \lambda = 1$$

Therefore the eigenvalues are 1, 0, 1.

Eigenvector for $\lambda = 0$

$|\mathbf{A} - \lambda\mathbf{I}|\ \mathbf{X} = 0$ which implies $\mathbf{AX} = 0$. That is

$$\begin{bmatrix} 1 & -6 & -4 \\ 0 & 4 & 2 \\ 0 & -6 & -3 \end{bmatrix} \begin{bmatrix} x_1 \\ x_2 \\ x_3 \end{bmatrix} = 0$$

$$
\begin{aligned}
x_1 - 6x_2 - 4x_3 &= 0 \\
4x_2 + 2x_3 &= 0 \\
2x_3 &= -4x_2 \\
x_3 &= -2x_2 \\
-6x_2 - 3x_3 &= 0
\end{aligned}
$$

Let $x_2 = k$, then

$$
\begin{aligned}
x_3 &= -2k \\
x_1 - 6k - 4(-2k) &= 0 \\
x_1 - 6k + 8k &= 0 \\
x_1 + 2k &= 0 \\
x_1 &= -2k
\end{aligned}
$$

Therefore the eigenvector is

$$
\begin{bmatrix} x_1 \\ x_2 \\ x_3 \end{bmatrix} = k \begin{bmatrix} -2 \\ +1 \\ -2 \end{bmatrix}
$$

Eigenvector for $\lambda = 1$

$$
\begin{bmatrix} 0 & -6 & -4 \\ 0 & 3 & 2 \\ 0 & -6 & -4 \end{bmatrix} \begin{bmatrix} x_1 \\ x_2 \\ x_3 \end{bmatrix} = 0
$$

The above matrix reduces to the following equations,

$$
\begin{aligned}
x_1 &= 0 \\
-6x_2 - 4x_3 &= 0 \\
3x_2 + 2x_3 &= 0 \\
x_3 &= \frac{-3}{2} x_2
\end{aligned}
$$

Let $x_2 = k$, therefore

$$
x_3 = \frac{-3}{2} k
$$

Therefore eigenvector for $\lambda = 1$ is

$$
\begin{bmatrix} 0 \\ 1 \\ -3/2 \end{bmatrix}
$$

Q 2. Find the characteristic roots of the matrix

$$A = \begin{bmatrix} 2 & 2 & 1 \\ 1 & 3 & 1 \\ 1 & 2 & 2 \end{bmatrix}$$

[JNTU May/June 2008]

OR

Find the eigenvalues and the corresponding eigenvectors of the matrix

$$\begin{bmatrix} 2 & 2 & 1 \\ 1 & 3 & 1 \\ 1 & 2 & 2 \end{bmatrix}$$

[JNTU Aug./Sep. 2007]

Answer: Let

$$A = \begin{bmatrix} 2 & 2 & 1 \\ 1 & 3 & 1 \\ 2 & 2 & 2 \end{bmatrix}$$

The characteristic equation of **A** is given by

$$|A - \lambda I| = 0$$

This implies

$$\begin{vmatrix} 2 & 2 & 1 \\ 1 & 3 & 1 \\ 1 & 2 & 2 \end{vmatrix} - \begin{vmatrix} \lambda & 0 & 0 \\ 0 & \lambda & 0 \\ 0 & 0 & \lambda \end{vmatrix} = 0$$

$$\begin{vmatrix} 2-\lambda & 2 & 1 \\ 1 & 3-\lambda & 1 \\ 1 & 2 & 2-\lambda \end{vmatrix} = 0$$

Expanding we get

$$(2 - \lambda)\,[(3 - \lambda)(2 - \lambda) - 2] - 2\,[(2 - \lambda) - 1] + [2 - (3 - \lambda)] = 0$$

$$(2 - \lambda)\,[6 - 3\lambda - 2\lambda + \lambda^2 - 2] - 4 + 2\lambda + 2 + 2 - 3 + \lambda = 0$$

$$(2 - \lambda)\,[\lambda^2 - 5\lambda + 4] + 3\lambda - 3 = 0$$

$$(2 - \lambda)\,[(\lambda - 1)(\lambda - 4)] + 3(\lambda - 1) = 0$$

$$(\lambda - 1)\,[(2 - \lambda)(\lambda - 4) + 3] = 0$$

$$(\lambda - 1)\,[6\lambda - \lambda^2 - 8 + 3] = 0$$

$$(\lambda - 1)\,[-\lambda^2 + 6\lambda - 5] = 0$$

$$(\lambda - 1)\,[\lambda^2 - 6\lambda + 5] = 0$$

$$(\lambda - 1)\,[(\lambda - 1)(\lambda - 5)] = 0$$

Therefore $\lambda = 1, 1, 5$ are the required eigenvalues. Now, we calculate the respective eigenvectors.

Eigenvectors corresponding to $\lambda = 1$

$$(\mathbf{A} - \lambda \mathbf{I})\,\mathbf{X} = 0$$

$$\begin{bmatrix} 1 & 2 & 1 \\ 1 & 2 & 1 \\ 1 & 2 & 1 \end{bmatrix}\begin{bmatrix} x_1 \\ x_2 \\ x_3 \end{bmatrix} = \begin{bmatrix} 0 \\ 0 \\ 0 \end{bmatrix}$$

$$x_1 + 2x_2 + x_3 = 0$$

Assume, $x_2 = k_1$, $x_3 = k_2$ [since $\rho(\mathbf{A} - \lambda_1) = 1$]; then, $x_1 = -2k_1 - k_2$

$$\begin{bmatrix} x_1 \\ x_2 \\ x_3 \end{bmatrix} = \begin{bmatrix} -2k_1 & -k_2 \\ k_1 & \\ & k_2 \end{bmatrix} = k_1\begin{bmatrix} -2 \\ 1 \\ 0 \end{bmatrix} + k_2\begin{bmatrix} -1 \\ 0 \\ 1 \end{bmatrix}$$

Therefore

$$\mathbf{X}_1 = \begin{bmatrix} -2 \\ 1 \\ 0 \end{bmatrix},\ \mathbf{X}_2 = \begin{bmatrix} -1 \\ 0 \\ 1 \end{bmatrix}$$

are eigenvectors corresponding to $\lambda = 1$.

Eigenvectors corresponding to $\lambda = 5$

$$\begin{bmatrix} 2-\lambda & 2 & 1 \\ 1 & 3-\lambda & 1 \\ 1 & 2 & 2-\lambda \end{bmatrix}\begin{bmatrix} x_1 \\ x_2 \\ x_3 \end{bmatrix} = \begin{bmatrix} 0 \\ 0 \\ 0 \end{bmatrix}$$

$$\begin{bmatrix} -3 & 2 & 1 \\ 1 & -2 & 1 \\ 1 & 2 & -3 \end{bmatrix}\begin{bmatrix} x_1 \\ x_2 \\ x_3 \end{bmatrix} = \begin{bmatrix} 0 \\ 0 \\ 0 \end{bmatrix}$$

Expanding we get

$$-3x_1 + 2x_2 + x_3 = 0 \qquad (2.41)$$

$$x_1 - 2x_2 + x_3 = 0 \qquad (2.42)$$

$$x_1 + 2x_2 - 3x_3 = 0 \qquad (2.43)$$

Solving (2.41) and (2.42) we get, $x_1 = x_3$. Assume $x_2 = k_3$ [since $\rho(\mathbf{A} - \lambda_1) = 2$]. Therefore

$$x_1 = x_3 = k_3$$

$$\mathbf{X}_3 = \begin{bmatrix} k_3 \\ k_3 \\ k_3 \end{bmatrix}$$

The simplest vector for $\lambda = 5$ is

$$\mathbf{X}_3 = k_3 \begin{bmatrix} 1 \\ 1 \\ 1 \end{bmatrix}$$

Therefore the eigenvectors for eigenvalues $\lambda = 1, 1, 5$ are as follows:

$$\mathbf{X}_1 = \begin{bmatrix} -2 \\ 1 \\ 0 \end{bmatrix}, \mathbf{X}_2 = \begin{bmatrix} -1 \\ 0 \\ 1 \end{bmatrix}, \mathbf{X}_3 = \begin{bmatrix} 1 \\ 1 \\ 1 \end{bmatrix}$$

Q 3. Determine the characteristic roots and the corresponding characteristic vectors of the matrix

$$\mathbf{A} = \begin{bmatrix} 8 & -6 & 2 \\ -6 & 7 & -4 \\ 2 & -4 & 3 \end{bmatrix}$$

[JNTU May/June 2008]

OR

Find the eigenvalues and the corresponding eigenvectors of the matrix

$$\mathbf{A} = \begin{bmatrix} 8 & -6 & 2 \\ -6 & 7 & -4 \\ 2 & -4 & 3 \end{bmatrix}$$

[JNTU May/June 2006, Nov. 2004]

Answer: Let λ be the eigenvalues of **A**. The characteristic equation of the matrix **A** is

$$|\mathbf{A} - \lambda\mathbf{I}| = 0$$

$$\begin{vmatrix} 8-\lambda & -6 & 2 \\ -6 & 7-\lambda & -4 \\ 2 & -4 & 3-\lambda \end{vmatrix} = 0$$

Expanding the matrix we get

$$(8-\lambda)\,[(3-\lambda)\,(7-\lambda) - 16] + 6[-6(3-\lambda) + 8] + 2[24 - 2(7-\lambda)] = 0$$

$$(8-\lambda)\,[21 - 10\lambda + \lambda^2 - 16] + 6[-18 + 6\lambda + 8] + 2[10 + 2\lambda] = 0$$

$$(8-\lambda)\,(\lambda^2 - 10\lambda + 5) + 6(6\lambda - 10) + 2(10 + 2\lambda) = 0$$

$$8\lambda^2 - 80\lambda + 40 - \lambda^3 + 10\lambda^2 - 5\lambda + 36\lambda - 60 + 20 + 4\lambda = 0$$

$$-\lambda^3 + 18\lambda^2 - 45\lambda = 0$$

$$-\lambda\,(\lambda^2 - 18\lambda + 45) = 0$$

$$\lambda\,(\lambda - 3)\,(\lambda - 15) = 0$$

Therefore $\lambda = 0$, 3, and 15 are the eigenvalues of the matrix **A**. Let

$$\mathbf{X} = \begin{bmatrix} x_1 \\ x_2 \\ x_3 \end{bmatrix}$$

be the required eigenvector then

$$(\mathbf{A} - \lambda \mathbf{I})\,\mathbf{X} = 0$$

$$\begin{bmatrix} 8-\lambda & -6 & 2 \\ -6 & 7-\lambda & -4 \\ 2 & -4 & 3-\lambda \end{bmatrix}\begin{bmatrix} x_1 \\ x_2 \\ x_3 \end{bmatrix} = \begin{bmatrix} 0 \\ 0 \\ 0 \end{bmatrix} \tag{2.44}$$

Eigenvectors corresponding to $\lambda = 0$

Putting $\lambda = 0$ in (2.44), we get

$$\begin{bmatrix} 8 & -6 & 2 \\ -6 & 7 & -4 \\ 2 & -4 & 3 \end{bmatrix}\begin{bmatrix} x_1 \\ x_2 \\ x_3 \end{bmatrix} = \begin{bmatrix} 0 \\ 0 \\ 0 \end{bmatrix}$$

$$\begin{bmatrix} 2 & -4 & 3 \\ -6 & 7 & -4 \\ 8 & -6 & 2 \end{bmatrix}\begin{bmatrix} x_1 \\ x_2 \\ x_3 \end{bmatrix} = \begin{bmatrix} 0 \\ 0 \\ 0 \end{bmatrix} \quad \text{Row 1} \leftrightarrow \text{Row 3}$$

$$\begin{bmatrix} 2 & -4 & 3 \\ 0 & -5 & 5 \\ 0 & 10 & -10 \end{bmatrix}\begin{bmatrix} x_1 \\ x_2 \\ x_3 \end{bmatrix} = \begin{bmatrix} 0 \\ 0 \\ 0 \end{bmatrix} \quad \begin{matrix} \text{Row 2} \to \text{Row 2} + 3\ \text{Row 1} \\ \text{Row 3} \to \text{Row 3} - 4\ \text{Row 1} \end{matrix}$$

$$\begin{bmatrix} 2 & -4 & 3 \\ 0 & -5 & 5 \\ 0 & 0 & 0 \end{bmatrix}\begin{bmatrix} x_1 \\ x_2 \\ x_3 \end{bmatrix} = \begin{bmatrix} 0 \\ 0 \\ 0 \end{bmatrix} \quad \text{Row 3} \to \text{Row 3} + 2\ \text{Row 2}$$

Rank of the coefficient matrix $(r) = 2$

Number of unknowns $(n) = 3$

Therefore, the number of free variables is given by $= n - r = 3 - 2 = 1$. The reduced system of equation is

$$2x_1 - 4x_2 + 3x_3 = 0$$
$$-5x_2 + 5x_3 = 0$$

Let $x_3 = c_1$, then $x_2 = c_1$. Hence, the first equation gives

$$2x_1 - 4c_1 + 3c_1 = 0$$
$$2x_1 - c_1 = 0$$
$$x_1 = \frac{c_1}{2}$$

Hence the given vectors corresponding to $\lambda = 0$ are given as follows:

$$\mathbf{X}_1 = \begin{bmatrix} c_1/2 \\ c_1 \\ c_1 \end{bmatrix} = c_1 \begin{bmatrix} 1/2 \\ 1 \\ 1 \end{bmatrix}$$

Eigenvectors corresponding to $\lambda = 3$

Putting $\lambda = 3$ in (2.44), we get

$$\begin{bmatrix} 5 & -6 & 2 \\ -6 & 4 & -4 \\ 2 & -4 & 0 \end{bmatrix} \begin{bmatrix} x_1 \\ x_2 \\ x_3 \end{bmatrix} = \begin{bmatrix} 0 \\ 0 \\ 0 \end{bmatrix}$$

$$\begin{bmatrix} 5 & -6 & 2 \\ -6 & 4 & -4 \\ 1 & -2 & 0 \end{bmatrix} \begin{bmatrix} x_1 \\ x_2 \\ x_3 \end{bmatrix} = \begin{bmatrix} 0 \\ 0 \\ 0 \end{bmatrix} \quad \text{Row } 3 \rightarrow \text{Row } 3 \times (1/2)$$

$$\begin{bmatrix} 1 & -2 & 0 \\ -6 & 4 & -4 \\ 5 & -6 & 2 \end{bmatrix} \begin{bmatrix} x_1 \\ x_2 \\ x_3 \end{bmatrix} = \begin{bmatrix} 0 \\ 0 \\ 0 \end{bmatrix} \quad \text{Row } 1 \leftrightarrow \text{Row } 3$$

$$\begin{bmatrix} 1 & -2 & 0 \\ 0 & -8 & -4 \\ 0 & 4 & 2 \end{bmatrix} \begin{bmatrix} x_1 \\ x_2 \\ x_3 \end{bmatrix} = \begin{bmatrix} 0 \\ 0 \\ 0 \end{bmatrix} \quad \begin{matrix} \text{Row } 2 \rightarrow \text{Row } 2 + 6 \text{ Row } 1 \\ \text{Row } 3 \rightarrow \text{Row } 3 - 5 \text{ Row } 1 \end{matrix}$$

$$\begin{bmatrix} 1 & -2 & 0 \\ 0 & -8 & -4 \\ 0 & 0 & 0 \end{bmatrix} \begin{bmatrix} x_1 \\ x_2 \\ x_3 \end{bmatrix} = \begin{bmatrix} 0 \\ 0 \\ 0 \end{bmatrix} \quad \text{Row } 3 \rightarrow 2 \text{ Row } 3 + \text{Row } 2$$

$$\begin{bmatrix} 1 & -2 & 0 \\ 0 & 2 & 1 \\ 5 & 0 & 0 \end{bmatrix} \begin{bmatrix} x_1 \\ x_2 \\ x_3 \end{bmatrix} = \begin{bmatrix} 0 \\ 0 \\ 0 \end{bmatrix} \quad \text{Row } 2 \rightarrow \text{Row } 2 \times (-1/4)$$

$r = \text{Rank } (\mathbf{A} - \lambda\mathbf{I}) = 2$

Number of unknowns $(n) = 3$

Number of free variables $= n - r = 3 - 2 = 1$

The reduced system of equations equivalent to (2.44) is

$$x_1 - 2x_2 = 0 \Rightarrow x_1 = 2x_2$$

$$2x_2 + x_3 = 0 \Rightarrow 2x_2 = -x_3$$

Let $x_2 = c_2$. Then

$$x_3 = -2c_2 \text{ and } x_1 = 2c_2$$

Therefore

$$\mathbf{X}_2 = \begin{bmatrix} x_1 \\ x_2 \\ x_3 \end{bmatrix} = \begin{bmatrix} 2c_2 \\ c_2 \\ -2c_2 \end{bmatrix} = c_2 \begin{bmatrix} 2 \\ 1 \\ -2 \end{bmatrix}$$

Eigenvectors corresponding to $\lambda = 15$
Putting $\lambda = 15$ in (2.44) we get,

$$\begin{bmatrix} -7 & -6 & 2 \\ -6 & -8 & -4 \\ 2 & -4 & -12 \end{bmatrix} \begin{bmatrix} x_1 \\ x_2 \\ x_3 \end{bmatrix} = \begin{bmatrix} 0 \\ 0 \\ 0 \end{bmatrix}$$

$$\begin{bmatrix} 1 & -2 & -6 \\ -6 & -8 & -4 \\ -7 & -6 & 2 \end{bmatrix} \begin{bmatrix} x_1 \\ x_2 \\ x_3 \end{bmatrix} = \begin{bmatrix} 0 \\ 0 \\ 0 \end{bmatrix} \quad \text{Row 3} \rightarrow \text{Row 3} \times (1/2), \ \text{Row 1} \leftrightarrow \text{Row 3}$$

$$\begin{bmatrix} 1 & -2 & -6 \\ 0 & -20 & -40 \\ 0 & -20 & -40 \end{bmatrix} \begin{bmatrix} x_1 \\ x_2 \\ x_3 \end{bmatrix} = \begin{bmatrix} 0 \\ 0 \\ 0 \end{bmatrix} \quad \begin{matrix} \text{Row 2} \rightarrow \text{Row 2} + 6 \text{ Row 1} \\ \text{Row 3} \rightarrow \text{Row 3} + 7 \text{ Row 1} \end{matrix}$$

$$\begin{bmatrix} 1 & -2 & -6 \\ 0 & -20 & -40 \\ 0 & 0 & 0 \end{bmatrix} \begin{bmatrix} x_1 \\ x_2 \\ x_3 \end{bmatrix} = \begin{bmatrix} 0 \\ 0 \\ 0 \end{bmatrix} \quad \text{Row 3} \rightarrow \text{Row 3} - \text{Row 2}$$

$$\begin{bmatrix} 1 & -2 & -6 \\ 0 & 1 & 2 \\ 0 & 0 & 0 \end{bmatrix} \begin{bmatrix} x_1 \\ x_2 \\ x_3 \end{bmatrix} = \begin{bmatrix} 0 \\ 0 \\ 0 \end{bmatrix} \quad \text{Row 2} \rightarrow \text{Row 2} \times (-1/20)$$

The reduced system of equation is

$$x_1 - 2x_2 - 6x_3 = 0$$
$$x_2 + 2x_3 = 0$$

$r = \text{Rank}\ (\mathbf{A} - \lambda \mathbf{I}) = 2$
$n = \text{Number of unknowns} = 3$
Number of free variables $= 3 - 2 = 1$
Taking $x_3 = c_3$ we get $x_2 = -2c_3$ and hence,

(a) $x_1 - 2(-2c_3) - 6c_3 = 0 \Rightarrow x_1 + 4c_3 - 6c_3 = 0 \Rightarrow x_1 = 2c_3$
(b) $-2c_3 + 2c_3 = 0$

Hence the eigenvectors corresponding to $\lambda = 15$ are

$$\mathbf{X}_3 = \begin{bmatrix} 2c_3 \\ -2c_3 \\ c_3 \end{bmatrix} = c_3 \begin{bmatrix} 2 \\ -2 \\ 1 \end{bmatrix}$$

The eigenvectors for eigenvalues $\lambda = 0, 3, 15$ are as follows:

$$\mathbf{X}_1 = \begin{bmatrix} 1/2 \\ 1 \\ 1 \end{bmatrix}, \mathbf{X}_2 = \begin{bmatrix} 2 \\ 1 \\ -2 \end{bmatrix}, \mathbf{X}_3 = \begin{bmatrix} 2 \\ -2 \\ 1 \end{bmatrix}$$

Q 4. If λ be an eigenvalue of the nonsingular matrix **A**, show that

(a) λ^{-1} is an eigenvalue of $\mathbf{A}^{-1}$ and

(b) $\dfrac{|\mathbf{A}|}{\lambda}$ is an eigenvalue of adj **A**. [JNTU May/June 2006]

Answer: (a) λ^{-1} is an eigenvalue of $\mathbf{A}^{-1}$
λ is an eigenvalue of the matrix **A** which is a nonsingular matrix. Therefore there exists a nonzero matrix **X** such that $\mathbf{AX} = \lambda\mathbf{X}$. Multiplying both sides by $\mathbf{A}^{-1}$ (since **A** is nonsingular $\mathbf{A}^{-1}$ exists), we get

$$\mathbf{X} = \mathbf{A}^{-1}(\lambda\mathbf{X})$$
$$\mathbf{X} = \lambda(\mathbf{A}^{-1}\mathbf{X})$$
$$\frac{1}{\lambda}\mathbf{X} = \mathbf{A}^{-1}\mathbf{X}$$
$$\mathbf{A}^{-1}\mathbf{X} = \lambda^{-1}\mathbf{X}$$

λ^{-1} is an eigenvalue of $\mathbf{A}^{-1}$.

(b) $\dfrac{|\mathbf{A}|}{\lambda}$ is an eigenvalue of adj A

$$\mathbf{AX} = \lambda\mathbf{X}$$
$$(\text{adj } \mathbf{A})(\mathbf{AX}) = (\text{adj } \mathbf{A})(\lambda\mathbf{X})$$
$$(\text{adj } \mathbf{A})\,\mathbf{AX} = \lambda(\text{adj } \mathbf{A})(\mathbf{X})$$

Since **A** is nonsingular (adj **A**) **A** = $|\mathbf{A}|$. Therefore

$$|\mathbf{A}|\ \mathbf{IX} = \lambda(\text{adj } \mathbf{A})\,\mathbf{X}$$
$$|\mathbf{A}|\ \mathbf{X} = \lambda(\text{adj } \mathbf{A})\,\mathbf{X}$$
$$\frac{|\mathbf{A}|}{\lambda}\mathbf{X} = (\text{adj } \mathbf{A})\,\mathbf{X}$$
$$(\text{adj } \mathbf{A})\,\mathbf{X} = \frac{|\mathbf{A}|}{\lambda}\mathbf{X}$$

That is, $|\mathbf{A}|/\lambda$ is an eigenvalue of adj **A**.

Q 5. Show that if $\lambda_1, \lambda_2, \ldots, \lambda_n$ are the latent roots of a matrix **A**, then $\mathbf{A}^3$ has the latent roots (a) $\lambda_1^3, \lambda_2^3, \ldots, \lambda_n^3$ and (b) $\lambda_1, \lambda_2, \lambda_3, \ldots, \lambda_n$. **[JNTU May 2005, Nov. 2004]**

Answer: (a) Let λ be any eigenvalue of **A**, then there exists an eigenvector X, such that

$$\mathbf{AX} = \lambda\mathbf{X} \tag{2.45}$$

Multiplying (2.45) with **A** we get

$$\mathbf{A}(\mathbf{AX}) = \mathbf{A}(\lambda\mathbf{X})$$

$$\mathbf{A}(\mathbf{AX}) = \lambda(\mathbf{AX})$$

$$\mathbf{A}^2\mathbf{X} = \lambda(\lambda\mathbf{X})$$

Therefore $\mathbf{AX} = \lambda\mathbf{X}$ from (2.45)

$$\mathbf{A}^2\mathbf{X} = \lambda^2\mathbf{X} \tag{2.46}$$

Multiplying (2.46) with **A** we get

$$\mathbf{A}(\mathbf{A}^2\mathbf{X}) = \mathbf{A}(\lambda^2\mathbf{X})$$

$$\mathbf{A}(\mathbf{A}^2\mathbf{X}) = \lambda^2(\mathbf{AX})$$

$$\mathbf{A}^3\mathbf{X} = \lambda^3\mathbf{X}$$

Thus, if $\lambda_1, \lambda_2, \lambda_3, \ldots, \lambda_n$ be eigenvalues of **A** then the matrix $\mathbf{A}^3$ has eigenvalues $\lambda_1^3, \lambda_2^2, \lambda_3^3, \ldots, \lambda_n^3$.

(b) Multiplying with k on both sides of (2.45), we get

$$k(\mathbf{AX}) = k(\lambda\mathbf{X})$$

$$k(\lambda\mathbf{X}) = (k\lambda)\mathbf{X}$$

Therefore from (2.45), $\mathbf{A} = \lambda$. Hence

$$(k\mathbf{A})\mathbf{X} = (k\lambda)\mathbf{X}$$

Therefore if $k\lambda$ is the eigenvalue of $k\mathbf{A}$, then the matrix $k\mathbf{A}$ has the eigenvalues of $k\lambda_1, k\lambda_2, k\lambda_3, \ldots, k\lambda_n$.

Q 6. Show that **A** and $\mathbf{A}^\mathrm{T}$ has same eigenvalues but different eigenvectors.

[JNTU Nov. 2004, May/June 2004]

Answer: Let

$$\mathbf{A} = [a_{ij}]_{n\times n} \text{ and } \mathbf{A}^\mathrm{T} = [a_{ji}]_{n\times n};\ i, j = 1, 2, \ldots, n$$

Consider the characteristic equation of **A** and $\mathbf{A}^\mathrm{T}$:

$$|\mathbf{A} - \lambda\mathbf{I}| = 0$$

$$\begin{vmatrix} a_{11}-\lambda & a_{12} & \cdots & a_{1n} \\ a_{21} & a_{22}-\lambda & \cdots & a_{2n} \\ . & \cdots & \cdots & \cdots \\ . & \cdots & \cdots & \cdots \\ . & \cdots & \cdots & \cdots \\ a_{n1} & a_{n2} & \cdots & a_{nn}-\lambda \end{vmatrix} = 0 \tag{2.47}$$

Now we calculate $|A^T - \lambda I| = 0$:

$$\begin{vmatrix} a_{11}-\lambda & a_{21} & \cdots & a_{n1} \\ a_{12} & a_{22}-\lambda & \cdots & a_{n2} \\ \cdot & \cdots & \cdots & \cdots \\ \cdot & \cdots & \cdots & \cdots \\ \cdot & \cdots & \cdots & \cdots \\ a_{1n} & a_{2n} & \cdots & a_{nn}-\lambda \end{vmatrix} = 0 \tag{2.48}$$

Since the interchange of rows and columns does not alter the determinant, therefore (2.47) and (2.48) are the same. That is

$$|A - \lambda I| = |A^T - \lambda I|$$

Therefore the characteristic roots of A and A^T are the same.

Q 7. If **A** and **B** are *n*-rowed square matrices and if **A** is invertible, show that $A^{-1}B$ and BA^{-1} have same eigenvalues. **[JNTU Aug./Sep. 2007, May 2005]**

Answer: It is given that A is invertible. That is

$$A^{-1}B = B^{-1}(BA^{-1})B \quad (\text{since } B^{-1}B = I)$$

Let $A^{-1}B = P$ and $BA^{-1} = Q$. Then

$$\begin{aligned} P &= B^{-1}(Q)B \\ P - \lambda I &= B^{-1}QB - \lambda I \\ &= B^{-1}QB - \lambda B^{-1}B \\ &= B^{-1}QB - B^{-1}\lambda B \\ &= B^{-1}(QB - \lambda B) \\ &= B^{-1}(Q - \lambda I)B \end{aligned}$$

This implies that

$$|P - \lambda I| = |Q - \lambda I|BB^{-1}$$

Since $BB^{-1} = I$, therefore

$$|P - \lambda I| = |Q - \lambda I|$$

Therefore characteristic equations of **P** and **Q** are same. This implies that eigenvalues of $A^{-1}B$ and BA^{-1} are the same.

Q 8. Verify Cayley–Hamilton theorem for the matrix

$$A = \begin{bmatrix} 8 & -8 & 2 \\ 4 & -3 & -2 \\ 3 & -4 & 1 \end{bmatrix}$$

{JNTU Aug./Sep. 2006, May 2005, Nov. 2004}

Answer: It is given that

$$\mathbf{A} = \begin{bmatrix} 8 & -8 & 2 \\ 4 & -3 & -2 \\ 3 & -4 & 1 \end{bmatrix}$$

Then its characteristic equation is

$$|\mathbf{A} - \lambda \mathbf{I}| = 0$$

$$\begin{vmatrix} 8-\lambda & -8 & 2 \\ 4 & -3-\lambda & -2 \\ 3 & -4 & 1-\lambda \end{vmatrix} = 0$$

$$(8 - \lambda)\,[(-3 - \lambda)(1 - \lambda) - 8] + 8[4(1 - \lambda)] + 6 + 2\,[-16 - 3(-3 - \lambda)] = 0$$

$$(8 - \lambda)(-3 + 3\lambda - \lambda + \lambda^2 - 8) + 8[4 - 4\lambda + 6] + 2\,[-16 + 9 + 3\lambda] = 0$$

$$(8 - \lambda)(\lambda^2 + 2\lambda - 11) + 8(10 - 4\lambda) + 2(3\lambda - 7) = 0$$

$$8(\lambda^2 + 2\lambda - 11) - \lambda(\lambda^2 + 2\lambda - 11) + 80 - 32\lambda + 6\lambda - 14 = 0$$

$$8\lambda^2 + 16\lambda - 88 - \lambda^3 - 2\lambda^2 + 11\lambda + 80 - 32\lambda + 6\lambda - 14 = 0$$

$$-\lambda^3 + 6\lambda^2 + \lambda - 22 = 0$$

$$\lambda^3 - 6\lambda^2 - \lambda + 22 = 0$$

By Cayley–Hamilton theorem, **A** should satisfy its characteristic equation. To verify,

$$\mathbf{A}^3 - 6\mathbf{A}^2 - \mathbf{A} + 22 = 0 \qquad (2.49)$$

We first calculate $\mathbf{A}^2$ and $\mathbf{A}^3$.

$$\mathbf{A}^2 = \begin{bmatrix} 8 & -8 & 2 \\ 4 & -3 & -2 \\ 3 & -4 & 1 \end{bmatrix}\begin{bmatrix} 8 & -8 & 2 \\ 4 & -3 & -2 \\ 3 & -4 & 1 \end{bmatrix} = \begin{bmatrix} 64-32+6 & -64+24-8 & 16+16+2 \\ 32-12-6 & -32+9+8 & 8+6-2 \\ 24-16+3 & -24+12-4 & 6+8+1 \end{bmatrix}$$

$$= \begin{bmatrix} 38 & -48 & 234 \\ 14 & -15 & 12 \\ 11 & -16 & 15 \end{bmatrix}$$

$$\mathbf{A}^3 = \mathbf{A}^2 \times \mathbf{A} = \begin{bmatrix} 38 & -48 & 34 \\ 14 & -15 & 12 \\ 11 & -16 & 15 \end{bmatrix}\begin{bmatrix} 8 & -8 & 2 \\ 4 & -3 & -2 \\ 3 & -4 & 1 \end{bmatrix}$$

$$= \begin{bmatrix} (38\times8-48\times8+34\times3) & (-8\times38+48\times3-34\times4) & (38\times2+48\times2+34\times1) \\ (14\times8-15\times4+12\times3) & (14\times-8+15\times3-12\times4) & (14\times2+15\times2+12\times1) \\ (11\times8-16\times4+15\times3) & (11\times-8+16\times3-15\times4) & (11\times2+16\times2+15\times1) \end{bmatrix}$$

$$= \begin{bmatrix} 214 & -269 & 206 \\ 88 & -115 & 70 \\ 69 & -100 & 69 \end{bmatrix}$$

Substituting the respective values in the LHS of (2.49) we get

$$\begin{bmatrix} 214 & -269 & 206 \\ 88 & -115 & 70 \\ 69 & -100 & 69 \end{bmatrix} - 6\begin{bmatrix} 38 & -48 & 34 \\ 14 & -15 & 12 \\ 11 & -16 & 15 \end{bmatrix} - \begin{bmatrix} 8 & -8 & 2 \\ 4 & -3 & -2 \\ 3 & -4 & 1 \end{bmatrix} + 22\begin{bmatrix} 1 & 0 & 0 \\ 0 & 1 & 0 \\ 0 & 0 & 1 \end{bmatrix}$$

$$= \begin{bmatrix} 214 & -269 & 206 \\ 88 & -115 & 70 \\ 69 & -100 & 69 \end{bmatrix} - \begin{bmatrix} 228 & -288 & 204 \\ 84 & -90 & 72 \\ 66 & -96 & 90 \end{bmatrix} - \begin{bmatrix} 8 & -8 & 2 \\ 4 & -3 & -2 \\ 3 & -4 & 1 \end{bmatrix} + \begin{bmatrix} 22 & 0 & 0 \\ 0 & 22 & 0 \\ 0 & 0 & 22 \end{bmatrix}$$

$$= \begin{bmatrix} 214-228-8+22 & -269-288+8+0 & 206-204-2+0 \\ 88-84-4+0 & -115+90+3+22 & 70-72+2+0 \\ 69-66-3+0 & -100+96+4+0 & 69-90-1+22 \end{bmatrix}$$

$$= \begin{bmatrix} 0 & 0 & 0 \\ 0 & 0 & 0 \\ 0 & 0 & 0 \end{bmatrix}$$

Therefore matrix **A** satisfies its own characteristic equation. Hence, Cayley–Hamilton theorem is proved.

Q 9. Diagonalize the matrix

$$\begin{bmatrix} 8 & -6 & 2 \\ -6 & 7 & -4 \\ 2 & -4 & 3 \end{bmatrix}$$

[JNTU Aug./Sep. 2007]

Answer: Let **A** denote the given matrix. Then

$$\mathbf{A} = \begin{bmatrix} 8 & -6 & 2 \\ -6 & 7 & -4 \\ 2 & -4 & 3 \end{bmatrix}$$

Assume λ be the eigenvalue of **A**. The characteristic equation of the matrix **A** is

$$|\mathbf{A} - \lambda\mathbf{I}| = 0$$

$$\begin{vmatrix} 8-\lambda & -6 & 2 \\ -6 & 7-\lambda & -4 \\ 2 & -4 & 3-\lambda \end{vmatrix} = 0$$

$$(8-\lambda)\,[(7-\lambda)(3-\lambda)-(-4\times -4)]-(-6)\,[-6\,(3-\lambda)$$

$$-(2\times -4)]+2[(-6\times -4)-2(7-\lambda)]=0$$

$$(8-\lambda)\,(21-10\lambda+\lambda^2-16)+6(8+6\lambda-18)+2(10+2\lambda)=0$$

$$(8-\lambda)\,(\lambda^2-10\lambda+5)+6\,(6\lambda-10)+2(10+2\lambda)=0$$

$$8\lambda^2-80\lambda+40-\lambda^3+10\lambda^2-5\lambda+36\lambda-60+20+4\lambda=0$$

$$-\lambda^3+18\lambda^2-45\lambda=0$$

$$-\lambda(\lambda^2-18\lambda+45)=0$$

$$\lambda(\lambda-3)\,(\lambda-15)=0$$

Therefore $\lambda = 0, 3, 15$ are the eigenvalues. Let **X** be the given vector corresponding to eigenvalues λ.

$$\mathbf{AX} = \lambda\mathbf{X}$$

$$(\mathbf{A} - \lambda\mathbf{I})\mathbf{X} = \mathbf{0}$$

$$\begin{bmatrix} 8-\lambda & -6 & 2 \\ -6 & 7-\lambda & -4 \\ 2 & -4 & 3-\lambda \end{bmatrix}\begin{bmatrix} x_1 \\ x_2 \\ x_3 \end{bmatrix} = \begin{bmatrix} 0 \\ 0 \\ 0 \end{bmatrix}$$

Eigenvector corresponding to $\lambda = 0$

$$\begin{bmatrix} 8 & -6 & 2 \\ -6 & 7 & -4 \\ 2 & -4 & 3 \end{bmatrix}\begin{bmatrix} x_1 \\ x_2 \\ x_3 \end{bmatrix} = \begin{bmatrix} 0 \\ 0 \\ 0 \end{bmatrix}$$

$$8x_1 - 6x_2 + 2x_3 = 0$$

$$-6x_1 + 7x_2 - 4x_3 = 0$$

$$2x_1 - 4x_2 + 3x_3 = 0$$

Let $x_2 = k_1$, $x_1 = (1/2)\,k_1$, $x_3 = x_2 = k_1$. Therefore eigenvector for $\lambda = 0$ is

$$\mathbf{X}_1 = \begin{bmatrix} (1/2)k_1 \\ k_1 \\ k_1 \end{bmatrix}$$

The simplest vector for $\lambda = 0$ is

$$\mathbf{X}_1 = k_1\begin{bmatrix} 1/2 \\ 1 \\ 1 \end{bmatrix} = \begin{bmatrix} 1 \\ 2 \\ 2 \end{bmatrix} \text{ [multiplying with 2]}$$

Eigenvector corresponding to $\lambda = 3$

$$\begin{bmatrix} 5 & -6 & 2 \\ -6 & 4 & -4 \\ 2 & -4 & 0 \end{bmatrix}\begin{bmatrix} x_1 \\ x_2 \\ x_3 \end{bmatrix} = \begin{bmatrix} 0 \\ 0 \\ 0 \end{bmatrix}$$

$$5x_1 - 6x_2 + 2x_3 = 0$$

$$-6x_1 + 4x_2 - 4x_3 = 0$$

$$2x_1 - 4x_2 = 0 \Rightarrow x_1 = 2x_2$$

Let $x_2 = k_2$ then $x_1 = 2k_2$ and $x_3 = -2k_2$. Therefore eigenvector for $\lambda = 3$ is

$$\mathbf{X}_2 = \begin{bmatrix} 2k_2 \\ 1k_2 \\ -2k_2 \end{bmatrix} = k_2\begin{bmatrix} 2 \\ 1 \\ -2 \end{bmatrix}$$

Eigenvector corresponding to $\lambda = 15$

$$-7x_1 - 6x_2 + \ \ 2x_3 = 0$$

$$-6x_1 - 8x_2 - \ \ 4x_3 = 0$$

$$2x_1 - 4x_2 - 12x_3 = 0$$

Let $x_3 = k_3$, then $x_2 = -2k_3$, $x_1 = 2k_3$. Hence, the eigenvectors corresponding to $\lambda = 15$ are given as follows:

$$\mathbf{X}_3 = \begin{bmatrix} 2k_3 \\ -2k_3 \\ k_3 \end{bmatrix} = k_3\begin{bmatrix} 2 \\ -2 \\ 1 \end{bmatrix}$$

Therefore modal matrix, is given by $\mathbf{P} = [\mathbf{X}_1, \mathbf{X}_2, \mathbf{X}_3]$. That is

$$\mathbf{P} = \begin{bmatrix} 1 & 2 & 2 \\ 2 & 1 & -2 \\ 2 & -2 & 1 \end{bmatrix}$$

Now determinant of **P** is

$$|\mathbf{P}| = 1(1-4) - 2(2+4) + 2(-4-2) = -3 - 12 - 12 = -27$$

Since $|\mathbf{P}| \neq 0$, therefore **P** is a nonsingular matrix and so $\mathbf{P}^{-1}$ exists. Now

$$\mathbf{P}^{-1} = \frac{1}{|\mathbf{P}|} \text{adj } \mathbf{P} = \frac{-1}{27} \begin{vmatrix} -3 & -6 & -6 \\ -6 & -3 & 6 \\ -6 & 6 & -3 \end{vmatrix}$$

$$\mathbf{AP} = \begin{bmatrix} 8 & -6 & 2 \\ -6 & 7 & -4 \\ 2 & -4 & 3 \end{bmatrix} \begin{bmatrix} 1 & 2 & 2 \\ 2 & 1 & -2 \\ 2 & -2 & 1 \end{bmatrix} = \begin{bmatrix} 0 & 6 & 30 \\ 0 & 3 & -30 \\ 0 & -6 & 15 \end{bmatrix}$$

Therefore diagonal matrix is $\mathbf{P}^{-1}\ \mathbf{AP}$. Solving we get

$$\mathbf{P}^{-1}\mathbf{AP} = \frac{-1}{27} \begin{bmatrix} -3 & -6 & -6 \\ -6 & -3 & 6 \\ -6 & 6 & -3 \end{bmatrix} \begin{bmatrix} 8 & -6 & 2 \\ -6 & 7 & 4 \\ 2 & -4 & 3 \end{bmatrix} \begin{bmatrix} 1 & 2 & 2 \\ 2 & 1 & -2 \\ 2 & -2 & 1 \end{bmatrix} = \frac{-1}{27} \begin{bmatrix} 0 & 0 & 0 \\ 0 & -81 & 0 \\ 0 & 0 & -405 \end{bmatrix}$$

$$\mathbf{P}^{-1}\mathbf{AP} = \mathbf{D} = \begin{bmatrix} 0 & 0 & 0 \\ 0 & 3 & 0 \\ 0 & 0 & 15 \end{bmatrix}$$

Q 10. Diagonalize the matrix

$$\mathbf{A} = \begin{bmatrix} 8 & -8 & -2 \\ 4 & -3 & -2 \\ 3 & -4 & 1 \end{bmatrix}$$

{JNTU Aug./Sep. 2007, Aug./Sep. 2006}

Answer: The characteristic equation of **A** is $|\mathbf{A} - \lambda\mathbf{I}| = 0$. That is

$$\begin{vmatrix} 8-\lambda & -8 & -2 \\ 4 & -3-\lambda & -2 \\ 3 & -4 & 1-\lambda \end{vmatrix} = 0$$

$$(8-\lambda)\,[(-3-\lambda)\,(1-\lambda) - 8] + 8\,[4(1-\lambda) + 6] - 2\,[-16 - 3\,(-3-\lambda)] = 0$$

$$(8-\lambda)\,(-3 + 3\lambda - \lambda + \lambda^2 - 8) + 8(\,4 - 4\lambda + 6) - 2(-16 + 9 + 3\lambda) = 0$$

$$(8-\lambda)(\lambda^2 + 2\lambda - 11) + 8(-4\lambda + 10) - 2(3\lambda - 7) = 0$$

$$8\lambda^2 + 16\lambda - 88 - \lambda^3 - 2\lambda^2 + 11\lambda - 32\lambda + 80 - 6\lambda + 14 = 0$$

$$-\lambda^3 + 6\lambda^2 - 11\lambda + 6 = 0$$

$$\lambda^3 - 6\lambda^2 + 11\lambda - 6 = 0$$

Therefore $\lambda = 3, 1, 2$. The diagonal matrix is

$$\begin{bmatrix} 3 & 0 & 0 \\ 0 & 1 & 0 \\ 0 & 0 & 2 \end{bmatrix}$$

We now find the eigenvectors when $\lambda = 3, 1, 2$.

Eigenvector corresponding to $\lambda = 3$

$$\begin{bmatrix} 8-3 & -8 & -2 \\ 4 & -3-3 & -2 \\ 3 & -4 & 1-3 \end{bmatrix}\begin{bmatrix} x_1 \\ x_2 \\ x_3 \end{bmatrix} = \begin{bmatrix} 0 \\ 0 \\ 0 \end{bmatrix}$$

$$\begin{bmatrix} 5 & -8 & -2 \\ 4 & -6 & -2 \\ 3 & -4 & -2 \end{bmatrix}\begin{bmatrix} x_1 \\ x_2 \\ x_3 \end{bmatrix} = \begin{bmatrix} 0 \\ 0 \\ 0 \end{bmatrix}$$

Expanding we get

$$5x_1 - 8x_2 - 2x_3 = 0$$
$$3x_1 - 4x_2 - 2x_3 = 0$$
$$4x_1 - 6x_2 - 2x_3 = 0$$

Solving, we have

$$x_1 = 2x_2$$
$$x_2 = x_3 = k_1 \text{ (say)}$$

Therefore $x_1 = 2k_1$. Hence

$$\mathbf{X}_1 = \begin{bmatrix} 2k_1 \\ k_1 \\ k_1 \end{bmatrix} = k_1\begin{bmatrix} 2 \\ 1 \\ 1 \end{bmatrix}$$

Now

$$|\mathbf{X}_1| = \sqrt{4+1+1} = \sqrt{6}$$

Therefore

$$\frac{\mathbf{X}_1}{|\mathbf{X}_1|} = \begin{bmatrix} 2/\sqrt{6} \\ 1/\sqrt{6} \\ 1/\sqrt{6} \end{bmatrix}$$

Eigenvector corresponding to $\lambda = 1$

$$\begin{bmatrix} 8-1 & -8 & -2 \\ 4 & -3-1 & -2 \\ 3 & -4 & 1-1 \end{bmatrix}\begin{bmatrix} x_1 \\ x_2 \\ x_3 \end{bmatrix} = \begin{bmatrix} 0 \\ 0 \\ 0 \end{bmatrix}$$

$$\begin{bmatrix} 7 & -8 & -2 \\ 4 & -4 & -2 \\ 3 & -4 & 0 \end{bmatrix}\begin{bmatrix} x_1 \\ x_2 \\ x_3 \end{bmatrix} = \begin{bmatrix} 0 \\ 0 \\ 0 \end{bmatrix}$$

Expanding we get

$$7x_1 - 8x_2 - 2x_3 = 0 \quad (2.50)$$

$$4x_1 - 4x_2 - 2x_3 = 0 \quad (2.51)$$

$$3x_1 = 4x_2 \Rightarrow x_1 = \frac{4}{3}x_2 \quad (2.52)$$

Substituting this value of x_1 in (2.51) we get

$$4\left(\frac{4}{3}x_2\right) - 4x_2 - 2x_3 = 0$$

$$16x_2 - 12x_2 - 6x_3 = 0$$

$$4x_2 = 6x_3$$

$$x_3 = \frac{4}{6}x_2 = \frac{2}{3}x_2$$

Let $x_2 = k_2$. Therefore

$$\mathbf{X}_2 = \begin{bmatrix} (4/3)k_2 \\ k_2 \\ (2/3)k_2 \end{bmatrix} = k_2 \begin{bmatrix} 4/3 \\ 1 \\ 2/3 \end{bmatrix}$$

Therefore

$$|\mathbf{X}_2| = \sqrt{\frac{16}{9} + 1 + \frac{4}{9}} = \sqrt{\frac{16+9+4}{9}} = \sqrt{\frac{29}{9}} = \frac{\sqrt{29}}{3}$$

$$\frac{\mathbf{X}_2}{|\mathbf{X}_2|} = \begin{bmatrix} (4/3)\times(3/\sqrt{29}) \\ 1\times(3/\sqrt{29}) \\ (2/3)\times(3/\sqrt{29}) \end{bmatrix} = \begin{bmatrix} 4/\sqrt{29} \\ 3/\sqrt{29} \\ 2/\sqrt{29} \end{bmatrix}$$

Eigenvector corresponding to $\lambda = 2$

$$\begin{bmatrix} 8-2 & -8 & -2 \\ 4 & -3-2 & -2 \\ 3 & -4 & 1-2 \end{bmatrix} \begin{bmatrix} x_1 \\ x_2 \\ x_3 \end{bmatrix} = \begin{bmatrix} 0 \\ 0 \\ 0 \end{bmatrix}$$

Expanding we get

$$6x_1 - 8x_2 - 2x_3 = 0$$

$$4x_1 - 5x_2 - 2x_3 = 0$$

$$3x_1 - 4x_2 - x_3 = 0$$

We know solve the set of equations. Let $x_2 = k_3$. Then substituting and solving we get

$$x_1 = \frac{3}{2}x_2 = \frac{3}{2}k_3$$

$$x_3 = \frac{1}{2}x_2 = \frac{1}{2}k_3$$

$$\mathbf{X}_3 = \begin{bmatrix} (3/2)k_3 \\ k_3 \\ (1/2)k_3 \end{bmatrix} = k_3\begin{bmatrix} 3/2 \\ 1 \\ 1/2 \end{bmatrix}$$

$$|\mathbf{X}_3| = \sqrt{\frac{9}{4}+1+\frac{1}{4}} = \sqrt{\frac{9+4+1}{4}} = \sqrt{\frac{14}{4}} = \frac{\sqrt{14}}{2}$$

Therefore

$$\frac{\mathbf{X}_3}{|\mathbf{X}_3|} = \begin{bmatrix} (3/2)\times(2/\sqrt{14}) \\ (1)\times 2/\sqrt{14} \\ (3/2)\times(2/\sqrt{14}) \end{bmatrix} = \begin{bmatrix} 3/\sqrt{14} \\ 2/\sqrt{14} \\ 1/\sqrt{14} \end{bmatrix}$$

Now the modal matrix **P** is given by

$$\mathbf{P} = \left[\frac{\mathbf{X}_1}{|\mathbf{X}_1|}\frac{\mathbf{X}_2}{|\mathbf{X}_2|}\frac{\mathbf{X}_3}{|\mathbf{X}_3|}\right] = \begin{bmatrix} 2/\sqrt{6} & 4/\sqrt{29} & 3/\sqrt{14} \\ 1/\sqrt{6} & 3/\sqrt{29} & 3/\sqrt{14} \\ 1/\sqrt{6} & 2/\sqrt{29} & 3/\sqrt{14} \end{bmatrix}$$

Therefore

$$\mathbf{P}^{-1}\mathbf{AP} = \begin{bmatrix} 2.44 & -4.89 & 2.44 \\ -5.38 & 5.38 & 5.38 \\ 3.74 & 0 & -7.48 \end{bmatrix}\begin{bmatrix} 8 & -8 & -2 \\ 4 & -3 & -2 \\ 3 & -4 & 1 \end{bmatrix}\begin{bmatrix} 2/\sqrt{6} & 4/\sqrt{29} & 3/\sqrt{14} \\ 1/\sqrt{6} & 3/\sqrt{29} & 2/\sqrt{14} \\ 1/\sqrt{6} & 2/\sqrt{29} & 1/\sqrt{14} \end{bmatrix}$$

$$\mathbf{P}^{-1}\mathbf{AP} = \mathbf{D} = \begin{bmatrix} 3 & 0 & 0 \\ 0 & 1 & 0 \\ 0 & 0 & 2 \end{bmatrix}$$

Q 11. Diagonalize the matrix

$$\mathbf{A} = \begin{bmatrix} 1 & 1 & 1 \\ 0 & 2 & 1 \\ -4 & 4 & 3 \end{bmatrix}$$

and hence find $\mathbf{A}^4$. **[JNTU April/May 2007, May/June 2006]**

Answer: The characteristic equation of **A** is

$$|\mathbf{A} - \lambda\mathbf{I}| = 0$$

That is

$$\begin{vmatrix} 1 & 1 & 1 \\ 0 & 2 & 1 \\ -4 & 4 & 3 \end{vmatrix} - \begin{vmatrix} \lambda & 0 & 0 \\ 0 & \lambda & 0 \\ 0 & 0 & \lambda \end{vmatrix} = 0$$

$$\begin{vmatrix} 1-\lambda & 1 & 1 \\ 0 & 2-\lambda & 1 \\ -4 & 4 & 3-\lambda \end{vmatrix} = 0$$

Solving we get

$$(1-\lambda)(6-2\lambda-3\lambda+\lambda^2-4)-1(0+4)+1(8-4\lambda)=0$$
$$(1-\lambda)(\lambda^2-5\lambda+2)-4+8-4\lambda=0$$
$$(1-\lambda)(\lambda^2-5\lambda+2)+4(1-\lambda)=0$$
$$(1-\lambda)(\lambda^2-5\lambda+6)=0$$
$$(1-\lambda)(\lambda-2)(\lambda-3)=0$$

Therefore the eigenvalues are $\lambda = 1, 2, 3$.

Eigenvector corresponding to $\lambda = 1$

$$\begin{bmatrix} 1-1 & 1 & 1 \\ 0 & 2-1 & 1 \\ -4 & 4 & 3-1 \end{bmatrix}\begin{bmatrix} x_1 \\ x_2 \\ x_3 \end{bmatrix} = \begin{bmatrix} 0 \\ 0 \\ 0 \end{bmatrix}$$

$$\begin{bmatrix} 0 & 1 & 1 \\ 0 & 1 & 1 \\ -4 & 4 & 2 \end{bmatrix}\begin{bmatrix} x_1 \\ x_2 \\ x_3 \end{bmatrix} = \begin{bmatrix} 0 \\ 0 \\ 0 \end{bmatrix}$$

Solving we get

$$x_2 + x_3 = 0 \Rightarrow x_2 = -x_3 \tag{2.53}$$
$$-4x_1 + 4x_2 + 2x_3 = 0 \tag{2.54}$$

Substituting (2.53) in (2.54) we get

$$-4x_1 - 4x_3 + 2x_3 = 0$$
$$-4x_1 - 2x_3 = 0$$
$$2x_1 = -x_3$$

Let $x_3 = k_1$. Then $x_1 = (-1/2)\,k_1$ and $x_2 = -k_1$. Therefore the eigenvector for λ is

$$\mathbf{X}_1 = \begin{bmatrix} -(1/2)k_1 \\ -k_1 \\ k_1 \end{bmatrix} = k_1 \begin{bmatrix} -1/2 \\ -1 \\ 1 \end{bmatrix}$$

The simplest vector for $\lambda = 1$ is

$$\mathbf{X}_1 = \begin{bmatrix} -1/2 \\ -1 \\ 1 \end{bmatrix} = \begin{bmatrix} 1 \\ 2 \\ -2 \end{bmatrix}$$

If $\lambda = 2$, then

$$\begin{bmatrix} -1 & 1 & 1 \\ 0 & 0 & 1 \\ -4 & 4 & 1 \end{bmatrix} \begin{bmatrix} x_1 \\ x_2 \\ x_3 \end{bmatrix} = \begin{bmatrix} 0 \\ 0 \\ 0 \end{bmatrix}$$

Expanding and solving we get

$$\begin{aligned} -x_1 + x_2 + x_3 &= 0 \\ x_3 &= 0 \\ -4x_1 + 4x_2 + x_3 &= 0 \end{aligned}$$

From these equations we get

$$x_1 = x_2, \quad x_3 = 0$$

So, the simplest vector for $\lambda = 2$

$$\mathbf{X}_2 = \begin{bmatrix} 1 \\ 1 \\ 0 \end{bmatrix}$$

If $\lambda = 3$, then

$$\begin{bmatrix} -2 & 1 & 1 \\ 0 & -1 & 1 \\ -4 & 4 & 0 \end{bmatrix} \begin{bmatrix} x_1 \\ x_2 \\ x_3 \end{bmatrix} = \begin{bmatrix} 0 \\ 0 \\ 0 \end{bmatrix}$$

In equation form we get

$$\begin{aligned} -2x_1 + x_2 + x_3 &= 0 \\ -x_2 + x_3 &= 0 \\ -4x_1 + 4x_2 &= 0 \\ x_1 = x_2 &= x_3 \end{aligned}$$

Therefore, the simplest vector for $\lambda = 3$ is

$$\mathbf{X}_3 = \begin{bmatrix} 1 \\ 1 \\ 1 \end{bmatrix}$$

Thus, we have

$$\mathbf{X}_1 = \begin{bmatrix} 1 \\ 2 \\ -2 \end{bmatrix}, \ \mathbf{X}_2 = \begin{bmatrix} 1 \\ 1 \\ 0 \end{bmatrix}, \ \mathbf{X}_3 = \begin{bmatrix} 1 \\ 1 \\ 1 \end{bmatrix}$$

Therefore the modal matrix, $\mathbf{P} = [\mathbf{X}_1 \ \mathbf{X}_2 \ \mathbf{X}_3]$, that is

$$\mathbf{P} = \begin{bmatrix} 1 & 1 & 1 \\ 2 & 1 & 1 \\ -2 & 0 & 1 \end{bmatrix}$$

Hence

$$|\mathbf{P}| = 1\,(1-0) - 1\,(2+2) + 1\,(0+2)$$
$$= 1 - 4 + 2 = -1 \neq 0$$

$\mathbf{P}$ is a nonsingular matrix, so $\mathbf{P}^{-1}$ exists. Therefore

$$\mathbf{P}^{-1} = \frac{1}{|\mathbf{P}|} \text{ adj } \mathbf{P} = -\begin{bmatrix} 1 & -1 & 0 \\ -4 & 3 & 1 \\ 2 & -2 & -1 \end{bmatrix} = \begin{bmatrix} -1 & 1 & 0 \\ 4 & -3 & -1 \\ -2 & 2 & 1 \end{bmatrix}$$

The required similarity transformation is, $\mathbf{P}^{-1}\mathbf{AP} = \mathbf{D}(1, 2, 3)$. We now have to find $\mathbf{A}^4$. Consider

$$\mathbf{A}^4 = \mathbf{PD}^4\mathbf{P}^{-1} \tag{2.55}$$

We first find $\mathbf{D}^4\mathbf{P}^{-1}$. It is given by

$$\mathbf{D}^4\mathbf{P}^{-1} = \begin{bmatrix} 1^4 & 0 & 0 \\ 0 & 2^4 & 0 \\ 0 & 0 & 3^4 \end{bmatrix} \begin{bmatrix} -1 & 1 & 0 \\ 4 & -3 & -1 \\ -2 & 2 & 1 \end{bmatrix} = \begin{bmatrix} 1 & 0 & 0 \\ 0 & 16 & 0 \\ 0 & 0 & 81 \end{bmatrix} \begin{bmatrix} -1 & 1 & 0 \\ 4 & -3 & -1 \\ -2 & 2 & 1 \end{bmatrix}$$

$$= \begin{bmatrix} -1+0+0 & 1+0+0 & 0+0+0 \\ 0+(16\times 4)+0 & 0(16\times -3)+0 & 0+(16\times -1)+0 \\ 0+0+(81\times -2) & 0+0+(81\times 2) & 0+0+(81\times 1) \end{bmatrix}$$

$$= \begin{bmatrix} -1 & 1 & 0 \\ 64 & -48 & -16 \\ -162 & 162 & 81 \end{bmatrix}$$

Substituting this value in (2.55) we get

$$\mathbf{A}^4 = \mathbf{P}\begin{bmatrix} -1 & 1 & 0 \\ 64 & -48 & -16 \\ -162 & 162 & 81 \end{bmatrix} = \begin{bmatrix} 1 & 1 & 12 \\ 1 & 1 & 1 \\ -2 & 0 & 1 \end{bmatrix}\begin{bmatrix} -1 & 1 & 0 \\ 64 & -48 & -16 \\ -162 & 162 & 81 \end{bmatrix}$$

$$= \begin{bmatrix} -1+64-162 & 1-48+162 & 0+16+81 \\ -2+64-162 & 2-48+162 & 0-16+81 \\ 2+0-162 & -2-0+162 & 0+0+81 \end{bmatrix}$$

$$= \begin{bmatrix} -99 & 115 & 65 \\ -100 & 116 & 65 \\ -160 & 160 & 81 \end{bmatrix}$$

Summary

The practical importance of matrix eigenvalue problems can hardly be overrated. The problems are defined by the vector equation

$$\mathbf{Ax} = \lambda\mathbf{x} \tag{2.56}$$

$\mathbf{A}$ is a given square matrix. All matrices in this chapter are ***square.*** λ is a scalar. To *solve* the problem (2.56) means to determine values of λ, called **eigenvalues** (or **characteristic values**) of $\mathbf{A}$, such that (2.56) has a nontrivial solution $\mathbf{x}$ (that is, $\mathbf{x} \neq 0$), called an **eigenvector** of $\mathbf{A}$ corresponding to that λ. An $n \times n$ matrix has at least one and at most n numerically different eigenvalues. These are the solutions of the **characteristic equation**:

$D(\lambda) = \det(\mathbf{A} - \lambda\mathbf{I})$

$$= \begin{bmatrix} a_{11}-\lambda & a_{12} & \cdots & a_{1n} \\ a_{21} & a_{22}-\lambda & \cdots & a_{2n} \\ \cdot & \cdot & \cdots & \cdot \\ a_{n1} & a_{n2} & \cdots & a_{nn}-\lambda \end{bmatrix} = 0 \tag{2.57}$$

$D(\lambda)$ is called the **characteristic determinant** of $\mathbf{A}$. By expanding it we get the **characteristic polynomial** of $\mathbf{A}$, which is of degree n in λ.

We also discuss the diagonalization of matrices and the transformation of quadratic forms to principal axes and its relation to eigenvalues.

Eigenvalues λ (values λ for which $\mathbf{Ax} = \lambda\mathbf{x}$ has a solution $\mathbf{x} \neq \mathbf{0}$, called an **eigenvector**) can be characterized by inequalities, e.g., in **Gerschgorin's theorem**, which gives n circular disks which contain the whole spectrum (all eigenvalues) of $\mathbf{A}$, of centers a_{jj} and radii $\Sigma|a_{jk}|$ (sum over k from 1 to n, $k \neq j$).

Approximations of eigenvalues can be obtained by iteration, starting from an $\mathbf{x}_0 \neq \mathbf{0}$ and computing $\mathbf{x}_1 = \mathbf{Ax}_0$, $\mathbf{x}_2 = \mathbf{Ax}_1$, ..., $\mathbf{x}_n = \mathbf{Ax}_{n-1}$. In this **power method** the **Rayleigh quotient**

$$q = \frac{(\mathbf{Ax})^{\mathrm{T}}\mathbf{x}}{\mathbf{x}^{\mathrm{T}}\mathbf{x}} \qquad (\mathbf{x} = \mathbf{x}_n) \tag{2.58}$$

gives an approximation of an eigenvalue (usually that of the greatest absolute value) and, if $\mathbf{A}$ is symmetric, an error bound is

$$|\epsilon| \leqq \sqrt{\frac{(\mathbf{Ax})^{\mathrm{T}}\mathbf{Ax}}{\mathbf{x}^{\mathrm{T}}\mathbf{x}} - q^2} \tag{2.59}$$

Convergence may be slow but can be improved by a *spectral shift.*

In the last section we define the concept of Modal and Spectral Matrices.

Review Questions and Problems

1. Find the following expressions or give reasons why they are undefined.
   a. In solving an eigenvalue problem, what is given and what is sought?
   b. Do there exist square matrices without eigenvalues? Eigenvectors corresponding to more than one eigenvalue of a given matrix?
   c. What is the defect? Why is it important? Give examples.
   d. Can a complex matrix have real eigenvalues? Real eigenvectors? Give reasons.
   e. What is diagonalization of a matrix? Transformation of a form to principal axes?
   f. What is an eigenbasis? When does it exist? Why is it important?
   g. Does a 3 × 3 matrix always have a real eigenvalue?
   h. Give a few typical applications in which eigenvalue problems occur.
2. Find an eigenbasis and diagonalize. (Show the details.)

   a. $\begin{bmatrix} 101 & 72 \\ -144 & -103 \end{bmatrix}$

   b. $\begin{bmatrix} 14.4 & -11.2 \\ -11.2 & 102.6 \end{bmatrix}$

   c. $\begin{bmatrix} -14 & 10 \\ -10 & 11 \end{bmatrix}$.

   d. $\begin{bmatrix} 15 & 4 & -4 \\ 6 & 10 & 8 \\ -12 & -2 & -7 \end{bmatrix}, \lambda = 18$

   e. $\begin{bmatrix} 5 & 8/3 & -2/3 \\ 2 & 2/3 & 4/3 \\ -4 & -4/3 & -8/3 \end{bmatrix}$

   f. $\begin{bmatrix} -6 & 11 & 3 \\ 4 & 1 & 3 \\ -4 & 10 & 8 \end{bmatrix}, \lambda = 2$

3. Verify that **A** and $\hat{\mathbf{A}} = \mathbf{P}^{-1}\mathbf{AP}$ have the same spectrum. Here, **A**, **P** are:

   a. $\begin{bmatrix} 3.8 & 2.4 \\ 2.4 & 0.2 \end{bmatrix}, \begin{bmatrix} 1 & 2 \\ 2 & 1 \end{bmatrix}$

   b. $\begin{bmatrix} -22 & 20 & 10 \\ -4 & 20 & -8 \\ 28 & -14 & 29 \end{bmatrix}, \begin{bmatrix} 1 & 0 & 2 \\ 0 & 2 & 4 \\ 2 & 8 & 0 \end{bmatrix}$

   c. $\begin{bmatrix} 2 & 2 & -2 \\ 3 & 1 & -3 \\ 1 & -1 & -1 \end{bmatrix}, \begin{bmatrix} 1 & 2 & 3 \\ 2 & 0 & 2 \\ 3 & 2 & 4 \end{bmatrix}$

4. **Transformation to Canonical Form.** Reduce the quadratic form to principal axes.
   a. $11.56x_1^2 + 20.16x_1x_2 + 17.44x_2^2 = 100$
   b. $1.09x_1^2 - 0.06x_1x_2 + 1.01x_2^2 = 1$
   c. $14x_1^2 + 24x_1x_2 - 4x_2^2 = 20$
5. What are the main problem areas in numeric linear algebra?
6. What is pivoting? When and how would you apply it?
7. What happens if you apply Gauss elimination to a system that has no solutions?
8. What is Doolittle's method? Its connection to Gauss elimination?
9. What is Cholesky's method? When would you apply it?
10. What do you know about the convergence of the Gauss–Seidel method?
11. What is ill-conditioning? What is the condition number and its significance?
12. What is least squares approximation? What are the normal equations?

13. What is an eigenvalue of a matrix? Why are eigenvalue problems important? Give typical examples.
14. Why are similarity transformations of matrices important in designing numeric methods? Give examples.
15. What is the power method for eigenvalues? What are its advantages and disadvantages?
16. State Gerschgorin's theorem from memory. Can you remember its proof?
17. State Schur's inequality and give some applications of it.
18. What is tridiagonalization? When would you apply it?
19. What is the idea of the QR-method? When would you apply the method?
20. Solve:

a.
$$\begin{aligned} 4x_2 - 3x_3 &= 11.8 \\ 5x_1 + 3x_2 + x_3 &= 34.2 \\ 6x_1 - 7x_2 + 2x_3 &= -3.1 \end{aligned}$$

b.
$$\begin{aligned} x_1 + x_2 + x_3 &= 5 \\ x_1 + 2x_2 + 2x_3 &= 6 \\ x_1 + 2x_2 + 3x_3 &= 8 \end{aligned}$$

c.
$$\begin{aligned} 5x_1 + x_2 - 3x_3 &= 17 \\ -5x_2 + 15x_3 &= -10 \\ 2x_1 - 3x_2 + 9x_3 &= 0 \end{aligned}$$

d.
$$\begin{aligned} 2x_1 \qquad + 3x_3 &= 15 \\ 4x_2 - x_3 &= -13 \\ 3x_1 - x_2 + 5x_3 &= 26 \end{aligned}$$

21. Solve Prob. 3(c) by Doolittle's method.
22. Solve Prob. 3(c) by Cholesky's method.
23. Compute the inverse of:

a. $\begin{bmatrix} 1.0 & 2.0 & 0.5 \\ 0.5 & 1.0 & 0.5 \\ 1.5 & 2.0 & 1.0 \end{bmatrix}$

b. $\begin{bmatrix} 1.5 & 2.0 & 1.0 \\ 2.0 & 3.5 & 1.5 \\ 1.0 & 1.5 & 9.0 \end{bmatrix}$

c. $\begin{bmatrix} 5 & 1 & 1 \\ 1 & 6 & 0 \\ 1 & 0 & 8 \end{bmatrix}$

3 Linear Transformations

LEARNING OBJECTIVES

After reading this chapter, you will be able to understand the following:

- Symmetric, skew-symmetric, and orthogonal matrices.
- Quadratic and canonical form of matrices.
- Sylvester's law of inertia.
- Complex matrices (Hermitian, skew-Hermitian and unitary).
- Eigenvalues and eigenvectors of complex matrices.

In this chapter we shall discuss symmetric, skew-symmetric, and orthogonal matrices. We would also discuss their complex counterparts – Hermitian, skew-Hermitian, and unitary matrices – which play a role in modern physics.

3.1 Symmetric, Skew-Symmetric, Orthogonal, and Quadratic Matrices

We consider three classes of real square matrices that occur quite frequently in applications because they have several remarkable properties which we shall now discuss. The first two of these classes have already been mentioned in Chapter 1.

DEFINITIONS (Symmetric, Skew-Symmetric, and Orthogonal Matrices)
A real square matrix $\mathbf{A} = [a_{jk}]$ *is called*

1. **symmetric** if transposition leaves it unchanged, that is

$$\mathbf{A}^{\mathrm{T}} = \mathbf{A}, \quad \text{thus} \quad a_{kj} = a_{jk} \tag{3.1}$$

2. **skew-symmetric** if transposition gives the negative of **A**, that is

$$\mathbf{A}^{\mathrm{T}} = -\mathbf{A}, \quad \text{thus} \quad a_{kj} = -a_{jk} \tag{3.2}$$

3. **orthogonal** if transposition gives the inverse of **A**, that is

$$\mathbf{A}^{\mathrm{T}} = \mathbf{A}^{-1} \tag{3.3}$$

Example 1 Symmetric, Skew-Symmetric, and Orthogonal Matrices

The matrices

$$\begin{bmatrix} -3 & 1 & 5 \\ 1 & 0 & -2 \\ 5 & -2 & 4 \end{bmatrix}, \begin{bmatrix} 0 & 9 & -12 \\ -9 & 0 & 20 \\ 12 & -20 & 0 \end{bmatrix}, \begin{bmatrix} 2/3 & 1/3 & 2/3 \\ -2/3 & 2/3 & 1/3 \\ 1/3 & 2/3 & -2/3 \end{bmatrix}$$

are symmetric, skew-symmetric, and orthogonal, respectively, as you should verify. Every skew-symmetric matrix has all main diagonal entries zero. (Can you prove this? See Chapter 1.)

THEOREM 1: *Any real square matrix* **A** *may be written as the sum of a symmetric matrix* **R** *and a skew-symmetric matrix* S, *where*

$$R = \frac{1}{2}(A + A^T) \quad \text{and} \quad S = \frac{1}{2}(A - A^T) \tag{3.4}$$

PROOF: Let A be any square matrix. We can write **[JNTU 2005, 2004]**

$$A = \frac{1}{2}(A + A^T) + \frac{1}{2}(A - A^T) = R + S$$

where

$$R = \frac{1}{2}(A + A^T) \quad \text{and} \quad S = \frac{1}{2}(A - A^T)$$

We have

$$R^T = \left[\frac{1}{2}(A + A^T)\right]^T = \frac{1}{2}(A + A^T)^T \quad [\text{since } (kA)^T = kA^T\,]$$

$$= \frac{1}{2}\left[A^T + (A^T)^T\right] = \frac{1}{2}(A + A^T) = R$$

Therefore **R** is a symmetric matrix. Now

$$S^T = \left[\frac{1}{2}(A - A^T)\right]^T = \frac{1}{2}(A - A^T)^T = \frac{1}{2}[A^T - (A^T)^T]$$

$$= \frac{1}{2}(A^T - A) = -\frac{1}{2}(A - A^T) = -S$$

Therefore S is a skew-symmetric matrix. Thus the theorem is proved. Also this sum is unique. We leave this as an exercise. ■

Example 2 **llustration of Formula (3.4)**

$$A = \begin{bmatrix} 9 & 5 & 2 \\ 2 & 3 & -8 \\ 5 & 4 & 3 \end{bmatrix} = R + S = \begin{bmatrix} 9.0 & 3.5 & 3.5 \\ 3.5 & 3.0 & -2.0 \\ 3.5 & -2.0 & 3.0 \end{bmatrix} + \begin{bmatrix} 0 & 1.5 & -1.5 \\ -1.5 & 0 & -6.0 \\ 1.5 & 6.0 & 0 \end{bmatrix}$$

THEOREM 2 (Eigenvalues of Symmetric and Skew-Symmetric Matrices):

(a) *The eigenvalues of a symmetric matrix are real.*

(b) *The eigenvalues of a skew-symmetric matrix are pure imaginary or zero.*

This basic theorem (and an extension of it) will be proved in Sec. 3.3.

Example 3 **Eigenvalues of Symmetric and Skew-Symmetric Matrices**

The matrices

$$\begin{bmatrix} 5 & 3 \\ 3 & 5 \end{bmatrix} \quad \text{and} \quad \begin{bmatrix} -5 & 2 \\ 2 & -5 \end{bmatrix}$$

are symmetric and have real eigenvalues. The skew-symmetric matrix in Example 1 has the eigenvalues 0, $-25i$, and $25i$. (Verify this.) The following matrix has the real eigenvalues 1 and 5 but is not symmetric. Does this contradict Theorem 2?

$$\begin{bmatrix} 3 & 4 \\ 1 & 3 \end{bmatrix}$$

Orthogonal Transformations and Orthogonal Matrices

Orthogonal transformations are transformations

$$\mathbf{y} = \mathbf{Ax} \tag{3.5}$$

where **A** is an orthogonal matrix. With each vector **x** in R^n such a transformation assigns a vector y in R^n. For instance, the plane rotation through an angle θ

$$\mathbf{y} = \begin{bmatrix} y_1 \\ y_2 \end{bmatrix} = \begin{bmatrix} \cos\theta & -\sin\theta \\ \sin\theta & \cos\theta \end{bmatrix} \begin{bmatrix} x_1 \\ x_2 \end{bmatrix} \tag{3.6}$$

is an orthogonal transformation. It can be shown that any orthogonal transformation in the plane or in three-dimensional space is a **rotation** (possibly combined with a reflection in a straight line or a plane, respectively).

The main reason for the importance of orthogonal matrices is as follows.

THEOREM 3 (Invariance of Inner Product): *An orthogonal transformation preserves the value of the **inner product** of vectors* **a** *and* **b** *in* R^n, *defined by*

$$\mathbf{a} \times \mathbf{b} = \mathbf{a}^T\mathbf{b} = [a_1 \cdots a_n] \begin{bmatrix} b_1 \\ \vdots \\ b_n \end{bmatrix} \tag{3.7}$$

That is, for any **a** *and* **b** *in* R^n, *orthogonal* $n \times n$ *matrix* **A**, *and* **u** = **Aa**, **v** = **Ab** *we have* **u** × **v** = **a** × **b**. *Hence the transformation also preserves the **length** or **norm** of any vector* **a** *in* R^n *given by*

$$\|\mathbf{a}\| = \sqrt{\mathbf{a} \times \mathbf{a}} = \sqrt{\mathbf{a}^T\mathbf{a}} \tag{3.8}$$

PROOF: Let **A** be orthogonal. Let **u** = **Aa** and **v** = **Ab**. We must show that **u** × **v** = **a** × **b**. Now $(\mathbf{Aa})^T = \mathbf{a}^T\mathbf{A}^T$ by the rules of transposition discussed in Chapter 1 and $\mathbf{A}^T\mathbf{A} = \mathbf{A}^{-1}\mathbf{A} = \mathbf{I}$ by (3.3). Hence

$$\mathbf{u} \times \mathbf{v} = \mathbf{u}^T\mathbf{v} = (\mathbf{Aa})^T\mathbf{Ab} = \mathbf{a}^T\mathbf{A}^T\mathbf{Ab} = \mathbf{a}^T\mathbf{Ib} = \mathbf{a}^T\mathbf{b} = \mathbf{a} \times \mathbf{b} \tag{3.9}$$

From this the invariance of $\|\mathbf{a}\|$ follows if we set **b** = **a**. ■

Orthogonal matrices have further interesting properties as follows.

THEOREM 4 (Orthonormality of Column and Row Vectors): *A real square matrix is orthogonal if and only if its column vectors* $\mathbf{a}_1, \ldots, \mathbf{a}_n$ *(and also its row vectors) form an* ***orthonormal system****, that is,*

$$\mathbf{a}_j \times \mathbf{a}_k = \mathbf{a}_j^{\mathrm{T}} \mathbf{a}_k = \begin{cases} 0 & \text{if } j \neq k \\ 1 & \text{if } j = k \end{cases} \tag{3.10}$$

PROOF:

(a) Let **A** be orthogonal. Then $\mathbf{A}^{-1}\mathbf{A} = \mathbf{A}^{\mathrm{T}}\mathbf{A} = \mathbf{I}$, in terms of column vectors $\mathbf{a}_1, \ldots, \mathbf{a}_n$,

$$\mathbf{I} = \mathbf{A}^{-1}\mathbf{A} = \mathbf{A}^{\mathrm{T}}\mathbf{A} = \begin{bmatrix} \mathbf{a}_1^{\mathrm{T}} \\ \vdots \\ \mathbf{a}_n^{\mathrm{T}} \end{bmatrix} [\mathbf{a}_1 \cdots \mathbf{a}_n] = \begin{bmatrix} \mathbf{a}_1^{\mathrm{T}}\mathbf{a}_1 & \mathbf{a}_1^{\mathrm{T}}\mathbf{a}_2 & \cdots & \mathbf{a}_1^{\mathrm{T}}\mathbf{a}_n \\ \cdot & \cdot & \cdots & \cdot \\ \mathbf{a}_n^{\mathrm{T}}\mathbf{a}_1 & \mathbf{a}_n^{\mathrm{T}}\mathbf{a}_2 & \cdots & \mathbf{a}_n^{\mathrm{T}}\mathbf{a}_n \end{bmatrix} \tag{3.11}$$

The last equality implies (3.10), by the definition of the $n \times n$ unit matrix **I**. From (3.3) it follows that the inverse of an orthogonal matrix is orthogonal. Now the column vectors of $\mathbf{A}^{-1}$ $(= \mathbf{A}^{\mathrm{T}})$ are the row vectors of **A**. Hence the row vectors of **A** also form an orthonormal system.

(b) Conversely, if the column vectors of **A** satisfy (3.10), the off-diagonal entries in (3.11) must be 0 and the diagonal entries 1. Hence $\mathbf{A}^{\mathrm{T}}\mathbf{A} = \mathbf{I}$, as (3.11) shows. Similarly, $\mathbf{A}\mathbf{A}^{\mathrm{T}} = \mathbf{I}$. This implies $\mathbf{A}^{\mathrm{T}} = \mathbf{A}^{-1}$ because also $\mathbf{A}^{-1}\mathbf{A} = \mathbf{A}\mathbf{A}^{-1} = \mathbf{I}$ and the inverse is unique. Hence **A** is orthogonal. Similarly when the row vectors of **A** form an orthonormal system, by what has been said at the end of part (a). ■

THEOREM 5 (Determinant of an Orthogonal Matrix): *The determinant of an orthogonal matrix has the value* +1 *or* −1.

PROOF: From the results det **AB** = det **A** det **B** and det $\mathbf{A}^{\mathrm{T}}$ = det **A**, we get for an orthogonal matrix

$$1 = \det \mathbf{I} = \det(\mathbf{A}\mathbf{A}^{-1}) = \det(\mathbf{A}\mathbf{A}^{\mathrm{T}}) = \det \mathbf{A} \det \mathbf{A}^{\mathrm{T}} = (\det \mathbf{A})^2 \Rightarrow \det \mathbf{A} = \pm 1$$ ■

Example 4 Illustration of Theorems 4 and 5

The last matrix in Example 1 and the matrix in (3.6) illustrate Theorems 4 and 5 because their determinants are −1 and +1, as you should verify.

THEOREM 6 (Eigenvalues of an Orthogonal Matrix): *The eigenvalues of an orthogonal matrix* **A** *are real or complex conjugates in pairs and have absolute value* 1.

PROOF: The first part of the statement holds for any real matrix **A** because its characteristic polynomial has real coefficients, so that its zeros (the eigenvalues of **A**) must be as indicated. The claim that $|\lambda| = 1$ will be proved in Sec. 3.3. ■

Example 5 Eigenvalues of an Orthogonal Matrix

The orthogonal matrix in Example 1 has the characteristic equation

$$-\lambda^3 + \frac{2}{3}\lambda^2 + \frac{2}{3}\lambda - 1 = 0$$

Now one of the eigenvalues must be real (why?), hence +1 or −1. Trying, we find −1. Division by $\lambda + 1$ gives $-(\lambda^2 - 5\lambda/3 + 1) = 0$ and the two eigenvalues $(5 + i\sqrt{11})/6$ and $(5 - i\sqrt{11})/6$, which have absolute value 1. Verify all of this.

Looking back at this section, you will find that the numerous basic results it contains have relatively short, straightforward proofs. This is typical of large portions of matrix eigenvalue theory.

CONCEPT CHECK

1. **(Verification)** Verify the statements in Example 1.
2. Verify the statements in Examples 3 and 4.
3. Are the eigenvalues of **A** + **B** of the form $\lambda_j + \mu_j$, where λ_j and μ_j are the eigenvalues of **A** and **B**, respectively?
4. **(Orthogonality)** Prove that eigenvectors of a symmetric matrix corresponding to different eigenvalues are orthogonal. Give an example.
5. **(Skew-symmetric matrix)** Show that the inverse of a skew-symmetric matrix is skew-symmetric.
6. Do there exist nonsingular skew-symmetric $n \times n$ matrices with odd n?
7. **(Orthogonal matrix)** Do there exist skew-symmetric orthogonal 3×3 matrices?
8. **(Symmetric matrix)** Do there exist non-diagonal symmetric 3×3 matrices that are orthogonal?
9. Are the following matrices symmetric, skew-symmetric, or orthogonal? Find their spectrum (thereby illustrating Theorems 1 and 5). (Show the details of your work.)

(a) $\begin{bmatrix} 0.96 & -0.28 \\ 0.28 & 0.96 \end{bmatrix}$ (b) $\begin{bmatrix} a & b \\ -b & a \end{bmatrix}$

(c) $\begin{bmatrix} 3 & 1 \\ -1 & 1 \end{bmatrix}$ (d) $\begin{bmatrix} \cos\theta & -\sin\theta & 0 \\ \sin\theta & \cos\theta & 0 \\ 0 & 0 & 1 \end{bmatrix}$

(e) $\begin{bmatrix} 14 & 4 & -2 \\ 4 & 14 & 2 \\ -2 & 2 & 17 \end{bmatrix}$

(f) $\begin{bmatrix} 0 & -6 & -12 \\ 6 & 0 & -12 \\ 12 & 12 & 0 \end{bmatrix}$

(g) $\begin{bmatrix} 0 & 0 & 1 \\ 0 & 1 & 0 \\ -1 & 0 & 0 \end{bmatrix}$

(h) $\begin{bmatrix} 4/9 & 8/9 & 1/9 \\ -7/9 & 4/9 & -4/9 \\ -4/9 & 1/9 & 8/9 \end{bmatrix}$

(i) $\begin{bmatrix} a & b & b \\ b & a & b \\ b & b & a \end{bmatrix}$

10. **(Rotation in space)** Give a geometric interpretation of the transformation $\mathbf{y} = \mathbf{Ax}$ with **A** as in Prob. 9(d) and **x** and **y** referred to a Cartesian coordinate system.
11. **WRITING PROJECT. Section Summary.** Summarize the main concepts and facts in this section, with illustrative examples of your own.

3.2 Quadratic Forms

The study of quadratic forms finds more use in Electrical and Electronics Engineering problems. They aid in deciding the stability of systems. By definition, a **quadratic form Q** in the components $x_1, \ldots, x_n$ of a vector **x** is a sum of n^2 terms, namely,

$$
\begin{aligned}
Q = \mathbf{x}^{T}\mathbf{A}\mathbf{x} &= \sum_{j=1}^{n}\sum_{k=1}^{n} a_{jk}x_j x_k \\
&= a_{11}x_1^{\,2} + a_{12}x_1x_2 + \cdots + a_{1n}x_1x_n \\
&\quad + a_{21}x_2x_1 + a_{22}x_2^{\,2} + \cdots + a_{2n}x_2x_n \\
&\quad + \cdots\cdots\cdots\cdots\cdots\cdots\cdots\cdots \\
&\quad + a_{n1}x_nx_1 + a_{n2}x_nx_2 + \cdots + a_{nn}x_n^{\,3}
\end{aligned}
\tag{3.12}
$$

$\mathbf{A} = [a_{jk}]$ is called the **coefficient matrix** of the form. We may assume that $\mathbf{A}$ is ***symmetric***, because we can take off-diagonal terms together in pairs and write the result as a sum of two equal terms; see the following example.

Example 6 Quadratic Form. Symmetric Coefficient Matrix

Let

$$\mathbf{x}^{T}\mathbf{A}\mathbf{x} = [x_1 \quad x_2]\begin{bmatrix} 3 & 4 \\ 6 & 2 \end{bmatrix}\begin{bmatrix} x_1 \\ x_2 \end{bmatrix} = 3x_1^{\,2} + 4x_1x_2 + 6x_2x_1 + 2x_2^{\,2} = 3x_1^{\,2} + 10x_1x_2 + 2x_2^{\,2}$$

Here $4 + 6 = 10 = 5 + 5$. From the corresponding ***symmetric*** matrix $\mathbf{C} = [c_{jk}]$, where $c_{jk} = (1/2)(a_{jk} + a_{kj})$, thus $c_{11} = 3$, $c_{12} = c_{21} = 5$, $c_{22} = 2$, we get the same result; indeed,

$$\mathbf{x}^{T}\mathbf{C}\mathbf{x} = [x_1 \quad x_2]\begin{bmatrix} 3 & 5 \\ 5 & 2 \end{bmatrix}\begin{bmatrix} x_1 \\ x_2 \end{bmatrix} = 3x_1^{\,2} + 5x_1x_2 + 5x_2x_1 + 2x_2^{\,2} = 3x_1^{\,2} + 10x_1x_2 + 2x_2^{\,2}$$

Quadratic forms occur in physics and geometry, for instance, in connection with conic sections (ellipses $x_1^{\,2}/a^2 + x_2^{\,2}/b^2 = 1$, etc.) and quadratic surfaces (cones, etc.). Their transformation to principal axes is an important practical task related to the diagonalization of matrices, as follows.

By the theorem "A symmetric matrix has an orthonormal basis of eigenvectors for R^n" (Chapter 2) the *symmetric* coefficient matrix $\mathbf{A}$ of (3.12) has an orthonormal basis of eigenvectors. Hence if we take these as column vectors, we obtain a matrix $\mathbf{X}$ that is orthogonal, so that $\mathbf{X}^{-1} = \mathbf{X}^{T}$. From the result (Chapter 2; theorem on "Diagonalization of a Matrix")

$$\mathbf{D} = \mathbf{X}^{-1}\mathbf{A}\mathbf{X}$$

we thus have $\mathbf{A} = \mathbf{X}\mathbf{D}\mathbf{X}^{-1} = \mathbf{X}\mathbf{D}\mathbf{X}^{T}$. Substitution into (3.12) gives

$$Q = \mathbf{x}^{T}\mathbf{X}\mathbf{D}\mathbf{X}^{T}\mathbf{x} \tag{3.13}$$

If we set $\mathbf{X}^{T}\mathbf{x} = \mathbf{y}$, then, since $\mathbf{X}^{T} = \mathbf{X}^{-1}$, we get

$$\mathbf{x} = \mathbf{X}\mathbf{y} \tag{3.14}$$

Furthermore, in (3.13) we have $\mathbf{x}^{T}\mathbf{X} = (\mathbf{X}^{T}\mathbf{x})^{T} = \mathbf{y}^{T}$ and $\mathbf{X}^{T}\mathbf{x} = \mathbf{y}$, so that Q becomes simply

$$Q = \mathbf{y}^{T}\,\mathbf{D}\mathbf{y} = \lambda_1 y_1^{\,2} + \lambda_2 y_2^{\,2} + \cdots + \lambda_n y_n^{\,2} \tag{3.15}$$

This proves the following basic theorem.

THEOREM 7 (Principal Axes Theorem): *The substitution* (3.14) *transforms a quadratic form*

$$Q = \mathbf{x}^T A\mathbf{x} = \sum_{j=1}^{n}\sum_{k=1}^{n} a_{jk}x_j x_k \quad (a_{kj} = a_{jk})$$

to the principal axes form or ***canonical form*** (3.15), *where* $\lambda_1, \ldots, \lambda_n$ *are the (not necessarily distinct) eigenvalues of the (symmetric!) matrix* **A**, *and* **X** *is an orthogonal matrix with corresponding eigenvectors* $\mathbf{x}_1, \ldots, \mathbf{x}_n$, *respectively, as column vectors.*

Example 7 Transformation to Principal Axes. Conic Sections

Find out what type of conic section the following quadratic form represents and transform it to principal axes:

$$Q = 17x_1^{\,2} - 30x_1x_2 + 17x_2^{\,2} = 128$$

Solution: We have $Q = \mathbf{x}^T\mathbf{A}\mathbf{x}$, where

$$\mathbf{A} = \begin{bmatrix} 17 & -15 \\ -15 & 17 \end{bmatrix}, \quad \mathbf{x} = \begin{bmatrix} x_1 \\ x_2 \end{bmatrix}$$

This gives the characteristic equation $(17 - \lambda)^2 - 15^2 = 0$. It has the roots $\lambda_1 = 2$, $\lambda_2 = 32$. Hence (3.15) becomes

$$Q = 2y_1^{\,2} + 32y_2^{\,2}$$

We see that $Q = 128$ represents the ellipse $2y_1^{\,2} + 32y_2^{\,2} = 128$, that is,

$$\frac{y_1^{\,2}}{8^2} + \frac{y_2^{\,2}}{2^2} = 1$$

If we want to know the direction of the principal axes in the x_1x_2-coordinates, we have to determine normalized eigenvectors from $(\mathbf{A} - \lambda\mathbf{I})\,\mathbf{x} = \mathbf{0}$ with $\lambda = \lambda_1 = 2$ and $\lambda = \lambda_2 = 32$ and then use (3.14). We get

$$\begin{bmatrix} 1/\sqrt{2} \\ 1/\sqrt{2} \end{bmatrix} \quad \text{and} \quad \begin{bmatrix} -1/\sqrt{2} \\ 1/\sqrt{2} \end{bmatrix}$$

hence

$$\mathbf{x} = \mathbf{X}\mathbf{y} = \begin{bmatrix} 1/\sqrt{2} & -1/\sqrt{2} \\ 1/\sqrt{2} & 1/\sqrt{2} \end{bmatrix}\begin{bmatrix} y_1 \\ y_2 \end{bmatrix}$$

In equation form we get

$$x_1 = (y_1/\sqrt{2}) - (y_2/\sqrt{2})$$
$$x_2 = (y_1/\sqrt{2}) + (y_2/\sqrt{2})$$

This is a 45° rotation.

Positive, Negative, Definite, and Semi-Definite Quadratic Forms

DEFINITION If a quadratic form with real coefficients has the property that it is equal to or greater than zero for all real values of its variables, it is said to be ***positive***.

DEFINITION A positive form which is zero only for the values $x_1 = x_2 = \cdots = x_n = 0$ is said to be ***positive definite***.

DEFINITION A positive form which is zero for real values other than $x_1 = x_2 = \cdots = x_n = 0$ is said to be ***positive semi-definite***.

DEFINITION A quadratic form with real coefficients which is equal to or less than zero for all real values of its variables is said to be ***negative***.

DEFINITION A negative form which is zero only for the values $x_1 = x_2 = \cdots = x_n = 0$ is said to be ***negative definite***.

DEFINITION A negative form which is zero for real values other than $x_1 = x_2 = \cdots = x_n = 0$ is said to be ***negative semi-definite***.

Clearly a negative definite or a negative semi-definite form can be converted into a positive form of the corresponding type by multiplying it by –1. A quadratic form which can take both positive and negative values for real values of its variables is said to be ***indefinite***. Table 1 provides a better understanding of the concepts explained above.

Note: (i) The real quadratic form $\mathbf{x}^T\mathbf{A}\mathbf{x}$ is positive definite if every principal minor of **A** is positive. (ii) The real quadratic form $\mathbf{x}^T\mathbf{A}\mathbf{x}$ is negative definite if every principal minor of odd order is negative and every principal minor of even order is positive.

Table 1 Examples of Various Types of Quadratic Forms

Type of quadratic form	*Example*
Positive definite	$x_1^2 + x_2^2$
Negative definite	$-(x_1^2 + x_2^2)$
Positive semi-definite	$(x_1 + x_2)^2$
Indefinite	$x_1^2 - x_2^2$

The matrix **A** of the quadratic form $Q(x) = \mathbf{x}^T$ is said to be positive definite, negative definite, positive semi-definite, negative semi-definite or indefinite according to the nature of $Q(x)$. Correspondingly $Q(x)$ is said to be singular or nonsingular according as its matrix **A** is singular or nonsingular, that is, depending on whether $|\mathbf{A}|$ is equal to or different from zero.

The symmetric matrix $\mathbf{A} = (a_{ij})$ is called the matrix of the quadratic form and the rank of **A** is called the rank of the form. If the rank is r (say) is less than n (the number of variables), the quadratic form is called singular, otherwise nonsingular.

Transformation of Quadratic Form

The linear transformation $\mathbf{X} = \mathbf{BY}$ carries the quadratic form with symmetric matrix **A** into the quadratic form $(\mathbf{BY})^T\mathbf{A}(\mathbf{BY}) = \mathbf{Y}^T(\mathbf{B}^T\mathbf{AB})\mathbf{Y}$ with symmetric matrix $\mathbf{B}^T\mathbf{AB}$.

Canonical Form

A quadratic form of rank r can be reduced to form

$$h_1y_1^2 + h_2y_2^2 + \cdots + h_ry_r^2, \quad h_i \neq 0 \tag{3.16}$$

which is called the ***canonical form***. In general the relation

$$\mathbf{Y} = \mathbf{AX} \tag{3.17}$$

where

$$\mathbf{Y} = \begin{bmatrix} y_1 \\ y_2 \\ \cdot\cdot \\ \cdot\cdot \\ \cdot\cdot \\ y_n \end{bmatrix}, \quad \mathbf{A} = \begin{bmatrix} a_{11}a_{12}\cdots\cdots\cdots a_n \\ a_{21}a_{22}\cdots\cdots\cdots a_{2n} \\ \cdots\cdots\cdots\cdots\cdots\cdots \\ \cdots\cdots\cdots\cdots\cdots\cdots \\ a_{n1}a_{n2}\cdots\cdots\cdots a_{nn} \end{bmatrix}, \quad \mathbf{X} = \begin{bmatrix} x_1 \\ x_2 \\ x_3 \\ \cdot\cdot \\ \cdot\cdot \\ x_n \end{bmatrix}$$

gives a linear transformation from n variables $x_1, x_2, \ldots, x_n$ to the variables $y_1, y_2, \ldots, y_n$, that is the transformation of the vector **x** to the vector **y**. The linear transformation (3.17) is said to be orthogonal if it transforms ${y_1}^2 + {y_2}^2 + \cdots + {y_n}^2$ into ${x_1}^2 + {x_2}^2 + \cdots + {x_n}^2$.

Index of a Quadratic Form

Every real quadratic form (of rank say r) can be reduced by a real nonsingular transformation to the canonical form (3.16). The number of positive terms, p, in such a form is called the ***index***.

Signature of a Quadratic Form

The difference between the number of positive (p) and negative terms in (3.16) is called the ***signature*** of the quadratic form. If r is the rank of a quadratic form, then the signature is given by

$$p - (r - p)$$

Example 8

Find a symmetric matrix **A** such that $\mathbf{Q} = \mathbf{x}^{\mathrm{T}}\mathbf{A}\mathbf{x}$ where

$$\mathrm{Q} = {x_1}^2 + 2x_1x_2 + 3{x_2}^2 + 6x_2x_3 + 2{x_3}^2$$

Solution:

$$\mathrm{A} = \begin{bmatrix} \text{Coeff. of } {x_1}^2 & (1/2)\text{ Coeff. of } x_1x_2 & (1/2)\text{ Coeff. of } x_1x_3 \\ (1/2)\text{ Coeff. of } x_2x_1 & \text{Coeff. of } {x_2}^2 & (1/2)\text{ Coeff. of } x_2x_3 \\ (1/2)\text{ Coeff. of } x_3x_1 & (1/2)\text{ Coeff. of } x_3x_2 & \text{Coeff. of } {x_3}^2 \end{bmatrix}$$

$$= \begin{bmatrix} 1 & 1 & 0 \\ 1 & 3 & 3 \\ 0 & 3 & 2 \end{bmatrix}$$

Example 9

Reduce

$$\mathbf{Q} = \mathbf{x}^{\mathrm{T}} \begin{bmatrix} 1 & -2 & 4 \\ -2 & 2 & 0 \\ 4 & 0 & -7 \end{bmatrix} \mathbf{x}$$

to sum of squares (i.e., canonical form).

Solution: Consider

$$\left[\begin{array}{ccc|ccc} 1 & -2 & 4 & 1 & 0 & 0 \\ -2 & 2 & 0 & 0 & 1 & 0 \\ 4 & 0 & -7 & 0 & 0 & 1 \end{array}\right]$$

$$[\mathbf{A}|\mathbf{I}] = \begin{bmatrix} 1 & 2 & 4 \\ 0 & 1 & 4 \\ 0 & 0 & 1 \end{bmatrix} \mathbf{Y}$$

Perform the operations Row 2 → Row 2 + 2 Row 1, Column 2 → Column 2 + 2 Column 1, Row 3 → Row 3 – 4 Row 1, and Column 3 → Column 3 – 4 Column 1 on the left-side matrix of partitioning and simultaneously perform the operations Row 2 → Row 2 + 2 Row 1 and Row 3 → Row 3 – 4 Row 1 on the right-side matrix of the partitioning. We get

$$\left[\begin{array}{ccc|ccc} 1 & 0 & 0 & 1 & 0 & 0 \\ 0 & -2 & 8 & 2 & 1 & 0 \\ 0 & 8 & -23 & -4 & 0 & 1 \end{array}\right]$$

Perform Row 3 → Row 3 + 4 Row 2, Column 3 → Column 3 + 4 Column 2 at a time on the left-side matrix of the partitioning and perform Row 3 → Row 3 + 4 Row 2 on the right-side matrix of the partitioning. We get

$$\left[\begin{array}{ccc|ccc} 1 & 0 & 0 & 1 & 0 & 0 \\ 0 & -2 & 0 & 2 & 1 & 0 \\ 0 & 0 & 9 & 4 & 4 & 1 \end{array}\right] = [\mathbf{D}|\mathbf{B}^{\mathrm{T}}]$$

Thus the transformation

$$\begin{bmatrix} 1 & 2 & 4 \\ 0 & 1 & 4 \\ 0 & 0 & 1 \end{bmatrix} \mathbf{Y}$$

reduces $\mathbf{Q}$ to $y_1^2 - 2y_2^2 + 9y_3^2$.

Lagrange's reduction: For the above problem

$$\begin{aligned} Q &= x_1^2 + 2x_2^2 - 7x_3^2 - 4x_1x_2 + 8x_1x_3 \\ &= x_1^2 - 4x_1(x_2 - 2x_3) + 2x_2^2 - 7x_3^2 \\ &= [x_1^2 - 4x_1(x_2 - 2x_3) + 4(x_2 - 2x_3)] + 2x_2^2 - 7x_3^2 - 4(x_2 - 2x_3)^2 \\ &= (x_1 - 2x_2 + 4x_3)^2 - 2(x_2^2 - 8x_2x_3) - 23x_3^2 \\ &= (x_1 - 2x_2 + 4x_3)^2 - 2(x_2^2 - 8x_2x_3 + 16x_3^2) + 9x_3^2 \\ &= (x_1 - 2x_2 + 4x_3)^2 - 2(x_2^2 - 4x_3)^2 + 9x_3^2 \end{aligned}$$

Thus

$$\begin{aligned} y_1 &= x_1 - 2x_2 + 4x_3 \quad \text{or} \quad x_1 = y_1 + y_2 + 4y_3 \\ y_2 &= x_2 - 4x_3 \quad \text{or} \quad x_2 = y_2 + 4y_3 \\ y_3 &= x_3 \quad \text{or} \quad x_3 = y_3 \end{aligned}$$

reduces Q to $y_1^2 - 2y_2^2 + 9y_3^2$.

Example 10

Classify the following quadratic form

$$3x_1^2 + 3x_2^2 + 6x_3^2 - 2x_1x_2 - 4x_1x_3$$

and find the rank, index, and signature of the quadratic form.

Solution:

$$A = \begin{bmatrix} 3 & -1 & -2 \\ -1 & 3 & 0 \\ -2 & 0 & 6 \end{bmatrix}$$

$$= \begin{bmatrix} 3 & 0 & -2 \\ 0 & 8/3 & -2/3 \\ -2 & -2/3 & 16/3 \end{bmatrix} \quad \begin{array}{l} \text{Row } 2 \to \text{Row } 2 - (1/3) \text{ Row } 1 \\ \text{Column } 2 \to \text{Column } 2 - (1/3) \text{ Column } 1 \end{array}$$

$$= \begin{bmatrix} 3 & 0 & 0 \\ 0 & 8/3 & -2/3 \\ 0 & -2/3 & 4 \end{bmatrix} \quad \begin{array}{l} \text{Row } 3 \to \text{Row } 3 - (2/3) \text{ Row } 1 \\ \text{Column } 3 \to \text{Column } 3 - (2/3) \text{ Column } 1 \end{array}$$

$$= \begin{bmatrix} 3 & 0 & 0 \\ 0 & 8/3 & 0 \\ 0 & 0 & 23/6 \end{bmatrix} \quad \begin{array}{l} \text{Row } 3 \to \text{Row } 3 - (1/4) \text{ Row } 2 \\ \text{Column } 3 \to \text{Column } 3 - (1/4) \text{ Column } 2 \end{array}$$

Sum of squares is

$$3y_1^{\,2} + \frac{8}{3}y_2^{\,2} + \frac{23}{6}y_3^{\,2}$$

which shows that the given quadratic form is positive definite.

Number of positive signs = 3
Number of negative signs = 0
Rank = 3
Index of the quadratic form = 3
Signature of the quadratic form = 3 − 0 = 3

Example 11

Classify the following quadratic form:

$$x_1^{\,2} + 2x_2^{\,2} + 2x_3^{\,2} + 4x_4^{\,2} - 2x_1x_2 + 2x_2x_3 + 6x_3x_4$$

Solution:

$$A = \begin{bmatrix} 1 & -1 & 0 & 0 \\ -1 & 2 & 1 & 0 \\ 0 & 1 & 2 & 3 \\ 0 & 0 & 3 & 4 \end{bmatrix}$$

$$= \begin{bmatrix} 1 & 0 & 0 & 0 \\ 0 & 1 & 1 & 0 \\ 0 & 1 & 2 & 3 \\ 0 & 1 & 3 & 4 \end{bmatrix} \begin{matrix} \text{Row } 2 \to \text{Row } 2 - \text{Row } 1 \\ \text{Column } 2 \to \text{Column } 2 - \text{Column } 1 \end{matrix}$$

$$= \begin{bmatrix} 1 & 0 & 0 & 0 \\ 0 & 1 & 0 & 0 \\ 0 & 0 & 1 & 3 \\ 0 & 0 & 3 & 4 \end{bmatrix} \begin{matrix} \text{Row } 3 \to \text{Row } 3 - \text{Row } 2 \\ \text{Column } 3 \to \text{Column } 3 - \text{Column } 2 \end{matrix}$$

$$= \begin{bmatrix} 1 & 0 & 0 & 0 \\ 0 & 1 & 0 & 0 \\ 0 & 0 & 1 & 0 \\ 0 & 0 & 0 & -5 \end{bmatrix} \begin{matrix} \text{Row } 4 \to \text{Row } 4 - 3\text{Row } 3 \\ \text{Column } 4 \to \text{Column } 4 - 3\text{ Column } 3 \end{matrix}$$

Sum of squares is

$$y_1^{\,2} + y_2^{\,2} + y_3^{\,2} - 5y_4^{\,2}$$

which shows that the given quadratic form is indefinite.

Example 12

Classify the following quadratic form:

$$2x_1^{\,2} + 2x_2^{\,2} + x_3^{\,2} + 2x_1x_3 + 2x_2x_3$$

Solution:

$$A = \begin{bmatrix} 2 & 0 & 1 \\ 0 & 2 & 1 \\ 1 & 1 & 1 \end{bmatrix}$$

$$= \begin{bmatrix} 2 & 0 & 0 \\ 0 & 2 & 1 \\ 0 & 1 & 1/2 \end{bmatrix} \quad \begin{matrix} \text{Row } 3 \to \text{Row } 3 - (1/2)\,\text{Row } 2 \\ \text{Column } 3 \to \text{Column } 3 - (1/2)\text{ Column } 2 \end{matrix}$$

$$= \begin{bmatrix} 2 & 0 & 0 \\ 0 & 2 & 0 \\ 0 & 0 & 0 \end{bmatrix} \quad \begin{matrix} \text{Row } 3 \to \text{Row } 3 - (1/2)\,\text{Row } 2 \\ \text{Column } 3 \to \text{Column } 3 - (1/2)\text{ Column } 2 \end{matrix}$$

Sum of the squares is

$$2y_1^{\,2} + 2y_2^{\,2}$$

which shows that the given quadratic form is semi-definite.

CONCEPT CHECK

1. Reduce the quadratic form $7x^2 + 6y^2 + 5z^2 - 4xy - 4yz$ to the canonical form. [JNTU May 2007]
2. Find the eigenvectors of the matrix

$$\begin{bmatrix} 6 & -2 & 2 \\ -2 & 3 & -1 \\ 2 & -1 & 3 \end{bmatrix}$$

   and hence reduce $6x^2 + 3y^2 + 3z^2 - 2yz + 4zx - 4xy$ to a sum of squares. [JNTU 2000]
3. Reduce the quadratic form $3x^2 + 5y^2 + 3z^2 - 2yz - 2zx - 2xy$ to the canonical form. [JNTU 2006, 2005]
4. Reduce the quadratic form $3x_1^{\,2} + 3x_2^{\,2} + 3x_3^{\,2} + 2x_1x_2 + 2x_1x_3 - 2x_2x_3$ into sum of squares form by an orthogonal transformation and give the matrix of transformation.

3.3 Sylvester's Law of Inertia

Let

$$Q(A) = C_0A^n + C_1A^{n-1} + C_2A^{n-2} + \cdots + C_{n-1}A + C_nI$$

and

$$|\lambda \mathbf{I} - \mathbf{A}| = f(\lambda)$$

Let the adjoint matrix of $[\lambda\mathbf{I} - \mathbf{A}] = [f(\lambda)]$ be

$$\mathbf{Z}(\lambda) = \frac{|f(\lambda)|}{f'(\lambda)} = \frac{\text{Adjoint matrix } [\lambda\mathbf{I} - \mathbf{A}]}{f'(\lambda)}$$

Then according to Sylvester's theorem

$$\mathbf{Q}(\mathbf{A}) = \mathbf{Q}(\lambda_1)\mathbf{Z}(\lambda_1) + \mathbf{Q}(\lambda_2)\mathbf{Z}(\lambda_2) + \cdots + \mathbf{Q}(\lambda_n)\mathbf{Z}(\lambda_n) = r = \sum_{i=1}^{n} \mathbf{Q}(\lambda_i)\mathbf{Z}(\lambda_i)$$

Example 13

If

$$\mathbf{A} = \begin{bmatrix} 3 & 0 \\ 0 & 1 \end{bmatrix}$$

find $\mathbf{A}^{200}$.

Solution:

$$f(\lambda) = |\lambda\mathbf{I} - \mathbf{A}| = \left| \begin{bmatrix} \lambda & 0 \\ 0 & \lambda \end{bmatrix} - \begin{bmatrix} 3 & 0 \\ 0 & 1 \end{bmatrix} \right| = \begin{vmatrix} \lambda - 3 & 0 \\ 0 & \lambda - 1 \end{vmatrix}$$

$$f(\lambda) = (\lambda - 3)(\lambda - 1) = 0$$

Using $f(\lambda) = 0$ we get two roots $\lambda_1 = 1$, $\lambda_2 = 3$. Now

$$f(\lambda) = (\lambda - 3)(\lambda - 1) = \lambda^2 - 4\lambda + 3$$

Differentiating we get

$$f'(\lambda) = 2\lambda - 4$$

For $\lambda_1 = 1$ and $\lambda_2 = 3$ we get

$$f'(1) = -2,\ f'(3) = 2$$

Also $[f(\lambda)]$ = adjoint matrix of the matrix $[\lambda\mathbf{I} - \mathbf{A}]$. Therefore

$$\mathbf{Z}(\lambda_1) = \mathbf{Z}(1) = \frac{f(1)}{f'(1)} = \begin{bmatrix} \lambda - 1 & 0 \\ 0 & \lambda - 3 \end{bmatrix} = \begin{bmatrix} 0 & 0 \\ 0 & 1 \end{bmatrix}$$

$$\mathbf{Z}(\lambda_2) = \mathbf{Z}(3) = \frac{f(3)}{f'(3)} = \frac{1}{2}\begin{bmatrix} 2 & 0 \\ 0 & 0 \end{bmatrix} = \begin{bmatrix} 1 & 0 \\ 0 & 1 \end{bmatrix}$$

By Sylvester theorem

$$Q(A) = Q(\lambda_1)Z(\lambda_1) + Q(\lambda_2)Z(\lambda_2)$$

$$\begin{aligned} A^{200} &= Q(\lambda_1)Z(\lambda_1) + Q(\lambda_2)Z(\lambda_2) \\ &= (\lambda_1)^{200}\begin{bmatrix} 0 & 0 \\ 0 & 1 \end{bmatrix} + (\lambda_2)^{200}\begin{bmatrix} 1 & 0 \\ 0 & 1 \end{bmatrix} \\ &= 1^{200}\begin{bmatrix} 0 & 0 \\ 0 & 1 \end{bmatrix} + 3^{200}\begin{bmatrix} 1 & 0 \\ 0 & 0 \end{bmatrix} \\ &= \begin{bmatrix} 0 & 0 \\ 0 & 1 \end{bmatrix} + \begin{bmatrix} 3^{200} & 0 \\ 0 & 0 \end{bmatrix} \\ &= \begin{bmatrix} 3^{200} & 0 \\ 0 & 1 \end{bmatrix} \end{aligned}$$

CONCEPT CHECK

1. Find A^{100} where

$$A = \begin{bmatrix} 2 & 0 \\ 0 & 1 \end{bmatrix}$$

2. Find A^{50} where

$$A = \begin{bmatrix} 1 & 0 \\ 0 & 3 \end{bmatrix}$$

3.4 Complex Matrices

The three classes of real matrices in Sec. 3.1 have complex counterparts that are of practical interest in certain applications, mainly because of their spectra (see Theorem 8 in this section), for instance, in quantum mechanics. To define these classes, we need the following standard notations:

NOTATIONS: $\bar{A} = [\,\overline{a_{jk}}\,]$ is obtained from $A = [a_{jk}]$ by replacing each entry $a_{jk} = \alpha + i\beta$ (α, β real) with its complex conjugate $\overline{a_{jk}} = \alpha - i\beta$. Also, $(\bar{A})^T = [\,\overline{a_{kj}}\,]$ is the transpose of $(\bar{A})^T$, hence the conjugate transpose of A.

Example 14 Notations

If $A = \begin{bmatrix} 3+4i & 1-i \\ 6 & 2-5i \end{bmatrix}$, then $\bar{A} = \begin{bmatrix} 3-4i & 1+i \\ 6 & 2+5i \end{bmatrix}$ and $(\bar{A})^T = \begin{bmatrix} 3-4i & 6 \\ 1+i & 2+5i \end{bmatrix}$

DEFINITION (Hermitian, Skew-Hermitian, and Unitary Matrices)

A square matrix $\mathbf{A} = [a_{kj}]$ is called

(a) **Hermitian** if $(\bar{\mathbf{A}})^{\mathrm{T}} = \mathbf{A}$, that is, $\overline{a_{kj}} = a_{jk}$

(b) **Skew-Hermitian** if $(\bar{\mathbf{A}})^{\mathrm{T}} = -\mathbf{A}$, that is, $\overline{a_{kj}} = -a_{jk}$

(c) **Unitary** if $(\bar{\mathbf{A}})^{\mathrm{T}} = \mathbf{A}^{-1}$

The first two classes are named after Hermite. From the definitions we see the following. If **A** is Hermitian, the entries on the main diagonal must satisfy $\overline{a_{jj}} = a_{jj}$; that is, they are real. Similarly, if **A** is skew-Hermitian, then $\overline{a_{jj}} = -a_{jj}$. If we set $a_{jj} = \alpha + i\beta$, this becomes $\alpha - i\beta = -(\alpha + i\beta)$. Hence $\alpha = 0$, so that a_{jj} must be pure imaginary or 0.

Example 15 Hermitian, Skew-Hermitian, and Unitary Matrices

$$\mathbf{A} = \begin{bmatrix} 4 & 1-3i \\ 1+3i & 7 \end{bmatrix}, \quad \mathbf{B} = \begin{bmatrix} 3i & 2+i \\ -2+i & -i \end{bmatrix}, \quad \mathbf{C} = \begin{bmatrix} \frac{1}{2}i & \frac{1}{2}\sqrt{3} \\ \frac{1}{2}\sqrt{3} & \frac{1}{2}i \end{bmatrix}$$

are Hermitian, skew-Hermitian, and unitary matrices, respectively, as you may verify by using the definitions.

Note the following:

1. If a Hermitian matrix is real, then $(\bar{\mathbf{A}})^{\mathrm{T}} = \mathbf{A}^{\mathrm{T}} = \mathbf{A}$. Hence a real Hermitian matrix is a symmetric matrix (Sec. 3.1).
2. Similarly, if a skew-Hermitian matrix is real, then $(\bar{\mathbf{A}})^{\mathrm{T}} = (\bar{\mathbf{A}})^{\mathrm{T}} = -\mathbf{A}$. Hence a real skew-Hermitian matrix is a skew-symmetric matrix.
3. Finally, if a unitary matrix is real, then $(\bar{\mathbf{A}})^{\mathrm{T}} = \mathbf{A}^{\mathrm{T}} = \mathbf{A}^{-1}$. Hence a real unitary matrix is an orthogonal matrix.

This shows that Hermitian, skew-Hermitian, and unitary matrices generalize symmetric, skew-symmetric, and orthogonal matrices, respectively.

Eigenvalues

It is quite remarkable that the matrices under consideration have spectra (sets of eigenvalues; see Chapter 2) that can be characterized in a general way as follows (see Figure 1).

THEOREM 8 (Eigenvalues):

(a) *The eigenvalues of a Hermitian matrix (and thus of a symmetric matrix) are real.*

(b) *The eigenvalues of a skew-Hermitian matrix (and thus of a skew-symmetric matrix) are pure imaginary or zero.*

(c) *The eigenvalues of a unitary matrix (and thus of an orthogonal matrix) have absolute value* 1.

PROOF: We prove Theorem 8. Let λ be an eigenvalue and $\mathbf{x}$ an eigenvector of **A**. Multiply $\mathbf{Ax} = \lambda\mathbf{x}$ from the left by $(\bar{\mathbf{x}})^{\mathrm{T}}$, thus $(\bar{\mathbf{x}})^{\mathrm{T}}\mathbf{Ax} = \lambda(\bar{\mathbf{x}})^{\mathrm{T}}\mathbf{x}$, and divide by $(\bar{\mathbf{x}})^{\mathrm{T}}\mathbf{x} = \overline{x_1}x_1 + \cdots + \overline{x_n}x_n = |x_1|^2 + \cdots + |x_n|^2$, which is real and not 0 because $\mathbf{x} \neq \mathbf{0}$. This gives

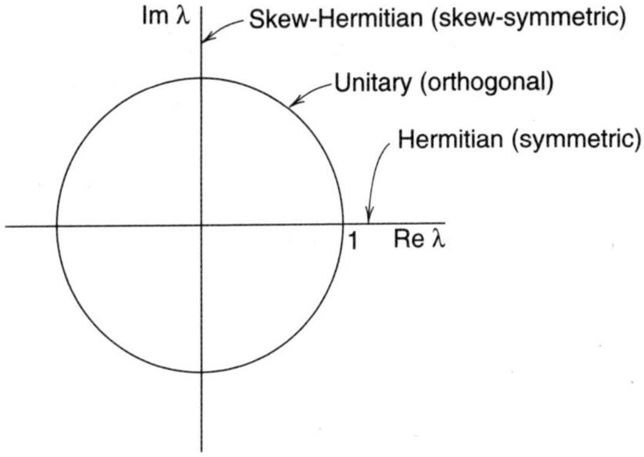

Figure 1 Location of the eigenvalues of Hermitian, skew-Hermitian, and unitary matrices in the complex λ plane.

$$\lambda = \frac{(\bar{\mathbf{x}})^{\mathrm{T}} \mathbf{A}\mathbf{x}}{(\bar{\mathbf{x}})^{\mathrm{T}} \mathbf{x}} \tag{3.18}$$

1. If $\mathbf{A}$ is Hermitian, $(\bar{\mathbf{A}})^{\mathrm{T}} = \mathbf{A}$ or $\mathbf{A}^{\mathrm{T}} = \bar{\mathbf{A}}$ and we show that then the numerator in (3.18) is real, which makes λ real. $(\bar{\mathbf{x}})^{\mathrm{T}}\mathbf{A}\mathbf{x}$ is a scalar; hence taking the transpose has no effect. Thus

$$(\bar{\mathbf{x}})^{\mathrm{T}} \mathbf{A}\mathbf{x} = [(\bar{\mathbf{x}})^{\mathrm{T}} (\mathbf{A}\mathbf{x})^{\mathrm{T}}] = \mathbf{x}^{\mathrm{T}}\mathbf{A}^{\mathrm{T}}\bar{\mathbf{x}} = \mathbf{x}^{\mathrm{T}}\bar{\mathbf{A}}\bar{\mathbf{x}} = \overline{[(\bar{\mathbf{x}})^{\mathrm{T}} (\mathbf{A}\mathbf{x})^{\mathrm{T}}]} \tag{3.19}$$

   Hence, $(\bar{\mathbf{x}})^{\mathrm{T}}\mathbf{A}\mathbf{x}$ equals its complex conjugate, so that it must be real. ($a + ib = a - ib$ implies $b = 0$.)

2. If $\mathbf{A}$ is skew-Hermitian, $\mathbf{A}^{\mathrm{T}} = -\bar{\mathbf{A}}$ and instead of (3.19) we obtain

$$(\bar{\mathbf{x}})^{\mathrm{T}} \mathbf{A}\mathbf{x} = -\overline{[(\bar{\mathbf{x}})^{\mathrm{T}} (\mathbf{A}\mathbf{x})^{\mathrm{T}}]} \tag{3.20}$$

   so that $(\bar{\mathbf{x}})^{\mathrm{T}}\mathbf{A}\mathbf{x}$ equals minus its complex conjugate and is pure imaginary or 0. [$a + ib = -(a - ib)$ implies $a = 0$.]

3. Let $\mathbf{A}$ be unitary. We take $\mathbf{A}\mathbf{x} = \lambda\mathbf{x}$ and its conjugate transpose

$$(\bar{\mathbf{A}}\bar{\mathbf{x}})^{\mathrm{T}} = (\bar{\lambda}\bar{\mathbf{x}})^{\mathrm{T}} = \bar{\lambda}(\bar{\mathbf{x}})^{\mathrm{T}}$$

   and multiply the two left sides and the two right sides,

$$(\bar{\mathbf{A}}\bar{\mathbf{x}})^{\mathrm{T}} \mathbf{A}\mathbf{x} = \bar{\lambda}\lambda(\bar{\mathbf{x}})^{\mathrm{T}}\mathbf{x} = |\lambda|^2 (\bar{\mathbf{x}})^{\mathrm{T}}\mathbf{x}$$

   But $\mathbf{A}$ is unitary, $(\bar{\mathbf{A}})^{\mathrm{T}} = \mathbf{A}^{-1}$, so that on the left we obtain

$$(\bar{\mathbf{A}}\bar{\mathbf{x}})^{\mathrm{T}} \mathbf{A}\mathbf{x} = (\bar{\mathbf{x}})^{\mathrm{T}}(\bar{\mathbf{A}})^{\mathrm{T}}\mathbf{A}\mathbf{x} = (\bar{\mathbf{x}})^{\mathrm{T}}\mathbf{A}^{-1}\mathbf{A}\mathbf{x} = (\bar{\mathbf{x}})^{\mathrm{T}}\mathbf{I}\,\mathbf{x} = (\bar{\mathbf{x}})^{\mathrm{T}}\mathbf{x}$$

Together, $(\bar{\mathbf{x}})^{\mathrm{T}}\mathbf{x} = |\lambda|^2(\bar{\mathbf{x}})^{\mathrm{T}}\mathbf{x}$. We now divide by $(\bar{\mathbf{x}})^{\mathrm{T}}\mathbf{x}$ ($\neq 0$) to get $|\lambda|^2 = 1$. Hence $|\lambda| = 1$. This proves Theorem 8 as well as Theorems 2 and 6 in Sec. 3.1. ■

Key properties of orthogonal matrices (invariance of the inner product, orthonormality of rows and columns) generalize to unitary matrices in a remarkable way. To see this, instead of R^n we now use the **complex vector space** C^n of all complex vectors with n complex numbers as components and complex numbers as scalars. For such complex vectors the **inner product** is defined by (note the overbar for the complex conjugate)

$$\mathbf{a} \times \mathbf{b} = (\overline{\mathbf{a}})^{\mathrm{T}}\mathbf{b} \tag{3.21}$$

The **length** or **norm** of such a complex vector is a *real* number defined by

$$\|\mathbf{a}\| = \sqrt{\mathbf{a} \times \mathbf{a}} = \sqrt{(\overline{\mathbf{a}})^{\mathrm{T}}\mathbf{a}} = \sqrt{\overline{a}_1 a_1 + \cdots + \overline{a}_n a_n} = \sqrt{|a_1|^2 + \cdots + |a_n|^2} \tag{3.22}$$

Example 16 Illustration of Theorem 3

For the matrices in Example 15 we find by direct calculation

Matrix		*Characteristic equation*	*Eigenvalues*
A	Hermitian	$\lambda^2 - 11\lambda + 18 = 0$	$9, \quad 2$
B	Skew-Hermitian	$\lambda^2 - 2i\lambda + 8 = 0$	$4i, \quad -2i$
C	Unitary	$\lambda^2 - i\lambda - 1 = 0$	$\frac{1}{2}\sqrt{3} + \frac{1}{2}i, \quad -\frac{1}{2}\sqrt{3} + \frac{1}{2}i$

and

$$\left| \pm\frac{1}{2}\sqrt{3} + \frac{1}{2}i \right|^2 = \frac{3}{4} + \frac{1}{4} = 1$$

THEOREM 9 (Invariance of Inner Product): *A unitary transformation, that is,* $\mathbf{y} = \mathbf{A}\mathbf{x}$ *with a unitary matrix* $\mathbf{A}$, *preserves the value of the inner product* (3.21), *hence also the norm* (3.22).

PROOF: The proof is the same as that of Theorem 3.21 which the theorem generalizes. In the analog of (3.9), we now have bars,

$$\mathbf{u} \times \mathbf{v} = (\overline{\mathbf{u}})^{\mathrm{T}}\mathbf{v} = (\overline{\mathbf{A}}\overline{\mathbf{a}})^{\mathrm{T}}\mathbf{A}\mathbf{b} = (\overline{\mathbf{a}})^{\mathrm{T}}(\overline{\mathbf{A}})^{\mathrm{T}}\mathbf{A}\mathbf{b} = (\overline{\mathbf{a}})^{\mathrm{T}}\mathbf{I}\mathbf{b} = (\overline{\mathbf{a}})^{\mathrm{T}}\mathbf{b} = \mathbf{a} \times \mathbf{b}$$ ■

The complex analog of an orthonormal systems of real vectors (see Sec. 3.1) is defined as follows.

DEFINITION **(Unitary System)** A *unitary system* is a set of complex vectors satisfying the relationships

$$\mathbf{a}_j \times \mathbf{b}_k = (\overline{\mathbf{a}_j})^{\mathrm{T}}\mathbf{a}_k = \begin{cases} 0 & \text{if} \quad j \neq k \\ 1 & \text{if} \quad j = k \end{cases} \tag{3.23}$$

Theorem 4 in Sec. 3.1 extends to complex as follows.

THEOREM 10 (Unitary Systems of Column and Row Vectors): *A complex square matrix is unitary if and only if its column vectors (and also its row vectors) form a unitary system.*

PROOF: The proof is the same as that of Theorem 4 except for the bars required in $(\overline{\mathbf{A}})^{\mathrm{T}} = \mathbf{A}^{-1}$ and in (3.21) and (3.23). ■

THEOREM 11 (Determinant of a Unitary Matrix): *Let* **A** *be a unitary matrix. Then its determinant has absolute value one, that is,* $|\det \mathbf{A}| = 1$.

PROOF: Similarly as in Sec. 3.1 we obtain

$$1 = \det(\mathbf{A}\mathbf{A}^{-1}) = \det \mathbf{A}\,(\bar{\mathbf{A}})^{\mathrm{T}} = \det \mathbf{A}\ \det\ (\bar{\mathbf{A}})^{\mathrm{T}} = \det \mathbf{A}\ \det\ (\bar{\mathbf{A}})^{\mathrm{T}}$$
$$= \det \mathbf{A} + \overline{\det \mathbf{A}} = |\det \mathbf{A}|^2$$

Hence $|\det \mathbf{A}| = 1$ (where det **A** may now be complex). ■

Example 17 Unitary Matrix Illustrating Theorems 1c and 2–4

For the vectors $\mathbf{a}^{\mathrm{T}} = [2 \quad -i]$ and $\mathbf{b}^{\mathrm{T}} = [1+i \quad 4i]$ we get $(\bar{\mathbf{a}})^{\mathrm{T}} = [2 \quad i]^{\mathrm{T}}$ and $(\bar{\mathbf{a}})^{\mathrm{T}}\mathbf{b} = 2(1+i) - 4 = -2 + 2i$ and with

$$\mathbf{A} = \begin{bmatrix} 0.8i & 0.6 \\ 0.6 & 0.8i \end{bmatrix} \quad \text{also} \quad \mathbf{A}\mathbf{a} = \begin{bmatrix} i \\ 2 \end{bmatrix} \quad \text{and} \quad \mathbf{A}\mathbf{b} = \begin{bmatrix} -0.8+3.2i \\ -2.6+0.6i \end{bmatrix}$$

as one can readily verify. This gives $(\overline{\mathbf{A}\mathbf{a}})^{\mathrm{T}}\mathbf{A}\mathbf{b} = -2 + 2i$, illustrating Theorem 2. The matrix is unitary. Its columns form a unitary system,

$$(\bar{\mathbf{a}}_1)^{\mathrm{T}}\mathbf{a}_1 = -0.8i \times 0.8i + 0.6^2 = 1, \qquad (\bar{\mathbf{a}}_1)^{\mathrm{T}}\mathbf{a}_2 = -0.8i \times 0.6 + 0.6 \times 0.8i = 0$$
$$(\bar{\mathbf{a}}_2)^{\mathrm{T}}\mathbf{a}_2 = 0.6^2 + (-0.8i)0.8i = 1$$

and so do its rows. Also, $\det \mathbf{A} = -1$. The eigenvalues are $0.6 + 0.8i$ and $-0.6 + 0.8i$, with eigenvectors $[1 \quad 1]^{\mathrm{T}}$ and $[1 \quad -1]^{\mathrm{T}}$, respectively.

THEOREM 12 (Basis of Eigenvectors): *A Hermitian, skew-Hermitian, or unitary matrix has a basis of eigenvectors for* C^n *that is a unitary system.*

Example 18 Unitary Eigenbases

The matrices **A, B, C** in Example 15 have the following unitary systems of eigenvectors, as you should verify.

$$\mathbf{A}:\quad \frac{1}{\sqrt{35}}|1-3i \quad 5|^{\mathrm{T}} \quad (\lambda = 9), \quad \frac{1}{\sqrt{14}}|1-3i \quad -2|^{\mathrm{T}} \quad (\lambda = 2)$$

$$\mathbf{B}:\quad \frac{1}{\sqrt{30}}|1-2i \quad -5|^{\mathrm{T}} \ (\lambda = -2i), \quad \frac{1}{\sqrt{30}}|5 \quad 1+2i|^{\mathrm{T}} \quad (\lambda = 4i)$$

$$\mathbf{C}:\quad \frac{1}{\sqrt{2}}|1 \quad 1|^{\mathrm{T}} \quad \left[\lambda = \frac{1}{2}(i+\sqrt{3})\right], \quad \frac{1}{\sqrt{2}}|1 \quad -1|^{\mathrm{T}} \quad \left[\lambda = \frac{1}{2}(i-\sqrt{3})\right]$$

Hermitian and Skew-Hermitian Forms

The concept of a quadratic form (Sec. 3.2) can be extended to complex. We call the numerator $(\bar{\mathbf{x}})^{\mathrm{T}}\mathbf{A}\mathbf{x}$ in (3.15) a **form** in the components $x_1, \ldots, x_n$ of **x**, which may now be complex. This form is again a sum of n^2 terms:

$$(\overline{\mathbf{x}})^{\mathrm{T}}\mathbf{A}\mathbf{x} = \sum_{j=1}^{n}\sum_{k=1}^{n} a_{jk}\overline{x_j}\,x_k = a_{11}\,\overline{x_1}x_1 + \cdots + a_{1n}\overline{x_1}x_n$$
$$+ a_{21}\overline{x_2}\,x_1 + \cdots + a_{2n}\overline{x_2}\,x_n \tag{3.24}$$
$$+ \cdots\cdots\cdots\cdots\cdots\cdots$$
$$+ a_{n1}\overline{x_n}\,x_1 + \cdots + a_{nn}\overline{x_n}\,x_n$$

A is called its **coefficient matrix.** The form is called a **Hermitian** or **skew-Hermitian form** if **A** is Hermitian or skew-Hermitian, respectively. *The value of a Hermitian form is real, and that of a skew-Hermitian form is pure imaginary or zero.* This can be seen directly from (3.19) and (3.20) and accounts for the importance of these forms in physics. Note that (3.19) and (3.20) are valid for any vectors because in the proof of (3.19) and (3.20) we did not use that **x** is an eigenvector but only that $(\overline{\mathbf{x}})^{\mathrm{T}}\mathbf{x}$ is real and not 0.

Example 19 Hermitian Form

For **A** in **Example** 15 and, say, $\mathbf{x} = [1+i \quad 5i]^{\mathrm{T}}$ we get

$$(\overline{\mathbf{x}})^{\mathrm{T}}\mathbf{A}\mathbf{x} = [(1-i) \quad -5i]\begin{bmatrix} 4 & 1-3i \\ 1+3i & 7 \end{bmatrix}\begin{bmatrix} 1+i \\ 5i \end{bmatrix} = [(1-i) \quad -5i]\begin{bmatrix} 4(1+i)+(1-3i)\times 5i \\ (1+3i)(1+i)+7\times 5i \end{bmatrix} = 223$$

Clearly, if **A** and **x** in (3.18) are real, then (3.21) reduces to a quadratic form.

CONCEPT CHECK

1. **(Verification)** Verify the statements in Examples 14 and 15.
2. **(Product)** Show $(\overline{\mathbf{BA}})^{\mathrm{T}} = -\mathbf{AB}$ for **A** and **B** in Example 15. For any $n \times n$ Hermitian **A** and skew-Hermitian **B**.
3. Show that $(\overline{\mathbf{ABC}})^{\mathrm{T}} = -\mathbf{C}^{-1}\mathbf{BA}$ for any $n \times n$ Hermitian **A**, skew-Hermitian **B**, and unitary **C**.
4. **(Eigenvectors)** Find eigenvectors of **A, B, C** in Examples 15 and 16.
5. Are the following matrices Hermitian? Skew-Hermitian? Unitary? Find their eigenvalues (thereby verifying Theorem 8) and eigenvectors.

(a) $\begin{bmatrix} 4 & i \\ -i & 2 \end{bmatrix}$ (b) $\begin{bmatrix} 0 & 2i \\ 2i & 0 \end{bmatrix}$

(c) $\begin{bmatrix} 1/\sqrt{2} & i/\sqrt{2} \\ -i/\sqrt{2} & -1/\sqrt{2} \end{bmatrix}$ (d) $\begin{bmatrix} 0 & i \\ i & 0 \end{bmatrix}$

(e) $\begin{bmatrix} 5i & 0 & 0 \\ 0 & 0 & 5i \\ 0 & 5i & 0 \end{bmatrix}$ (f) $\begin{bmatrix} 0 & 1+i & 0 \\ 1-i & 0 & 1+i \\ 0 & 1-i & 0 \end{bmatrix}$

(g) $\begin{bmatrix} 0 & 0 & i \\ 0 & i & 0 \\ i & 0 & 0 \end{bmatrix}$

6. **PROJECT. Complex Matrices**

   (a) **Decomposition**. Show that any square matrix may be written as the sum of a Hermitian and a skew-Hermitian matrix. Give examples.

   (b) **Normal matrix.** This important concept denotes a matrix that commutes with its conjugate transpose, $\mathbf{A}(\overline{\mathbf{A}})^{\mathrm{T}} = (\overline{\mathbf{A}})^{\mathrm{T}}\mathbf{A}$. Prove that Hermitian, skew-Hermitian, and unitary matrices are normal. Give corresponding examples of your own.

(c) **Normality criterion.** Prove that **A** is normal if and only if the Hermitian and skew-Hermitian matrices in (a) commute.

(d) Find a simple matrix that is not normal. Find a normal matrix that is not Hermitian, skew-Hermitian, or unitary.

(e) **Unitary matrices.** Prove that the product of two unitary $n \times n$ matrices and the inverse of a unitary matrix are unitary. Give examples.

(f) **Powers of unitary matrices** in applications may sometimes be very simple. Show that $\mathbf{C}^{12} = \mathbf{I}$ in Example 15. Find further examples.

7. Is the given matrix (call it **A**) Hermitian or skew-Hermitian? Find $(\overline{\mathbf{x}})^{\mathrm{T}}\mathbf{A}\mathbf{x}$. (Show all the details.) a, b, c, k are real.

(a) $\begin{bmatrix} 0 & -3i \\ -3i & 0 \end{bmatrix}, \mathbf{x} = \begin{bmatrix} 4+i \\ 3-i \end{bmatrix}$

(b) $\begin{bmatrix} a & b+ic \\ b-ic & k \end{bmatrix}, \mathbf{x} = \begin{bmatrix} x_1 \\ x_2 \end{bmatrix}$

(c) $\begin{bmatrix} 2 & 1+i \\ 1-i & 1 \end{bmatrix}, \mathbf{x} = \begin{bmatrix} i \\ 2i \end{bmatrix}$

8. **(Pauli spin matrices)** Find the eigenvalues and eigenvectors of the so-called *Pauli spin matrices* and show that $\mathbf{S}_x\mathbf{S}_y = i\mathbf{S}_z$, $\mathbf{S}_y\mathbf{S}_x = -i\mathbf{S}_z$, $\mathbf{S}_x^{\,2} = \mathbf{S}_x^{\,2} = \mathbf{S}_z^{\,2} = \mathbf{I}$, where

$$\mathbf{S}_x = \begin{bmatrix} 0 & 1 \\ 1 & 0 \end{bmatrix}, \quad S_y = \begin{bmatrix} 0 & -i \\ i & 0 \end{bmatrix},$$

$$\mathbf{S}_z = \begin{bmatrix} 1 & 0 \\ 0 & -1 \end{bmatrix}$$

9. Find the eigenvalues of the following matrices:

(a) $$\mathbf{A} = \begin{bmatrix} 4 & 1-3i \\ 1+3i & 7 \end{bmatrix}$$

(b) $$\mathbf{B} = \begin{bmatrix} 3i & 2+i \\ -2+i & -i \end{bmatrix}$$

(c) $$\mathbf{C} = \begin{bmatrix} \frac{1}{2}i & \frac{\sqrt{3}}{2} \\ \frac{\sqrt{3}}{2} & \frac{1}{2}i \end{bmatrix}$$

(d) $$\mathbf{D} = \begin{bmatrix} 4 & 1-3i \\ 1+3i & 7 \end{bmatrix}$$

[JNTU May 2006]

10. Find the eigenvalues and eigenvectors of the following Hermitian matrix:

$$\begin{bmatrix} 2 & 3+4i \\ 3-4i & 2 \end{bmatrix}$$

[JNTU 2006]

11. Show that

$$\mathbf{A} = \begin{bmatrix} i & 0 & 0 \\ 0 & 0 & i \\ 0 & i & 0 \end{bmatrix}$$

is a skew-Hermitian matrix and also unitary. Find eigenvalues and the corresponding eigenvectors of **A**. [JNTU May 2006]

12. Prove that every Hermitian matrix can be written as $\mathbf{A} + i\mathbf{B}$, where **A** is real and symmetric and **B** is real and skew-symmetric.

[JNTU 2005]

Solved Problems

Q 1. Show that any square matrix **A** = **B** + **C**, where **B** is symmetric and **C** is skew-symmetric matrices.

[JNTU Nov. 2005, May/June 2004]

Answer: Let A be any square matrix, such that

$$A = \frac{1}{2}(A + A^T) + \frac{1}{2}(A - A^T)$$
$$= B + C$$

where

$$B = \frac{1}{2}(A + A^T)$$

$$C = \frac{1}{2}(A - A^T)$$

Consider

$$B^T = \left[\frac{1}{2}(A + A^T)\right]^T = \frac{1}{2}(A + A^T)^T$$

$$B^T = \frac{1}{2}\left[A^T + (A^T)^T\right] = \frac{1}{2}(A + A^T) = B$$

Therefore $B^T = B$ or **B** is symmetric. Again,

$$C^T = \left[\frac{1}{2}(A - A^T)\right]^T = \frac{1}{2}(A - A^T)^T = \frac{1}{2}[A^T - (A^T)^T] = -\frac{1}{2}(A - A^T) = -C$$

Therefore $C^T = -C$, or **C** is skew-symmetric. Thus, **A** can be expressed as the sum of symmetric and skew-symmetric matrices.

Q 2. Show that every square matrix can be expressed uniquely as a sum of a symmetric and skew-symmetric matrices.

[JNTU Aug./Sep. 2007]

OR

Prove that every square matrix can be uniquely expressed as a sum of symmetric and skew-symmetric matrices.

[JNTU April/May 2007]

Answer: Let **A** be any square matrix, such that,

$$A = \frac{1}{2}(A + A^T) + \frac{1}{2}(A - A^T)\ \{\text{say}\} = M + N$$

where

$$M = \frac{1}{2}(A + A^T) \tag{3.25}$$

and

$$N = \frac{1}{2}(A - A^T) \tag{3.26}$$

We have

$$M^T = \left[\frac{1}{2}(A + A^T)\right]^T$$

since $(kA)^T = kA^T$ we get

$$M^T = \frac{1}{2}[A^T + (A^T)^T]$$

$$= \frac{1}{2}(A^T + A) = M$$

Therefore **M** is a symmetric matrix. Now

$$N = \frac{1}{2}(A - A^T)$$

$$N^T = \left[\frac{1}{2}(A - A^T)\right]^T$$

$$= \frac{1}{2}[A^T - (A^T)]^T$$

$$= \frac{1}{2}(A^T - A)$$

$$= \frac{-1}{2}(A - A^T) = -N$$

Therefore **N** is a skew-symmetric matrix. **A** is a sum of a symmetric and skew-symmetric matrices.

Now, to prove uniqueness. Suppose, **A** = **P** + Q, where, **P** is symmetric matrix and Q is skew-symmetric matrix. Therefore $P^T = P$ and $Q^T = -Q$. Then,

$$A^T = (P + Q)^T = P^T + Q^T = P - Q$$

Now $A = P + Q$ and $A^T = P - Q$. Therefore

$$\frac{1}{2}(A + A^T) = \frac{1}{2}(P + Q + P - Q) = \frac{1}{2} \times 2P = P$$

$$\frac{1}{2}(A - A^T) = \frac{1}{2}(P + Q - P + Q) = \frac{1}{2} \times 2Q = Q$$

Therefore $M = P$ and $N = Q$ and hence they are unique also.

Q 3. Show that

$$A = \frac{1}{2}\begin{bmatrix} -1 & 1 & 1 & 1 \\ 1 & -1 & 1 & 1 \\ 1 & 1 & -1 & 1 \\ 1 & 1 & 1 & -1 \end{bmatrix}$$

is orthogonal. **[JNTU Aug./Sep. 2006, May/June 2006, Nov. 2005]**

Answer: The given matrix is

$$A = \frac{1}{2}\begin{bmatrix} -1 & 1 & 1 & 1 \\ 1 & -1 & 1 & 1 \\ 1 & 1 & -1 & 1 \\ 1 & 1 & 1 & -1 \end{bmatrix}$$

A is orthogonal matrix if $AA^T = I_4$. First calculate A^T

$$A^T = \frac{1}{2}\begin{bmatrix} -1 & 1 & 1 & 1 \\ 1 & -1 & 1 & 1 \\ 1 & 1 & -1 & 1 \\ 1 & 1 & 1 & -1 \end{bmatrix}$$

We now calculate AA^T

$$AA^T = \frac{1}{4}\begin{bmatrix} -1 & 1 & 1 & 1 \\ 1 & -1 & 1 & 1 \\ 1 & 1 & -1 & 1 \\ 1 & 1 & 1 & -1 \end{bmatrix}\begin{bmatrix} -1 & 1 & 1 & 1 \\ 1 & -1 & 1 & 1 \\ 1 & 1 & -1 & 1 \\ 1 & 1 & 1 & -1 \end{bmatrix}$$

$$= \frac{1}{4}\begin{bmatrix} 1+1+1+1 & -2+2 & -2+2 & -2+2 \\ -2+2 & 1+1+1+1 & -2+2 & 1-2+2 \\ -2+2 & -2+2 & 1+1+1+1 & -2+2 \\ -2+2 & -2+2 & -2+2 & 1+1+1+1 \end{bmatrix}$$

$$= \frac{1}{4}\begin{bmatrix} 4 & 0 & 0 & 0 \\ 0 & 4 & 0 & 0 \\ 0 & 0 & 4 & 0 \\ 0 & 0 & 0 & 4 \end{bmatrix}$$

$$= \frac{4}{4}\begin{bmatrix} 1 & 0 & 0 & 0 \\ 0 & 1 & 0 & 0 \\ 0 & 0 & 1 & 0 \\ 0 & 0 & 0 & 1 \end{bmatrix} = I_4$$

Hence $AA^T = I_4$. Therefore **A** is orthogonal.

Q4. Find the orthogonal transformation which transforms the quadratic form $x_1^{\ 2} + 3x_2^{\ 2} + 3x_3^{\ 2} - 2x_2x_3$ to canonical form. **[JNTU May 2005, May/June 2004]**

Answer: The symmetric matrix **A** corresponding to the given quadratic form **Q** is

$$A = \begin{bmatrix} 1 & 0 & 0 \\ 0 & 3 & -1 \\ 0 & -1 & 3 \end{bmatrix}$$

The characteristic equation of **A** is $|\mathbf{A} - \lambda\mathbf{I}| = 0$.

$$\begin{vmatrix} 1-\lambda & 0 & 0 \\ 0 & 3-\lambda & -1 \\ 0 & -1 & 3-\lambda \end{vmatrix} = 0$$

Applying the transformation Column 2 $\rightarrow$ Column 2 + Column 3 we get

$$\begin{vmatrix} 1-\lambda & 0 & 0 \\ 0 & 2-\lambda & -\lambda \\ 0 & 2-\lambda & 3-\lambda \end{vmatrix} = 0$$

$$(1-\lambda)(2-\lambda)\begin{vmatrix} 1 & 0 & 0 \\ 0 & 1 & -1 \\ 0 & 1 & 3-\lambda \end{vmatrix} = 0$$

Applying Row 3 $\rightarrow$ Row 3 − Row 2 we have

$$(1-\lambda)(2-\lambda)\begin{vmatrix} 1 & 0 & 0 \\ 0 & 1 & -1 \\ 0 & 0 & 4-\lambda \end{vmatrix} = 0$$

Applying Column 3 $\rightarrow$ Column 3 + Column 2 we get

$$(1-\lambda)(2-\lambda)\begin{vmatrix} 1 & 0 & 0 \\ 0 & 1 & -1 \\ 0 & 0 & 4-1 \end{vmatrix} = 0$$

Solving we get

$$(1-\lambda)(2-\lambda)(4-\lambda) = 0$$

Hence $\lambda = 1$, 2 and 4 are the eigenvalues of **A**.

Case (i): Eigenvectors corresponding to $\lambda = 1$

The eigenvectors corresponding to an eigenvalue λ are nontrivial solutions of the system:

$$(\mathbf{A} - \lambda\mathbf{I})\mathbf{X} = 0 \tag{3.27}$$

Take $\lambda = 1$ in (3.27). Then, since $\lambda = 1$

$$(\mathbf{A} - \mathbf{I})\mathbf{X} = 0$$

$$\begin{bmatrix} 0 & 0 & 0 \\ 0 & 2 & -1 \\ 0 & -1 & 2 \end{bmatrix}\begin{bmatrix} x_1 \\ x_2 \\ x_3 \end{bmatrix} = \begin{bmatrix} 0 \\ 0 \\ 0 \end{bmatrix}$$

Applying Row 3 $\rightarrow$ 2 Row 3 + Row 2 we get

$$\begin{bmatrix} 0 & 0 & 0 \\ 0 & 2 & -1 \\ 0 & 0 & 3 \end{bmatrix}\begin{bmatrix} x_1 \\ x_2 \\ x_3 \end{bmatrix} = \begin{bmatrix} 0 \\ 0 \\ 0 \end{bmatrix}$$

In equation form

$$2x_2 - x_3 = 0 \quad \text{and} \quad 3x = 0$$

$$x_2 = 0 \quad \text{and} \quad x_3 = 0$$

Let, $x_1 = c_1$, arbitrary. Then, the eigenvectors corresponding to $\lambda = 1$ are

$$\mathbf{X}_1 = \begin{bmatrix} x_1 \\ x_2 \\ x_3 \end{bmatrix} = \begin{bmatrix} c_1 \\ 0 \\ 0 \end{bmatrix} = c_1 \begin{bmatrix} 1 \\ 0 \\ 0 \end{bmatrix}$$

$$\mathbf{X}_1 = \begin{bmatrix} 1 \\ 0 \\ 0 \end{bmatrix}$$

Case (ii): Eigenvectors corresponding to $\lambda = 2$
Take $\lambda = 2$ in (3.27). Then,

$$(\mathbf{A} - 2\mathbf{I})\mathbf{X} = 0$$

$$\begin{bmatrix} -1 & 0 & 0 \\ 0 & 1 & -1 \\ 0 & -1 & 1 \end{bmatrix} \begin{bmatrix} x_1 \\ x_2 \\ x_3 \end{bmatrix} = \begin{bmatrix} 0 \\ 0 \\ 0 \end{bmatrix}$$

Applying Row 3 $\rightarrow$ Row 3 + Row 2 we get

$$\begin{bmatrix} -1 & 0 & 0 \\ 0 & 1 & -1 \\ 0 & 0 & 0 \end{bmatrix} \begin{bmatrix} x_1 \\ x_2 \\ x_3 \end{bmatrix} = \begin{bmatrix} 0 \\ 0 \\ 0 \end{bmatrix}$$

Expanding we obtain

$$-x_1 = 0 \Rightarrow x_1 = 0$$

$$x_2 - x_3 = 0 \Rightarrow x_2 = x_3$$

Write $x_2 = c_2$ arbitrary so that $x_3 = c_2$. Hence, all the eigenvectors corresponding to $\lambda = 2$ are

$$\mathbf{X}_2 = \begin{bmatrix} x_1 \\ x_2 \\ x_3 \end{bmatrix} = \begin{bmatrix} 0 \\ c_2 \\ c_2 \end{bmatrix} = c_2 \begin{bmatrix} 0 \\ 1 \\ 1 \end{bmatrix}$$

$$\mathbf{X}_2 = \begin{bmatrix} 0 \\ 1 \\ 1 \end{bmatrix}$$

Case (iii): Eigenvectors corresponding to $\lambda = 4$
Take $\lambda = 4$ in (3.29). Then,

$$(\mathbf{A} - 4\mathbf{I})\mathbf{X} = 0$$

$$\begin{bmatrix} -3 & 0 & 0 \\ 0 & -1 & -1 \\ 0 & -1 & -1 \end{bmatrix} \begin{bmatrix} x_1 \\ x_2 \\ x_3 \end{bmatrix} = \begin{bmatrix} 0 \\ 0 \\ 0 \end{bmatrix}$$

Applying Row 1 → Row 1 × (–1/3) and Row 3 → Row 3 – Row 2 we get

$$\begin{bmatrix} -1 & 0 & 0 \\ 0 & -1 & -1 \\ 0 & 0 & 0 \end{bmatrix} \begin{bmatrix} x_1 \\ x_2 \\ x_3 \end{bmatrix} = \begin{bmatrix} 0 \\ 0 \\ 0 \end{bmatrix}$$

Expanding we get

$$x_1 = 0$$
$$-x_2 - x_3 = 0 \Rightarrow x_2 = -x_3$$

Put $x_3 = c_3$, arbitrary. Then, $x_2 = -c_3$. Hence, all the eigenvectors corresponding to $\lambda = 4$ are

$$\mathbf{X}_3 = \begin{bmatrix} x_1 \\ x_2 \\ x_3 \end{bmatrix} = \begin{bmatrix} 0 \\ -c_3 \\ c_3 \end{bmatrix} = c_3 \begin{bmatrix} 0 \\ -1 \\ 1 \end{bmatrix}$$

$$\mathbf{X}_3 = \begin{bmatrix} 0 \\ -1 \\ 1 \end{bmatrix}$$

Thus,

$$\mathbf{X}_1 = \begin{bmatrix} 1 \\ 0 \\ 0 \end{bmatrix}, \ \mathbf{X}_2 = \begin{bmatrix} 0 \\ 1 \\ 1 \end{bmatrix} \ \text{and} \ \mathbf{X}_3 = \begin{bmatrix} 0 \\ -1 \\ 1 \end{bmatrix}$$

are eigenvectors corresponding to distinct eigenvalues ($\lambda = 1$, 2, and 4) of the symmetric matrix **A**. Hence $\mathbf{X}_1$, $\mathbf{X}_2$ and $\mathbf{X}_3$ are pairwise orthogonal. The inner products,

$$\mathbf{X}_1^{\mathrm{T}}\mathbf{X}_2 = \mathbf{X}_2^{\mathrm{T}}\mathbf{X}_3 = \mathbf{X}_3^{\mathrm{T}}\mathbf{X}_1 = 0$$

Normalizing these, we get

$$\hat{\mathbf{e}}_1 = \frac{\mathbf{X}_1}{|\mathbf{X}_1|} = \begin{bmatrix} 1 \\ 0 \\ 0 \end{bmatrix}$$

where $|\mathbf{X}_1| = \sqrt{1+0+0} = 1$. Similarly

$$\hat{\mathbf{e}}_2 = \frac{\mathbf{X}_2}{|\mathbf{X}_2|} = \begin{bmatrix} 0 \\ 1/\sqrt{2} \\ 1/\sqrt{2} \end{bmatrix}, \ \hat{\mathbf{e}}_3 = \frac{\mathbf{X}_3}{|\mathbf{X}_3|} = \begin{bmatrix} 0 \\ -1/\sqrt{2} \\ 1/\sqrt{2} \end{bmatrix}$$

where $|\mathbf{X}_2| = |\mathbf{X}_3| = \sqrt{1+1+0} = \sqrt{2}$. Hence, $\mathbf{P} = [\hat{e}_1\ \hat{e}_2\ \hat{e}_3]$ is the modal matrix and

$$\mathbf{P}^{-1}\mathbf{AP} = \mathbf{D} = \text{Diagonal form}\begin{bmatrix} 1 & 0 & 0 \\ 0 & 2 & 0 \\ 0 & 0 & 4 \end{bmatrix}\text{of } \mathbf{A}$$

Now $\mathbf{X} = \mathbf{PY}$, where

$$\mathbf{Y} = \begin{bmatrix} y_1 \\ y_2 \\ y_3 \end{bmatrix}$$

Therefore $\mathbf{Y}^{\mathrm{T}} = [y_1\ y_2\ y_3]$, that is the quadratic form is reduced into the canonical form. Now

$$\mathbf{Y}^{\mathrm{T}}\mathbf{DY} = [y_1 y_2 y_3]\begin{bmatrix} 1 & 0 & 0 \\ 0 & 2 & 0 \\ 0 & 0 & 4 \end{bmatrix}\begin{bmatrix} y_1 \\ y_2 \\ y_3 \end{bmatrix} = y_1{}^2 + 2y_2{}^2 + 4y_3{}^2$$

The orthogonal transformation of the reduction is $\mathbf{X} = \mathbf{PY}$. That is

$$\begin{bmatrix} x_1 \\ x_2 \\ x_3 \end{bmatrix} = \begin{bmatrix} 1 & 0 & 0 \\ 0 & 1/\sqrt{2} & -1/\sqrt{2} \\ 0 & 1/\sqrt{2} & 1/\sqrt{2} \end{bmatrix}\begin{bmatrix} y_1 \\ y_2 \\ y_3 \end{bmatrix}$$

In equation form we get

$$x_1 = y_1$$
$$x_2 = \frac{1}{\sqrt{2}}y_2 - \frac{1}{\sqrt{2}}y_3$$
$$x_3 = \frac{1}{\sqrt{2}}y_2 + \frac{1}{\sqrt{2}}y_3$$

Q 5. Prove that,

(a) The eigenvalues of a Hermitian matrix are real.

(b) The eigenvalues of a skew-Hermitian matrix are purely imaginary or zero.

[JNTU Nov. 2005, May/June 2004, Nov. 2003]

Answer: (a) Eigenvalues of a Hermitian matrix are real

Let **A** be a Hermitian matrix such that $(\bar{\mathbf{A}})^{\mathrm{T}} = \mathbf{A}$. Let λ be an eigenvalue of **A** and **X** be corresponding eigenvector. Then

$$\bar{\mathbf{A}}\bar{\mathbf{X}} = \bar{\lambda}\bar{\mathbf{X}}$$
$$(\bar{\mathbf{A}}\bar{\mathbf{X}})^{\mathrm{T}} = (\bar{\lambda}\bar{\mathbf{X}})^{\mathrm{T}}$$
$$(\bar{\mathbf{X}})^{\mathrm{T}}(\bar{\mathbf{A}})^{\mathrm{T}} = \bar{\lambda}(\bar{\mathbf{X}})^{\mathrm{T}}$$
$$(\bar{\mathbf{X}})^{\mathrm{T}}\bar{\mathbf{A}} = \bar{\lambda}(\bar{\mathbf{X}})^{\mathrm{T}}$$

Postmultiplying by **X**, we get

$$(\bar{\mathbf{X}})^{\mathrm{T}}\mathbf{AX} = \bar{\lambda}(\bar{\mathbf{X}})^{\mathrm{T}}\mathbf{X}$$

$$(\bar{\mathbf{X}})^{\mathrm{T}}\lambda\mathbf{X} = \bar{\lambda}\,[(\bar{\mathbf{X}})^{\mathrm{T}}\mathbf{X}] \quad \text{[since } \lambda \text{ is an eigenvalue of } \mathbf{A}]$$

$$\lambda[(\bar{\mathbf{X}})^{\mathrm{T}}\mathbf{X}] = \bar{\lambda}\,[(\bar{\mathbf{X}})^{\mathrm{T}}\mathbf{X}]$$

$$(\lambda - \bar{\lambda})[(\bar{\mathbf{X}})^{\mathrm{T}}\mathbf{X}] = 0$$

$$\lambda - \bar{\lambda} = 0 \qquad \text{[since } (\bar{\mathbf{X}})^{\mathrm{T}}\mathbf{X} \neq 0]$$

That is λ is real. The eigenvalues of a Hermitian matrix are real.

(ii) Eigenvalues of a skew-Hermitian matrix are purely imaginary or zero
Let **A** be a skew-Hermitian matrix so that

$$(\bar{\mathbf{A}})^{\mathrm{T}} = -\mathbf{A}$$

Let λ be an eigenvalue of **A** and **X** be the corresponding eigenvector. Then

$$\mathbf{AX} = \lambda\mathbf{X}$$

$$\bar{\mathbf{A}}\bar{\mathbf{X}} = \bar{\lambda}\bar{\mathbf{X}}$$

$$(\bar{\mathbf{A}}\bar{\mathbf{X}})^{\mathrm{T}} = (\bar{\lambda}\bar{\mathbf{X}})^{\mathrm{T}}$$

$$(\bar{\mathbf{X}})^{\mathrm{T}}(\bar{\mathbf{A}})^{\mathrm{T}} = \bar{\lambda}(\bar{\mathbf{X}})^{\mathrm{T}}$$

$$-(\bar{\mathbf{X}})^{\mathrm{T}}\mathbf{A} = \bar{\lambda}(\bar{\mathbf{X}})^{\mathrm{T}} \qquad \text{[since } (\bar{\mathbf{A}})^{\mathrm{T}} = -\mathbf{A}]$$

Postmultiplying by **X**, we get

$$-(\bar{\mathbf{X}})^{\mathrm{T}}\mathbf{AX} = \bar{\lambda}\,(\bar{\mathbf{X}})^{\mathrm{T}}\mathbf{X}$$

$$-(\bar{\mathbf{X}})^{\mathrm{T}}\lambda\mathbf{X} = \bar{\lambda}\,[(\bar{\mathbf{X}})^{\mathrm{T}}\mathbf{X}] \qquad \text{[since } \lambda \text{ is an eigenvalue of } \mathbf{A}]$$

$$-\lambda[(\bar{\mathbf{X}})^{\mathrm{T}}\mathbf{X}] = \bar{\lambda}\,[(\bar{\mathbf{X}})^{\mathrm{T}}\mathbf{X}]$$

$$(\bar{\lambda} + \lambda)[(\bar{\mathbf{X}})^{\mathrm{T}}\mathbf{X}] = 0$$

$$\bar{\lambda} + \lambda = 0 \qquad \text{[since } (\bar{\mathbf{X}})^{\mathrm{T}}\mathbf{X} \neq 0]$$

Therefore λ is purely imaginary or zero.

Q 6. Show that

$$\mathbf{A} = \begin{bmatrix} i & 0 & 0 \\ 0 & 0 & i \\ 0 & i & 0 \end{bmatrix}$$

is a skew-Hermitian matrix and also unitary. Find eigenvalues and the corresponding eigenvectors of **A**.

[JNTU Aug./Sep. 2006, May/June 2006, Nov. 2004]

Answer: Given that

$$A = \begin{bmatrix} i & 0 & 0 \\ 0 & 0 & i \\ 0 & i & 0 \end{bmatrix}$$

To show that A is a skew-Hermitian matrix and is unitary. To find the given values corresponding to the eigenvectors of A. Any matrix A is said to be skew-Hermitian if $(\bar{A})^T = -A$. Now

$$\bar{A} = \begin{bmatrix} -i & 0 & 0 \\ 0 & 0 & -i \\ 0 & -i & 0 \end{bmatrix}$$

Taking transpose we get

$$(\bar{A})^T = \begin{bmatrix} -i & 0 & 0 \\ 0 & 0 & -i \\ 0 & -i & 0 \end{bmatrix} = -\begin{bmatrix} i & 0 & 0 \\ 0 & 0 & i \\ 0 & i & 0 \end{bmatrix} = -A$$

Therefore A is a skew-Hermitian. To check whether A is unitary that is to prove that $A(\bar{A})^T = (\bar{A})^T A = I$

$$A(\bar{A})^T = \begin{bmatrix} i & 0 & 0 \\ 0 & 0 & i \\ 0 & i & 0 \end{bmatrix}\begin{bmatrix} -i & 0 & 0 \\ 0 & 0 & -i \\ 0 & -i & 0 \end{bmatrix} = \begin{bmatrix} -i^2 & 0 & 0 \\ 0 & -i^2 & 0 \\ 0 & 0 & -i^2 \end{bmatrix} = \begin{bmatrix} 1 & 0 & 0 \\ 0 & 1 & 0 \\ 0 & 0 & 1 \end{bmatrix} \tag{3.28}$$

Now

$$(\bar{A})^T A = \begin{bmatrix} -i & 0 & 0 \\ 0 & 0 & -i \\ 0 & -i & 0 \end{bmatrix}\begin{bmatrix} i & 0 & 0 \\ 0 & 0 & i \\ 0 & i & 0 \end{bmatrix} = \begin{bmatrix} -i^2 & 0 & 0 \\ 0 & -i^2 & 0 \\ 0 & 0 & -i^2 \end{bmatrix} = \begin{bmatrix} 1 & 0 & 0 \\ 0 & 1 & 0 \\ 0 & 0 & 1 \end{bmatrix} \tag{3.29}$$

From (3.28) and (3.29), we get,

$$(\bar{A})^T A = A(\bar{A})^T = I$$

Therefore A is unitary matrix. We now find eigenvalues and eigenvectors. Therefore the characteristic equation of A is

$$|A - \lambda I| = 0$$

$$\left|\begin{bmatrix} i & 0 & 0 \\ 0 & 0 & i \\ 0 & i & 0 \end{bmatrix} - \lambda\begin{bmatrix} 1 & 0 & 0 \\ 0 & 1 & 0 \\ 0 & 0 & 1 \end{bmatrix}\right| = 0$$

$$\begin{vmatrix} i-\lambda & 0 & 0 \\ 0 & 0-\lambda & i \\ 0 & i & 0-\lambda \end{vmatrix} = 0$$

Expanding we get

$$(i-\lambda)\,(\lambda^2 - i^2) = 0$$
$$(i-\lambda)\,(i+\lambda)\,(i+\lambda) = 0$$
$$\lambda = -i, -i, i$$

Now, eigenvector corresponding to λ values, $(\mathbf{A} - \lambda\mathbf{I})\mathbf{X} = \mathbf{0}$, is given by

$$\begin{bmatrix} i-\lambda & 0 & 0 \\ 0 & 0-\lambda & i \\ 0 & i & 0-\lambda \end{bmatrix}\begin{bmatrix} x_1 \\ x_2 \\ x_3 \end{bmatrix} = \begin{bmatrix} 0 \\ 0 \\ 0 \end{bmatrix} \tag{3.30}$$

Eigenvector corresponding to $\lambda = -i$

$$\begin{bmatrix} 2i & 0 & 0 \\ 0 & i & i \\ 0 & i & i \end{bmatrix}\begin{bmatrix} x_1 \\ x_2 \\ x_3 \end{bmatrix} = \begin{bmatrix} 0 \\ 0 \\ 0 \end{bmatrix}$$

Solving we get

$$2ix_1 = 0 \Rightarrow x_1 = 0$$
$$x_2 i + x_3 i = 0 \Rightarrow i(x_2 + x_3) = 0 \Rightarrow x_2 + x_3 = 0 \Rightarrow x_2 = -x_3$$

Therefore eigenvector corresponding to $\lambda = -i$ is

$$\mathbf{X}_1 = \begin{bmatrix} x_1 \\ x_2 \\ x_3 \end{bmatrix} = \begin{bmatrix} 0 \\ 1 \\ -1 \end{bmatrix}$$

Eigenvector corresponding to $\lambda = i$

$$\begin{bmatrix} i-i & 0 & 0 \\ 0 & 0-i & i \\ 0 & i & 0-i \end{bmatrix}\begin{bmatrix} x_1 \\ x_2 \\ x_3 \end{bmatrix} = \begin{bmatrix} 0 \\ 0 \\ 0 \end{bmatrix}$$

$$\begin{bmatrix} 0 & 0 & 0 \\ 0 & -i & i \\ 0 & i & -i \end{bmatrix}\begin{bmatrix} x_1 \\ x_2 \\ x_3 \end{bmatrix} = \begin{bmatrix} 0 \\ 0 \\ 0 \end{bmatrix}$$

Expanding we get

$$-ix_2 + ix_3 = 0 \Rightarrow ix_2 - ix_3 = 0 \Rightarrow i(x_2 - x_3) = 0$$
$$x_2 - x_3 = 0 \Rightarrow x_2 = x_3$$

Let $x_1 = k$. Then we have two linearly independent eigenvectors when $\lambda = i$. The eigenvectors are given by

$$\mathbf{X}_2 = \begin{bmatrix} 0 \\ 1 \\ 1 \end{bmatrix} \quad \text{and} \quad \mathbf{X}_3 = \begin{bmatrix} 1 \\ 0 \\ 0 \end{bmatrix}$$

Q 7. Prove that every Hermitian matrix can be written as **A** + i**B** where **A** is real and symmetric and **B** is real and skew-symmetric. **[JNTU Aug./Sep. 2007, May 2005]**

Answer: Let **X** be a Hermitian matrix. Then $(\bar{\mathbf{X}})^{\mathrm{T}} = \mathbf{X}$. Let,

$$\mathrm{A} = \frac{1}{2}(\mathbf{X} + \bar{\mathbf{X}}) \quad \text{and} \quad \mathrm{B} = \frac{1}{2i}(\mathbf{X} - \bar{\mathbf{X}}) \tag{3.31}$$

Then both **A** and **B** are real matrices. The matrix **X** can be written as

$$\mathbf{X} = \frac{1}{2}(\mathbf{X} + \bar{\mathbf{X}}) + i\left[\frac{1}{2i}(\mathbf{X} - \bar{\mathbf{X}})\right] = \mathrm{A} + i\mathbf{B}$$

Now we have to prove that **A** is symmetric and **B** is skew-symmetric, that is

$$\mathrm{A}^{\mathrm{T}} = \left[\frac{1}{2}(\mathbf{X} + \bar{\mathbf{X}})\right]^{\mathrm{T}} = \frac{1}{2}(\mathbf{X} + \bar{\mathbf{X}})^{\mathrm{T}} = \frac{1}{2}[\mathbf{X}^{\mathrm{T}} + (\bar{\mathbf{X}})^{\mathrm{T}}] = \frac{1}{2}[\mathbf{X}^{\mathrm{T}} + (\mathbf{X})^{\mathrm{T}}] = \frac{1}{2}\ [\mathbf{X}^{\mathrm{T}} + \mathbf{X}] = \frac{1}{2}(\bar{\mathbf{X}} + \mathbf{X}) = \mathrm{A}$$

Therefore **A** is symmetric. Now

$$\mathbf{B}^{\mathrm{T}} = \left[\frac{1}{2i}(\mathbf{X} - \bar{\mathbf{X}})\right]^{\mathrm{T}} = \frac{1}{2i}(\mathbf{X} - \bar{\mathbf{X}})^{\mathrm{T}} = \frac{1}{2i}[\mathbf{X}^{\mathrm{T}} - (\bar{\mathbf{X}})^{\mathrm{T}}] = \frac{1}{2i}[\mathbf{X}^{\mathrm{T}} - \mathbf{X}] = -\frac{1}{2i}(\mathbf{X} - \bar{\mathbf{X}}) = -\mathbf{B}$$

Therefore **B** is skew-symmetric.

Q 8. Reduce the quadratic form $3x^2 + 5y^2 + 3z^2 - 2yz + 2zx - 2xy$ to the canonical form and specify the matrix of transformation. **[JNTU May/June 2008, Aug./Sep. 2006, May 2005]**

Answer: The symmetric matrix corresponding to the given quadratic form is $\mathrm{A} = [a_{ij}]_{n\times n}$, where

$$a_{ij} = \begin{cases} \frac{1}{2} \text{ coefficient of cross-product terms for } i \neq j \\ \text{coefficient of square terms for } i = j \end{cases}$$

Now

$$\mathrm{A} = \begin{bmatrix} 3 & -1 & 1 \\ -1 & 5 & -1 \\ 1 & -1 & 3 \end{bmatrix}$$

First, we find the diagonal form of **A** by the method of congruent reduction. Indeed, we write,

$$\mathbf{A} = \mathbf{I}_n \mathbf{A} \mathbf{I}_n$$

$$\begin{bmatrix} 3 & -1 & 1 \\ -1 & 5 & -1 \\ 1 & -1 & 3 \end{bmatrix} = \begin{bmatrix} 1 & 0 & 0 \\ 0 & 1 & 0 \\ 0 & 0 & 1 \end{bmatrix} \mathbf{A} \begin{bmatrix} 1 & 0 & 0 \\ 0 & 1 & 0 \\ 0 & 0 & 1 \end{bmatrix} \tag{3.32}$$

We apply row transformation to the LHS and to the pre-factor of **A** on RHS of (3.32). Similarly, we apply the corresponding column transformations to the LHS and to the post-factor of **A** on RHS of (3.32).
Applying Row 2 $\to$ 3 Row 2 + Row 1 and Row 3 $\to$ 3 Row 3 – Row 1 in (3.32), gives

$$\begin{bmatrix} 3 & -1 & 1 \\ 0 & 14 & -2 \\ 0 & -2 & 8 \end{bmatrix} = \begin{bmatrix} 1 & 0 & 0 \\ 1 & 3 & 0 \\ 1 & 0 & 3 \end{bmatrix} \mathbf{A} \begin{bmatrix} 1 & 0 & 0 \\ 0 & 1 & 0 \\ 0 & 0 & 1 \end{bmatrix}$$

Applying Column 2 $\to$ 3 Column 2 + Column 1 and Column 3 $\to$ 3 Column 3 – Column 1 we get

$$\begin{bmatrix} 3 & 0 & 0 \\ 0 & 42 & -6 \\ 0 & -6 & 24 \end{bmatrix} = \begin{bmatrix} 1 & 0 & 0 \\ 1 & 3 & 0 \\ 1 & 0 & 3 \end{bmatrix} \mathbf{A} \begin{bmatrix} 1 & 1 & -1 \\ 0 & 3 & 0 \\ 0 & 0 & 3 \end{bmatrix}$$

Applying Row 3 $\to$ 7 Row 3 + Row 2

$$\begin{bmatrix} 3 & 0 & 0 \\ 0 & 42 & -6 \\ 0 & 0 & -162 \end{bmatrix} = \begin{bmatrix} 1 & 0 & 0 \\ 1 & 3 & 0 \\ -6 & 0 & 21 \end{bmatrix} \mathbf{A} \begin{bmatrix} 1 & 1 & -1 \\ 0 & 3 & 0 \\ 0 & 0 & 3 \end{bmatrix}$$

Applying Column 3 $\to$ 7 Column 3 + Column 2

$$\begin{bmatrix} 3 & 0 & 0 \\ 0 & 42 & 0 \\ 0 & 0 & 1134 \end{bmatrix} = \begin{bmatrix} 1 & 0 & 0 \\ 1 & 3 & 0 \\ -6 & 0 & 21 \end{bmatrix} \mathbf{A} \begin{bmatrix} 1 & 1 & -6 \\ 0 & 3 & 0 \\ 0 & 0 & 21 \end{bmatrix} \tag{3.33}$$

Let

$$\mathbf{P} = \begin{bmatrix} 1 & 1 & -6 \\ 0 & 3 & 0 \\ 0 & 0 & 21 \end{bmatrix}$$

Then (3.33) gives

$$\mathbf{P}^{\mathrm{T}}\mathbf{A}\mathbf{P} = \mathbf{D} = \text{diagonal } (3, 42, 1134)$$

Thus, the linear transformation is $\mathbf{X} = \mathbf{P}\mathbf{Y}$ that is

$$\begin{bmatrix} x \\ y \\ z \end{bmatrix} = \begin{bmatrix} 1 & 1 & -6 \\ 0 & 3 & 0 \\ 0 & 0 & 21 \end{bmatrix} \begin{bmatrix} y_1 \\ y_2 \\ y_3 \end{bmatrix}$$

In equation form

$$x = y_1 + y_2 - 6y_3$$
$$y = 3y_2$$
$$z = 21y_3$$

Transforming the given quadratic form into the canonical form we get

$$\mathbf{C} = \mathbf{Y}^{\mathrm{T}}\,\mathbf{D}\mathbf{Y} = 3{y_1}^2 + 42{y_2}^2 + 1134{y_3}^2$$

Q 9. Reduce the quadratic form $8x^2 + 7y^2 + 3z^2 - 12xy - 8yz + 4zx$ into a "sum of squares" by an orthogonal transformation and give the matrix of transformation. Also state the nature.

[JNTU Aug./Sep. 2008]

Answer: Given quadratic form is $8x^2 + 7y^2 + 3z^2 - 12xy + 8yz + 4zx$. The matrix of the given quadratic form is

$$\mathbf{A} = \begin{bmatrix} 8 & -6 & 2 \\ -6 & 7 & -4 \\ 2 & -4 & 3 \end{bmatrix}$$

The characteristic equation of **A** is $|\mathbf{A} - \lambda\mathbf{I}| = 0$

$$\left|\begin{bmatrix} 8 & -6 & 2 \\ -6 & 7 & -4 \\ 2 & -4 & 3 \end{bmatrix} - \lambda\begin{bmatrix} 1 & 0 & 0 \\ 0 & 1 & 0 \\ 0 & 0 & 1 \end{bmatrix}\right| = 0$$

$$\left|\begin{bmatrix} 8 & -6 & 2 \\ -6 & 7 & -4 \\ 2 & -4 & 3 \end{bmatrix} - \begin{bmatrix} \lambda & 0 & 0 \\ 0 & \lambda & 0 \\ 0 & 0 & \lambda \end{bmatrix}\right| = 0$$

$$\begin{vmatrix} 8-\lambda & -6-0 & 2-0 \\ -6-0 & 7-\lambda & -4-0 \\ 2-0 & -4-0 & 3-\lambda \end{vmatrix} = 0$$

$$\begin{vmatrix} 8-\lambda & -6 & 2 \\ -6 & 7-\lambda & -4 \\ 2 & -4 & 3-\lambda \end{vmatrix} = 0$$

Expanding we get

$$(8-\lambda)[(7-\lambda)(3-\lambda) - (-4)(-4)] - (-6)[(-6)(3-\lambda) - (2)(-4)]$$
$$+ 2\,[(-6)(-4) - 2\,(7-\lambda)] = 0$$

$$(8-\lambda)(21 - 3\lambda - 7\lambda + \lambda^2 - 16) + 6(-18 + 6\lambda + 8) + 2(24 - 14 + 2\lambda) = 0$$

$$(8-\lambda)(\lambda^2 - 10\lambda + 5) + 6(6\lambda - 10) + 2(2\lambda + 10) = 0$$

$$(8\lambda^2 - \lambda^3 - 80\lambda + 10\lambda^2 + 40 - 5\lambda) + (36\lambda - 60) + (4\lambda + 20) = 0$$

$$-\lambda^3 + 18\lambda^2 - 45\lambda = 0$$
$$-(\lambda^3 - 18\lambda^2 + 45\lambda) = 0$$
$$\lambda^3 - 18\lambda^2 + 45\lambda = 0$$
$$\lambda(\lambda^2 - 18\lambda + 45) = 0$$

Therefore one root is $\lambda = 0$. We now solve the term in parentheses on the LHS

$$\lambda^2 - 18\lambda + 45 = 0$$
$$\lambda^2 - 3\lambda - 15\lambda + 45 = 0$$
$$\lambda(\lambda - 3) - 15(\lambda - 3) = 0$$
$$(\lambda - 3)(\lambda - 15) = 0$$
$$\lambda = 3, 15$$

Therefore the three roots as $\lambda = 0, 3, 15$. Since, all the eigenvalues are positive, the nature of the quadratic form is positive definite. The eigenvectors are given by $(\mathbf{A} - \lambda\mathbf{I})\mathbf{X} = \mathbf{0}$.

Case (i): Eigenvector corresponding to $\lambda = 0$ is

$$(\mathbf{A} - 0\mathbf{I})\mathbf{X} = \mathbf{0}$$

$$\begin{bmatrix} 8-0 & -6 & 2 \\ -6 & 7-0 & -4 \\ 2 & -4 & 3-0 \end{bmatrix} \begin{bmatrix} x_1 \\ x_2 \\ x_3 \end{bmatrix} = \begin{bmatrix} 0 \\ 0 \\ 0 \end{bmatrix}$$

$$\begin{bmatrix} 8 & -6 & 2 \\ -6 & 7 & -4 \\ 2 & -4 & 3 \end{bmatrix} \begin{bmatrix} x_1 \\ x_2 \\ x_3 \end{bmatrix} = \begin{bmatrix} 0 \\ 0 \\ 0 \end{bmatrix}$$

Expanding we get

$$8x_1 - 6x_2 + 2x_3 = 0$$
$$-6x_1 + 7x_2 - 4x_3 = 0$$
$$2x_1 - 4x_2 + 3x_3 = 0$$

On solving first two equations, we get

$$\frac{x_1}{24-14} = \frac{x_2}{-12-32} = \frac{x_3}{56-36}$$

$$\frac{x_1}{10} = \frac{x_2}{20} = \frac{x_3}{20}$$

$$\frac{x_1}{1} = \frac{x_2}{2} = \frac{x_3}{2}$$

Thus, for eigenvalue $\lambda = 0$, the eigenvector is

$$\mathbf{X}_1 = \begin{bmatrix} 1 \\ 2 \\ 2 \end{bmatrix}$$

Case (ii): Eigenvector corresponding to $\lambda = 3$ is

$$(\mathbf{A} - 3\mathbf{I})\mathbf{X} = \mathbf{0}$$

$$\begin{bmatrix} 8-3 & -6 & 2 \\ -6 & 7-3 & -4 \\ 2 & -4 & 3-3 \end{bmatrix}\begin{bmatrix} x_1 \\ x_2 \\ x_3 \end{bmatrix} = \begin{bmatrix} 0 \\ 0 \\ 0 \end{bmatrix}$$

$$\begin{bmatrix} 5 & -6 & 2 \\ -6 & 4 & -4 \\ 2 & -4 & 0 \end{bmatrix}\begin{bmatrix} x_1 \\ x_2 \\ x_3 \end{bmatrix} = \begin{bmatrix} 0 \\ 0 \\ 0 \end{bmatrix}$$

Expanding we get

$$5x_1 - 6x_2 + 2x_3 = 0$$
$$-6x_1 + 4x_2 - 4x_3 = 0$$
$$2x_1 - 4x_2 = 0$$

On solving first two equations, we get

$$\frac{x_1}{24-8} = \frac{x_2}{-12-20} = \frac{x_3}{20-36}$$

$$\frac{x_1}{16} = \frac{x_2}{8} = \frac{x_3}{-16}$$

$$\frac{x_1}{2} = \frac{x_2}{1} = \frac{x_3}{-2}$$

Thus, for eigenvalue $\lambda = 3$, the eigenvector is

$$\mathbf{X}_2 = \begin{bmatrix} 2 \\ 1 \\ -2 \end{bmatrix}$$

Case (iii): Eigenvector corresponding to $\lambda = 15$ is

$$(\mathbf{A} - 15\mathbf{I})\mathbf{X} = \mathbf{0}$$

$$\begin{bmatrix} 8-15 & -6 & 2 \\ -6 & 7-15 & -4 \\ 2 & -4 & 3-15 \end{bmatrix}\begin{bmatrix} x_1 \\ x_2 \\ x_3 \end{bmatrix} = \begin{bmatrix} 0 \\ 0 \\ 0 \end{bmatrix}$$

$$\begin{bmatrix} -7 & -6 & 2 \\ -6 & -8 & -4 \\ 2 & -4 & -12 \end{bmatrix}\begin{bmatrix} x_1 \\ x_2 \\ x_3 \end{bmatrix} = \begin{bmatrix} 0 \\ 0 \\ 0 \end{bmatrix}$$

Expanding we get

$$-7x_1 - 6x_2 + 2x_3 = 0$$
$$-6x_1 - 8x_2 - 4x_3 = 0$$
$$2x_1 - 4x_2 - 12x_3 = 0$$

On solving first two equations, we get

$$\frac{x_1}{24+16} = \frac{x_2}{-12-28} = \frac{x_3}{56-36}$$

$$\frac{x_1}{40} = \frac{x_2}{-40} = \frac{x_3}{20}$$

$$\frac{x_1}{2} = \frac{x_2}{-2} = \frac{x_3}{1}$$

Thus, for eigenvalue $\lambda = 15$, the eigenvector is

$$\mathbf{X}_3 = \begin{bmatrix} 2 \\ -2 \\ 1 \end{bmatrix}$$

The model matrix is $\mathbf{P} = [\mathbf{X}_1\ \mathbf{X}_2\ \mathbf{X}_3]$. Therefore

$$\mathbf{P} = \begin{bmatrix} 1 & 2 & 2 \\ 2 & 1 & -2 \\ 2 & -2 & 1 \end{bmatrix}$$

The normalized matrix is

$$\mathbf{B} = \begin{bmatrix} 1/\sqrt{1^2+2^2+2^2} & 2/\sqrt{2^2+1^2+(-2)^2} & 2/\sqrt{2^2+(-2)^2+1^2} \\ 2/\sqrt{1^2+2^2+2^2} & 1/\sqrt{2^2+1^2+(-2)^2} & -2/\sqrt{2^2+(-2)^2+1^2} \\ 2/\sqrt{1^2+2^2+2^2} & -2/\sqrt{2^2+1^2+(-2)^2} & 1/\sqrt{2^2+(-2)^2+1^2} \end{bmatrix}$$

$$\mathbf{B} = \begin{bmatrix} 1/3 & 2/3 & 2/3 \\ 2/3 & 1/3 & -2/3 \\ 2/3 & -2/3 & 1/3 \end{bmatrix}$$

The above matrix is an orthogonal matrix. The diagonalized matrix, $\mathbf{D} = \mathbf{B}^{-1}\mathbf{AB} = \mathbf{B}^{T}\mathbf{AB}$ [since $\mathbf{B}^{T}\mathbf{B} = \mathbf{I}$]. Now

$$\mathbf{B}^{T} = \begin{bmatrix} 1/3 & 2/3 & 2/3 \\ 2/3 & 1/3 & -2/3 \\ 2/3 & -2/3 & 1/3 \end{bmatrix}$$

Therefore

$$\mathbf{D} = \begin{bmatrix} 1/3 & 2/3 & 2/3 \\ 2/3 & 1/3 & -2/3 \\ 2/3 & -2/3 & 1/3 \end{bmatrix}\begin{bmatrix} 8 & -6 & 2 \\ -6 & 7 & -4 \\ 2 & -4 & 3 \end{bmatrix}\begin{bmatrix} 1/3 & 2/3 & 2/3 \\ 2/3 & 1/3 & -2/3 \\ 2/3 & -2/3 & 1/3 \end{bmatrix} = \mathbf{B\,AB}^{T} \tag{3.34}$$

On multiplying (3.34), we get,

$$\mathbf{D} = \begin{bmatrix} 0 & 0 & 0 \\ 0 & 3 & 0 \\ 0 & 0 & 15 \end{bmatrix}$$

Therefore $\mathbf{Q} = \mathbf{Y}^{T}\mathbf{DY}$

$$\mathbf{Q} = [y_1\ y_2\ y_3]\begin{bmatrix} 0 & 0 & 0 \\ 0 & 3 & 0 \\ 0 & 0 & 15 \end{bmatrix}\begin{bmatrix} y_1 \\ y_2 \\ y_3 \end{bmatrix}$$

$$= [y_1\ y_2\ y_3]\begin{bmatrix} 0xy_1 & 0xy_2 & 0xy_3 \\ 0xy_1 & 3xy_2 & 0xy_3 \\ 0xy_1 & 0xy_2 & 15xy_3 \end{bmatrix}$$

$$= 3{y_2}^2 + 15{y_3}^2$$

This is the required canonical form.

Q 10. Find the transformation that will transform $10x^2 + 2y^2 + 5z^2 + 4xy - 10xz + 6yz$ into a sum of square and find its reduced form. **[JNTU Aug./Sep. 2008]**

OR

Find the nature of the quadratic form index and signature $10x^2 + 2y^2 + 5z^2 + 4xy - 10xz + 6yz$.

[JNTU April/May 2007]

Answer: The symmetric matrix corresponding to the given quadratic form is $\mathbf{A} = [a_{ij}]_{m\times n}$, where

$$a_{ij} = \begin{cases} \dfrac{1}{2}\ \text{(coefficient of cross-product terms)} & \text{for } i \neq j \\ \text{coefficient of square terms} & \text{for } i = j \end{cases}$$

Therefore

$$\mathbf{A} = \begin{bmatrix} 10 & -2 & -5 \\ -2 & 2 & 3 \\ -5 & 3 & 5 \end{bmatrix}$$

First, we find the diagonal form of **A**, by the method of congruent reduction. Indeed, we write,

$$\mathbf{A} = \mathbf{I}_3 \, \mathbf{A} \, \mathbf{I}_3$$

$$\begin{bmatrix} 10 & -2 & -5 \\ -2 & 2 & 3 \\ -5 & 3 & 5 \end{bmatrix} = \begin{bmatrix} 1 & 0 & 0 \\ 0 & 1 & 0 \\ 0 & 0 & 1 \end{bmatrix} \mathbf{A} \begin{bmatrix} 1 & 0 & 0 \\ 0 & 1 & 0 \\ 0 & 0 & 1 \end{bmatrix} \tag{3.35}$$

We apply row transformation to the LHS and to the pre-factor of **A** on RHS of (3.35). Similarly, we apply the corresponding column transformation to the LHS and to the post-factor of **A** on (3.35).

$$\begin{bmatrix} 1 & -1/5 & -1/2 \\ -2 & 2 & 3 \\ -5 & 3 & 5 \end{bmatrix} = \begin{bmatrix} 1/10 & 0 & 0 \\ 0 & 1 & 0 \\ 0 & 0 & 1 \end{bmatrix} \mathbf{A} \begin{bmatrix} 1 & 0 & 0 \\ 0 & 1 & 0 \\ 0 & 0 & 1 \end{bmatrix} \quad \text{Row 1} \to \text{Row 1} \times (1/10)$$

$$\begin{bmatrix} 1 & -1/5 & -1/2 \\ 0 & 8/5 & 2 \\ 0 & 2 & 5/2 \end{bmatrix} = \begin{bmatrix} 1/10 & 0 & 0 \\ 1/5 & 1 & 0 \\ 1/2 & 0 & 1 \end{bmatrix} \mathbf{A} \begin{bmatrix} 1 & 0 & 0 \\ 0 & 1 & 0 \\ 0 & 0 & 1 \end{bmatrix} \quad \begin{matrix} \text{Row 2} \to \text{Row 2} + 2 \text{ Row 1} \\ \text{Row 3} \to \text{Row 3} + 5 \text{ Row 1} \end{matrix}$$

$$\begin{bmatrix} 1 & -1/8 & -1/2 \\ 0 & 1 & 2 \\ 0 & 5/4 & 5/2 \end{bmatrix} = \begin{bmatrix} 1/10 & 0 & 0 \\ 1/5 & 1 & 0 \\ 1/2 & 0 & 1 \end{bmatrix} \mathbf{A} \begin{bmatrix} 1 & 0 & 0 \\ 0 & 5/8 & 0 \\ 0 & 0 & 1 \end{bmatrix} \quad \text{Column 2} \to \text{Column 2} \times (5/8)$$

$$\begin{bmatrix} 1 & 0 & -1/4 \\ 0 & 1 & 2 \\ 0 & 0 & 0 \end{bmatrix} = \begin{bmatrix} 1/8 & 1/8 & 0 \\ 1/5 & 1 & 0 \\ 1/4 & -5/4 & 1 \end{bmatrix} \mathbf{A} \begin{bmatrix} 1 & 0 & 0 \\ 0 & 5/8 & 0 \\ 0 & 0 & 1 \end{bmatrix} \quad \begin{matrix} \text{Row 1} \to \text{Row 1} + (1/8) \text{Row 2} \\ \text{Row 3} \to \text{Row 3} + (5/4) \text{Row 2} \end{matrix}$$

$$\begin{bmatrix} 1 & 0 & -1/4 \\ 0 & 1 & 0 \\ 0 & 0 & 0 \end{bmatrix} = \begin{bmatrix} 1/8 & 1/8 & 0 \\ 1/5 & 1 & 0 \\ 1/4 & -5/4 & 1 \end{bmatrix} \mathbf{A} \begin{bmatrix} 1 & 0 & 0 \\ 0 & 5/8 & -5/4 \\ 0 & 0 & 1 \end{bmatrix} \quad \text{Column 3} \to \text{Column 3} - 2 \text{ Column 2}$$

$$\begin{bmatrix} 1 & 0 & 0 \\ 0 & 1 & 0 \\ 0 & 0 & 0 \end{bmatrix} = \begin{bmatrix} 1/8 & 1/8 & 0 \\ 1/5 & 1 & 0 \\ 1/4 & -5/4 & 1 \end{bmatrix} \mathbf{A} \begin{bmatrix} 1 & 0 & 1/4 \\ 0 & 5/8 & -5/4 \\ 0 & 0 & 1 \end{bmatrix} \quad \text{Column 3} \to \text{Column 3} + (1/4) \text{Column 2}$$

Therefore the quadratic form is reduced to normal form.

Nature of the quadratic form is positive, where, $r = 2$, $n = 3$, and $s = 2$

Index = Number of positive terms in normal form, that is 2

Signature $= 2s - r = 2(2) - 2 = 2$

Q 11. State whether the quadratic form $x_1^2 + 4x_2^2 + x_3^2 - 4x_2x_3 + 2x_3x_1 + 4x_2x_3$ is definite, semi-definite, or indefinite. **[JNTU Nov. 2005]**

Answer: The given quadratic form is

$$x_1^2 + 4x_2^2 + x_3^2 - 4x_2x_3 + 2x_2x_1 - 4x_2x_3$$

Let the matrix form be

$$\mathbf{A} = \begin{bmatrix} 1 & -2 & 1 \\ -2 & 4 & -2 \\ 1 & -2 & 1 \end{bmatrix}$$

The characteristic equation of **A** is $|\mathbf{A} - \lambda \mathbf{I}| = 0$, that is

$$\begin{vmatrix} 1-\lambda & -2 & 1 \\ -2 & 4-\lambda & -2 \\ 1 & -2 & 1-\lambda \end{vmatrix} = 0$$

Applying Column 1 → Column 1 + Column 2 + Column 3 we get

$$\begin{vmatrix} \lambda & -2 & 1 \\ -\lambda & 4-\lambda & -2 \\ \lambda & -2 & 1-\lambda \end{vmatrix} = 0$$

$$(-\lambda)\begin{vmatrix} 1 & -2 & 1 \\ 1 & 4-\lambda & -2 \\ 1 & -2 & 1-\lambda \end{vmatrix} = 0$$

Applying the row transformations Row 2 → Row 2 − Row 1 and Row 3 → Row 3 − Row 1 we get

$$(-\lambda)\begin{vmatrix} 1 & -2 & 1 \\ 0 & 6-\lambda & -3 \\ 0 & 0 & -\lambda \end{vmatrix} = 0$$

$$(-\lambda)(-\lambda)\begin{vmatrix} 1 & -2 & 1 \\ 0 & 6-\lambda & -3 \\ 0 & 0 & 1 \end{vmatrix} = 0$$

$$(-\lambda)(-\lambda)(6-\lambda) = 0$$

The roots are $\lambda = 0, 0, 6$. The eigenvalues of **A** are 0, 0, 6. As they are all nonnegative and **A** has at least one nonzero eigenvalue, the given quadratic form is positive, semi-definite.

Q 12. Use Lagrange's reduction method to reduce the quadratic form $x_1^2 + 6x_2^2 + 18x_3^2 + 4x_1x_2 + 8x_1x_2 - 4x_2x_3$ to the sum of squares and find the corresponding linear transformation. Also find the index and the signature. **[JNTU May/June 2006, Nov. 2005]**

Answer: It is given that

$$\mathbf{A} = \begin{bmatrix} 1 & 2 & 4 \\ 2 & 6 & -2 \\ 4 & -2 & 18 \end{bmatrix}$$

The quadratic form is $\mathbf{X}^T\mathbf{A}\mathbf{X}$. This gives

$$[x_1\, x_2\, x_3] \begin{bmatrix} 1 & 2 & 4 \\ 2 & 6 & -2 \\ 4 & -2 & 18 \end{bmatrix} \begin{bmatrix} x_1 \\ x_2 \\ x_3 \end{bmatrix} = [x_1\, x_2\, x_3] \begin{bmatrix} x_1 & + & 2x_2 & + & 4x_3 \\ 2x_1 & + & 6x_2 & + & 2x_3 \\ 4x_1 & - & 2x_2 & + & 18x_3 \end{bmatrix}$$

Expanding we get

$$\begin{aligned} x_1^2 + 6x_2^2 + 18x_3^2 + 4x_1x_2 + 8x_1x_3 - 4x_2x_3 &= [x_1^2 + 4x_1(x_2 + 2x_3)] + 6x_2^2 + 18x_3^2 - 4x_2x_3 \\ &= [x_1^2 + 4x_1(x_2 + 2x_3) + 2^2(x_2 + 2x_3)^2] + 2^2(x_2 + 2x_3) + 6x_2^2 + 18x_3^2 - 4x_2x_3 \\ &= [x_1 + 2(x_2 + 2x_3)]^2 + 2x_2^2 + 2x_3^2 - 20x_2x_3 \\ &= [x_1 + 2(x_2 + 2x_3)]^2 + 2(x_2^2 - 10x_2x_3) + 2x_3^2 \\ &= [x_1 + 2(x_2 + 2x_3)]^2 + 2(x_2^2 - 10x_2x_3 + 5^2x_3^2) - 2(5^2x_3^2) + 2x_3^2 \\ &= [x_1 + 2(x_2 + 2x_3)]^2 + 2(x_2 - 5x_3)^2 - 48x_3^2 \\ &= y_1^2 + 2y_2^2 - 48y_3^2 \end{aligned}$$

where $y_1 = [x_1 + 2(x_2 + 2x_3)]$, $y_2 = x_2 - 5x_3$ and $y_3 = x_3$. Now $\mathbf{A} = \mathbf{I}_3\mathbf{A}\mathbf{I}_3$. In matrix form it is

$$\begin{bmatrix} 1 & 2 & 4 \\ 2 & 6 & -2 \\ 4 & -2 & 18 \end{bmatrix} = \begin{bmatrix} 1 & 0 & 0 \\ 0 & 1 & 0 \\ 0 & 0 & 1 \end{bmatrix} \mathbf{A} \begin{bmatrix} 1 & 0 & 0 \\ 0 & 1 & 0 \\ 0 & 0 & 1 \end{bmatrix}$$

Applying row transformations Row 2 $\to$ Row 2 $-$ 2 Row 1 and Row 3 $\to$ Row 3 $-$ 4 Row 1 we get

$$\begin{bmatrix} 1 & 2 & 4 \\ 0 & 2 & -10 \\ 0 & -10 & 2 \end{bmatrix} = \begin{bmatrix} 1 & 0 & 0 \\ -2 & 1 & 0 \\ -4 & 0 & 1 \end{bmatrix} \mathbf{A} \begin{bmatrix} 1 & 0 & 0 \\ 0 & 1 & 0 \\ 0 & 0 & 1 \end{bmatrix}$$

Applying Column 2 → Column 2 − 2 Column 1 and Column 3 → Column 3 − 4 Column 1 we get

$$\begin{bmatrix}1 & 0 & 0\\0 & 2 & -10\\0 & -10 & 2\end{bmatrix}=\begin{bmatrix}1 & 0 & 0\\-2 & 1 & 0\\-4 & 0 & 1\end{bmatrix}\mathbf{A}\begin{bmatrix}1 & -2 & -4\\0 & 1 & 0\\0 & 0 & 1\end{bmatrix}$$

Applying Row 2 → Row 2 × (1/2) and Row 3 → Row 3 × (1/2) we get

$$\begin{bmatrix}1 & 0 & 0\\0 & 1 & -5\\0 & -5 & 1\end{bmatrix}=\begin{bmatrix}1 & 0 & 0\\-1 & 1/2 & 0\\-2 & 0 & 1/2\end{bmatrix}\mathbf{A}\begin{bmatrix}1 & -2 & -4\\0 & 1 & 0\\0 & 0 & 1\end{bmatrix}$$

Applying Row 3 → Row 3 + 5 Row 2 we get

$$\begin{bmatrix}1 & 0 & 0\\0 & 1 & -5\\0 & 0 & -24\end{bmatrix}=\begin{bmatrix}1 & 0 & 0\\-1 & 1/2 & 0\\-7 & 5/2 & 1/2\end{bmatrix}\mathbf{A}\begin{bmatrix}1 & -2 & -4\\0 & 1 & 0\\0 & 0 & 1\end{bmatrix}$$

Applying Column 3 → Column 3 + 5 Column 2 we get

$$\begin{bmatrix}1 & 0 & 0\\0 & 1 & 0\\0 & 0 & -24\end{bmatrix}=\begin{bmatrix}1 & 0 & 0\\-1 & 1/2 & 0\\-7 & 5/2 & 1/2\end{bmatrix}\mathbf{A}\begin{bmatrix}1 & -2 & -14\\0 & 1 & 5\\0 & 0 & 1\end{bmatrix}$$

Therefore

$\rho(\mathrm{A}) = 3$

Index = 2 (number of positive terms)

Signature = 2 − 1 = 1

Linear transformation is

$$\mathbf{X} = \mathbf{PY}$$

$$\mathbf{X}=\begin{bmatrix}x_1\\x_2\\x_3\end{bmatrix},\ \mathbf{P}=\begin{bmatrix}1 & -1 & -7\\0 & 1/2 & 5/2\\0 & 0 & 1/2\end{bmatrix},\ \mathbf{Y}=\begin{bmatrix}y_1\\y_2\\y_3\end{bmatrix}$$

$$\mathbf{X}=\begin{bmatrix}1 & -1 & -7\\0 & 1/2 & 5/2\\0 & 0 & 1/2\end{bmatrix}\begin{bmatrix}y_1\\y_2\\y_3\end{bmatrix}$$

$$\begin{bmatrix}x_1\\x_2\\x_3\end{bmatrix}=\begin{bmatrix}y_1 - y_2 - 7y_3\\(1/2)y_2 + (5/2)y_3\\(1/2)y_3\end{bmatrix}$$

$$x_1 = y_1 - y_2 - 7y_3$$
$$x_2 = (1/2)y_2 + (5/2)y_3$$
$$x_3 = (1/2)y_3$$

Summary

The practical importance of matrix eigenvalue problems can hardly be overrated. The problems are defined by the vector equation

$$\mathbf{Ax} = \lambda\mathbf{x} \tag{3.36}$$

A is a given square matrix. All matrices in this chapter are ***square***. λ is a scalar. We have discussed this is Chapter 2.

The present chapter is devoted to eigenvalue problems for **symmetric** ($\mathbf{A}^T = \mathbf{A}$), **skew-symmetric** ($\mathbf{A}^T = -\mathbf{A}$), and **orthogonal matrices** ($\mathbf{A}^T = \mathbf{A}^{-1}$). It concerns the transformation of quadratic forms to principal axes and its relation to eigenvalues as also Sylvester's law of Intertia.

The last section extends Sec. 3.1 to the complex analogs of those real matrices, called **Hermitian** [$(\bar{\mathbf{A}})^T = \mathbf{A}$], **skew-Hermitian** [$(\bar{\mathbf{A}})^T = -\mathbf{A}$], and **unitary matrices** [$(\bar{\mathbf{A}})^T = \mathbf{A}^{-1}$]. All the eigenvalues of a Hermitian matrix (and a symmetric one) are real. For a skew-Hermitian (and a skew-symmetric) matrix they are pure imaginary or zero. For a unitary (and an orthogonal) matrix they have absolute value 1.

Review Questions and Problems

1. In solving an eigenvalue problem, what is given and what is sought?
2. Do there exist square matrices without eigenvalues? Eigenvectors corresponding to more than one eigenvalue of a given matrix?
3. What is the defect? Why is it important? Give examples.
4. Can a complex matrix have real eigenvalues? Real eigenvectors? Give reasons.
5. Does a 3 × 3 matrix always have a real eigenvalue?
6. Give a few typical applications in which eigenvalue problems occur.
7. Verify that **A** and $\hat{\mathbf{A}} = \mathbf{P}^{-1}\mathbf{AP}$ have the same spectrum. Here, **A**, **P** are:

   a. $\begin{bmatrix} 3.8 & 2.4 \\ 2.4 & 0.2 \end{bmatrix}, \begin{bmatrix} 1 & 2 \\ 2 & 1 \end{bmatrix}$

   b. $\begin{bmatrix} -22 & 20 & 10 \\ -4 & 20 & -8 \\ 28 & -14 & 29 \end{bmatrix}, \begin{bmatrix} 1 & 0 & 2 \\ 0 & 2 & 4 \\ 2 & 8 & 0 \end{bmatrix}$

   c. $\begin{bmatrix} 2 & 2 & -2 \\ 3 & 1 & -3 \\ 1 & -1 & -1 \end{bmatrix}, \begin{bmatrix} 1 & 2 & 3 \\ 2 & 0 & 2 \\ 3 & 2 & 4 \end{bmatrix}$

8. Reduce the quadratic form to principal axes.

   a. $11.56x_1^2 + 20.16x_1x_2 + 17.44x_2^2 = 100$

   b. $1.09x_1^2 - 0.06x_1x_2 + 1.01x_2^2 = 1$

   c. $14x_1^2 + 24x_1x_2 - 4x_2^2 = 20$

9. Define (a) quadratic form and (b) cononical form.

   [JNTU Aug./Sep. 2006, May 2005, Nov. 2003]

4 Solution of Nonlinear Systems

LEARNING OBJECTIVES

After reading this chapter, you will be able to understand the following:

- Bisection method.
- Method of false position or regula falsi method.
- Iteration method.
- Newton–Raphson method.
- Interpolation and error in interpolation.
- Finite and divided differences.
- Newton forward and Newton backward interpolation formulas.
- Central, Gauss forward, Gauss backward difference formulas.
- Newton divided difference interpolation.
- Lagrange's interpolation.
- Spline interpolation.

Numeric analysis, more briefly also called **numerics**, concerns **numeric methods**, that is, methods for solving problems in terms of numbers or corresponding graphical representations. It also includes the investigation of the range of applicability and of the accuracy and stability of these methods.

Typical tasks for numerics are the evaluation of definite integrals, the solution of equations and linear systems, the solution of differential or integral equations for which there are no solution formulas, and the evaluation of experimental data for which we want to obtain, for example, an approximating polynomial.

Numerics has increased in importance to the engineer more than any other field of mathematics owing to the ongoing development of powerful software resulting from great research activity in numerics: new methods are invented, existing methods are improved and adapted, and old methods – impractical in precomputer times – are rediscovered. A main goal in these activities is the development of well-structured software. And in large-scale work – millions of equations or steps of iteration – even small algorithmic improvements may have a large effect on computing time, storage demand, accuracy, and stability.

This chapter on numerics begins with an explanation of some general concepts, such as floating point, roundoff errors, and general numeric errors and their propagation. We discuss methods for solving equations, interpolation methods, including splines. The purpose of this chapter is twofold. First, for all these tasks the student should become familiar with the most basic (but not too complicated) numeric solution methods. These are indispensable for the engineer, because for many problems there is no solution formula (think of a complicated integral or a polynomial of high degree or the interpolation of values obtained by measurements). In other cases a complicated solution formula may exist but may be practically useless. Second, the student should learn to understand some basic ideas and concepts that are important throughout numerics, such as the practical form of algorithms, the estimation of errors, and the order of convergence.

4.1 Basic Concepts

Numeric methods are used to solve problems on computers or calculators by numeric calculations, resulting in a table of numbers and/or graphical representations (figures). The steps from a given situation (in engineering, economics, etc.) to the final answer are usually as follows.

1. **Modeling.** We set up a mathematical model of our problem, such as an integral, a system of equations, or a differential equation.
2. **Choosing a numeric method** and parameters (e.g., step size), perhaps with a preliminary error estimation.
3. **Programming.** We use the algorithm to write a corresponding program in a CAS, such as Maple, Mathematica, Matlab, or Mathcad, or, say, in Fortran, C, or C^{++}, selecting suitable routines from a software system as needed.
4. **Doing the computation.**
5. **Interpreting the results** in physical or other terms; also deciding to rerun if further results are needed.

Steps 1 and 2 are related. A slight change of the model may often admit of a more efficient method. To choose methods, we must first get to know them. In Step 3 the program consists of the given data and a sequence of instructions to be executed by the computer in a certain order for producing the answer in numeric or graphic form.

To create a good understanding of the nature of numeric work, we continue in this section with some simple general remarks.

Floating-Point Form of Numbers

We know that in decimal notation, every real number is represented by a finite or an infinite sequence of decimal digits. Now most computers have two ways of representing numbers, called *fixed point* and *floating point.* In a **fixed-point** system all numbers are given with a fixed number of decimals after the decimal point; for example, numbers given with 3 decimals are 62.358, 0.014, 1.000. In a text we would write, say, 3 decimals as 3D. Fixed-point representations are impractical in most scientific computations because of their limited range (explain!) and will not concern us.

In a **floating-point** system we write, for instance,

$$0.6247 \times 10^3, \qquad 0.1735 \times 10^{-13}, \qquad -0.2000 \times 10^{-1}$$

or sometimes also

$$6.247 \times 10^2, \qquad 1.735 \times 10^{-14}, \qquad -2.000 \times 10^{-2}$$

We see that in this system the number of significant digits is kept fixed, whereas the decimal point is "floating." Here, a **significant digit** of a number c is any given digit of c, except possibly for zeros to the left of the first nonzero digit; these zeros serve only to fix the position of the decimal point. (Thus any other zero is a significant digit of c.) For instance, each of the numbers

$$1360, \quad 1.360, \quad 0.001360$$

has 4 significant digits. In a text we indicate, say, 4 significant digits, by 4S.

The use of exponents permits us to represent very large and very small numbers. Indeed, theoretically any nonzero number a can be written as

$$a = \pm m \times 10^n, \qquad 0.1 \leqq |m| < 1, \qquad n \text{ integer} \tag{4.1}$$

On the computer, m is limited to k digits (e.g., $k = 8$) and n is limited, giving representations (for finitely many numbers only!)

$$\bar{a} = \pm \bar{m} \times 10^n, \qquad \bar{m} = 0 \times d_1 \times d_2 \times \cdots \times d_k, \qquad d_1 > 0 \tag{4.2}$$

These numbers $\bar{a}$ are often called *k-digit decimal machine numbers*. Their fractional part m (or $\bar{m}$) is called the *mantissa*. This has nothing to do with "mantissa" as used for logarithms. n is called the *exponent* of $\bar{a}$.

Underflow and Overflow

The range of exponents that a typical computer can handle is very large. The IEEE (Institute of Electrical and Electronics Engineers) floating-point standard for **single precision** (the usual number of digits in calculations) is about $-38 < n < 38$ (about $-125 < n^* < 125$ for the exponent in **binary representations**, i.e., representations in base 2). [For so-called **double precision** it is about $-308 < n < 308$ (about $-1020 < n^* < 1020$ for binary).] If in a computation, a number outside that range occurs, this is called **underflow** when the number is smaller and **overflow** when it is larger. In the case of underflow the result is usually set to zero and computation continues. Overflow causes the computer to halt. Standard codes (by IMSL, NAG, etc.) are written to avoid overflow. Error messages on overflow may then indicate programming errors (incorrect input data, etc.).

Roundoff

An error is caused by **chopping** (= discarding all decimals from some decimal on) or **rounding**. This error is called **roundoff error**, regardless of whether we chop or round. The rule for rounding off a number to k decimals is as follows. (The rule for rounding off to k significant digits is the same, with "decimal" replaced by "significant digit.")

Roundoff Rule

Discard the $(k + 1)$th and all subsequent decimals. (a) If the number thus discarded is less than half a unit in the kth place, leave the kth decimal unchanged (*"rounding down"*). (b) If it is greater than half a unit in the kth place, add one to the kth decimal (*"rounding up"*). (c) If it is exactly half a unit, round off to the nearest *even* decimal. (Example: Rounding off 3.45 and 3.55 to 1 decimal gives 3.4 and 3.6, respectively.) The last part of the rule is supposed to ensure that in discarding exactly half a decimal, rounding up and rounding down happens about equally often, on the average. If we round off 1.2535 to 3, 2, 1 decimals, we get 1.254, 1.25, 1.3, but if 1.25 is rounded off to one decimal, without further information, we get 1.2. Chopping is not recommended because the corresponding error can be larger than that in rounding, and is systematic. (Nevertheless, some computers use it because it is simpler and faster. On the other hand, some computers and calculators improve accuracy of results by doing intermediate calculations using one or more extra digits, called *guarding digits*.)

Error in Rounding

Let $\bar{a} = fl(a)$ in (4.2) be the floating-point computer approximation of a in (4.1) obtained by rounding, where fl suggests **floating**. Then the roundoff rule gives (by dropping exponents) $|m - \bar{m}| \leqq 10^{-k}/2$. Since $|m| \geqq 0.1$, this implies (when $a \neq 0$)

$$\left|\frac{a-\bar{a}}{a}\right| \approx \left|\frac{m-\bar{m}}{m}\right| \leqq \frac{1}{2} \times 10^{1-k} \tag{4.3}$$

The right side $u = 10^{(1-k)}/2$ is called the **rounding unit**. If we write $\bar{a} = a(1+\delta)$, we have by algebra $(\bar{a} - a)/a = \delta$, hence $|\delta| \leqq u$ by (4.3). *This shows that the rounding unit u is an error bound in rounding.*

Rounding errors may ruin a computation completely, even a small computation. In general, these errors become the more dangerous the more arithmetic operations (perhaps several millions!) we have to perform. It is therefore important to analyze computational programs for expected rounding errors and to find an arrangement of the computations such that the effect of rounding errors is as small as possible.

The arithmetic in a computer is not exact either and causes further errors; however, these will not be relevant to our discussion.

Algorithm and Stability

Numeric methods can be formulated as algorithms. An **algorithm** is a step-by-step procedure that states a numeric method in a form (a "**pseudocode**") understandable to humans. The algorithm is then used to write a program in a programming language that the computer can understand so that it can execute the numeric method. Important algorithms follow in the next sections.

Stability

To be useful, an algorithm should be **stable;** that is, small changes in the initial data should cause only small changes in the final results. However, if small changes in the initial data can produce large changes in the final results, we call the algorithm **unstable**.

This "*numeric instability*," which in most cases can be avoided by choosing a better algorithm, must be distinguished from "*mathematical instability*" of a problem, which is called "*ill-conditioning*," a concept we have discussed in a previous chapter. Some algorithms are stable only for certain initial data, so that one must be careful in such a case.

Errors of Numeric Results

Final results of computations of unknown quantities generally are **approximations**; that is, they are not exact but involve errors. Such an error may result from a combination of the following effects. **Roundoff errors** result from rounding. **Experimental errors** are errors of given data (probably arising from measurements). **Truncating errors** result from truncating (prematurely breaking off), for instance, if we replace a Taylor series with the sum of its first few terms. These errors depend on the computational method used and must be dealt with individually for each method. ["Truncating" is sometimes used as a term for chopping off (see before), a terminology that is not recommended.]

Formulas for Errors

If $\tilde{a}$ is an approximate value of a quantity whose exact value is a, we call the difference

$$\epsilon = a - \tilde{a} \tag{4.4}$$

the **error** of $\tilde{a}$. Hence

$$a = \tilde{a} + \epsilon, \qquad \text{True value} = \text{Approximation} + \text{Error} \tag{4.5}$$

For instance, if $\tilde{a} = 10.5$ is an approximation of $a = 10.2$, its error is $\epsilon = -0.3$. The error of an approximation $\tilde{a} = 1.60$ of $a = 1.82$ is $\epsilon = 0.22$.

CAUTION! In the literature $|a - \tilde{a}|$ ("absolute error") or $\tilde{a} - a$ are sometimes also used as definitions of error.

The **relative error** ϵ_r of $\tilde{a}$ is defined by

$$\epsilon_r = \frac{\epsilon}{a} = \frac{a - \tilde{a}}{a} = \frac{\text{Error}}{\text{True value}} \qquad (a \neq 0) \tag{4.6}$$

This looks useless because a is unknown. But if $|\epsilon|$ is much less than $|\tilde{a}|$, then we can use $\tilde{a}$ instead of a and get

$$\epsilon_r \approx \frac{\epsilon}{\tilde{a}} \tag{4.7}$$

This still looks problematic because ϵ is unknown – if it were known, we could get $a = \tilde{a} + \epsilon$ from (4.4) and we would be done. But what one often can obtain in practice is an **error bound** for $\tilde{a}$, that is, a number β such that

$$|\epsilon| \leqq \beta, \qquad \text{hence} \qquad |a - \tilde{a}| \leqq \beta$$

This tells us how far away from our computed $\tilde{a}$ the unknown a can at most lie. Similarly, for the relative error, an error bound is a number β_r such that

$$|\epsilon_r| \leqq \beta_r, \qquad \text{hence} \qquad \left|\frac{a - \tilde{a}}{a}\right| \leqq \beta_r$$

Error Propagation

This is an important matter. It refers to how errors at the beginning and in later steps (roundoff, for example) propagate into the computation and affect accuracy, sometimes very drastically. We state here what happens to error bounds. Namely, bounds for the *error* add under addition and subtraction, whereas bounds for the *relative error* add under multiplication and division. You do well to keep this in mind.

THEOREM 1 (Error Propagation):

(a) *In addition and subtraction, an error bound for the results is given by the sum of the error bounds for the terms.*

(b) *In multiplication and division, an error bound for the relative error of the results is given (approximately) by the sum of the bounds for the relative errors of the given numbers.*

Basic Error Principle

Every numeric method should be accompanied by an error estimate. If such a formula is lacking, is extremely complicated, or is impractical because it involves information (for instance, on derivatives) that is not available, the following may help.

Error Estimation by Comparison

Do a calculation twice with different accuracy. Regard the difference $\tilde{a}_2 - \tilde{a}_1$ of the results $\tilde{a}_1, \tilde{a}_2$ as a (perhaps crude) estimate of the error ϵ_1 of the inferior result $\tilde{a}_1$. Indeed, $\tilde{a}_1 + \epsilon_1 = \tilde{a}_2 + \epsilon_2$ by formula (4.5). This implies $\tilde{a}_2 - \tilde{a}_1 = \epsilon_1 - \epsilon_2 \approx \epsilon_1$ because $\tilde{a}_2$ is generally more accurate than $\tilde{a}_1$, so that $|\epsilon_2|$ is small compared to $|\epsilon_1|$.

Loss of Significant Digits

This means that a result of a calculation has fewer correct digits than the numbers from which it was obtained. This happens if we subtract two numbers of about the same size, for example, 0.1439 – 0.1426 ("subtractive cancellation"). It may occur in simple problems, but it can be avoided in most cases by simple changes of the algorithm – if one is aware of it! Let us illustrate this with the following basic problem.

Example 1 Quadratic Equation. Loss of Significant Digits

Find the roots of the equation

$$x^2 - 40x + 2 = 0$$

using 4 significant digits (abbreviated 4S) in the computation.

Solution: A formula for the roots x_1, x_2 of a quadratic equation $ax^2 + bx + c = 0$ is

$$x_1 = \frac{1}{2a}(-b + \sqrt{b^2 - 4ac}), \quad x_2 = \frac{1}{2a}(-b - \sqrt{b^2 - 4ac}) \tag{4.8}$$

Furthermore, since $x_1 x_2 = c/a$, another formula for those roots is

$$x_1 \quad \text{as before,} \qquad x_2 = \frac{c}{ax_1} \tag{4.9}$$

From (4.8) we obtain $x = 20 \pm \sqrt{398} = 20.00 \pm 19.95$. This gives $x_1 = 20.00 + 19.95 = 39.95$, involving no difficulty, whereas $x_2 = 20.00 - 19.95 = 0.05$ is poor because it involves loss of significant digits.

In contrast, (4.9) gives $x_1 = 39.95$, $x_2 = 2.000/39.95 = 0.05006$, in error by less than one unit of the last digit, as a computation with more digits shows. (The 10S-value is 0.05006265674.)

Comment. To avoid misunderstandings: 4S was used for convenience; (4.9) is better than (4.8) regardless of the number of digits used. For instance, the 8S-computation by (4.8) is $x_1 = 39.949937$, $x_2 = 0.050063$, which is poor, and by (4.9) it is x_1 as before, $x_2 = 2/x_1 = 0.050062657$. In a quadratic equation with real roots, if x_2 is absolutely largest (because $b > 0$), use (4.8) for x_2 and then $x_1 = c/(ax_2)$.

CONCEPT CHECK

1. **(Floating point)** Write 98.17, −100.987, 0.0057869, −13600 in floating-point form, rounded to 4S (4 significant digits).
2. Write −0.0286403, 11.25845, −31681.55 in floating-point form rounded to 6S.
3. Small differences of large numbers may be particularly strongly affected by rounding errors. Illustrate this by computing 0.36443/(17.862 − 17.798) as given with 5S, then rounding stepwise to 4S, 3S, and 2S, where "stepwise" means: round the *rounded* numbers, not the given ones.
4. Do the work in Prob. 3 with numbers of your choice that give even more drastically different results. How can you avoid such difficulties?
5. The quotient in Prob. 3 is of the form $a/(b - c)$. Write it as $a(b + c)/(b^2 - c^2)$. Compute it first with 5S, then rounding numerator 12.996 and denominator 2.28 stepwise as in Prob. 3. Compare and comment.
6. **(Quadratic equation)** Solve $x^2 - 20x + 1 = 0$ by (4.8) and by (4.9), using 6S in the computation. Compare and comment.
7. Do the computations in Prob. 6 with 4S and 2S.
8. Solve $x^2 + 100x + 2 = 0$ by (4.8) and (4.9) with 5S and compare.
9. Calculate $1/e = 0.367879$ (6S) from the partial sums of 5 to 10 terms of the Maclaurin series (a) of e^{-x} with $x = 1$, (b) of e^x with $x = 1$ and then taking the reciprocal. Which is more accurate?
10. Addition with a fixed number of significant digits depends on the order in which you add the numbers. Illustrate this with an example. Find an empirical rule for the best order.
11. **Approximations of** $\pi = 3.14159265358979\ldots$ are 22/7 and 355/113. Determine the corresponding errors and relative errors to three significant digits.
12. **(Rounding and adding)** Let $a_1, \ldots, a_n$ be numbers with a_j correctly rounded to D_j decimals. In calculating the sum $a_1 + \cdots + a_n$, retaining $D = \min D_j$ decimals, is it essential that we first add and then round the result or that we first round each number to D decimals and then add?
13. Show that in Example 1 the absolute value of the error of $x_2 = 2.000/39.95 = 0.05006$ is less than 0.00001.
14. **Overflow and underflow** can sometimes be avoided by simple changes in a formula. Explain this in terms of

$$\sqrt{x^2 + y^2} = x\sqrt{1 + (y/x)^2}$$

with $x^2 \geqq y^2$ and x so large that x^2 would cause overflow. Invent examples of your own.

15. **(Nested form)** Evaluate

$$\begin{aligned} f(x) &= x^3 - 7.5x^2 + 11.2x + 2.8 \\ &= [(x - 7.5)x + 11.2]x + 2.8 \end{aligned}$$

at $x = 3.94$ using 3S arithmetic and rounding, in both of the given forms. The latter, called the *nested form*, is usually preferable since it minimizes the number of operations and thus the effect of rounding.

16. **WRITING PROJECT. Numerics.** In your own words write about the overall role of numeric methods in applied mathematics, why they are important, where and when they must be used or can be used, and how they are influenced by the use of the computer in engineering and other work.

4.2 Roots of Algebraic and Transcendental Equations

From here on, each section will be devoted to some basic kind of problem and corresponding solution methods. We start with numerical methods for solving the equations of the form

$$f(x) = 0 \tag{4.10}$$

where $f(x)$ is in general, a *nonlinear function* of a real variable x. $f(x)$ could be linear [e.g., $f(x) = ax + b$] or quadratic [e.g., $f(x) = ax^2 + bx + c$], or in general any algebraic function or a polynomial of degree n expressed as

$$f(x) = x^n + a_1 x^{n-1} + \cdots + a_n \tag{4.11}$$

n being positive and finite, and all coefficients $a_1, a_2, \ldots, a_n$ are assumed to be real. These are examples of **algebraic equations** and in such cases the solutions are also called **roots of the equations.** The function $f(x)$ could also be a transcendental equation, that is, an expression involving transcendental functions, such as

$$\tan x - x\,e^x + 2\cos x, \quad ax + b\log x = c, \quad 1 + \cos x - 5\sinh x \tag{4.12}$$

If $f(x)$ is a linear function of x expressed as $f(x) = ax + b = 0$, then its solution is given by

$$x = -b/a \tag{4.13}$$

Similarly for a quadratic function given by $f(x) = ax^2 + bx + c$, the solutions are

$$x_1 = \frac{-b + \sqrt{b^2 - 4ac}}{2a}, \quad x_2 = \frac{-b - \sqrt{b^2 - 4ac}}{2a} \tag{4.14}$$

Equations (4.13) and (4.14) are known as closed form solutions, and a mere substitution of the coefficients will give us the answers. Such closed form solutions are not possible for higher order polynomials or transcendental type of equations, and therefore solutions are obtained using numerical techniques only. These techniques are popularly known as **iterative** methods, that is, methods in which we start from an initial guess x_0 and compute step-by-step (in general better and better) approximations x_1, x_2, ... of an unknown solution of (4.10).

Solving $f(x) = 0$ is nothing but finding **the roots** of the equation $f(x) = 0$ or evaluating **the zero** of the function $f(x) = 0$. If we call the function as

$$y = f(x)$$

and plot $f(x)$ versus x, the intersections of this curve with x axis will give the required solutions, that is at these values of x, $y = f(x) = 0$. The solutions or roots of $f(x) = 0$ are thus nothing but the **zero crossings** of the function $f(x)$.

4.3 Iterative Methods

In this section we will discuss the iterative methods for solving algebraic and transcendental equations. **Iterative methods are trial and error type of methods**, and are totally amenable to computer programming. In these methods:

1. A sequence of steps is repeated over and over again (in each iteration) – they are thus natural methods to use with computers.
2. A generalized formulation is possible for a particular set of problems – for example, a general algorithm to solve polynomial equations of any order n may be written.
3. Rounding errors are negligible in trial and error procedures compared to closed form solution based procedures – this is a distinct advantage.

There are five basic questions that one should seek an answer to before using an iterative method:

1. ***How to start an iterative cycle?*** – To start, one has to choose a "guess value" for the root expected; an educated guess is always useful and it also ensures a quick solution.
2. ***How much computational effort is required for each iteration?*** – The evaluations common for each iteration must be logically described.

3. ***How to proceed from one guess (or trial) value to the next subsequent value?*** – This process must be logical (and mathematical) but not random.
4. ***How fast is the convergence?*** – All successive approximations for the roots, must finally lead (or converge on) to the actual root value as close as possible. This closeness is given in terms of quick "accuracy" or "error limit" (and is included in the program).
5. ***When to stop the iterative cycle?*** – The number of iterations must be finite for obtaining the required solution or even if there is no solution.

All these aspects are going to be considered in developing the algorithms and writing the programs. We shall be focussing our attention on five or six important iterative methods in these discussions.

Fixed-Point Iteration for Solving Equations f(x) = 0

Our present use of the word "fixed point" has absolutely nothing to do with that described in Sec. 4.1. In one way or another we transform (4.10) *algebraically* into the form

$$x = g(x) \tag{4.15}$$

Then we choose an x_0 and compute $x_1 = g(x_0)$, $x_2 = g(x_1)$, and in general

$$x_{n+1} = g(x_n) \qquad (n = 0, 1, \ldots) \tag{4.16}$$

A solution of (4.15) is called a **fixed point** of g, motivating the name of the method. This is a solution of (4.10), since from $x = g(x)$ we can return to the original form $f(x) = 0$. From (4.10) we may get several different forms of (4.15). The behavior of corresponding iterative sequences $x_0, x_1, \ldots$ may differ, in particular, with respect to their speed of convergence. Indeed, some of them may not converge at all. Let us illustrate these facts with a simple example.

Example 2 An Iteration Process (Fixed-Point Iteration)

Set up an iteration process for the equation $f(x) = x^2 - 3x + 1 = 0$. Since we know the solutions

$$x = 1.5 \pm \sqrt{1.25} = 2.618034 \quad \text{and} \quad 0.381966$$

we can watch the behavior of the error as the iteration proceeds.

Solution: The equation may be written as

$$x = g_1(x) = \frac{1}{3}(x^2 + 1), \quad \text{thus} \quad x_{n+1} = \frac{1}{3}(x_n{}^2 + 1) \tag{4.17}$$

If we choose $x_0 = 1$, we obtain the sequence (Figure 1a; computed with 6S and then rounded)

$$x_0 = 1.000, \qquad x_1 = 0.667, \qquad x_2 = 0.481, \qquad x_3 = 0.411, \qquad x_4 = 0.390, \ldots$$

which seems to approach the smaller solution. If we choose $x_0 = 2$, the situation is similar. If we choose $x_0 = 3$, we obtain the sequence (Figure 1a, upper part)

$$x_0 = 3.000, \qquad x_1 = 3.333, \qquad x_2 = 4.037, \qquad x_3 = 5.766, \qquad x_4 = 11.415, \ldots$$

which diverges. Our equation may also be written (divide by x) as

$$x = g_2(x) = 3 - \frac{1}{x}, \quad \text{thus} \quad x_{n+1} = 3 - \frac{1}{x_n} \tag{4.18}$$

and if we choose $x_0 = 1$, we obtain the sequence (Figure 1b)

$$x_0 = 1.000, \qquad x_1 = 2.000, \qquad x_2 = 2.500, \qquad x_3 = 2.600, \qquad x_4 = 2.615, \ldots$$

which seems to approach the larger solution. Similarly, if we choose $x_0 = 3$, we obtain the sequence (Figure 1b)

$$x_0 = 3.000, \qquad x_1 = 2.667, \qquad x_2 = 2.625, \qquad x_3 = 2.619, \qquad x_4 = 2.618, \ldots$$

Our figure shows the following. In the lower part of Figure 1a the slope of $g_1(x)$ is less than the slope of $y = x$, which is 1, thus $|g_1'(x)| < 1$, and we seem to have convergence. In the upper part, $g_1(x)$ is steeper ($g_1'(x) > 1$) and we have divergence. In Figure 1b the slope of $g_2(x)$ is less near the intersection point [$x = 2.618$, fixed point of g_2, solution of $f(x) = 0$], and both sequences seem to converge. From all this we conclude that convergence seems to depend on the fact that in a neighborhood of a solution the curve of $g(x)$ is less steep than the straight line $y = x$, and we shall now see that this condition $|g'(x)| < 1$ (= slope of $y = x$) is sufficient for convergence.

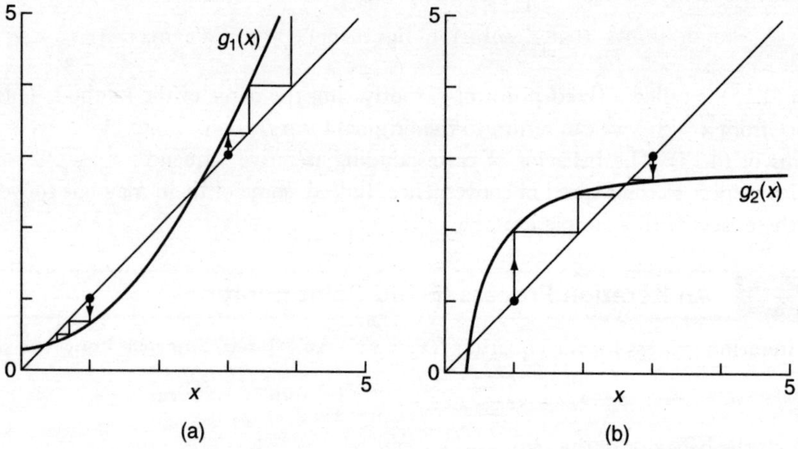

Figure 1 Example 2, iterations (4.17) and (4.18).

An iteration process defined by (4.16) is called **convergent** for an x_0 if the corresponding sequence x_0, x_1, ... is convergent. A sufficient condition for convergence is given in the following theorem, which has various practical applications.

THEOREM 2 (Convergence of Fixed-Point Iteration): *Let $x = s$ be a solution of $x = g(x)$ and suppose that g has a continuous derivative in some interval J containing s. Then if $|g'(x)| \leqq K < 1$ in J, the iteration process defined by* (4.16) *converges for any x_0 in J, and the limit of the sequence $\{x_n\}$ is s.*

PROOF: By the mean value theorem of differential calculus there is a t between x and s such that

$$g(x) - g(s) = g'(t)\,(x - s)$$

Since $g(s) = s$ and $x_1 = g(x_0), x_2 = g(x_1), \ldots$, we obtain from this and the condition on $|g'(x)|$ in the theorem

$$|x_n - s| = |g(x_{n-1}) - g(s)| = |g'(t)|\,|x_{n-1} - s| \leqq K|x_{n-1} - s|$$

Applying this inequality n times, for $n, n-1, \ldots, 1$ gives

$$|x_n - s| \leqq K|x_{n-1} - s| \leqq K^2|x_{n-2} - s| \leqq \cdots \leqq K^n|x_0 - s|$$

Since $K < 1$, we have $K^n \to 0$; hence $|x_n - s| \to 0$ as $n \to \infty$. ■

We mention that a function g satisfying the condition in Theorem 2 is called a **contraction** because $|g(x) - g(v)| \leqq K|x - v|$, where $K < 1$. Furthermore, K gives information on the speed of convergence. For instance, if $K = 0.5$, then the accuracy increases by at least 2 digits in only 7 steps because $0.5^7 < 0.01$.

Example 3 An Iteration Process. Illustration of Theorem 2

Find a solution of $f(x) = x^3 + x - 1 = 0$ by iteration.

Solution: A sketch shows that a solution lies near $x = 1$. We may write the equation as $(x^2 + 1)x = 1$ or

$$x = g_1(x) = \frac{1}{1+x^2}$$

so that

$$x_{n+1} = \frac{1}{1+x_n^{\,2}}$$

Also

$$|g_1'(x)| = \frac{2|x|}{(1+x^2)^2} < 1$$

for any x because $4x^2/(1+x^2)^4 = 4x^2/(1+4x^2+\cdots) < 1$, so that by Theorem 2 we have convergence for any x_0. Choosing $x_0 = 1$, we obtain (Figure 2)

$$x_1 = 0.500, \quad x_2 = 0.800, \quad x_3 = 0.610, \quad x_4 = 0.729, \quad x_5 = 0.653, \quad x_6 = 0.701, \ldots$$

The solution exact to 6D is $s = 0.682328$. The given equation may also be written as

$$x = g_2(x) = 1 - x^3$$

Then

$$|g_2'(x)| = 3x^2$$

and this is greater than 1 near the solution, so that we cannot apply Theorem 2 and assert convergence. Try $x_0 = 1$, $x_0 = 0.5$, $x_0 = 2$ and see what happens. The example shows that the transformation of a given $f(x) = 0$ into the form $x = g(x)$ with g satisfying $|g'(x)| \leqq K < 1$ may need some experimentation.

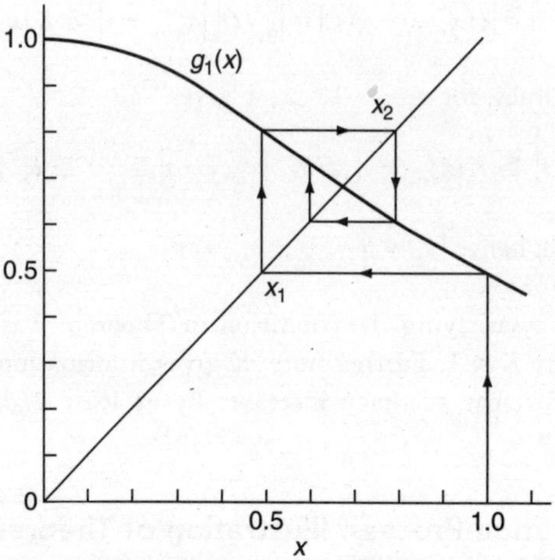

Figure 2 Iteration in Example 3.

Bisection Method

This is one of the simplest methods and convergence is guaranteed but slow; that is, solution of $f(x)$ can always be obtained. This method is also known as the "Methods of successive bisection." This method is based on the principle that if $f(x)$ is continuous between x_0 and x_1, and $f(x_0)$, $f(x_1)$ are of opposite signs then there exists at least one root between $x = x_0$ and $x = x_1$.

We begin the iterative cycle by choosing two trial points x_0 and x_1 which enclose the actual root. Then $f(x_0)$ and $f(x_1)$ are of opposite signs. The interval (x_0, x_1) is bisected and its midpoint x_2 is obtained as

$$x_2 = (x_1 + x_0)/2$$

If $f(x_2) = 0$, then x_2 itself is the root. If not, the root lies either between x_0 and x_2 *or* between x_1 and x_2. That is, if $f(x_0)$ and $f(x_2)$ are of opposite signs, then root lies between x_0 and x_2; *or* else if $f(x_1)$ and $f(x_2)$ are of opposite signs, the root will lie between x_1 and x_2. The new interval for searching the root is therefore (x_0, x_2) in the first case and (x_1, x_2) in the second case, which is much less compared to the first interval (x_0, x_1). This process is illustrated in Figure 3. This new interval will be the beginning interval for the next iteration, and the processes of bisectioning and finding another such x_2 are repeated. [For computation sake, the same set of designations x_0, x_1 are used as the starting points for each iteration; their values being different and the range of interval is half of the previous interval. This is evident in the algorithm for Bisection Method.]

The search interval gets halved in each successive iteration, and the range gets reduced by a factory of 2. A total of 10 steps will reduce the interval size by 2^{10} (= 1024). The iterations are stopped as soon as this search interval becomes smaller than the precision or accuracy (or error limit) with which we want the answer. Thus the root is calculated to the desired accuracy, the general iteration formula being

$$x_{i+1} = (x_i + x_{i-1})/2 \quad (i = 1, 2, 3, \ldots)$$

for each iteration. This required precision or accuracy can be given in two ways – either absolute precision (or absolute error measure) or to prescribe the measure of relative error (the safer way) – a measure of this relative precision is described by the value of

$$\left|\frac{x_1 - x_0}{x_1}\right| \quad \text{or} \quad \left|\frac{x_1 - x_0}{x_0}\right|$$

If the answer is required with n significant digits, then the iterative process is continued till

$$\left|\frac{x_1 - x_0}{x_1}\right| \leqq 10^{-n}$$

(Students can choose a value of $n = 4$ as a typical value, which in turn declares an accuracy of 10^{-4} or 0.0001.)

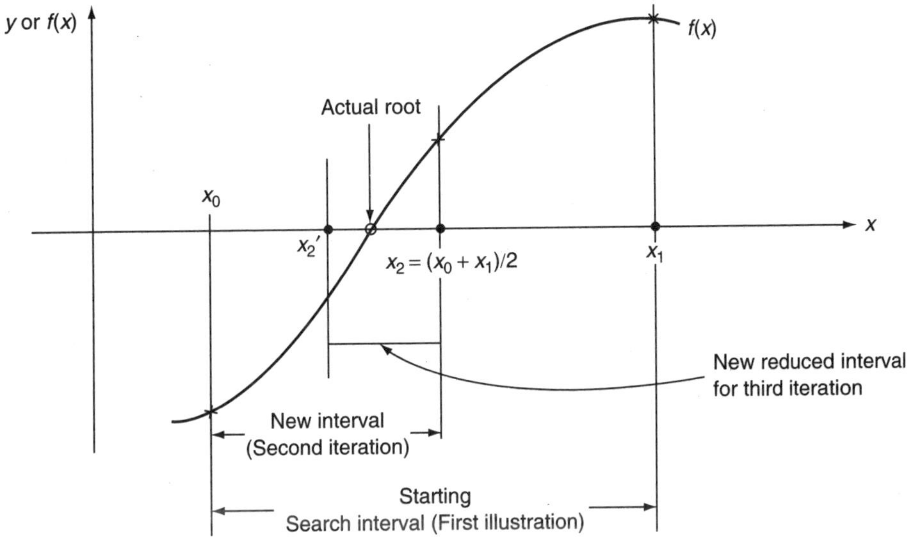

Figure 3 Illustration of bisection method.

Algorithm for the bisection method and the associated computer program can easily be developed following the logic explained in this subsection. The starting values x_0, x_1 are so chosen that $f(x_0)$ and $f(x_1)$ are of opposite signs, $f(x)$ being given. Note that e is the accuracy and can be taken as 10^{-4} (or any suitable value).Thus, the inputs are: x_0, x_1, e, N (where N is the maximum permissible number of iterations to be specified by user), and the outputs will be: root, number of iterations taken.

ALGORITHM BISECTION METHOD

INPUTS: x_0, x_1, e, N (N is the maximum permissible number of iterations specified by user.)

OUTPUTS: Roots, number of iterations taken

Step 1: Read x_0, x_1, e, N

Step 2: FX0 $\leftarrow f(x_0)$

Step 3: FX1 $\leftarrow f(x_1)$ [remarks: the function $f(x)$ is given and can be separately listed as a subprogram in source code. FX0, FX1 are values of the function at x_0, x_1].

Step 4: For IT = 1 to N, in steps of 1, do up to step 16
Step 5: err ← $|(x_1-x_0)/x_1|$ (remarks : if $x_1 = 0$, use $(x_1 - x_0)/x_0$ as err)
Step 6: $x_2 \leftarrow (x_1 + x_0)/2$
Step 7: If (err ≤ e) then
Step 8: Write "Solution converges," x_2, IT
Step 9: Stop (end if)
Step 10: FX2 ← $f(x_2)$
Step 11: FP ← FX0 * FX2
Step 12: If (FP < 0.0) then
Step 13: $x_1 \leftarrow x_2$
Step 14: FX1 ← FX2
Step 15: Else $x_0 \leftarrow x_2$
Step 16: FX0 ← FX2
End of FOR loop of step 4
Step 17: Write "no solution in," N, "iterations"
Step 18: Stop

Example 4

Find a real root of the equation $f(x) = x^3 - x - 1 = 0$ numerically.

Solution: Computationally, the root can be evaluated using the above program. The root can also be calculated as follows: Choose $x_0 = 1$, $x_1 = 2$, $e = 0.0001$ (for manual calculations), therefore

$$f(x_0) = -1, \; f(x_1) = 5$$

They are of opposite signs. So there is at least one root in the range (1, 2). The iterative procedure is tabulated below (for six iterations).

	No. of iterations	x_0	x_1	$f(x_0)$	$f(x_1)$	$\|(x_1 - x_0)/x_1\|$	$x_2 = (x + x_0)/2$	$f(x_2)$
	1	1.0	2.0	−1.0	5.0	0.5	1.5	+0.875
$x_2 \to x_1$	2	1.0	1.5	−1.0	0.875	1/3 = 0.3333	1.25	−19/64 = −0.2969
$x_2 \to x_0$	3	1.25	1.5	−19/64	0.875	1/6	1.375	+0.2246
$x_2 \to x_1$	4	1.25	1.375	−19/64	0.2246	0.09091	1.3125	−0.0515
$x_2 \to x_0$	5	1.3125	1.375	−0.0515	0.2246	0.04545	1.34375	0.0826
	6						1.328125	

The actual expected root value is ≈1.3247180, which can be obtained only after about 20 computations, if the accuracy limit is kept at 10^{-6} in the above program. Convergence is assured and solution can be obtained, but only after 20 iterations. This implies, the rate of convergence is slow.

Method of False Position

This is also known as Regula Falsi method. It is better than the previous method in the sense that it also guarantees convergence at a faster rate (i.e., solution can be obtained in less number of iterations, all other parameters being same).

As before, the iterative procedure is started by choosing two values x_0 and x_1 such that $f(x_0)$ and $f(x_1)$ are of opposite signs. Then the two points $[x_0, f(x_0)]$ and $[x_1, f(x_1)]$ are joined by a straight line. The intersection of this line with the x axis gives x_2. If $f(x_2)$ and $f(x_0)$ are of opposite signs, then, as before, we replace x_1 by x_2; or else, replace x_0 by x_2. This yields a new set of values for x_0 and x_1; the present range will be much smaller than the range or interval between the first chosen set of x_0 and x_1. The convergence is thus established, and the iterations are carried over with the new set of x_0 and x_1. Another x_2 is found by the intersection of the straight line joining the new $f(x_0)$ and $f(x_1)$ points with x axis. Each new or successive interval of search is smaller than the previous interval and it is guaranteed to converge to the root.

The convergence can be checked by checking the relative error in terms of x_2 *or* by checking the value of $f(x_2)$, which is nothing but checking for absolute error. Since we have established the relative error measure for bisection method, let us use the measure of absolute error here. Thus, for getting the final computed value of the root, the condition should be

$$|f(x_2)| < e$$

where e is specified by the user. This is acceptable provided that the slope $f(x)$ near the root is **not too small**. If the slope is too small, there can be a large error in the root calculation, even though $f(x)$ at x_2 is small. Otherwise, relative error measure can always be used as suggested earlier. The procedure is illustrated in Figure 4.

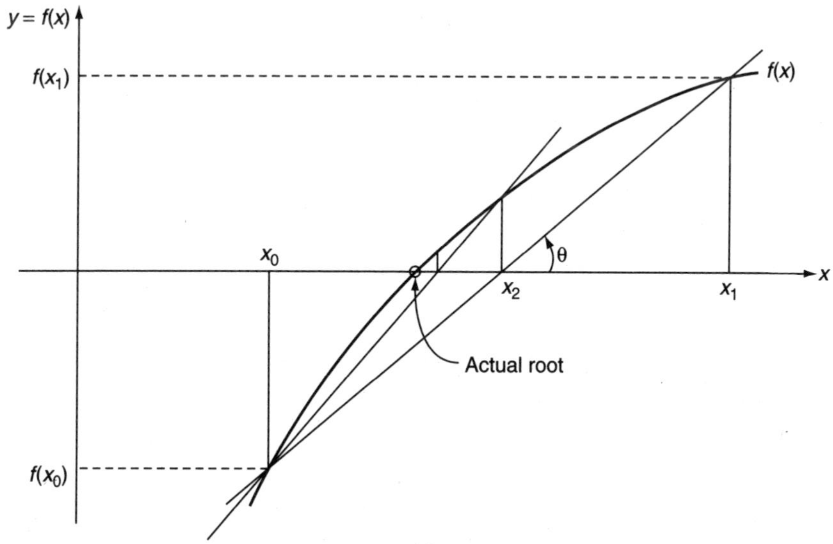

(a)

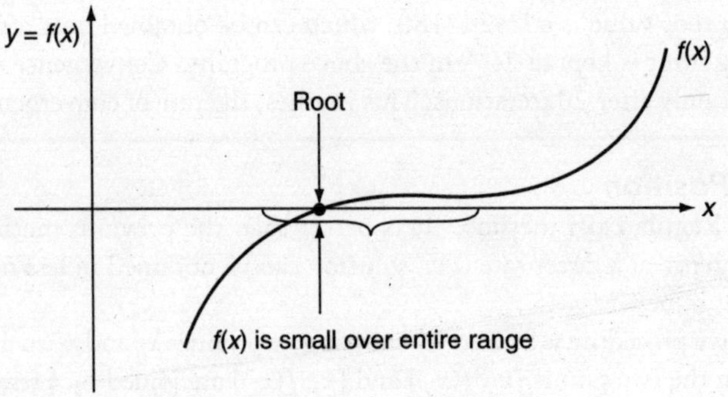

(b)

Figure 4 Illustration of regula falsi method: (a) $|f(x_2)| < e$ acceptable; (b) slope is too small near the root $|f(x_2)| < e$.

From this figure it can be seen (from the equation of the straight line) that

$$\frac{f(x_1) - f(x_0)}{x_1 - x_0} = \tan\theta = \frac{f(x_1) - 0}{x_1 - x_2} \tag{4.19}$$

where

$$x_2 = \frac{x_0 f(x_1) - x_1 f(x_0)}{f(x_1) - f(x_0)}$$

The algorithm for bisection method is written as follows:

ALGORITHM REGULA FALSI METHOD

Step 1: Read x_0, x_1, e, N
Step 2: F0 $\leftarrow f(x_0)$
Step 3: F1 $\leftarrow f(x_1)$
Step 4: For IT = 1 to N, in steps of 1, do up to step 14
Step 5: $x_2 \leftarrow$ (F1 * x_0 – F0 * x_1) / (F1 – F0)
Step 6: F2 $\leftarrow f(x_2)$
Step 7: If |F2| < e, then
Step 8: Write "Solution convergent," x_2, F2, IT
Step 9: Stop
Step 10: If (F2 * F0 < 0.0) then
Step 11: $x_1 \leftarrow x_2$
Step 12: F1 $\leftarrow$ F2
Step 13: Else $x_0 \leftarrow x_2$

Step 14: F0 ← F2

Step 15: Write "no convergence in," N, "iterations"

Step 16: Write x_2, F2

Step 17: Stop

Example 5

Evaluate the real root of the equation $x^3 - 2x = 5$, correct to 4 decimals places.

Solution: Choose $x_0 = 2$ and $x_1 = 3$, such that $f(x_0) = x^3 - 2x - 5$. At $x = 2, f(x_0) = -1$ and $f(x_1) = x^3 - 2x - 5$. At $x = 3, f(x_1) = 16$ are of opposite signs. Therefore we can expect atleast one root between $x = 2$ and $x = 3$. Since four decimal accuracy is specified, the number of significant digits in the expected value would be 5 (i.e., 2.0945). For 5 significant digit accuracy, error or accuracy limit would be 10^{-5}.

Thus inputs are: $x_0 = 2$, $x_1 = 3$, $e = 10^{-5}$, and let $N = 25$ be the maximum number of iterations permissible. The real root of the equation is 2.0945....

Example 6

Calculate the real root of the equation $x^2 = 25$ by the regula falsi method.

Solution: Choose $x_0 = 2$, $x_1 = 7$, $e = 10^{-4}$, therefore

$$f(x_0) = x^2 - 25$$

At $x = 2$, we get

$$f(x_0) = -21$$

Also

$$f(x_1) = x^2 - 25$$

At $x = 7$, we get

$$f(x_1) = 24$$

Now $f(x_0)$ and $f(x_1)$ are of opposite signs. So there is at least one root between 2 and 7. The tabulations are given below:

Iteration no.	x_0	x_1	x_2	$f(x_0)$	$f(x_1)$	$f(x_2)$	
1	2	7	4.3333	−21	24	−6.2225	$x_2 \to x_0$
2	4.3333	7	4.8824	−6.2225	24	−1.1622	$x_2 \to x_0$
3	4.8824	7	4.9802	−1.1622	24	−0.1976	$x_2 \to x_0$
4	4.9802	7	4.9967	−0.1976	24	−0.0330	$x_2 \to x_0$
5	4.9967	7	4.9995	−0.0330	24	−0.005	$x_2 \to x_0$
6	4.9995	7	4.9999	−0.005	24	−0.001	$x_2 \to x_0$

Calculations (manual) for six iterations are shown in the above table. After six iterations the approximate root is (given by x_2) 4.9999, if an error limit of 0.001 is acceptable. For higher precision some more iterations would be required and the root converges to 5.0 which is the actual root. The number of iterations required in this case would be much less than those in bisection method. Regula falsi method provides convergent solution with a faster rate of convergence than the bisection method.

Students are advised to calculate (manually and computationally) Examples 5 and 6 by both bisection and regular falsi methods using same precision limit (e) and compare the results.

Newton's Method

Newton's method, also known as **Newton–Raphson's method,** is another iteration method for solving equations $f(x) = 0$, where f is assumed to have a continuous derivative f'. The method is commonly used because of its simplicity and great speed. The underlying idea is that we approximate the graph of f by suitable tangents. Using an approximate value x_0 obtained from the graph of f, we let x_1 be the point of intersection of the x-axis and the tangent to the curve of f at x_0 (see Figure 5). Then

$$\tan\beta = f'(x_0) = \frac{f(x_0)}{x_0 - x_1}$$

hence

$$x_1 = x_0 - \frac{f(x_0)}{f'(x_0)}$$

In the second step we compute $x_2 = x_1 - f(x_1)/f'(x_1)$, in the third step x_3 from x_2 again by the same formula, and so on. We thus have the algorithm shown next. Formula (4.21) in this algorithm can also be obtained if we algebraically solve Taylor's formula

$$f(x_{n+1}) \approx f(x_n) + (x_{n+1} - x_n)f'(x_n) = 0 \tag{4.20}$$

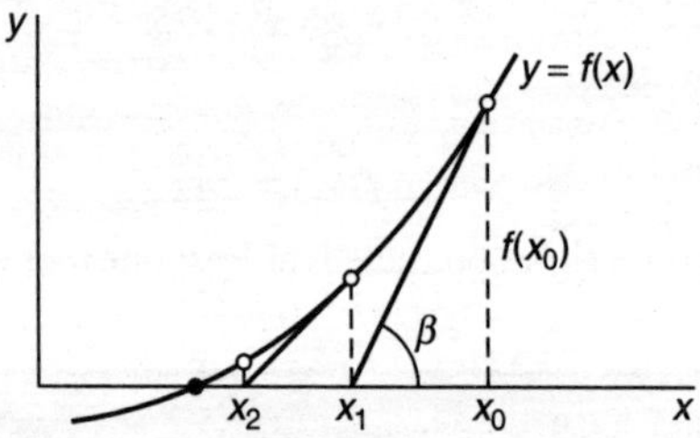

Figure 5 Newton's method.

ALGORITHM NEWTON (f, f', x_0, ϵ, N)

This algorithm computes a solution of $f(x) = 0$ given an initial approximation x_0 (starting value of the iteration). Here the function $f(x)$ is continuous and has a continuous derivative $f'(x)$.

INPUT: f, f', initial approximation x_0, tolerance $\epsilon > 0$, maximum number of iterations N.

OUTPUT: Approximate solution x_n ($n \leqq N$) or message of failure.

For $n = 0, 1, 2, \ldots, N-1$ do:

1 Compute $f'(x_n)$.

2 If $f'(x_n) = 0$ then OUTPUT "Failure". Stop.

[Procedure completed unsuccessfully]

3 Else compute

$$x_{n+1} = x_n - \frac{f(x_n)}{f'(x_n)} \tag{4.21}$$

4 If $|x_{n+1} - x_n| \leqq \epsilon |x_n|$ then OUTPUT x_{n+1}. Stop.

[Procedure completed successfully]

End

5 OUTPUT "Failure". Stop.

[Procedure completed unsuccessfully after N iterations]

End NEWTON

If it happens that $f'(x_n) = 0$ for some n (see line 2 of the algorithm), then try another starting value x_0. Line 3 is the heart of Newton's method. The inequality in line 4 is a **termination criterion**. If the sequence of the x_n converges and the criterion holds, we have reached the desired accuracy and stop. In this line the factor $|x_n|$ is needed in the case of zeros of very small (or very large) absolute value because of the high density (or of the scarcity) of machine numbers for those x.

WARNING! The criterion by itself does not imply convergence. For example, the harmonic series diverges, although its partial sums $x_n = \Sigma_{k=1}^{n} 1/k$ satisfy the criterion because $\lim (x_{n+1} - x_n) = \lim(1/(n+1)) = 0$.

Line 5 gives another termination criterion and is needed because Newton's method may diverge or, due to a poor choice of x_0, may not reach the desired accuracy by a reasonable number of iterations. Then we may try another x_0. If $f(x) = 0$ has more than one solution, different choices of x_0 may produce different solutions. Also, an iterative sequence may sometimes converge to a solution different from the expected one.

Example 7 Square Root

Set up a Newton iteration for computing the square root x of a given positive number c and apply it to $c = 2$.

Solution: We have $x = \sqrt{c}$, hence $f(x) = x^2 - c = 0$, $f'(x) = 2x$, and (4.21) takes the form

$$x_{n+1} = x_n - \frac{x_n^{\;2} - c}{2x_n} = \frac{1}{2}\left(x_n + \frac{c}{x_n}\right)$$

For $c = 2$, choosing $x_0 = 1$, we obtain

$$x_1 = 1.500000, \quad x_2 = 1.416667, \quad x_3 = 1.414216, \quad x_4 = 1.414214, \ldots$$

x_4 is exact to 6D.

Example 8 Iteration for a Transcendental Equations

Find the positive solution of $2 \sin x = x$.

Solution: Setting $f(x) = x - 2\sin x$, we have $f'(x) = 1 - 2\cos x$, and (4.21) gives

$$x_{n+1} = x_n - \frac{x_n - 2\sin x_n}{1 - 2\cos x_n} = \frac{2(\sin x_n - x_n \cos x_n)}{1 - 2\cos x_n} = \frac{N_n}{D_n}$$

From the graph of f we conclude that the solution is near $x_0 = 2$. We compute:

n	x_n	N_n	D_n	x_{n+1}
0	2.00000	3.48318	1.83229	1.90100
1	1.90100	3.12470	1.64847	1.89552
2	1.89552	3.10500	1.63809	1.89550
3	1.89550	3.10493	1.63806	1.89549

$x_4 = 1.89549$ is exact to 5D since the solution to 6D is 1.895494.

Example 9 Newton's Method Applied to an Algebraic Equation

Apply Newton's method to the equation $f(x) = x^3 + x - 1 = 0$.

Solution: From (4.21) we have

$$x_{n+1} = x_n - \frac{x_n^{\;3} + x_n - 1}{3x_n^{\;2} + 1} = \frac{2x_n^{\;3} + 1}{3x_n^{\;2} + 1}$$

Starting from $x_0 = 1$, we obtain

$$x_1 = 0.750000, \quad x_2 = 0.686047, \quad x_3 = 0.682340, \quad x_4 = 0.682328, \ldots$$

where x_4 has the error -1×10^{-6}. A comparison with Example 3 shows that the present convergence is much more rapid. This may motivate the concept of the *order of an iteration process,* to be discussed next.

Secant Method

Newton's method is very powerful but has the disadvantage that the derivative f' may sometimes be a far more difficult expression than f itself and its evaluation therefore computationally expensive. This situation suggests the idea of replacing the derivative with the difference quotient

$$f'(x_n) \approx \frac{f(x_n) - f(x_{n-1})}{x_n - x_{n-1}}$$

Then instead of (4.21) we have the formula of the popular secant method:

$$x_{n+1} = x_n - f(x_n)\frac{x_n - x_{n-1}}{f(x_n) - f(x_{n-1})} \tag{4.22}$$

Geometrically, we intersect the x axis at x_{n+1} with the secant of $f(x)$ passing through P_{n-1} and P_n in Figure 6. We need two starting values x_0 and x_1. Evaluation of derivatives is now avoided. It can be shown that convergence is **superlinear**, almost quadratic like Newton's method. The algorithm is similar to that of Newton's method, as the student may show.

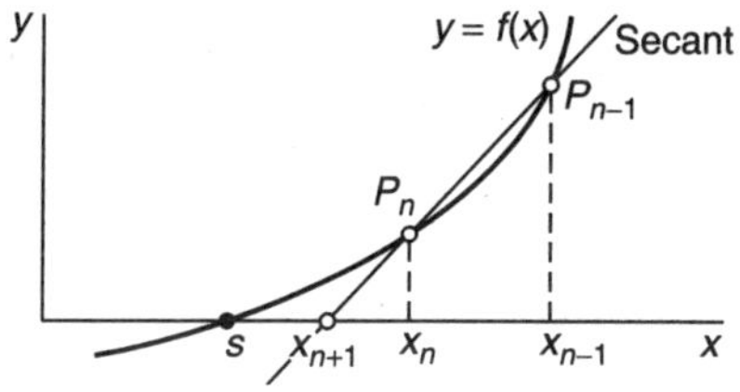

Figure 6 Secant method.

CAUTION! It is ***not*** good to write (4.22) as

$$x_{n+1} = \frac{x_{n-1}f(x_n) - x_n f(x_{n-1})}{f(x_n) - f(x_{n-1})}$$

Here this may lead to loss of significant digits if x_n and x_{n-1} are about equal. (Can you see this from the formula?)

Example 10 Secant Method

Find the positive solution of $f(x) = x - 2\sin x = 0$ by the secant method, starting from $x_0 = 2$, $x_1 = 1.9$.

Solution: Here (4.22) is

$$x_{n+1} = x_n - \frac{(x_n - 2\sin x_n)(x_n - x_{n-1})}{x_n - x_{n-1} + 2(\sin x_{n-1} - \sin x_n)} = x_n - \frac{N_n}{D_n}$$

The numerical values are:

n	x_{n-1}	x_n	N_n	D_n	$x_{n+1} - x_n$
1	2.000000	1.900000	−0.000740	−0.174005	−0.004253
2	1.900000	1.895747	−0.000002	−0.006986	−0.000252
3	1.895747	1.895494	0		0

$x_3 = 1.895494$ is exact to 6D.

Successive Approximation Method

This is one of the simplest methods of finding the roots of the equation $f(x) = 0$. It does not involve evaluation of the derivative nor does it have the minimum slope restriction problem. The method is quite simple and very easy to program. However, there is no guarantee of convergence.

In this method, function $f(x) = 0$, whose roots or zeros are to be found, is expressed in the form

$$x = g(x) \tag{4.23}$$

The roots of the equation $f(x) = 0$ are now same as the points of intersection of the straight line representing x and the function $g(x)$. For example, consider the solution of the function

$$f(x) = e^{-x} - 2x + 1 = 0$$

This can be written as

$$x = \frac{e^{-x} + 1}{2}$$

and thus

$$g(x) = \frac{e^{-x} + 1}{2}$$

The roots of $f(x) = 0$ are nothing but points of intersection of the line representing x and the curve $(e^{-x} + 1)/2$. Once the function $g(x)$ is identified, the iterative technique is initiated with an initial guess value x_0. An appropriate value (no conditions here) for x_0 is chosen. The next successive value is obtained by substituting $x = x_0$ in (4.23) in $g(x)$, and $g(x_0)$ is now taken as the new guess x_1. The value $g(x_1)$ will become another iterative value x_2 and so on. The iterations are carried until two successive values of x are close enough, that is within the acceptable error limit. This implies the use of relative error measure for declaration of the root. Thus the iterative formula is

$$x_i = g(x_{i-1}) \tag{4.24}$$

That is, $x_1 = g(x_0)$, $x_2 = g(x_1)$, $x_3 = g(x_2)$, …, until the convergence is obtained or until the maximum prescribed iterations are over.

An algorithm implementing the above method is as follows:

ALGORITHM SUCCESSIVE APPROXIMATION METHOD

INPUT: x_0 – initial guess; e – error limit for root declaration (relative error measure used); N – Maximum number of iterations allowed

OUTPUT: Root, number of iterations taken

Step 1: Read x_0, e, N

Step 2: For IT = 1 to N, in steps of 1, do upto step 8

Step 3: $x_1 \leftarrow g(x_0)$

Step 4: rerr $\leftarrow (x_1 - x_0)/x_1$

Step 5: if (abs (err) < e) then

Step 6: Write "convergent solution," x_1, IT

Step 7: Top (endif)

Step 8: $x_0 \leftarrow x_1$

End of FOR loop for step 2

Step 9: Write "no convergence," x_0, x_1

Step 10: Stop

The only requirement here is to express the function $f(x)$ in $x = g(x)$ form. In this method, the order of convergence is 1 and evaluation of number of functions per iteration is also 1. There is no guarantee of convergence however.

Example 11

Find the real positive roots of the equation $x^3 - 4x + 1 = 0$ by the successive approximation method, through (i) computation and (ii) manual calculations.

Solution: Computation can be carried out using above program, where $g(x)$ is given by

$$x = g(x) = \frac{x^3 + 1}{4}$$

Students are advised to calculate the root (any one root) manually and prepare a table for iterative procedure as done earlier.

Answer: The roots are 0.2541 and 1.8608.

Order of an Iteration Method and Speed of Convergence

The quality of an iteration method may be characterized by the speed of convergence, as follows. Let $x_{n+1} = g(x_n)$ define an iteration method, and let x_n approximate a solution s of $x = g(x)$. Then $x_n = s - \epsilon_n$, where ϵ_n is the error of x_n. Suppose that g is differentiable a number of times, so that the Taylor formula gives

$$x_{n+1} = g(x_n) = g(s) + g'(s)(x_n - s) + \frac{1}{2}g''(s)(x_n - s)^2 + \cdots$$

$$= g(s) - g'(s)\epsilon_n + \frac{1}{2}g''(s)\epsilon_n{}^2 + \cdots \tag{4.25}$$

The exponent of ϵ_n in the first nonvanishing term after $g(s)$ is called the **order** of the iteration process defined by g. The order measures the speed of convergence.

To see this, subtract $g(s) = s$ on both sides of (4.25). Then on the left you get $x_{n+1} - s = -\epsilon_{n+1}$, where ϵ_{n+1} is the error of x_{n+1}. And on the right the remaining expression equals approximately its first nonzero term because $|\epsilon_n|$ is small in the case of convergence. Thus

$$\begin{aligned} &\text{(a)} \quad \epsilon_{n+1} \approx +g'(s)\epsilon_n && \text{in the case of first order} \\ &\text{(b)} \quad \epsilon_{n+1} \approx -\frac{1}{2}g''(s)\epsilon_n^2 && \text{in the case of second order, etc.} \end{aligned} \tag{4.26}$$

Thus if $\epsilon_n = 10^{-k}$ in some step, then for second order, $\epsilon_{n+1} = c \times (10^{-k})^2 = c \times 10^{-2k}$, so that the number of significant digits is about doubled in each step.

Convergence of Newton's Method

In Newton's method, $g(x) = x - f(x)/f'(x)$. By differentiation,

$$g'(x) = 1 - \frac{f'(x)^2 - f(x)f''(x)}{f'(x)^2} = \frac{f(x)f''(x)}{f'(x)^2} \tag{4.27}$$

Since $f(s) = 0$, this shows that also $g'(s) = 0$. Hence Newton's method is at least of second order. If we differentiate again and set $x = s$, we find that

$$g''(s) = \frac{f''(s)}{f'(s)} \tag{4.28}$$

which will not be zero in general. This proves the next the theorem.

THEOREM 3 (Second-Order Convergence of Newton's Method): *If $f(x)$ is three times differentiable and f' and f'' are not zero at a solution s of $f(x) = 0$, then for x_0 sufficiently close to s, Newton's method is of second order.*

Comments. For Newton's method, (4.26b) becomes, by (4.28),

$$\epsilon_{n+1} \approx \frac{f''(s)}{2f'(s)}\epsilon_n^2 \tag{4.29}$$

For the rapid convergence of the method indicated in Theorem 3 it is important that s be a *simple* zero of $f(x)$ [thus $f'(s) \neq 0$] and that x_0 be close to s, because in Taylor's formula we took only the linear term, assuming the quadratic term to be negligibly small. (With a bad x_0 the method may even diverge!)

Example 12 Prior Error Estimate of the Number of Newton Iteration Steps

Use $x_0 = 2$ and $x_1 = 1.901$ in Example 8 for estimating how many iteration steps we need to produce the solution to 5D accuracy. This is an **a priori estimate** or **prior estimate** because we can compute it after only one iteration, prior to further iterations.

Solution: We have $f(x) = x - 2\sin x = 0$. Differentiation gives

$$\frac{f''(s)}{2f'(s)} \approx \frac{f''(x_1)}{2f'(x_1)} = \frac{2\sin x_1}{2(1 - 2\cos x_1)} \approx 0.57$$

Hence (4.29) gives

$$|\epsilon_{n+1}| \approx 0.57\epsilon_n^{\ 2} \approx 0.57(0.57\epsilon_{n-1}^{\ 2})^2 = 0.57^3\epsilon_{n-1}^{\ 4} \approx \cdots \approx 0.57^M \epsilon_0^{\ m+1} \leqq 5\times 10^{-6}$$

where $M = 2^n + 2^{n-1} + \cdots + 2 + 1 = 2^{n+1} - 1$. We show below that $\epsilon_0 \approx -0.11$. Consequently, our condition becomes

$$0.57^M 0.11^{M+1} \leqq 5\times 10^{-6}$$

Hence $n = 2$ is the smallest possible n, according to this crude estimate, in good agreement with Example 8. $\epsilon_0 \approx -0.11$ is obtained from $\epsilon_1 - \epsilon_0 = (\epsilon_1 - s) - (\epsilon_0 - s) = -x_1 + x_0 \approx 0.10$, hence $\epsilon_1 = \epsilon_0 + 0.10 \approx -0.57\,\epsilon_0^{\ 2}$ or $0.57\,\epsilon_0^{\ 2} + \epsilon_0 + 0.10 \approx 0$, which gives $\epsilon_0 \approx -0.11$.

Difficulties in Newton's Method

Difficulties may arise if $|f'(x)|$ is very small near a solution s of $f(x) = 0$, for instance, if s is a zero of $f(x)$ of second (or higher) order (so that Newton's method converges only linearly). Geometrically, small $|f'(x)|$ means that the tangent of $f(x)$ near s almost coincides with the x axis (so that double precision may be needed to get $f(x)$ and $f'(x)$ accurately enough). Then for values $x = \tilde{s}$ far away from s we can still have small function values

$$R(\tilde{s}) = f(\tilde{s})$$

In this case we call the equation $f(x) = 0$ **ill-conditioned**. $R(\tilde{s})$ is called the **residual** of $f(x) = 0$ at $\tilde{s}$. Thus a small residual guarantees a small error of $\tilde{s}$ only if the equation is ***not*** ill-conditioned.

Example 13 An ILL-Conditioned Equation

$f(x) = x^5 + 10^{-4}x = 0$ is ill-conditioned. $x = 0$ is a solution. $f'(0) = 10^{-4}$ is small. At $\tilde{s} = 0.1$ the residual $f(0.1) = 2 \times 10^{-5}$ is small, but the error -0.1 is larger in absolute value by a factor 5000. Invent a more drastic example of your own.

Comparison of Iterative Methods

The main basis of comparison for the iterative methods are:

1. Rate of convergence.
2. Amount of computational efforts per iteration.
3. Sensitivity of the method to initial guess value and subsequent iterated values.

The details of each method described, with reference to the above parameters of comparison, have already been mentioned along with the description of that iterative method. For completeness sake, the comparison is also given in a tabulated form in Table 1.

Table 1 Comparison of Iterative Methods

Method	*Formula*	*Order of convergence*	*Evaluation of functions per iteration*	*Remarks*
Bisection method	$x_{i+1} = \frac{x_i + x_{i-1}}{2}$ x_i, x_{i-1} enclose the root.	Search interval reduces by a factor of 2 in each iteration.	1	Guaranteed convergence but slow. Not suitable for even order multiple or complex roots. Simple method.
Method of false position	$x_{i+1} = \frac{x_{i-1}f(x_i) - x_i f(x_{i-1})}{f(x_i) - f(x_{i-1})}$ x_i, x_{i-1} enclose the root.	1	1	Guaranteed convergence. Slope of $f(x)$ near the root should not be too small. Superior to bisection method.
Newton–Raphson method	$x_{i+1} = x_i - \frac{f(x_i)}{f'(x_i)}$	2	2	Convergence fast if starting point is near actual root. Sensitive to starting value. $f'(x)$ is required, slope should not be too small.
Secant method	$x_{i+1} = \frac{x_{i-1}f(x_i) - x_i f(x_{i-1})}{f(x_i) - f(x_{i-1})}$	1.62	1	Faster than bisection and false position if initial guess is near the actual root; if not, no guarantee of convergence. Economical method as it needs no derivative.
Successive approximation method	$x_i = g(x_{i-1})$	1	1	Easy to program, no guarantee of convergence. Simple, no derivative calculations or slope restrictions.

There are various other iterative methods also available. All methods are not suitable for all types of problems. If the solution is not possible by one method (after all checks), other methods could be tried.

Summary of Methods

The methods for computing solutions s of $f(x) = 0$ with given continuous (or differentiable) $f(x)$ start with an initial approximation x_0 of s and generate a sequence $x_1, x_2, \ldots$ by **iteration**. **Fixed-point methods** solve $f(x) = 0$ written as $x = g(x)$, so that s is a *fixed point* of g, that is, $s = g(s)$. For $g(x) = x - f(x)/f'(x)$ this is **Newton's method**, which for good x_0 and simple zeros converges quadratically (and for multiple zeros linearly). From Newton's method the **secant method** follows by replacing $f'(x)$ by a difference quotient. The **bisection method** and the **method of false position** always converge, but often slowly.

CONCEPT CHECK

1. Apply fixed-point iteration and answer related questions where indicated. Show details of your work.
   (a) $x = 1.4 \sin x$, $x_0 = 1.4$.
   (b) Do the iterations indicated at the end of Example 2. Sketch a figure.
   (c) Why do we obtain a monotone sequence in Example 1, but not in Example 2?
   (d) $f = x^4 - x + 0.2 = 0$, the root near 1, $x_0 = 1$.
   (e) f as in Part (d), the root near 0, $x_0 = 0$.
   (f) Find the smallest positive solution of $\sin x = e^{-0.5x}$, $x_0 = 1$.
   (g) (Bessel functions, drumhead) A partial sum of the Maclaurin series of $J_0(x)$ is $f(x) = 1 - (x^2/4) + (x^4/64) - (x^6/2304)$. Conclude from a sketch that $f(x) = 0$ near $x = 2$. Write $f(x) = 0$ as $x = g(x)$ (by dividing $f(x)$ by $x/4$ and taking the resulting x term to the other side). Find the zero.
2. Apply Newton's method (6D accuracy). First sketch the function(s) to see what is going on.
   (a) $\sin x = \cot x$, $x_0 = 1$
   (b) $x = \cos x$, $x_0 = 1$
   (c) $x^3 - 5x + 3 = 0$, $x_0 = 2$
   (d) $x + \ln x = 2$, $x_0 = 2$
3. **(Vibrating beam)** Find the solution of $\cos x \cosh x = 1$ near $x = 3\pi/2$.
4. **(Heating, cooling)** At what time x (4S-accuracy only) will the processes governed by $f_1(x) = 100(1 = e^{20.2x})$ and $f_2(x) = 40e^{-0.01x}$ reach the same temperature? Also find the latter.
5. **(Associated Legendre functions)** Find the smallest positive zero of
   $$P_4^{\,2} = (1 - x^2)P_4'' = (15/2)(-7x^4 + 8x^2 - 1)$$
   (a) by Newton's method, **(b)** exactly, by solving a quadratic equation.
6. **(Legendre polynomials)** Find the largest root of the Legendre polynomial $P_5(x)$ given by $P_5(x) = (63x^5 - 70x^3 + 15x)/8$ **(a)** by Newton's method, **(b)** from a quadratic equation.
7. Design a Newton iteration for cube roots and compute $\sqrt[3]{7}$ (6D, $x_0 = 2$).
8. Design a Newton iteration for $\sqrt[k]{c}$ $(c > 0)$. Use it to compute $\sqrt{2}, \sqrt[3]{2}, \sqrt[4]{2}, \sqrt[5]{2}$ (6D, $x_0 = 1$).
9. **TEAM PROJECT. Bisection Method.** This simple but slowly convergent method for finding a solution of $f(x) = 0$ with continuous f is based on the **intermediate value theorem**, which states that if a continuous function f has opposite signs at some $x = a$ and $x = b$ $(> a)$, that is, either $f(a) < 0$, $f(b) > 0$ or $f(a) > 0$, $f(b) < 0$, then f must be 0 somewhere on $[a, b]$. The solution is found by repeated bisection of the interval and in each iteration picking that half which also satisfies that sign condition.
   (a) **Algorithm.** Write an algorithm for the method.
   (b) **Comparison.** Solve $x = \cos x$ by Newton's method and by bisection method. Compare.
   (c) Solve $e^{-x} = \ln x$ and $e^x + x^4 + x = 2$ by bisection method.

10. **TEAM PROJECT. Method of False Position (Regula falsi).** Figure 7 shows the idea. We assume that f is continuous. We compute the x-intercept c_0 of the line through $[a_0, f(a_0), b_0, f(b_0)]$. If $f(c_0) = 0$, we are done. If $f(a_0)f(c_0) < 0$ (as in Figure 7), we set $a_1 = a_0$, $b_1 = c_0$ and repeat to get c_1, etc. If $f(a_0)f(c_0) > 0$, then $f(c_0)f(b_0) < 0$ and we set $a_1 = c_0$, $b_1 = b_0$, etc.

    (a) **Algorithm.** Show that

    $$c_0 = \frac{a_0 f(b_0) - b_0 f(a_0)}{f(b_0) - f(a_0)}$$

    and write an algorithm for the method.

    (b) **Comparison.** Solve $x^3 = 5x + 6$ by Newton's method, the secant method, and the method of false position. Compare.

    (c) Solve $x^4 = 2$, $\cos x = \sqrt{x}$, and $x + \ln x = 2$ by the method of false position.

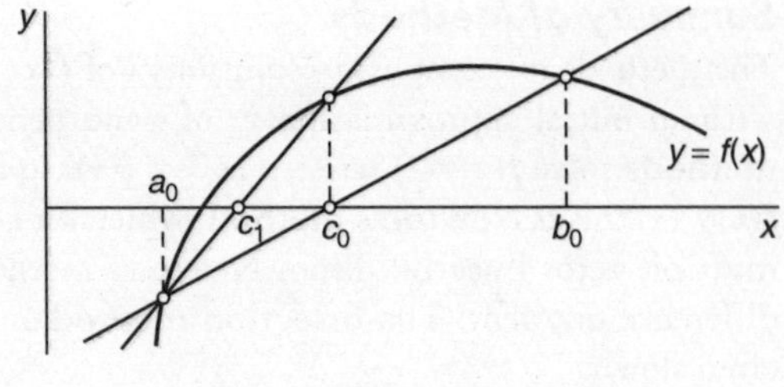

Figure 7 Method of false position.

11. Solve, using x_0 and x_1 as indicated.

    (a) Prob. 2(c), $x_0 = 0.5$, $x_1 = 2.0$

    (b) $e^{-x} - \tan x = 0$, $x_0 = 1$, $x_1 = 0.7$

    (c) Prob. 2(a), $x_0 = 1$, $x_1 = 0.5$

    (d) Prob. 2(b), $x_0 = 0.5$, $x_1 = 1$

12. **WRITING PROJECT. Solution of Equations.** Compare the methods in this section and problem set, discussing advantages and disadvantages using examples of your own.

4.4 Interpolation

Interpolation means finding (approximate) values of a function $f(x)$ for an x between ***different*** x values $x_0, x_1, \ldots, x_n$ at which the values of $f(x)$ are given. These values may come from a "mathematical" function, such as a logarithm or a Bessel function, or, perhaps more frequently, they may be measured or automatically recorded values of an "empirical" function, such as the air resistance of a car or an airplane at different speeds, or the yield of a chemical process at different temperatures, or the size of the U.S. population as it appears from censuses taken at 10-year intervals. We write these given values of a function f in the form

$$f_0 = f(x_0), \quad f_1 = f(x_1), \quad \ldots, \quad f_n = f(x_n)$$

or as ordered pairs

$$(x_0, f_0), \quad (x_1, f_1), \quad \ldots, \quad (x_n, f_n)$$

A standard idea in interpolation now is to find a polynomial $p_n(x)$ of degree n (or less) that assumes the given values; thus

$$p_n(x_0) = f_0, \quad p_n(x_1) = f_1, \quad \ldots, \quad p_n(x_n) = f_n \tag{4.30}$$

We call this p_n an **interpolation polynomial** and $x_0, \ldots, x_n$ the **nodes.** And if $f(x)$ is a mathematical function, we call p_n an **approximation** of f (or a **polynomial approximation,** because there are other kinds of approximations, as we shall see later). We use p_n to get (approximate) values of f for x's between x_0 and x_n (**"interpolation"**) or sometimes outside this interval $x_0 \leqq x \leqq x_n$ (**"extrapolation"**).

Motivation: Polynomials are convenient to work with because we can readily differentiate and integrate them, again obtaining polynomials. Moreover, they approximate *continuous* functions with any desired accuracy. That is, for any continuous $f(x)$ on an interval J: $a \leqq x \leqq b$ and error bound $\beta > 0$, there is a polynomial $p_n(x)$ (of sufficiently high degree n) such that

$$|f(x) - p_n(x)| < \beta \quad \text{for all } x \text{ on } J$$

This is the famous **Weierstrass approximation theorem.**

Existence and Uniqueness: p_n satisfying (4.30) for given data exists – we give formulas for it below. p_n is unique. Indeed, if another polynomial q_n also satisfies $q_n(x_0) = f_0, \ldots, q_n(x_n) = f_n$, then $p_n(x) - q_n(x) = 0$ at $x_0, \ldots, x_n$, but a polynomial $p_n - q_n$ of degree n (or less) with $n + 1$ roots must be identically zero, as we know from algebra; thus $p_n(x) = q_n(x)$ for all x, which means uniqueness.

How to Find p_n? This is the important practical question. We answer it by explaining several standard methods. For given data, these methods give the same polynomial, by the uniqueness just proved (which is thus of practical interest!), but expressed in several forms suitable for different purposes.

Error in Interpolation

If f is itself a polynomial of degree n (or less), it must coincide with p_n because the $n + 1$ data (x_0, f_0), $\ldots$, (x_n, f_n) determine a polynomial uniquely, so the error is zero. Now the special f has its $(n + 1)$th derivative identically zero. This makes it plausible that for a *general* f its $(n + 1)$th derivative $f^{(n+1)}$ should measure the error

$$\epsilon_n(x) = f(x) - p_n(x)$$

It can be shown that this is true if $f^{(n+1)}$ exists and is continuous. Then, with a suitable t between x_0 and x_n (or between x_0, x_n, and x if we extrapolate),

$$\epsilon_n(x) = f(x) - p_n(x) = (x - x_0)(x - x_1) \ldots (x - x_n) \frac{f^{(n+1)}(t)}{(n+1)!} \tag{4.31}$$

Thus $|\epsilon_n(x)|$ is 0 at the nodes and small near them, because of continuity. The product $(x - x_0) \cdots (x - x_n)$ is large for x away from the nodes. This makes extrapolation risky. And interpolation at an x will be best if we choose nodes on both sides of that x. Also, we get error bounds by taking the smallest and the largest value of $f^{(n+1)}(t)$ in (4.31) on the interval $x_0 \leqq t \leqq x_n$ (or on the interval also containing x if we *extrapolate*).

Most importantly, since p_n is unique, as we have shown, we have the next theorem.

THEOREM 4 (Error of Interpolation): *Formula* (4.31) *gives the error for* ***any*** *polynomial interpolation method if $f(x)$ has a continuous* $(n + 1)$*th derivative.*

Practical error estimate: If the derivative in (4.31) is difficult or impossible to obtain, apply the Error Principle, that is, take another node and the Lagrange polynomial $p_{n+1}(x)$ and regard $p_{n+1}(x) - p_n(x)$ as a (crude) error estimate for $p_n(x)$.

An example of error estimate of a linear interpolation is given in Example 15.

Lagrange Interpolation

Given $(x_0, f_0), (x_1, f_1), \ldots, (x_n, f_n)$ with arbitrarily spaced x_j, Lagrange had the idea of multiplying each f_j by a polynomial that is 1 at x_j and 0 at the other n nodes and then taking the sum of these $n + 1$ polynomials. Clearly, this gives the unique interpolation polynomial of degree n or less. Beginning with the simplest case, let us see how this works.

Linear interpolation is interpolation by the straight line through (x_0, f_0), (x_1, f_1); see Figure 8. Thus the linear Lagrange polynomial p_1 is a sum $p_1 = L_0 f_0 + L_1 f_1$ with L_0 the linear polynomial that is 1 at x_0 and 0 at x_1; similarly, L_1 is 0 at x_0 and 1 at x_1. Obviously,

$$L_0(x) = \frac{x - x_1}{x_0 - x_1}, \quad L_1(x) = \frac{x - x_0}{x_1 - x_0}$$

This gives the linear Lagrange polynomial

$$p_1(x) = L_0(x)f_0 + L_1(x)f_1 = \frac{x - x_1}{x_0 - x_1} \times f_0 + \frac{x - x_0}{x_1 - x_0} \times f_1 \tag{4.32}$$

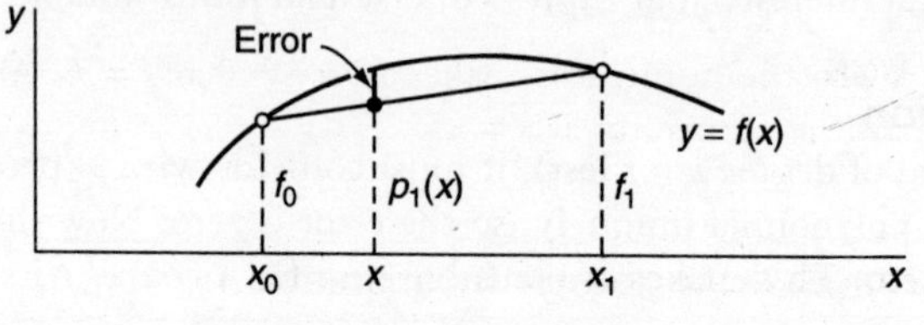

Figure 8 Linear interpolation.

Note that Lagrange interpolation is for unevenly spaced points.

Example 14 Linear Lagrange Interpolation

Compute a 4D-value of ln 9.2 from ln 9.0 = 2.1972, ln 9.5 = 2.2513 by linear Lagrange interpolation and determine the error, using ln 9.2 = 2.2192 (4D).

Solution: Here $x_0 = 9.0, x_1 = 9.5, f_0 = \ln 9.0, f_1 = \ln 9.5$. In (4.32) we need

$$L_0(x) = \frac{x - 9.5}{-0.5} = -2.0(x - 9.5), \quad L_0(9.2) = -2.0(-0.3) = 0.6$$

$$L_1(x) = \frac{x - 9.0}{0.5} = 2.0(x - 9.0), \quad L_1(9.2) = 2 \times 0.2 = 0.4$$

(see Figure 9) and obtain the answer

$$\ln 9.2 \approx p_1(9.2) = L_0(9.2)\,f_0 + L_1(9.2)f_1 = 0.6 \times 2.1972 + 0.4 \times 2.2513 = 2.2188$$

The error is $\epsilon = a - \tilde{a} = 2.2192 - 2.2188 = 0.0004$. Hence linear interpolation is not sufficient here to get 4D-accuracy; it would suffice for 3D-accuracy.

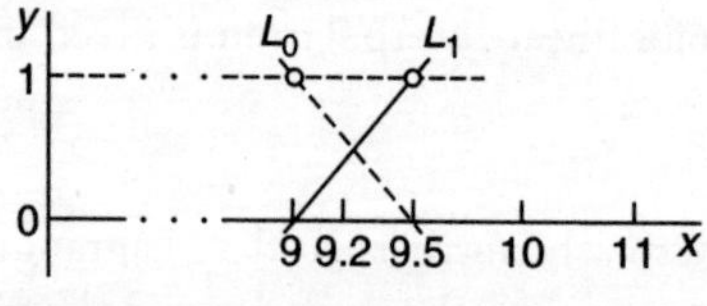

Figure 9 L_0 and L_1 in Example 14.

Quadratic interpolation is interpolation of given $(x_0, f_0), (x_1, f_1), (x_2, f_2)$ by a second-degree polynomial $p_2(x)$, which by Lagrange's idea is

$$p_2(x) = L_0(x)f_0 + L_1(x)f_1 + L_2(x)f_2 \tag{4.33}$$

with $L_0(x_0) = 1, L_1(x_1) = 1, L_2(x_2) = 1$, and $L_0(x_1) = L_0(x_2) = 0$, etc. We claim that

$$L_0(x) = \frac{l_0(x)}{l_0(x_0)} = \frac{(x-x_1)(x-x_2)}{(x_0-x_1)(x_0-x_2)}$$
$$L_1(x) = \frac{l_1(x)}{l_1(x_1)} = \frac{(x-x_0)(x-x_2)}{(x_1-x_0)(x_1-x_2)} \tag{4.34}$$
$$L_2(x) = \frac{l_2(x)}{l_2(x_2)} = \frac{(x-x_0)(x-x_1)}{(x_2-x_0)(x_2-x_1)}$$

How did we get this? Well, the numerator makes $L_k(x_j) = 0$ if $j \neq k$. And the denominator makes $L_k(x_k) = 1$ because it equals the numerator at $x = x_k$.

Example 15 Quadratic Lagrange Interpolation

Compute ln 9.2 by using (4.33) and (4.34) from the data in Example 13 and the additional third value ln 11.0 = 2.3979.

Solution: In (4.33)

$$L_0(x) = \frac{(x-9.5)(x-11.0)}{(9.0-9.5)(9.0-11.0)} = x^2 - 20.5x + 104.5, \qquad L_0(9.2) = 0.5400$$

$$L_1(x) = \frac{(x-9.0)(x-11.0)}{(9.5-9.0)(9.5-11.0)} = -\frac{1}{0.75}(x^2 - 20x + 99), \qquad L_1(9.2) = 0.4800$$

$$L_2(x) = \frac{(x-9.0)(x-9.5)}{(11.0-9.0)(11.0-9.5)} = \frac{1}{3}(x^2 - 18.5x + 85.5), \qquad L_2(9.2) = -0.0200$$

(see Figure 10), so that (4.33) gives, exact to 4D,

$$\ln 9.2 \approx p_2(9.2) = 0.5400 \times 2.1972 + 0.4800 \times 2.2513 - 0.0200 \times 2.3979 = 2.2192$$

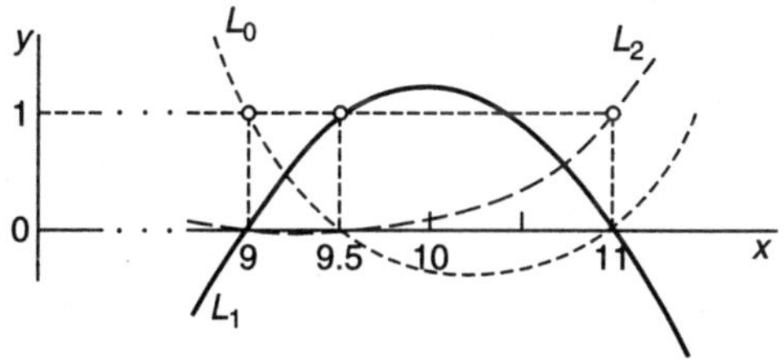

Figure 10 L_0, L_1, L_2 in Example 15.

General Lagrange Interpolation Polynomial

For general n we obtain

$$f(x) \approx p_n(x) = \sum_{k=0}^{n} L_k(x) f_k = \sum_{k=0}^{n} \frac{l_k(x)}{l_k(x_k)} f_k \tag{4.35}$$

where $L_k(x_k) = 1$ and L_k is 0 at the other nodes, and the L_k are independent of the function f to be interpolated. We get (4.35) if we take

$$\begin{aligned} l_0(x) &= (x-x_1)(x-x_2)\cdots(x-x_n) \\ l_k(x) &= (x-x_0)\cdots(x-x_{k-1})(x-x_{k+1})\cdots(x-x_n), \qquad 0<k<n \qquad (4.36) \\ l_n(x) &= (x-x_0)(x-x_1)\cdots(x-x_{n-1}) \end{aligned}$$

We can easily see that $p_n(x_k) = f_k$. Indeed, inspection of (4.36) shows that $l_k(x_j) = 0$ if $j \neq k$, so that for $x = x_k$, the sum in (4.35) reduces to the single term $[l_k(x_k)/l_k(x_k)]f_k = f_k$.

Example 16 Error Estimate (4.31) of Linear Interpolation. Damage by Roundoff. Error Principle

Estimate the error in Example 13 first by (4.31) directly and then by the Error Principle.

Solution:
(a) ***Estimation by*** (4.31). We have $n = 1, f(t) = \ln t, f'(t) = 1/t, f''(t) = -1/t^2$. Hence

$$\epsilon_1(x) = (x-9.0)(x-9.5)\,\frac{(-1)}{2t^2}, \quad \text{thus} \quad \epsilon_1(9.2) = \frac{0.03}{t^2}$$

$t = 9.0$ gives the maximum $0.03/9^2 = 0.00037$ and $t = 9.5$ gives the minimum $0.03/9.5^2 = 0.00033$, so that we get $0.00033 \leqq \epsilon_1(9.2) \leqq 0.00037$, or better, 0.00038 because $0.3/81 = 0.003703\ \ldots$.
But the error 0.0004 in Example 13 disagrees, and we can learn something! Repetition of the computation there with 5D instead of 4D gives

$$\ln 9.2 \approx p_1(9.2) = 0.6 \times 2.19722 + 0.4 \times 2.25129 = 2.21885$$

with an actual error $\epsilon = 2.21920 - 2.21885 = 0.00035$, which lies nicely near the middle between our two error bounds.
This shows that the discrepancy (0.0004 vs. 0.00035) was caused by rounding, which is not taken into account in (4.31).
(b) ***Estimation by the Error Principle.*** We calculate $p_1(9.2) = 2.21885$ as before and then $p_2(9.2)$ as in Example 14 but with 5D, obtaining

$$p_2(9.2) = 0.54 \times 2.19722 + 0.48 \times 2.25129 - 0.02 \times 2.39790 = 2.21916$$

The difference $p_2(9.2) - p_1(9.2) = 0.00031$ is the approximate error of $p_1(9.2)$ that we wanted to obtain; this is an approximation of the actual error 0.00035 given above.

CONCEPT CHECK

1. **(Linear interpolation)** Calculate $p_1(x)$ in Example 13. Compute from it $\ln 9.4 \approx p_1(9.4)$.
2. Estimate the error in Prob. 1 by (4.31).
3. **(Quadratic interpolation)** Calculate the Lagrange polynomial $p_2(x)$ for the 4D-values of the Gamma function $\Gamma(1.00) = 1.0000$, $\Gamma(1.02) = 0.9888$, $\Gamma(1.04) = 0.9784$, and

from it approximations of $\Gamma(x)$ for $x = 1.005$, 1.010, 1.015, 1.025, 1.030, 1.035.

4. **(Error bounds)** Derive error bounds for $p_2(9.2)$ in Example 14 from (4.31).
5. **(Error function)** Calculate the Lagrange polynomial $p_2(x)$ for the 5D-values of the error function

$$f(x) = \text{erf}\, x = (2\sqrt{\pi}) \int_0^x e^{-w^2}\, dw$$

namely, $f(0.25) = 0.27633$, $f(0.5) = 0.52050$, $f(1) = 0.84270$, and from p_2 an approximation of $f(0.75)$ (= 0.71116, 5D).

6. Derive an error bound in Prob. 5 from (4.31).
7. **(Sine integral)** Calculate the Lagrange polynomial $p_2(x)$ for the 4D-values of the sine integral Si(x), namely, Si(0) = 0, Si(1) = 0.9461, Si(2) = 1.6054, and from p_2 approximations of Si(0.5) (= 0.4931, 4D) and Si(1.5) (= 1.3247, 4D).
8. **(Linear and quadratic interpolation)** Find $e^{-0.25}$ and $e^{-0.75}$ by linear interpolation with $x_0 = 0$, $x_1 = 0.5$ and $x_0 = 0.5$, $x_1 = 1$, respectively. Then find $p_2(x)$ interpolating e^{-x} with $x_0 = 0$, $x_1 = 0.5$, $x_2 = 1$ and from it $e^{-0.25}$ and $e^{-0.75}$. Compare the errors of these linear and quadratic interpolations. Use 4D-values of e^{-x}.
9. **(Cubic Lagrange interpolation)** Calculate and sketch or graph L_0, L_1, L_2, L_3 for $x = 0$, 1, 2, 3 on common axes. Find $p_3(x)$ for the data

(0, 1)
(1, 0.765198)
(2, 0.223891)
(3, −0.260052)

[values of the Bessel function $J_0(x)$]. Find p_3 for $x = 0.5$, 1.5, 2.5 by interpolation.

10. **(Interpolation and extrapolation)** Calculate $p_2(x)$ in Example 14. Compute from it approximations of ln 9.4, ln 10, ln 10.5, ln 11.5, ln 12, compute the errors by using exact 4D-values, and comment.
11. **(Extrapolation)** Does a sketch or graph of the product of the $(x - x_j)$ in (4.31) for the data in Prob. 10 indicate that extrapolation is likely to involve larger errors than interpolation does?
12. **(Lower degree)** Find the degree of the interpolation polynomial for the data

(−2, 33)
(0, 5)
(2, 9)
(4, 45)
(6, 113)

4.5 Finite Differences and Their Applications

In this section, we shall discuss the various difference formulae and their applications. Let us consider a set of discrete data (x_i, y_i), $i = 0, 1, 2, \ldots, n$ of any function $y = f(x)$, $f(x)$ being unknown. All given x_i are equispaced, such that

$$x_i = x_0 + ih,\ i = 1, 2, \ldots, n$$

Thus $x_1 = x_0 + h$, $x_2 = x_0 + 2h = x_1 + h$, ..., $x_n = x_0 + nh$ or $x_n = x_{n-1} + h$. However y_i need not be equispaced.

The methods of interpolation are based on the concept of differences of a function – we have to define and formulate the differences and then tabulate the data. The differences that we define are "finite differences" in the sense that the difference between any pair of successive values of either variable in such difference tables is finite.

While formulating these finite difference tables, we deal with quantities whose values differ by small amounts, but we will not proceed to the limit in which these differences tend to zero. This means that the differences remain finite, and may become zero in some cases – but this is not a condition at all. There are different types of difference operators (as shown in Table 2). We shall be discussing a few such operators in this section.

Table 2 Different Types of Difference Operators

	Type	*Operation*	*Operator symbol*
(i)	Shift operator	$Ef(x) = f(x+h)$	E
(ii)	Forward difference operator	$\Delta f(x) = f(x+h) - f(x)$	Δ
(iii)	Backward difference operator	$\nabla f(x) = f(x) - f(x-h)$	∇
(iv)	Central difference operator	$\delta f(x) = f(x+h/2) - f(x-h/2)$	δ
(v)	Average operator	$\mu f(x) = (1/2)[f(x+h/2) + f(x-h/2)]$	μ
(vi)	Integral operator	$\mathrm{I}f(x) = \int_x^{x+h} f(x)dx$	I
(vii)	Differential operator	$\mathrm{D}f(x) = \frac{d}{dx}f(x)$	D

The last two operators, I and D, are not finite difference operations. They involve infinitesimal quantities and are not included in our present discussions. In this section, we deal with equally spaced points.

Newton's Divided Difference Interpolation

For given data $(x_0, f_0), \ldots, (x_n, f_n)$ the interpolation polynomial $p_n(x)$ satisfying (4.30) is unique, as we have shown. But for different purposes we may use $p_n(x)$ in different forms. **Lagrange's form** just discussed is useful for deriving formulas in numeric differentiation (approximation formulas for derivatives) and integration.

Practically more important are Newton's forms of $p_n(x)$, which we shall also use for solving ODEs. They involve fewer arithmetic operations than Lagrange's form. Moreover, it often happens that we have to increase the degree n to reach a required accuracy. Then in Newton's forms we can use all the previous work and just add another term, a possibility without counterpart for Lagrange's form. This also simplifies the application of the Error Principle (used in Example 15 for Lagrange). The details of these ideas are as follows.

Let $p_{n-1}(x)$ be the $(n-1)$th Newton polynomial (whose form we shall determine); thus $p_{n-1}(x_0) = f_0$, $p_{n-1}(x_1) = f_1, \ldots, p_{n-1}(x_{n-1}) = f_{n-1}$. Furthermore, let us write the nth Newton polynomial as

$$p_n(x) = p_{n-1}(x) + g_n(x) \tag{4.37}$$

hence

$$g_n(x) = p_n(x) - p_{n-1}(x) \tag{4.38}$$

Here $g_n(x)$ is to be determined so that $p_n(x_0) = f_0, p_n(x_1) = f_1, \ldots, p_n(x_n) = f_n$. Since p_n and p_{n-1} agree at $x_0, \ldots, x_{n-1}$, we see that g_n is zero there. Also, g_n will generally be a polynomial of nth degree because so is p_n, whereas p_{n-1} can be of degree $n-1$ at most. Hence g_n must be of the form

$$g_n(x) = a_n(x-x_0)(x-x_1)\cdots(x-x_{n-1}) \tag{4.39}$$

We determine the constant a_n. For this we set $x = x_n$ and solve (4.39) algebraically for a_n. Replacing $g_n(x_n)$ according to (4.38) and using $p_n(x_n) = f_n$, we see that this gives

$$a_n = \frac{f_n - p_{n-1}(x_n)}{(x_n - x_0)(x_n - x_1)\cdots(x_n - x_{n-1})} \tag{4.40}$$

We write a_k instead of a_n and show that a_k equals the **kth divided difference,** recursively denoted and defined as follows:

$$a_1 = f[x_0, x_1] = \frac{f_1 - f_0}{x_1 - x_0}$$

$$a_2 = f[x_0, x_1, x_2] = \frac{f[x_1, x_2] - f[x_0, x_1]}{x_2 - x_0}$$

and in general

$$a_k = f[x_0, \ldots, x_k] = \frac{f[x_1, \ldots, x_k] - f[x_0, \ldots, x_{k-1}]}{x_k - x_0} \tag{4.41}$$

If $n = 1$, then $p_{n-1}(x_n) = p_0(x_1) = f_0$ because $p_0(x)$ is constant and equal to f_0, the value of $f(x)$ at x_0. Hence (4.40) gives

$$a_1 = \frac{f_1 - p_0(x_1)}{x_1 - x_0} = \frac{f_1 - f_0}{x_1 - x_0} = f[x_0, x_1]$$

and (4.37) and (4.39) give the Newton interpolation polynomial of the first degree

$$p_1(x) = f_0 + (x - x_0)f[x_0, x_1]$$

If $n = 2$, then this p_1 and (4.40) give

$$a_2 = \frac{f_2 - p_1(x_2)}{(x_2 - x_0)(x_2 - x_1)} = \frac{f_2 - f_0 - (x_2 - x_0)f[x_0, x_1]}{(x_2 - x_0)(x_2 - x_1)} = f[x_0, x_1, x_2]$$

where the last equality follows by straightforward calculation and comparison with the definition of the right side. From (4.37) and (4.39) we thus obtain the second Newton polynomial

$$p_2(x) = f_0 + (x - x_0)f[x_0, x_1] + (x - x_0)(x - x_1)f[x_0, x_1, x_2]$$

For $n = k$, formula (4.37) gives

$$p_k(x) = p_{k-1}(x) + (x - x_0)(x - x_1)\cdots(x - x_{k-1})f[x_0, \ldots, x_k] \tag{4.42}$$

With $p_0(x) = f_0$ by repeated application with $k = 1, \ldots, n$ this finally gives **Newton's divided difference interpolation formula**

$$\begin{aligned} f(x) \approx f_0 &+ (x - x_0)f[x_0, x_1] + (x - x_0)(x - x_1)f[x_0, x_1, x_2] \\ &+ \cdots + (x - x_0)(x - x_1)\cdots(x - x_{n-1})f[x_0, \ldots, x_n] \end{aligned} \tag{4.43}$$

This is discussed in the next section. An algorithm is discussed next. The first do-loop computes the divided differences and the second the desired value $p_n(\hat{x})$.

Example 16 shows how to arrange differences near the values from which they are obtained; the latter always stand a half-line above and a half-line below in the preceding column. Such an arrangement is called a (divided) **difference table**.

ALGORITHM INTERPOL ($x_0, \ldots, x_N; f_0, \ldots, f_N; \hat{x}$)

This algorithm computes an approximation $p_n(\hat{x})$ of $f(\hat{x})$ at $\hat{x}$.

INPUT: Data $(x_0, f_0), (x_1, f_1), \ldots, (x_n, f_n)$; $\hat{x}$

OUTPUT: Approximation $p_n(\hat{x})$ of $f(\hat{x})$

Set $f[x_j] = f_j \quad (j = 0, \ldots, n)$.

For $m = 1, \ldots, n-1$ do:

1 For $j = 0, \ldots, n - m$ do:

$$f[x_j, \ldots, x_{j+m}] = \frac{f[x_{j+1}, \ldots, x_{j+m}] - f[x_j, \ldots, x_{j+m-1}]}{x_{j+m} - x_j}$$

End

End

Set $p_0(x) = f_0$.

For $k = 1, \ldots, n$ do:

$$p_k(\hat{x}) = p_{k-1}(\hat{x}) + (\hat{x} - x_0) \cdots (\hat{x} - x_{k-1}) f[x_0, \ldots, x_k]$$

End

OUTPUT $p_n(\hat{x})$

End INTERPOL

Example 17 Newton's Divided Difference Interpolation Formula

Compute $f(9.2)$ from the values shown in the first two columns of the following table.

x_j	$f_j = f(x_j)$	$f[x_j, x_{j+1}]$	$f[x_j, x_{j+1}, x_{j+2}]$	$f[x_j, \ldots, x_{j+3}]$
8.0	(2.079442)			
		(0.117783)		
9.0	2.197225		(−0.006433)	
		0.108134		(0.000411)
9.5	2.251292		−0.005200	
		0.097735		
11.0	2.397895			

Solution: We compute the divided differences as shown. Sample computation:

$$(0.097735 - 0.108134)/(11 - 9) = -0.005200$$

The values we need in (4.43) are circled. We have

$$\begin{aligned} f(x) \approx p_3(x) &= 2.079442 + 0.117783(x - 8.0) - 0.006433(x - 8.0)(x - 9.0) \\ &+ 0.000411(x - 8.0)(x - 9.0)(x - 9.5) \end{aligned}$$

At $x = 9.2$

$$f(9.2) \approx 2.079442 + 0.141340 - 0.001544 - 0.000030 = 2.219208$$

The value exact to 6D is $f(9.2) = \ln 9.2 = 2.219203$. Note that we can nicely see how the accuracy increases from term to term:

$$p_1(9.2) = 2.220782, \quad p_2(9.2) = 2.219238, \quad p_3(9.2) = 2.219208$$

Newton's Forward Difference Formula

Newton's formula (4.43) is valid for *arbitrarily spaced* nodes as they may occur in practice in experiments or observations. However, in many applications the x_j's are *regularly spaced* – for instance, in measurements taken at regular intervals of time. Then, denoting the distance by h, we can write

$$x_0, \quad x_1 = x_0 + h, \quad x_2 = x_0 + 2h, \ \ldots, \quad x_n = x_0 + nh \tag{4.44}$$

We show how (4.41) and (4.43) now simplify considerably! To get started, let us define the *first forward difference* of f at x_j by

$$\Delta f_j = f_{j+1} - f_j$$

the *second forward difference* of f at x_j by

$$\Delta^2 f_j = \Delta f_{j+1} - \Delta f_j$$

and, continuing in this way, the ***k*th forward difference** of f at x_j by

$$\Delta^k f_j = \Delta^{k-1} f_{j+1} - \Delta^{k-1} f_j \quad (k = 1, 2, \ldots) \tag{4.45}$$

Examples and an explanation of the name "forward" follow later. What is the point of this? We show that if we have regular spacing (4.44), then

$$f[x_0, \ldots, x_k] = \frac{1}{k!h^k} \Delta^k f_0 \tag{4.46}$$

We prove (4.46) by induction. It is true for $k = 1$ because $x_1 = x_0 + h$, so that

$$f[x_0, x_1] = \frac{f_1 - f_0}{x_1 - x_0} = \frac{1}{h}(f_1 - f_0) = \frac{1}{1!h} \Delta f_0$$

Assuming (4.46) to be true for all forward differences of order k, we show that (4.46) holds for $k + 1$. We use (4.41) with $k + 1$ instead of k; then we use $(k + 1)h = x_{k+1} - x_0$, resulting from (4.41), and finally (4.42) with $j = 0$, that is, $\Delta^{k+1} f_0 = \Delta^k f_1 - \Delta^k f_0$. This gives

$$\begin{aligned} f[x_0, \ldots, x_{k-1}] &= \frac{f[x_1, \ldots, x_{k+1}] - f[x_0, \ldots, x_k]}{(k+1)h} \\ &= \frac{1}{(k+1)h} \left[\frac{1}{k!h^k} \Delta^k f_1 - \frac{1}{k!h^k} \Delta^k f_0 \right] \\ &= \frac{1}{(k+1)!h^{k+1}} \Delta^{k+1} f_0 \end{aligned}$$

which is (4.46) with $k + 1$ instead of k. Formula (4.46) is proved.

Newton's Forward Difference Interpolation Formula

In (4.43) we finally set $x = x_0 + rh$. Then $x - x_0 = rh$, $x - x_1 = (r - 1)h$ since $x_1 - x_0 = h$, and so on. With this and (4.46), formula (4.43) becomes **Newton's** (or *Gregory–Newton's*) **forward difference interpolation formula**

$$f(x) \approx p_n(x) = \sum_{s=0}^{n} \binom{r}{s} \Delta^s f_0 \quad [x = x_0 + rh,\ r = (x - x_0)/h]$$
$$= f_0 + r\Delta f_0 + \frac{r(r-1)}{2!}\Delta^2 f_0 + \cdots + \frac{r(r-1)\cdots(r-n+1)}{n!}\Delta^n f_0 \tag{4.47}$$

where the **binomial coefficients** in the first line are defined by

$$\binom{r}{0} = 1, \quad \binom{r}{s} = \frac{r(r-1)(r-2)\cdots(r-s+1)}{s!} \quad (s > 0,\ \text{integer}) \tag{4.48}$$

and $s! = 1 \times 2 \times \cdots \times s$.

Error

From (4.31) we get, with $x - x_0 = rh$, $x - x_1 = (r - 1)h$, etc.,

$$\epsilon_n(x) = f(x) - p_n(x) = \frac{h^{n+1}}{(n+1)!} r(r-1)\cdots(r-n) f^{(n+1)}(t) \tag{4.49}$$

with t as characterized in (4.31). Formula (4.49) is an exact formula for the error, but it involves the unknown t. In Example 17 we shall show how to use (4.49) for obtaining an error estimate and an interval in which the true value of $f(x)$ must lie.

Comments on Accuracy

(a) The order of magnitude of the error $\epsilon_n(x)$ is about equal to that of the next difference not used in $p_n(x)$.

(b) One should choose $x_0, \ldots, x_n$ such that the x at which one interpolates is as well centered between $x_0, \ldots, x_n$ as possible.

The reason for (a) is that in (4.49),

$$f^{n+1}(t) \approx \frac{\Delta^{n+1} f(t)}{h^{n+1}}, \quad \frac{|r(r-1)\cdots(r-n)|}{1 \times 2 \times \cdots \times (n+1)} \leqq 1 \quad \text{if} \quad |r| \leqq 1$$

(and actually for any r as long as we do not *extrapolate*). The reason for (b) is that $|r(r - 1) \cdots (r - n)|$ becomes smallest for that choice.

Example 18 Newton's Forward Difference Formula. Error Estimation

Compute cosh 0.56 from (4.47) and the four values in the following table and estimate the error.

j	x_j	$f_j = \cosh x_j$	Δf_j	$\Delta^2 f_j$	$\Delta^3 f_j$
0	0.5	(1.127626)			
			(0.057839)		
1	0.6	1.185465		(0.011865)	
			0.069704		(0.000697)
2	0.7	1.255169		0.012562	
			0.082266		
3	0.8	1.337435			

Solution: We compute the forward differences as shown in the table. The values we need are circled. In (4.47) we have $r = (0.56 - 0.50)/0.1 = 0.6$, so that (4.47) gives

$$\cosh 0.56 \approx 1.127626 + 0.6 \times 0.057839 + \frac{0.6(-0.4)}{2} \times 0.011865 + \frac{0.6(-0.4)(-1.4)}{6} \times 0.000697$$

$$= 1.127626 + 0.034703 - 0.001424 + 0.000039$$

$$= 1.160944$$

Error estimate: From (4.49), since the fourth derivative is $\cosh^{(4)} t = \cosh t$,

$$\epsilon_3(0.56) = \frac{0.1^4}{4!} \times 0.6(-0.4)(-1.4)(-2.4) \cosh t = A \cosh t$$

where $A = -0.00000336$ and $0.5 \leqq t \leqq 0.8$. We do not know t, but we get an inequality by taking the largest and smallest $\cosh t$ in that interval:

$$A \cosh 0.8 \leqq \epsilon_3(0.62) \leqq A \cosh 0.5$$

Since

$$f(x) = p_3(x) + \epsilon_3(x)$$

this gives

$$p_3(0.56) + A \cosh 0.8 \leqq \cosh 0.56 \leqq p_3(0.56) + A \cosh 0.5$$

Numeric values are

$$1.160939 \leqq \cosh 0.56 \leqq 1.160941$$

The exact 6D-value is $\cosh 0.56 = 1.160941$. It lies within these bounds. Such bounds are not always so tight. Also, we did not consider roundoff errors, which will depend on the number of operations.

This example also explains the name "*forward* difference formula": we see that the differences in the formula slope forward in the difference table.

Newton's Backward Difference Formula

Instead of forward-sloping differences we may also employ backward-sloping differences. Table remains the same as before (same numbers, in the same positions), except for a very harmless change of the running subscript j (which we shall explain in Example 18). Nevertheless, purely for reasons of convenience it is standard to introduce a second name and notation for differences as follows. We define the first backward difference of f at x_j by

$$\nabla f_j = f_j - f_{j-1}$$

The *second backward difference* of f at x_j by

$$\nabla^2 f_j = \nabla f_j - \nabla f_{j-1}$$

and, continuing in this way, the ***k*th backward difference** of f at x_j by

$$\nabla^k f_j = \nabla^{k-1} f_j - \nabla^{k-1} f_{j-1} \qquad (k = 1, 2, \ldots) \qquad (4.50)$$

Newton's Backward Difference Interpolation Formula

A formula similar to (4.47) but involving backward differences is **Newton's** (or **Gregory–Newton's**) **backward difference interpolation formula**

$$f(x) \approx p_n(x) = \sum_{s=0}^{n} \binom{r+s-1}{s} \nabla^s f_0 \qquad [x = x_0 + rh, r = (x - x_0)/h]$$

$$= f_0 + r\nabla f_0 + \frac{r(r+1)}{2!}\nabla^2 f_0 + \cdots + \frac{r(r+1)\cdots(r+n-1)}{n!}\nabla^n f_0 \qquad (4.51)$$

Example 19 — Newton's Forward and Backward Interpolations

Compute a 7D-value of the Bessel function $J_0(x)$ for $x = 1.72$ from the four values in the following table, using (a) Newton's forward formula (4.47), (b) Newton's backward formula (4.51).

j_{for}	j_{back}	x_j	$J_0(x_j)$	*1st Diff.*	*2nd Diff.*	*3rd Diff.*
0	−3	1.7	0.3979849			
				−0.0579985		
1	−2	1.8	0.3399864		−0.0001693	
				−0.0581678		0.0004093
2	−1	1.9	0.2818186		0.0002400	
				−0.0579278		
3	0	2.0	0.2238908			

Solution: The computation of the differences is the same in both cases. Only their notation differs.

(a) ***Forward.*** In (4.47) we have $r = (1.72 - 1.70)/0.1 = 0.2$, and j goes from 0 to 3 (see first column). In each column we need the first given number, and (4.47) thus gives

$$J_0(1.72) \approx 0.3979849 + 0.2(-0.0579985) + \frac{0.2(-0.8)}{2}(-0.0001693) + \frac{0.2(-0.8)(-1.8)}{6} \times 0.0004093$$
$$= 0.3979849 - 0.0115997 + 0.0000135 + 0.0000196 = 0.3864183$$

which is exact to 6D, the exact 7D-value being 0.3864185.

(b) ***Backward.*** For (4.51) we use j shown in the second column, and in each column the last number. Since $r = (1.72 - 2.00)/0.1 = -2.8$, we thus get from (4.51)

$$J_0(1.72) \approx 0.2238908 - 2.8(-0.0579278) + \frac{-2.8(-1.8)}{2} \times 0.0002400 + \frac{-2.8(-1.8)(-0.8)}{6} \times 0.0004093$$
$$= 0.2238908 + 0.1621978 + 0.0006048 - 0.0002750 = 0.3864184$$

Central Difference Notation

This is a third notation for differences. The first central difference of $f(x)$ at x_j is defined by

$$\delta f_j = f_{j+1/2} - f_{j-1/2}$$

and the **kth central difference** of $f(x)$ at x_j by

$$\delta^k f_j = \delta^{k-1} f_{j+1/2} - \delta^{k-1} f_{j-1/2} \quad (j = 2, 3, \ldots) \tag{4.52}$$

Thus in this notation a difference table, for example, for f_{-1}, f_0, f_1, f_2, looks as follows:

x_{-1}	f_{-1}			
		$\delta f_{-1/2}$		
x_0	f_0		$\delta^2 f_0$	
		$\delta f_{1/2}$		$\delta^3 f_{1/2}$
x_1	f_1		$\delta^2 f_1$	
		$\delta f_{3/2}$		
x_2	f_2			

Central differences are used in numeric differentiation, differential equations, and centered interpolation formulas. These are formulas that use function values "symmetrically" located on both sides of the interpolation point x. Such values are available near the middle of a given table, where centered interpolation formulas tend to give better results than those of Newton's formulas, which do not have that "symmetry" property.

Gauss Interpolation Formula

These formulas are used to derive interpolation formulas that aid in finding the value of x which is around the middle to some specified values. In such cases, we take x_0 as one of the specified value of x that lies at the middle of the difference table. Table 3 illustrates the mid-portion of the forward difference table.

Table 3 Mid-Portion of the Forward Difference Table

x	y	Δy	$\Delta^2 y$	$\Delta^3 y$	$\Delta^4 y$	$\Delta^5 y$
x_{-4}	y_{-4}					
		Δy_{-4}				
x_{-3}	y_{-3}		$\Delta^2 y_{-4}$			
		Δy_{-3}		$\Delta^3 y_{-4}$		
x_{-2}	y_{-2}		$\Delta^2 y_{-3}$		$\Delta^4 y_{-4}$	
		Δy_{-2}		$\Delta^3 y_{-3}$		$\Delta^5 y_{-4}$
x_{-1}	y_{-1}		$\Delta^2 y_{-2}$		$\Delta^4 y_{-3}$	
		Δy_{-1}		$\Delta^3 y_{-2}$		$\Delta^5 y_{-3}$
x_0	y_0		$\Delta^2 y_{-1}$		$\Delta^4 y_{-2}$	
		Δy_0		$\Delta^3 y_{-1}$		$\Delta^5 y_{-2}$
x_1	y_1		$\Delta^2 y_0$		$\Delta^4 y_{-1}$	
		Δy_1		$\Delta^3 y_0$		$\Delta^5 y_{-1}$
x_2	y_2		$\Delta^2 y_1$		$\Delta^4 y_0$	
		Δy_2		$\Delta^3 y_1$		
x_3	y_3		$\Delta^2 y_2$			
		Δy_3				
x_4	y_4					

From the table, we note the following:

$$\begin{aligned}
\Delta y_0 &= \Delta y_{-1} + \Delta^2 y_{-1} \\
\Delta^2 y_0 &= \Delta^2 y_{-1} + \Delta^3 y_{-1} \\
\Delta^3 y_0 &= \Delta^3 y_{-1} + \Delta^4 y_{-1} \\
\Delta^4 y_0 &= \Delta^4 y_{-1} + \Delta^5 y_{-1}
\end{aligned} \tag{4.53}$$

and

$$\begin{aligned}
\Delta y_{-1} &= \Delta y_{-2} + \Delta^2 y_{-2} \\
\Delta^2 y_{-1} &= \Delta^2 y_{-2} + \Delta^3 y_{-2} \\
\Delta^3 y_{-1} &= \Delta^3 y_{-2} + \Delta^4 y_{-2} \\
\Delta^4 y_{-1} &= \Delta^4 y_{-2} + \Delta^5 y_{-2}
\end{aligned} \tag{4.54}$$

Using (4.53) and (4.54), we now obtain two versions of the following Newton's forward interpolation formula:

$$y_p = y_0 + p(\Delta y_0) + \frac{p(p-1)}{2!}(\Delta^2 y_0) + \frac{p(p-1)(p-2)}{3!}(\Delta^3 y_0) + \frac{p(p-1)(p-2)(p-3)}{4!}(\Delta^4 y_0) + \cdots \tag{4.55}$$

Here y_p is the value of y at $x = x_p = x_0 + ph$.

Gauss Forward Interpolation Formula

Substituting $\Delta^2 y_0$, $\Delta^3 y_0$, etc., from (4.53) in (4.55), we have

$$
\begin{aligned}
y_p &= y_0 + p(\Delta y_0) + \frac{p(p-1)}{2!}[\Delta^2 y_{-1} + \Delta^3 y_{-1}] \\
&\quad + \frac{p(p-1)(p-2)}{3!}[\Delta^3 y_{-1} + \Delta^4 y_{-1}] + \frac{p(p-1)(p-2)(p-3)}{4!}[\Delta^4 y_{-1} + \Delta^5 y_{-1}] + \cdots \\
&= y_0 + p\Delta y_0 + \frac{p(p-1)}{2!}(\Delta^2 y_{-1}) + \frac{(p+1)p(p-1)}{3!}(\Delta^3 y_{-1}) \\
&\quad + \frac{(p+1)(p-1)p(p-2)}{4!}(\Delta^4 y_{-1}) + \cdots
\end{aligned} \tag{4.56}
$$

Substituting $\Delta^4 y_{-1}$ from (4.54), (4.56) becomes

$$
\begin{aligned}
y_p &= y_0 + p(\Delta y_0) + \frac{p(p-1)}{2!}\Delta^2 y_{-1} + \frac{(p+1)p(p-1)}{3!}(\Delta^3 y_{-1}) \\
&\quad + \frac{(p+1)(p-1)p(p-2)}{4!}(\Delta^4 y_{-2}) + \cdots
\end{aligned}
$$

This is known as the Gauss forward interpolation formula.

Gauss Backward Interpolation Formula

Let us substitute Δy_0, $\Delta^2 y_0$, $\Delta^3 y_0$, … from (4.53) in (4.55). We get

$$
\begin{aligned}
y_p &= y_0 + p(\Delta y_{-1} + \Delta^2 y_{-1}) + \frac{p(p-1)}{2!}(\Delta^2 y_{-1} + \Delta^3 y_{-1}) + \frac{p(p-1)(p-2)}{3!}(\Delta^3 y_{-1} + \Delta^4 y_{-1}) \\
&\quad + \frac{p(p-1)(p-2)(p-3)}{4!}(\Delta^4 y_{-1} + \Delta^5 y_{-1}) + \cdots \\
&= y_0 + p(\Delta y_{-1}) + \frac{(p+1)p}{2!}(\Delta^2 y_{-1}) + \frac{(p+1)p(p-1)}{3!}(\Delta^3 y_{-1}) \\
&\quad + \frac{(p+1)p(p-1)(p-2)}{4!}(\Delta^4 y_{-1}) + \cdots
\end{aligned} \tag{4.57}
$$

Substituting for $\Delta^3 y_{-1}$ and $\Delta^4 y_{-1}$ from (4.54), (4.57) becomes

$$
\begin{aligned}
y_p &= y_0 + p(\Delta y_{-1}) + \frac{(p+1)p}{2!}\Delta^2 y_{-1} + \frac{(p+1)p(p-1)}{3!}(\Delta^3 y_{-1} + \Delta^4 y_{-2}) \\
&\quad + \frac{(p+1)p(p-1)(p-2)}{4!}(\Delta^4 y_{-2} + \Delta^5 y_{-2}) + \cdots \\
&= y_0 + p(\Delta y_{-1}) + \frac{(p+1)}{2!}(\Delta^2 y_{-1}) + \frac{(p+1)p(p-1)}{3!}(\Delta^3 y_{-2}) \\
&\quad + \frac{(p+1)p(p-1)(p-2)}{4!}(\Delta^4 y_{-2}) + \cdots
\end{aligned}
$$

This is the Newton's forward interpolation formula is known as the Gauss backward interpolation formula.

Example 20

Find $y(25)$, given that $y_{20} = 24, y_{24} = 32, y_{28} = 35, y_{32} = 40$, using Gauss forward difference formula.

[JNTU Sep. 2006]

Solution: Tabulating the given data we have

x	20	24	28	32
y	24	32	35	40

Using Gauss forward difference formula, we have

$$y(x) = y_0 + p\Delta y_0 + \frac{p(p-1)}{2!}\Delta^2 y_{-1} + \frac{(p+1)p(p-1)}{3!}\Delta^3 y_{-1} + \cdots$$

Now we take $x = 24$ as origin. Therefore $x_0 = 24$, $h = 4$, $x = 25$, and

$$p = \frac{x - x_0}{h} = \frac{25-24}{4} = 0.25$$

Therefore Gauss forward difference table is as follows:

x	p	y	Δy	$\Delta^2 y$	$\Delta^3 y$
20	–1	24			
			$8 = \Delta y_{-1}$		
24	0	32		$-5 = \Delta^2 y_{-1}$	
			$3 = \Delta y_0$		$7 = \Delta^3 y_{-1}$
28	1	35		$2 = \Delta^2 y_0$	
			$5 = \Delta y_1$		
32	2	40			

By Gauss Forward interpolation formula, we have

$$\begin{aligned} y(25) &= 32 + (0.25)^3 + \frac{(0.25)(0.25-1)}{6}(7) \\ &= 32 + 0.75 + 0.46875 - 0.2734 \\ &= 32.945 \end{aligned}$$

Example 21

Using Gauss backward difference formula, find $y(8)$ from the following table.

[JNTU May 2007, Sep. 2006]

x	0	5	10	15	20	25
y	7	11	14	18	24	32

Solution: By Gauss backward interpolation formula, we have

$$f(x) = y_p = y_0 + p\Delta y_{-1} + \frac{p(p+1)}{2!}\Delta^2 y_{-1} + \frac{(p+1)p(p-1)}{3!}\Delta^3 y_{-2}$$
$$+\frac{(p+1)p(p-1)(p-2)}{4!}\Delta^4 y_{-2} + \cdots$$

Here $y_p = y_8$, $y_0 = 14$, $x_0 = 10$, $y = 5$, and

$$p = \frac{x_1 - x_0}{h} = \frac{8-10}{5} = \frac{-2}{5} = -0.4$$

Therefore

$$y(8) = 14 - 0.4(3) + \frac{(-0.4)(-0.4+1)1}{2} + \frac{(-0.4+1)(-0.4)(-0.4-1)}{6}(2)$$
$$+\frac{(-0.4-2)(-0.4+1)(-0.4)(-0.4-1)}{24}(-1)$$
$$= 14 - 1.2 - 0.12 + 0.112 + 0.0336$$
$$= 12.8256$$

Detection of Errors by Use of Difference Tables

All difference tables must be formulated and filled up with care. Errors should not occur while calculating such simple differences. These tables can also be used to check errors in tabular values. Suppose that there is an error of +1 unit in a certain tabular value. As higher order differences are formed, the errors spread out, and get magnified as shown in Table 4. Here all zeros indicate that entries are correct, and nonzero quantities in x_i and y_i indicate the errors in units while entering x_i, y_i which get magnified and spread out as shown below.

Table 4 Spread of Errors and Their Magnification

x_i	y_i	Δ	Δ^2	Δ^3	Δ^4	Δ^5
0	0					
		0				
0	0		0			
		0		e		
0	0		e		$-4e$	
		e		$-3e$		$10e$
0	e		$-2e$		$6e$	
		$-e$		$3e$		$-10e$
0	0		e		$-4e$	
		0		$-e$		
0	0		0			
		0				
0	0					

This table reveals the following aspects

1. The effect of the error increases with the order of the difference.
2. The errors in any column are the binomial coefficients with alternate signs.
3. The algebraic sum of errors in any difference column is zero.
4. Maximum error occurs opposite the function value containing the error.

All these facts can be remembered and used to detect errors in difference tables.

CONCEPT CHECK

1. **(Newton's forward difference formula)** Set up (4.47) for the data in Prob. 7 of the previous concept check and derive $p_2(x)$ from (4.47).
2. Set up Newton's forward difference formula for the data in Prob. 3 of the previous concept check and compute $\Gamma(1.01)$, $\Gamma(1.03)$, $\Gamma(1.05)$.
3. **(Newton's divided difference formula)** Compute $f(0.8)$ and $f(0.9)$ from
$$f(0.5) = 0.479$$
$$f(1.0) = 0.841$$
$$f(2.0) = 0.909$$
by quadratic interpolation.
4. Compute $f(6.5)$ from
$$f(6.0) = 0.1506$$
$$f(7.0) = 0.3001$$
$$f(7.5) = 0.2663$$
$$f(7.7) = 0.2346$$
by cubic interpolation using (4.43).
5. **(Central differences)** Write the difference in the table in Example 17 in central difference notation.
6. **(Subtabulation)** Compute the Bessel function $J_1(x)$ for $x = 0.1, 0.3, \ldots, 0.9$ from $J_1(0) = 0$, $J_1(0.2) = 0.09950$, $J_1(0.4) = 0.19603$, $J_1(0.6) = 0.28670$, $J_1(0.8) = 0.36884$, $J_1(1.0) = 0.44005$. Use (4.47) with $n = 5$.
7. **(Notations)** Compute a difference table of $f(x) = x^3$ for $x = 0, 1, 2, 3, 4, 5$. Choose $x_0 = 2$ and write all occurring numbers in terms of the notations (a) for central differences, (b) for forward differences, (c) for backward differences.
8. **WRITING PROJECT. Interpolation: Comparison of Methods.** Make a list of 5–6 ideas that you feel are most basic in this section. Arrange them in the best logical order. Discuss them in a 2–3 page report.
9. **TEAM PROJECT. Interpolation and Extrapolation.**
   (a) **Lagrange practical error estimate.** Apply this to $p_1(9.2)$ and $p_2(9.2)$ for the data $x_0 = 9.0$, $x_1 = 9.5$, $x_2 = 11.0$, $f_0 = \ln x_0$, $f_1 = \ln x_1$, $f_2 = \ln x_2$ (6S-values).
   (b) **Extrapolation.** Given $(x_j, f(x_j)) = (0.2, 0.9980)$, $(0.4, 0.9686)$, $(0.6, 0.8443)$, $(0.8, 0.5358)$, $(1.0, 0)$. Find $f(0.7)$ from the quadratic interpolation polynomials based on (α) 0.6, 0.8, 1.0, (β) 0.4, 0.6, 0.8, (γ) 0.2, 0.4, 0.6. Compare the errors and comment. [Exact $f(x) = \cos(\pi x^2/2)$, $f(0.7) = 0.7181$ (4S).]
   (c) Graph the product of factors $(x - x_j)$ in the error formula (4.31) for $n = 2, \ldots, 10$ separately. What do these graphs show regarding accuracy of interpolation and extrapolation?
   (d) **Central differences.** Show that $\delta^2 f_m = f_{m+1} - 2f_m + f_{m-1}$, and, furthermore $\delta^3 f_{m+1/2} = f_{m+2} - 3f_{m+1} + 3f_m - f_{m-1}$, $\delta^n f_m = \Delta^n f_{m-n/2} = \nabla^n f_{m+n/2}$.
   (e) **Everett's interpolation formula**
$$f(x) \approx (1-r)f_0 + rf_1 + \frac{(2-r)(1-r)(-r)}{3!}\delta^2 f_0 + \frac{(r+1)r(r-1)}{3!}\delta^2 f_1$$
is an example of a formula involving only even-order differences. Use it to compute the Bessel function $J_0(x)$ for $x = 1.72$ from $J_0(1.60) = 0.4554022$ and $J_0(1.7)$, $J_0(1.8)$, $J_0(1.9)$ in Example 18.

4.6 Spline Interpolation

Given data (function values, points in the xy plane) $(x_0, f_0), (x_1, f_1), \ldots, (x_n, f_n)$ can be interpolated by a polynomial $P_n(x)$ of degree n or less so that the curve of $P_n(x)$ passes through these $n + 1$ points (x_j, f_j); here $f_0 = f(x_0), \ldots, f_n = f(x_n)$.

Now if n is large, there may be trouble: $P_n(x)$ may tend to oscillate for x between the **nodes** $x_0, \ldots, x_n$. Hence we must be prepared for **numeric instability**. Figure 11 shows a famous example by C. Runge for which the maximum error even approaches ∞ as $n \to \infty$ (with the nodes kept equidistant and their number increased). Figure 12 illustrates the increase of the oscillation with n for some other function that is piecewise linear.

Those undesirable oscillations are avoided by the method of splines initiated by I. J. Schoenberg in 1946. This method is widely used in practice. It also laid the foundation for much of modern **CAD (computer-aided design)**. Its name is borrowed from a *draftman's spline,* which is an elastic rod bent to pass through given points and held in place by weights. The mathematical idea of the method is as follows:

Instead of using a single high-degree polynomial P_n over the entire interval $a \leqq x \leqq b$ in which the nodes lie, that is,

$$a = x_0 < x_1 < \cdots < x_n = b \tag{4.58}$$

we use n low-degree, for example, cubic, polynomials

$$q_0(x), q_1(x), \ldots, q_{n-1}(x)$$

one over each subinterval between adjacent nodes, hence q_0 from x_0 to x_1, then q_1 from x_1 to x_2, and so on. From this we compose an interpolation function $g(x)$, called a **spline,** by fitting these polynomials together into a single continuous curve passing through the data points, that is,

$$g(x_0) = f(x_0) = f_0, g(x_1) = f(x_1) = f_1, \ldots, g(x_n) = f(x_n) = f_n \tag{4.59}$$

Note that $g(x) = q_0(x)$ when $x_0 \leqq x \leqq x_1$, then $g(x) = q_1(x)$ when $x_1 \leqq x \leqq x_2$, and so on, according to our construction of g. Thus **spline interpolation** *is piecewise polynomial interpolation.*

The simplest q_j's would be linear polynomials. However, the curve of a piecewise linear continuous function has corners and would be of little interest in general – think of designing the body of a car or a ship.

Cubic Splines

We shall consider cubic splines because these are the most important ones in applications. By definition, a **cubic spline** $g(x)$ interpolating given data $(x_0, f_0), \ldots, (x_n, f_n)$ is a continuous function on the interval $a = x_0 \leqq x \leqq x_n = b$ that has continuous first and second derivatives and satisfies the interpolation condition (4.59); furthermore, between adjacent nodes, $g(x)$ is given by a polynomial $q_j(x)$ of degree 3 or less.

We claim that there is such a cubic spline. And if in addition to (4.59) we also require that

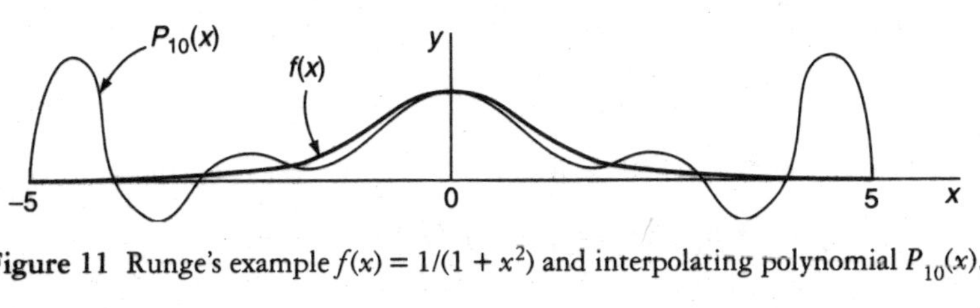

Figure 11 Runge's example $f(x) = 1/(1 + x^2)$ and interpolating polynomial $P_{10}(x)$.

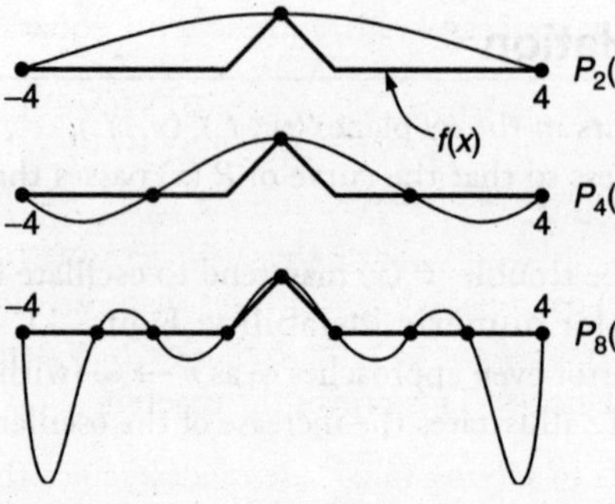

Figure 12 Piecewise linear function $f(x)$ and interpolation polynomials of increasing degrees.

$$g'(x_0) = k_0, \quad g'(x_n) = k_n \tag{4.60}$$

[given tangent directions of $g(x)$ at the two endpoints of the interval $a \leqq x \leqq b$], then we have a uniquely determined cubic spline. This is the content of the following existence and uniqueness theorem, whose proof will also suggest the actual determination of splines. [Condition (4.60) will be discussed after the proof.]

THEOREM 5 (Existence and Uniqueness of Cubic Splines): *Let* $(x_0, f_0), (x_1, f_1), \ldots, (x_n, f_n)$ *with rbitrarily spaced given* x_j *[see* (4.58)*] and given* $f_j = f(x_j), j = 5\ 0, 1, \ldots, n$*. Let* k_0 *and kn be any given numbers. Then there is one and only one cubic spline* $g(x)$ *corresponding to* (4.58) *and satisfying* (4.59) *and* (4.60)*.*

PROOF: By definition, on every subinterval I_j given by $x_j \leqq x \leqq x_{j+1}$ the spline $g(x)$ must agree with a polynomial $q_j(x)$ of degree not exceeding 3 such that

$$q_j(x_j) = f(x_j), \quad q_j(x_{j+1}) = f(x_{j+1}) \quad (j = 0, 1, \ldots, n-1) \tag{4.61}$$

For the derivatives we write

$$q_j'(x_j) = k_j, \quad q_j'(x_{j+1}) = k_{j+1} \quad (j = 0, 1, \ldots, n-1) \tag{4.62}$$

with k_0 and k_n given and $k_1, \ldots, k_{n-1}$ to be determined later. Equations (4.61) and (4.62) are four conditions for each $q_j(x)$. By direct calculation, using the notation

$$c_j = \frac{1}{h_j} = \frac{1}{x_{j+1} - x_j} \quad (j = 0, 1, \ldots, n-1) \tag{4.63}$$

we can verify that the unique cubic polynomial $q_j(x)$ $(j = 0, 1, \ldots, n-1)$ satisfying (4.61) and (4.62) is

$$\begin{aligned} q_j(x) = {} & f(x_j)c_j^2(x - x_{j+1})^2[1 + 2c_j(x - x_j)] + f(x_{j+1})c_j^2(x - x_j)^2[1 - 2c_j(x - x_{j+1})] \\ & + k_j c_j^2(x - x_j)(x - x_{j+1})^2 + k_{j+1}c_j^2(x - x_j)^2(x - x_{j+1}) \end{aligned} \tag{4.64}$$

Differentiating twice, we obtain

$$q_j''(x) = -6c_j^2 f(x_j) + 6c_j^2 f(x_{j+1}) - 4c_j k_j - 2c_j k_{j+1} \tag{4.65}$$

$$q_j''(x_{j+1}) = 6c_j^2 f(x_j) - 6c_j^2 f(x_{j+1}) + 2c_j k_j + 4c_j k_{j+1} \tag{4.66}$$

By definition, $g(x)$ has continuous second derivatives. This gives the conditions

$$q_{j-1}''(x_j) = q_j''(x_j) \quad (j = 1, \ldots, n-1)$$

If we use (4.66) with j replaced by $j-1$, and (4.65), these $n-1$ equations become

$$c_{j-1}k_{j-1}+2(c_{j-1}+c_j)k_j+c_jk_{j+1}=3[c_{j-1}{}^2\nabla f_j+c_j{}^2\nabla f_{j+1}] \tag{4.67}$$

where $\nabla f_j = f(x_j)-f(x_{j-1})$ and $\nabla f_{j+1}=f(x_{j+1})-f(x_j)$ and $j=1, \ldots, n-1$, as before. This linear system of $n-1$ equations has a un solution $k_1, \ldots, k_{n-1}$ since the coefficient matrix is strictly diagonally dominant (that is, in each row the (positive) diagonal entry is greater than the sum of the other (positive) entries). Hence the determinant of the matrix cannot be zero, so that we may determine unique values $k_1, \ldots, k_{n-1}$ of the first derivative of $g(x)$ at the nodes. This proves the theorem. ■

Storage and Time Demands in solving (4.67) are modest, since the matrix of (4.67) is **sparse** (has few nonzero entries) and **tridiagonal** (may have nonzero entries only on the diagonal and on the two adjacent "parallels" above and below it). Pivoting is not necessary because of that dominance. This makes splines efficient in solving large problems with thousands of nodes or more.

Condition (4.60) includes the **clamped conditions**

$$g'(x_0)=f'(x_0), \quad g'(x_n)=f'(x_n) \tag{4.68}$$

in which the tangent directions $f'(x_0)$ and $f'(x_n)$ at the ends are given. Other conditions of practical interest are the **free** or **natural conditions**

$$g''(x_0)=0, \quad g''(x_n)=0 \tag{4.69}$$

(geometrically: zero curvature at the ends, as for the draftman's spline), giving a **natural spline**.

Determination of Splines

Let k_0 and k_n be given. Obtain $k_1, \ldots, k_{n-1}$ by solving the linear system (4.67). Recall that the spline $g(x)$ to be found consists of n cubic polynomials $q_0, \ldots, q_{n-1}$. We write these polynomials in the form

$$q_j(x)=a_{j0}+a_{j1}(x-x_j)+a_{j2}(x-x_j)^2+a_{j3}(x-x_j)^3 \tag{4.70}$$

where $j=0, \ldots, n-1$. Using Taylor's formula, we obtain

$$a_{j0}=q_j(x_j)=f_j \qquad \text{[by (4.59)]}$$
$$a_{j1}=q_j'(x_j)=k_j \qquad \text{[by (4.62)]}$$
$$a_{j2}=\frac{1}{2}q_j''(x_j)=\frac{3}{h^2}(f_{j+1}-f_j)-\frac{1}{h}(k_{j+1}+2k_j) \qquad \text{[by (4.65)]}$$
$$a_{j3}=\frac{1}{6}q_j'''(x_j)=\frac{2}{h^3}(f_j-f_{j+1})+\frac{1}{h^2}(k_{j+1}+k_j) \tag{4.71}$$

with a_{j3} obtained by calculating $q_j''(x_{j+1})$ from (4.70) and equating the result to (4.66), that is,

$$q_j''(x_{j+1})=2a_{j2}+6a_{j3}h=\frac{6}{h^2}(f_j-f_{j+1})+\frac{2}{h}(k_j+2k_{j+1})$$

and now subtracting from this $2a_{j2}$ as given in (4.71) and simplifying.

Note that for ***equidistant nodes*** of distance $h_j = h$ we can write $c_j = c = 1/h$ in (4.63) and have from (4.67) simply

$$k_{j-1}+4k_j+k_{j+1}=\frac{3}{h}(f_{j+1}-f_{j-1}) \quad (j=1, \ldots, n-1) \tag{4.72}$$

Example 22 Spline Interpolation. Equidistant Nodes

Interpolate $f(x) = x^4$ on the interval $-1 \leqq x \leqq 1$ by the cubic spline $g(x)$ corresponding to the nodes $x_0 = -1$, $x_1 = 0$, $x_2 = 1$ and satisfying the clamped conditions $g'(-1) = f'(-1)$, $g'(1) = f'(1)$.

Solution: In our standard notation the given data are $f_0 = f(-1) = 1$, $f_1 = f(0) = 0$, $f_2 = f(1) = 1$. We have $h = 1$ and $n = 2$, so that our spline consists of $n = 2$ polynomials

$$q_0(x) = a_{00} + a_{01}(x+1) + a_{02}(x+1)^2 + a_{03}(x+1)^3 \qquad (-1 \leqq x \leqq 0)$$

$$q_1(x) = a_{10} + a_{11}x + a_{12}x^2 + a_{13}x^3 \qquad (0 \leqq x \leqq 1)$$

We determine the k_j from (4.72) (equidistance!) and then the coefficients of the spline from (4.71). Since $n = 2$, the system (4.72) is a single equation (with $j = 1$ and $h = 1$)

$$k_0 + 4k_1 + k_2 = 3(f_2 - f_0)$$

Here $f_0 = f_2 = 1$ (the value of x^4 at the ends) and $k_0 = -4$, $k_2 = 4$, the values of the derivative $4x^3$ at the ends -1 and 1. Hence

$$-4 + 4k_1 + 4 = 3(1-1) = 0, \quad k_1 = 0$$

From (4.71) we can now obtain the coefficients of q_0, namely, $a_{00} = f_0 = 1$, $a_{01} = k_0 = -4$, and

$$a_{02} = \frac{3}{1^2}(f_1 - f_0) - \frac{1}{1}(k_1 + 2k_0) = 3(0-1) - (0-8) = 5$$

$$a_{03} = \frac{2}{1^3}(f_0 - f_1) + \frac{1}{1^2}(k_1 + k_0) = 2(1-0) + (0-4) = -2$$

Similarly, for the coefficients of q_1 we obtain from (4.71) the values $a_{10} = f_1 = 0$, $a_{11} = k_1 = 0$, and

$$a_{12} = 3(f_2 - f_1) - (k_2 + 2k_1) = 3(1-0) - (4+0) = -1$$

$$a_{13} = 2(f_1 - f_2) + (k_2 + k_1) = 2(0-1) + (4+0) = 2$$

This gives the polynomials of which the spline $g(x)$ consists, namely,

$$g(x) = \begin{cases} q_0(x) = 1 - 4(x+1) + 5(x+1)^2 - 2(x+1)^3 = -x^2 - 2x^3 & \text{if } -1 \leqq x \leqq 0 \\ q_1(x) = -x^2 + 2x^3 & \text{if } \ 0 \leqq x \leqq 1 \end{cases}$$

Figure 13 shows $f(x)$ and this spline. Do you see that we could have saved over half of our work by using symmetry?

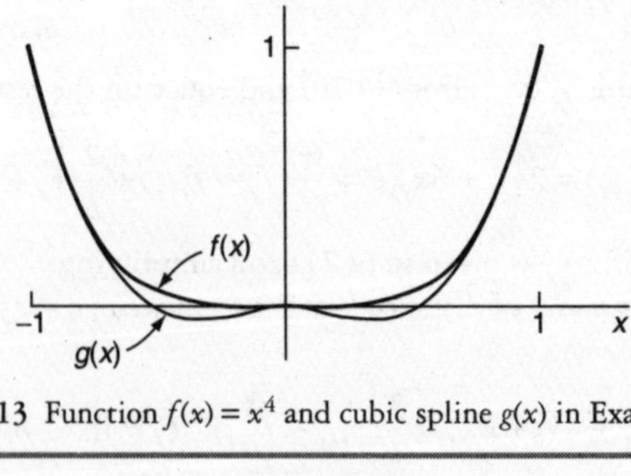

Figure 13 Function $f(x) = x^4$ and cubic spline $g(x)$ in Example 22.

Example 23 Natural Spline. Arbitrarily Spaced Nodes

Find a spline approximation and a polynomial approximation for the curve of the cross section of the circular-shaped Shrine of the Book in Jerusalem shown in Figure 14.

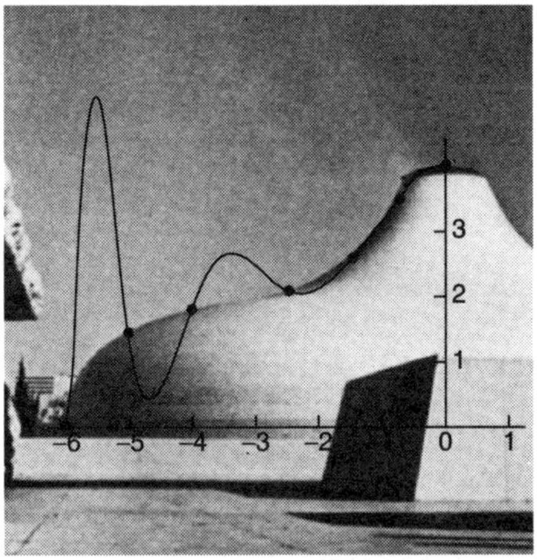

Figure 14 Shrine of the Book in Jerusalem (Architects F. Kissler and A. M. Bartus).

Solution: Thirteen points, about equally distributed along the contour (not along the x axis!), give these data:

x_j	−5.8	−5.0	−4.0	−2.5	−1.5	−0.8	0	0.8	1.5	2.5	4.0	5.0	5.8
f_j	0	1.5	1.8	2.2	2.7	3.5	3.9	3.5	2.7	2.2	1.8	1.5	0

The figure shows the corresponding interpolation polynomial of 12th degree, which is useless because of its oscillation. (Because of roundoff your software will also give you small error terms involving odd powers of x). The polynomial is

$$P_{12}(x) = 3.9000 - 0.65083x^2 + 0.033858x^4 + 0.011041x^6 - 0.0014010x^8 + 0.000055595x^{10} - 0.00000071867x^{12}$$

The spline follows practically the contour of the roof, with a small error near the nodes −0.8 and 0.8. The spline is symmetric. Its six polynomials corresponding to positive x have the following coefficients of their representations (4.70). [Note well that (4.70) is in terms of powers of $x - x_j$, not x!]

j	*x-interval*	a_{j0}	a_{j1}	a_{j2}	a_{j3}
0	0.0 ... 0.8	3.9	0.00	−0.61	−0.015
1	0.8 ... 1.5	3.5	−1.01	−0.65	0.66
2	1.5 ... 2.5	2.7	−0.95	0.73	−0.27
3	2.5 ... 4.0	2.2	−0.32	−0.091	0.084
4	4.0 ... 5.0	1.8	−0.027	0.29	−0.56
5	5.0 ... 5.8	1.5	−1.13	−1.39	0.58

CONCEPT CHECK

1. **WRITING PROJECT. Splines.** In your own words, and using as few formulas as possible, write a short report on spline interpolation, its motivation, a comparison with polynomial interpolation, and its applications.
2. **(Individual polynomial q_j)** Show that $q_j(x)$ in (4.64) satisfies the interpolation condition (4.61) as well as the derivative condition (4.62).
3. Verify the differentiations that give (4.65) and (4.66) from (4.64).
4. **(System for derivatives)** Derive the basic linear system (4.67) for $k_1, \ldots, k_{n-1}$ as indicated in the text.
5. **(Equidistant nodes)** Derive (4.72) from (4.67).
6. **(Coefficients)** Give the details of the derivation of a_{j2} and a_{j3} in (4.71).
7. Verify the computations in Example 21.
8. **(Comparison)** Compare the spline g in Example 1 with the quadratic interpolation polynomial over the whole interval. Find the maximum deviations of g and p_2 from f. Comment.
9. **(Natural spline condition)** Using the given coefficients, verify that the spline in Example 22 satisfies $g''(x) = 0$ at the ends.
10. Find the cubic spline $g(x)$ for the given data with k_0 and k_n as given.
    (a) $f(-2) = f(-1) = f(1) = f(2) = 0, f(0) = 1,$ $k_0 = k_4 = 0$
    (b) If we started from the piecewise linear function in Figure 15, we would obtain $g(x)$ in part (a) as the spline satisfying $g'(-2) = f'(-2) = 0, g'(2) = f'(2) = 0$. Find and sketch or graph the corresponding interpolation polynomial of fourth degree and compare it with the spline. Comment.
    (c) $f_0 = f(0) = 1, f_1 = f(2) = 9, f_2 = f(4) = 41,$ $f_3 = f(6) = 41, k_0 = 0, k_3 = -12$
    (d) $f_0 = f(-1) = 0, f_1 = f(0) = 4, f_2 = f(1) = 0,$ $k_0 = 0, k_2 = 0$. Is $g(x)$ even? (Give reason.)

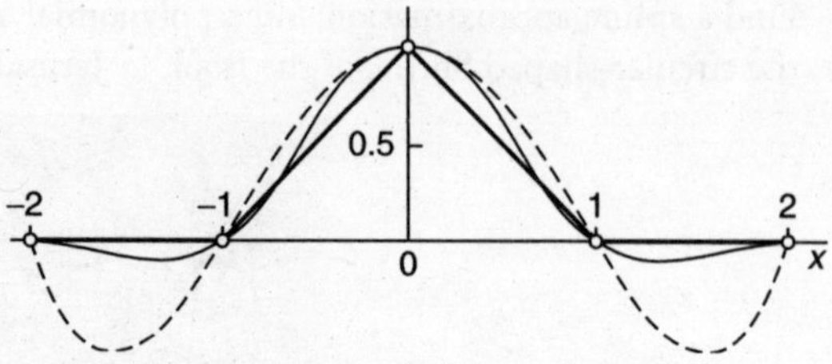

Figure 15 Spline and interpolation polynomial in Problems 10(a) and 10(b).

    (e) $f_0 = f(0) = 0, f_1 = f(1) = 1, f_2 = f(2) = 6,$ $f_3 = f(3) = 10, k_0 = 0, k_3 = 0$
    (f) $f_0 = f(0) = 1, f_1 = f(1) = 0, f_2 = f(2) = -1,$ $f_3 = f(3) = 0, k_0 = 0, k_3 = -6$
    (g) It can happen that a spline is given by the same polynomial in two adjacent subintervals. To illustrate this, find the cubic spline $g(x)$ for $f(x) = \sin x$ corresponding to the partition $x_0 = -\pi/2, x_1 = 0,$ $x_2 = \pi/2$ of the interval $-\pi/2 \leqq x \leqq \pi/2$ and satisfying $g'(-\pi/2) = f'(-\pi/2)$ and $g'(\pi/2) = f'(\pi/2)$.
11. If a cubic spline is three times continuously differentiable (that is, it has continuous first, second, and third derivatives), show that it must be a single polynomial.
12. **TEAM PROJECT. Hermite Interpolation and Bezier Curves.** In **Hermite interpolation** we are looking for a polynomial $p(x)$ (of degree $2n + 1$ or less) such that $p(x)$ and its derivative $p'(x)$ have given values at $n + 1$ nodes. (More generally, $p(x), p'(x), p''(x), \ldots$ may be required to have given values at the nodes.)
    (a) **Curves with given endpoints and tangents.** Let C be a curve in the xy plane parametrically represented by $\mathbf{r}(t) = [x(t), y(t)]$, $0 \leqq t \leqq 1$. Show that for given initial and terminal points of a curve and given initial and terminal tangents, say,

$$A: \quad \mathbf{r}_0 = [x(0), y(0)] = [x_0, y_0]$$
$$B: \quad \mathbf{r}_1 = [x(1), y(1)] = [x_1, y_1]$$

$$\mathbf{v}_0 = [x'(0), y'(0)] = [x_0', y_0'],$$
$$\mathbf{v}_1 = [x'(1), y'(1)] = [x_1', y_1']$$

we can find a curve C, namely,

$$\mathbf{r}(t) = \mathbf{r}_0 + \mathbf{v}_0 t + [3(\mathbf{r}_1 - \mathbf{r}_0) - (2\mathbf{v}_0 + \mathbf{v}_1)]t^2 + [2(\mathbf{r}_0 - \mathbf{r}_1) + \mathbf{v}_0 + \mathbf{v}_1]t^3;$$

in components,

$$x(t) = x_0 + x_0' t + (3(x_1 - x_0) - (2x_0' + x_1'))t^2 + (2(x_0 - x_1) + x_0' + x_1')t^3$$
$$y(t) = y_0 + y_0' t + (3(y_1 - y_0) - (2\,y_0' + y_1'))t^2 + (2(y_0 - y_1) + y_0' + y_1')t^3.$$

Note that this is a cubic Hermite interpolation polynomial, and $n = 1$ because we have two nodes (the endpoints of C). The two points

$$G_A\!: \mathbf{g}_0 = \mathbf{r}_0 + \mathbf{v}_0 = [x_0 + x_0', y_0 + y_0']$$

and

$$G_B\!: \mathbf{g}_1 = \mathbf{r}_1 - \mathbf{v}_1 = [x_1 - x_1', y_1 - y_1']$$

are called **guidepoints** because the segments AG_A and BG_B specify the tangents graphically. A, B, G_A, G_B determine C, and C can be changed quickly by moving the points. A curve consisting of such Hermite interpolation polynomials is called a **Bezier curve**, after the French engineer P. Bezier of the Renault Automobile Company, who introduced them in the early 1960s in designing car bodies. Bezier curves (and surfaces) are used in computer-aided design (CAD) and computer-aided manufacturing (CAM).

(b) Find and graph the Bezier curve and its guidepoints if A: [0, 0], B: [1, 0], $\mathbf{v}_0 = [1/2, 1/2]$, $\mathbf{v}_1 = [-1/2, -\sqrt{3}/4]$

(c) **Changing guidepoints** changes C. Moving guidepoints farther away makes C "staying near the tangents for a longer time." Confirm this by changing $\mathbf{v}_0$ and $\mathbf{v}_1$ in (b) to $2\mathbf{v}_0$ and $2\mathbf{v}_1$ (see Figure 16).

(d) Make experiments of your own. What happens if you change $\mathbf{v}_1$ in (b) to $-\mathbf{v}_1$. If you rotate the tangents? If you multiply $\mathbf{v}_0$ and $\mathbf{v}_1$ by positive factors less than 1?

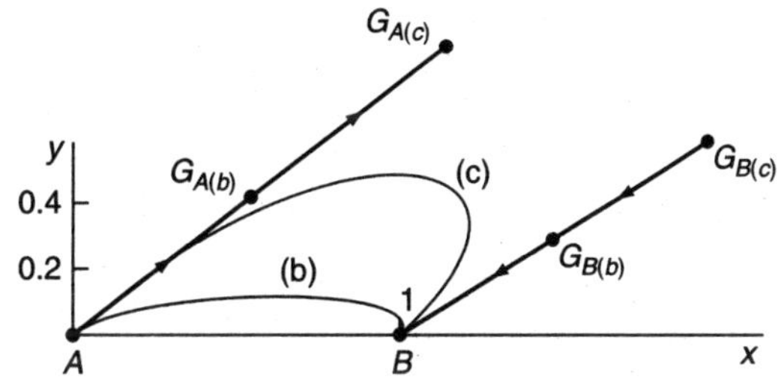

Figure 16 Team Project 18(b) and (c): Bezier curves.

Solved Problems

Q 1. If $f(x) = u(x)\,v(x)$, show that $f[x_0, x_1] = u[x_0]\,v[x_0, x_1] + u[x_0, x_1] \times v[x_1]$.

[JNTU May/June 2006, Nov./Dec. 2005, Nov. 2003]

Answer: The given equation is

$$f(x) = u(x)\,v(x)$$

The first-order divided difference between x_0 and x_1 is given by

$$y[x_0, x_1] = \frac{y_1 - y_0}{x_1 - x_0} = \frac{f(x_1) - f(x_0)}{x_1 - x_0}$$

$$f[x_0, x_1] = \frac{f[x_1] - f[x_0]}{x_1 - x_0}$$

Similarly,

$$u[x_0, x_1] = \frac{u[x_1] - u[x_0]}{x_1 - x_0}$$

$$v[x_0, x_1] = \frac{v[x_1] - v[x_0]}{x_1 - x_0}$$

Considering the right-hand side we get

$$u[x_0] \times v[x_0, x_1] + u[x_0, x_1]\, v[x_1] = u[x_0]\frac{v[x_1] - v[x_0]}{x_1 - x_0} + \frac{u[x_1] - u[x_0]}{x_1 - x_0} v[x_1]$$

$$= \frac{1}{x_1 - x_0}(u[x_0]v[x_1] - u[x_0]v[x_0] + u[x_1]v[x_1] - u[x_0]v[x_1])$$

$$= \frac{1}{x_1 - x_0} u[x_1]v[x_1] - u[x_0]v[x_0] \quad [\text{since } f[x] = u[x]v[x]]$$

$$= \frac{1}{x_1 - x_0}[f[x_1] - f[x_0]] = \frac{f[x_1] - f[x_0]}{x_1 - x_0} = f[x_0, x_1]$$

which is the left-hand side. Hence proved.

Q 2. Find the second difference of the polynomial $x^4 - 12x^3 + 42x^2 - 30x + 9$ with interval of differencing $h = 2$. **[JNTU Aug./Sep. 2008, May/June 2008]**

Answer: The given polynomial is

$$x^4 - 12x^3 + 42x^2 - 30x + 9; \qquad h = 2$$

Let

$$f(x) = x^4 - 12x^3 + 42x_2 - 30x + 9$$

Therefore

$$\Delta f(x) = f(x + h) - f(x)$$
$$= (x + h)^4 - 12(x + h)^3 + 42(x + h)^2 - 30(x + h) + 9 - x^4 + 12x^3 - 42x^2 + 30x - 9$$

Substituting the value of $h = 2$, we get

$$\Delta f(x) = (x + 2)^4 - 12(x + 2)^3 + 42(x + 2)^2 - 30(x + 2) + 9 - x^4 + 12x^3 - 42x^2 + 30x - 9$$

$$= (x^4 + 8x^3 + 24x^2 + 32x + 16) - 12(x^3 + 6x^2 + 12x + 8) + 42(x^2 + 4x + 4) - 30(x + 2) + 9 - x^4 + 12x^3 - 42x^2 + 30x - 9$$

$$= 8x^3 - 48x^2 + 56x + 28 = 4[2x^3 - 12x^2 + 14x + 7]$$

Now $\Delta^2 f(x) = \Delta[\Delta f(x)]$. Therefore

$$\Delta^2 f(x) = \Delta[4(2x^3 - 12x^2 + 14x + 7)]$$
$$= 4[2(x + 2)^3 - 12(x + 2)^2 + 14(x + 2) + 7 - 2x^3 + 12x^2 - 14x - 7]$$
$$= 4[12x^2 - 24x - 4] = 48x^2 - 96x - 16$$

Q 3. Find $f(22)$ from the following data using Newton's backward formula.

x	20	25	30	35	40	45
y	354	332	291	260	231	204

[JNTU Aug./Sep. 2007]

Answer: The backward difference table is

x	y	∇y	$\nabla^2 y$	$\nabla^3 y$	$\nabla^4 y$	$\nabla^5 y$
20	354					
		−22				
25	332		−19			
		−41		29		
30	291		10		−37	
		−31		−8		45
35	260		2		8	
		−29		0		
40	231		2			
		−27				
45	204					

Now $x = 22$. We have to find $f(22)$. Substituting the values of x, x_n, and h in

$$p = \frac{x - x_n}{h}$$

We get

$$p = \frac{22 - 45}{5} = \frac{-23}{5} = -4.6$$

Now

$$f(22) = y_n + p\nabla y_n + \frac{p(p+1)}{2}\nabla^2 y_n + \frac{p(p+1)(p+2)}{3!}\nabla^3 y_n + \frac{p(p+1)(p+2)(p+3)}{4!}\nabla^4 y_n$$

$$+\frac{p(p+1)(p+2)(p+3)(p+4)}{5!}\nabla^5 y_n \quad \text{(where } n = 5\text{)}$$

Substituting all values we get

$$f(22) = 204 - 4.6(-27) - 4.6\left[\frac{-4.6+1}{2}\right]\times 2 + 0 + \frac{(-4.6)(-4.6+1)(-4.6+2)(-4.6+3)}{24}\times 8$$

$$+\frac{(-4.6)(-4.6+1)(-4.6+2)(-4.6+3)(-4.6+4)}{120}\times 45$$

$$= 204 + 124.2 + 16.56 + 22.96 - 15.50$$

$$f(22) = y = 352.22$$

Q 4. Using Gauss backward difference formula, find $y(8)$ from the following table.

x	0	5	10	15	20	25
y	7	11	14	18	24	32

[JNTU April/May 2007, Aug./Sep. 2006]

Answer: To find $y(8)$ using Gauss backward difference formula, we first make the backward difference table.

x	y	Δy	$\Delta^2 y$	$\Delta^3 y$	$\Delta^4 y$	$\Delta^5 y$
0	$7 = y_{-2}$					
		$4 = \Delta y_{-2}$				
5	$11 = y_{-1}$		$-1 = \Delta^2 y_{-2}$			
		$3 = \Delta y_{-1}$		$2 = \Delta^3 y_{-2}$		
10	$14 = y_0$		$1 = \Delta^2 y_{-1}$		$-1 = \Delta^4 y_{-2}$	
		$4 = \Delta y_0$		$1 = \Delta^3 y_{-1}$		$0 = \Delta^5 y_{-2}$
15	$18 = y_1$		$2 = \Delta^2 y_0$		$-1 = \Delta^4 y_{-1}$	
		$6 = \Delta y_1$		$0 = \Delta^3 y_0$		
20	$24 = y_2$		$2 = \Delta^2 y_1$			
		$8 = \Delta y_2$				
25	$32 = y_3$					

Now $x_p = 8$. The Gauss backward difference formula is

$$f(x_p) = y_p = y_0 + p\Delta y_{-1} + \frac{p(p+1)}{2!}\Delta^2 y_{-1} + \frac{(p+1)p(p-1)}{3!}\Delta^3 y_{-2} + \frac{(p+1)p(p-1)(p-2)}{4!}\Delta^4 y_{-2} + \cdots$$

Now $y_p = y_{(8)}, y_0 = 14, x_0 = 10, h = 5$ and

$$p = \frac{x_p - x_0}{h} = \frac{8-10}{5} = \frac{-2}{5} = -0.4$$

Therefore substituting the values we get

$$f(8) = 14 - 0.4(3) + \frac{(-0.4)(-0.4+1)}{2}(1) + \frac{(-0.4+1)(-0.4)(-0.4-1)}{6}(2)$$

$$+\frac{(-0.4-2)(-0.4+1)(-0.4)(-0.4-1)}{24}(-1)$$

$$= 14 - 1.2 - 0.12 + 0.112 + 0.0336 = 12.8256$$

Therefore $f(8) = 12.8256$.

Q 5. Using Lagrange's interpolation formula, find $y(10)$ from the following table:

x	5	6	9	11
y	12	13	14	16

[JNTU Aug./Sep. 2008, May/June 2008]

Answer: The given values are $x_0 = 5, x_1 = 6, x_2 = 9, x_3 = 11$. Also $f(x_0) = 12, f(x_1) = 13, f(x_2) = 14$, $f(x_3) = 16$. The Lagrange's interpolation formula is

$$f(x) = \frac{(x-x_1)(x-x_2)(x-x_3)}{(x_0-x_1)(x_0-x_2)(x_0-x_3)}f(x_0) + \frac{(x-x_0)(x-x_2)(x-x_3)}{(x_1-x_0)(x_1-x_2)(x_1-x_3)}f(x_1)$$

$$+\frac{(x-x_0)(x-x_1)(x-x_3)}{(x_2-x_0)(x_2-x_1)(x_2-x_3)}f(x_2) + \frac{(x-x_0)(x-x_1)(x-x_2)}{(x_3-x_0)(x_3-x_1)(x_3-x_2)}f(x_3)$$

Substituting the values we get

$$f(10) = \frac{(10-6)(10-9)(10-11)}{(5-6)(5-9)(5-11)} \times 12 + \frac{(10-5)(10-9)(10-11)}{(6-5)(6-9)(6-11)} \times 13$$

$$+\frac{(10-5)(10-6)(10-11)}{(9-5)(9-6)(9-11)} \times 14 + \frac{(10-5)(10-6)(10-9)}{(11-5)(11-6)(11-9)} \times 16$$

$$= \frac{4\times1\times-1}{-1\times-4\times-6}\times12 + \frac{5\times1\times-1}{1\times-3\times-5}\times13 + \frac{5\times4\times-1}{4\times3\times-2}\times14 + \frac{5\times4\times1}{6\times5\times2}\times16$$

$$= 2 + \left(\frac{-13}{3}\right) + \left(\frac{35}{3}\right) + \frac{16}{3} = \frac{6-13+35+16}{3} = \frac{44}{3} = 14.6$$

Summary

In this chapter we discussed concepts that are relevant throughout numeric work as a whole and methods of a general nature, as opposed to methods for linear algebra or differential equations.

In scientific computations we use the **floating-point** representation of numbers; fixed-point representation is less suitable in most cases. Numeric methods give approximate values $\tilde{a}$ of quantities. The **error** ϵ of $\tilde{a}$ is

$$\epsilon = a - \tilde{a} \tag{4.73}$$

where a is the exact value. The *relative error* of $\tilde{a}$ is ϵ/a. Errors arise from rounding, inaccuracy of measured values, truncation (that is, replacement of integrals by sums, series by partial sums), and so on.

An algorithm is called **numerically stable** if small changes in the initial data give only correspondingly small changes in the final results. Unstable algorithms are generally useless because errors may become so large that results will be very inaccurate. Numeric instability of algorithms must not be confused with mathematical instability of problems.

Fixed-point iteration is a method for solving equations $f(x) = 0$ in which the equation is first transformed algebraically to $x = g(x)$, an initial guess x_0 for the solution is made, and then approximations $x_1, x_2, \ldots$, are successively computed by iteration from

$$x_{n+1} = g(x_n) \quad (n = 0, 1, \ldots) \tag{4.74}$$

Newton's method for solving equations $f(x) = 0$ is an iteration

$$x_{n+1} = x_n - \frac{f(x_n)}{f'(x_n)} \tag{4.75}$$

Here x_{n+1} is the x intercept of the tangent of the curve $y = f(x)$ at the point x_n. This method is of second order. If we replace f' in (4.75) by a difference quotient (geometrically: we replace the tangent by a secant), we obtain the **secant method**. We also discuss the *bisection method* (which converges slowly) and the *method of false position.*

Polynomial interpolation means the determination of a polynomial $p_n(x)$ such that $p_n(x_j) = f_j$, where $j = 0, \ldots, n$ and $(x_0, f_0), \ldots, (x_n, f_n)$ are measured or observed values, values of a function, etc. $p_n(x)$ is called an *interpolation polynomial.* For given data, $p_n(x)$ of degree n (or less) is unique. However, it can be written in different forms, notably in **Lagrange's form** or in **Newton's divided difference form** which requires fewer operations. For regularly spaced x_0, $x_1 = x_0 + h, \ldots, x_n = x_0 + nh$ the latter becomes **Newton's forward difference formula**

$$f(x) \approx p_n(x) = f_0 + r\Delta f_0 + \cdots + \frac{r(r-1)\cdots(r-n+1)}{n!}\Delta^n f_0$$

where $r = (x - x_0)/h$ and the forward differences are $\Delta f_j = f_{j+1} - f_j$ and

$$\Delta^k f_j = \Delta^{k-1} f_{j+1} - \Delta^{k-1} f_j \quad (k = 2, 3, \ldots)$$

A similar formula is *Newton's backward difference interpolation formula*.

Interpolation polynomials may become numerically unstable as n increases, and instead of interpolating and approximating by a single high-degree polynomial it is preferable to use a cubic **spline** $g(x)$, that is, a twice continuously differentiable interpolation function [thus, $g(x_j) = f_j$], which in each subinterval $x_j \leqq x \leqq x_{j+1}$ consists of a cubic polynomial $q_j(x)$.

Review Questions and Problems

1. What is a numeric method? How has the computer influenced numeric methods?
2. What is floating-point representation of numbers? Overflow and underflow?

3. How do error and relative error behave under addition? Under multiplication?
4. Why are roundoff errors important? State the rounding rules.
5. What is an algorithm? Which of its properties are important in software implementation?
6. Why is the selection of a good method at least as important on a large computer as it is on a small one?
7. Explain methods for solving equations, in particular fixed-point iteration and its convergence.
8. Can the Newton–Raphson method diverge? Is it fast? Same questions for the bisection method.
9. What is the advantage of Newton's interpolation formulas over Lagrange's?
10. What do you remember about errors in polynomial interpolation?
11. What is spline interpolation? Its advantage over polynomial interpolation?
12. Write −0.35287, 1274.799, −0.00614, 24.9482, 1/3, 85/7 in floating-point form with 5S (5 significant digits, properly rounded).
13. Compute (5.346 − 3.644)/(3.454 − 3.055) as given and then rounded stepwise to 3S, 2S, 1S. ("Stepwise" means rounding the four rounded numbers, not the given ones.) Comment on your results.
14. Compute 0.29731/(4.1232 − 4.0872) with the numbers as given and then rounded stepwise (that is, rounding the rounded numbers) to 4S, 3S, 2S. Comment.
15. Solve $x^2 - 50x + 1 = 0$ using 5S in the computation. Compare and comment.
16. Solve $x^2 - 200x + 4 = 0$ using 5S in the computation. Compare and comment.
17. Let 4.81 and 12.752 be correctly rounded to the number of digits shown. Determine the smallest interval in which the sum (using the true instead of the rounded values) must lie.
18. Answer the question in Prob. 17 for the difference 4.81 − 12.752.
19. What is the relative error of $n\bar{a}$ in terms of that of $\bar{a}$?
20. Show that the relative error of $\bar{a}^2$ is about twice that of $\bar{a}$.
21. Compute the solution of $x^5 = x + 0.2$ near $x = 0$ by transforming the equation into the form $x = g(x)$ and starting from $x_0 = 0$. (Use 6S.)
22. Solve $\cos x = x$ by iteration (6S, $x_0 = 1$), writing it as $x = (0.74x + \cos x)/1.74$, obtaining $x_4 = 0.739085$ (exact to 6S!). Why does this converge so rapidly?
23. Solve $x^4 - x^3 - 2x - 34 = 0$ by Newton's method with $x_0 = 3$ and 6S accuracy.
24. Solve $\cos x - x = 0$ by the method of false position.
25. Compute $f(1.28)$ from

$$f(1.0) = 3.00000$$
$$f(1.2) = 2.98007$$
$$f(1.4) = 2.92106$$
$$f(1.6) = 2.82534$$
$$f(1.8) = 2.69671$$
$$f(2.0) = 2.54030$$

by linear interpolation. By quadratic interpolation, using $f(1.2)$, $f(1.4)$, $f(1.6)$.
26. Find the cubic spline for the data

$$f(-1) = 3$$
$$f(1) = 1$$
$$f(3) = 23$$
$$f(5) = 45$$
$$k_0 = k_3 = 3$$

27. Find a real root of the equation $x\log_{10}x = 1.2$ which lies between 2 and 3 by bisection method. [JNTU 2008]
28. Find a real root of the equation $x^3 - 6x - 4 = 0$ by bisection method. [JNTU May 2006]
29. Find a real root of $x^3 - 5x + 3 = 0$ using bisection method. [JNTU May 2006]
30. Find a real root of the equation $x^3 - x - 11 = 0$ by bisection method. [JNTU April 2007]
31. Find the root of the equation $2x - \log_{10}x = 7$, which lies between 3.5 and by regula-falsi method.

OR

Find a real root of the equation $2x - \log x = 7$ by successive approximation method. [JNTU 2006]

32. Find a real root of $xe^x = 3$ using regula falsi method. [JNTU May 2006]
33. Evaluate $\sqrt{12}$ and $1/\sqrt{12}$ by the fixed-point iteration method.
34. Using Newton–Raphson method
    a. Find square root of a number
    b. Find reciprocal of a number

    [JNTU Sep. 2008]
35. Find a real root for $x\tan x + 1 = 0$ using Newton–Raphson method. [JNTU Sep. 2006]
36. Find a real root of $x + \log_{10}x - 2 = 0$ using Newton–Raphson method. [JNTU April 2007]
37. Find $f(22)$ from the Gauss forward formula. [JNTU May 2007]

x	20	25	30	35	40	45
y	354	332	291	260	231	204

5 Curve Fitting

LEARNING OBJECTIVES

After reading this chapter, you will be able to understand the following:

- Curve fitting and least squares method.
- Curvilinear regression.
- Weighted least squares approximation.
- Numerical differentiation and integration.
- Principal value integral.

In the previous chapter, we discussed how a function may be fitted to a set of tabulated values. It was assumed that the tabulated data had no errors. In such case the polynomial function was called an ***interpolating polynomial***. In many cases, it is likely that the tabulated data may have some errors, particularly when the data are obtained as a result of some experiment. These experimental errors are random quantities and therefore can be described only statistically. The errors may vary over a wide range and some may even be quite large. In such cases, we can only find an approximate fit of a mathematical function (or a polynomial) to the given set of data. However, we have to ensure that this function is "statistically the best fit" for that data. The question therefore is not just finding a fitting function, but to find a statistically best-fitted approximate polynomial. In general, the polynomial is not a simple linear function, it could be of nonlinear type (a curve and not a straight line) and hence the process is named "curve-fitting."

The method of least squares discussed in this chapter is probably the most systematic procedure to fit a unique curve (one and only one polynomial function) through a given set of data. This is widely used in practical computations and can also be easily implemented on personal computers.

5.1 Introduction to the Concept

In this section we introduce certain concepts that would be of use in the following sections. Let $x_1, x_2, \ldots, x_n$ be independent variables and $y_1, y_2, \ldots, y_n$ be the corresponding dependent variables. Let $\overline{y} = f(x)$ be an approximation to the function [i.e., $f(x)$ is the polynomial or curve to be fitted]. Then the errors between $\overline{y}_i$ (the approximate values) and y_i (the actual tabulated values) are

$$\begin{aligned} d_1 &= y_1 - \overline{y}_1 = y_1 - f(x_1) \\ d_2 &= y_2 - \overline{y}_2 = y_2 - f(x_2) \\ &\vdots \qquad\qquad \vdots \\ d_n &= y_n - \overline{y}_n = y_n - f(x_n) \end{aligned} \tag{5.1}$$

The function $f(x)$ is to be chosen in such a way that the errors $d_1, d_2, \ldots, d_n$ are small or minimized. The possible functions of the errors which one may consider for minimization are

$$\min \sum_i d_i = \min \sum_i (y_i - \overline{y}_i) \tag{5.2}$$

$$\min \sum_i |d_i| = \min \sum_i |y_i - \overline{y}_i| \tag{5.3}$$

$$\max|d_i| \leqq D \tag{5.4}$$

$$\min \sum_i d_i^2 = \min \sum_i (y_i - \bar{y}_i)^2 \tag{5.5}$$

These are called "metrics" or "distance functions" as they measure the distance between the tabulated values and the approximate values. The metric (5.2) is not applicable and is rejected as large deviations which are equal but of opposite signs will cancel one another. The metric (5.3) is a reasonable measure and minimizes the sum of absolute deviations; however, this is not amenable to analytic formulation and is again rejected because of statistical reasons. In metric (5.4) above, the maximum allowable deviation (or error) is to be kept below a pre-assigned value D – this formulation is termed Chebyshev approximation. The algorithms for these cases are quite complicated and hence not suggested at this level. The last metric (5.5) is of utmost importance because it minimizes the sum of squares of the deviations and has many desirable characteristics. It is generally agreed that the best representative curve is that for which "the sum of the squares of the deviations (or errors) is a minimum." This is the method of least squares and is the best known method (it was developed by Legendre). It leads to "regression" formulas (used in statistical computations) and its positive features are:

1. Errors of opposite sign do not cancel out.
2. It weighs large errors more than small errors.
3. Methods of calculus can be readily applied for analysis purposes.
4. The function $f(x)$, which is obtained by minimizing (5.5), can be shown to be the most likely one when the deviations d_i have a Gaussian or normal distribution.

We will examine this method in detail with a formulation called "regression" analysis later in the section. This can be further subdivided into linear and nonlinear types.

5.2 Curve Fitting and Least Squares Straight Line

Curve Fitting

Having discussed numerics for linear systems, we now turn to an important application, curve fitting, in which the solutions are obtained from linear systems. In **curve fitting** we are given n points (pairs of numbers) $(x_1, y_1), \ldots, (x_n, y_n)$ and we want to determine a function $f(x)$ such that

$$f(x_1) \approx y_1, \ldots, f(x_n) \approx y_n$$

approximately. The type of function (for example, polynomials, exponential functions, sine and cosine functions) may be suggested by the nature of the problem (the underlying physical law, for instance), and in many cases a polynomial of a certain degree will be appropriate.

Let us begin with a motivation. If we require strict equality $f(x_1) = y_1, \ldots, f(x_n) = y_n$ and use polynomials of sufficiently high degree, we may apply one of the methods in connection with interpolation. However, in certain situations this would not be the appropriate solution of the actual problem. For instance, to the four points

$$(-1.3, 0.103), \qquad (-0.1, 1.099), \qquad (0.2, 0.808), \qquad (1.3, 1.897) \tag{5.6}$$

there corresponds the interpolation polynomial $f(x) = x^3 - x + 1$ (Figure 1), but if we graph the points, we see that they lie nearly on a straight line. Hence if these values are obtained in an experiment and thus involve an experimental error, and if the nature of the experiment suggests a linear relation, we better fit a straight line through the points (Figure 1). Such a line may be useful for predicting

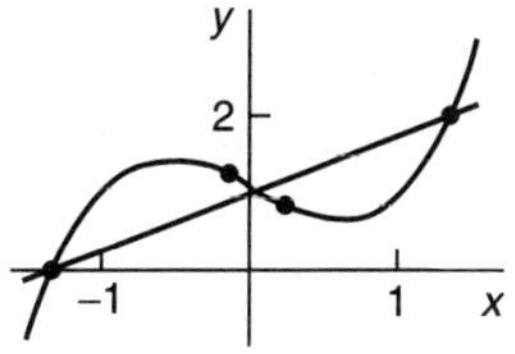

Figure 1 Approximate fitting of a straight line.

values to be expected for other values of x. A widely used principle for fitting straight lines is the **method of least squares** by Gauss and Legendre. We explain this method next.

Method of Least Squares

The straight line

$$y = a + bx$$

should be fitted through the given points $(x_1, y_1), \ldots, (x_n, y_n)$ so that the sum of the squares of the distances of those points from the straight line is minimum, where the distance is measured in the vertical direction (the y direction).

The point on the line with abscissa x_j has the ordinate $a + bx_j$. Hence its distance from (x_j, y_j) is $|y_j - a - bx_j|$ (Figure 2) and that sum of squares is

$$q = \sum_{j=1}^{n} (y_j - a - bx_j)^2$$

Here q depends on a and b. A necessary condition for q to be minimum is

$$\begin{aligned} \frac{\partial q}{\partial a} &= -2\sum (y_j - a - bx_j) = 0 \\ \frac{\partial q}{\partial b} &= -2\sum x_j(y_j - a - bx_j) = 0 \end{aligned} \tag{5.7}$$

(where we sum over j from 1 to n). Dividing by 2, writing each sum as three sums, and taking one of them to the right, we obtain the result

$$\begin{aligned} an + b\sum x_j &= \sum y_j \\ a\sum x_j + b\sum x_j^2 &= \sum x_j y_j \end{aligned} \tag{5.8}$$

These equations are called the **normal equations** of our problem.

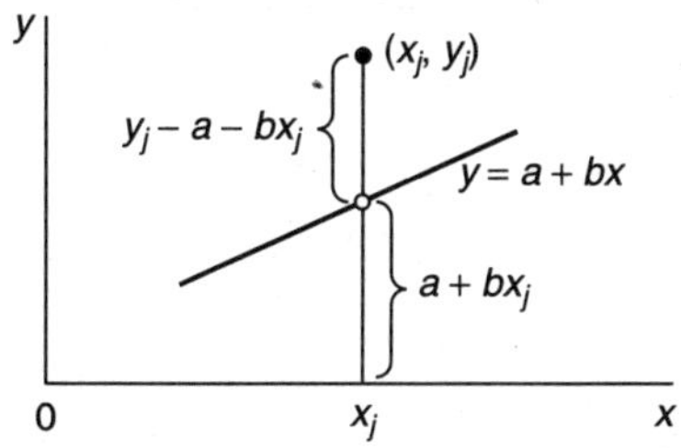

Figure 2 Vertical distance of a point (x_j, y_j) from a straight line $y = a + bx$.

Example 1 Straight Line

Using the method of least squares, fit a straight line to the four points given in formula (5.6).

Solution: We have $n = 4$, $\sum x_j = 0.1$, $\sum x_j^2 = 3.43$, $\sum y_j = 3.907$, $\sum x_j y_j = 2.3839$. Hence the normal equations are

$$4a + 0.10b = 3.9070$$
$$0.1a + 3.43b = 2.3839$$

The solution (rounded to 4D) is $a = 0.9601$, $b = 0.6670$, and we obtain the straight line (Figure 1)

$$y = 0.9601 + 0.6670x$$

Curve Fitting by Polynomials of Degree m

Our method of curve fitting can be generalized from a polynomial $y = a + bx$ to a polynomial of degree m

$$p(x) = b_0 + b_1 x + \cdots + b_m x^m \tag{5.9}$$

where $m \leqq n - 1$. Then q takes the form

$$q = \sum_{j=1}^{n} [y_j - p(x_j)]^2$$

and depends on $m + 1$ parameters $b_0, \ldots, b_m$. Instead of (5.7) we then have $m + 1$ conditions

$$\frac{\partial q}{\partial b_0} = 0, \qquad \ldots, \qquad \frac{\partial q}{\partial b_m} = 0 \tag{5.10}$$

which give a system of $m + 1$ normal equations. In the case of a quadratic polynomial

$$p(x) = b_0 + b_1 x + b_2 x^2 \tag{5.11}$$

the normal equations are (summation from 1 to n) as follows:

$$\begin{aligned} b_0 n + b_1 \sum x_j + b_2 \sum x_j^2 &= \sum y_j \\ b_0 \sum x_j + b_1 \sum x_j^2 + b_2 \sum x_j^3 &= \sum x_j y_j \\ b_0 \sum x_j^2 + b_1 \sum x_j^3 + b_2 \sum x_j^4 &= \sum x_j^2 y_j \end{aligned} \tag{5.12}$$

The derivation of (5.12) is left to the reader.

Example 2 Quadratic Parabola by Least Squares

Fit a parabola through the data (0, 5), (2, 4), (4, 1), (6, 6), (8, 7).

Solution: For the normal equations we need $n = 5$, $\sum x_j = 20$, $\sum x_j^2 = 120$, $\sum x_j^3 = 800$, $\sum x_j^4 = 5664$, $\sum y_j = 23$, $\sum x_j y_j = 104$, $\sum x_j^2 y_j = 696$. Hence these equations are

$$\begin{aligned} 5b_0 + 20b_1 + 120b_2 &= 23 \\ 20b_0 + 120b_1 + 800b_2 &= 104 \\ 120b_0 + 800b_1 + 5664b_2 &= 696 \end{aligned}$$

Solving them we obtain the quadratic least squares parabola (Figure 3):

$$y = 5.11429 - 1.41429x + 0.21429x^2$$

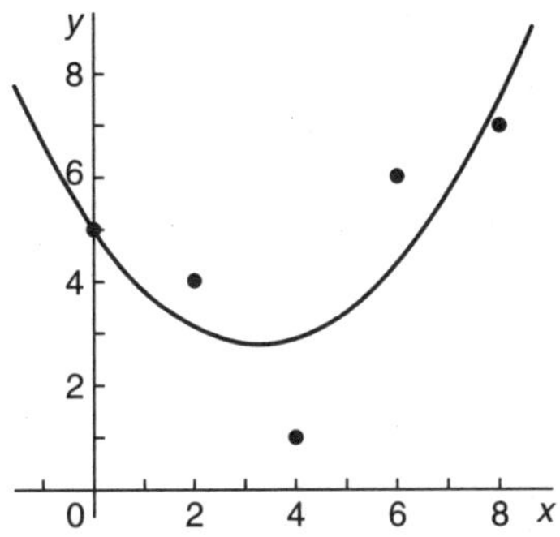

Figure 3 Least squares parabola in Example 2.

For a general polynomial (5.9) the normal equations form a linear system of equations in the unknowns $b_0, \ldots, b_m$. When its matrix **M** is nonsingular, we can solve the system by Cholesky's method (discussed in Chapter 1) because then **M** is positive definite (and symmetric). When the equations are nearly linearly dependent, the normal equations may become ill-conditioned and should be replaced by other methods. The least squares method also plays a role in statistics.

Curvilinear (or Nonlinear) Regression

In this method the given points (x_i, y_i), $i = 1, 2, 3, \ldots, n$, are fitted into a straight line utilizing the principle of least squares. This straight line fit leads to a linear regression formula. Let

$$f(x) = \bar{y} = a_1 x + a_0 \tag{5.13}$$

be the straight line to be fitted to the given data. We have to determine the unknown constants a_1, a_0 in such a way that they lead to the "best fit" straight line. We will use the least square errors (or deviations) criterion for achieving this.

From (5.1) we know that the errors or deviations at each x_i are given by d_i and from (5.5), the sum squares of deviations is written as

$$S = \sum_{i=1}^{n} d_i^{\,2} = \sum_{i=1}^{n} (y_i - \bar{y}_i)^2 = \sum_{i=1}^{n} [y_i - (a_1 x_i + a_0)]^2 \tag{5.14}$$

We have to minimize this sum S. Here x is the independent variable, (x_i, y_i) are the tabulated points (given) and $\bar{y}_i$ are approximate values at x_i. Equation (5.14) is called the "linear regression of y on x." That is given the data point (x_i, y_i), we have to identify the independent variable first and designate it as x (and all such points as x_i). By this, it is meant that x is the independent variable or control variable and has no errors in it. Variable y is the dependent variable (or measured one) and is likely to have errors. The error $y_i - \bar{y}_i$ implies the deviation between the measured (tabulated) values and the fitted values along the y direction.

We wish to minimize S given in (5.14). Mathematically we have

$$\min S = \min \sum_{i=1}^{n} [y_i - (a_1 x_i + a_0)]^2 \tag{5.15}$$

As this involves two unknowns a_0, a_1, we minimize (5.15) by taking the partial derivatives of S with a_0 and a_1 and setting them to zero (separately), that is, for S to be minimum

$$\frac{\partial S}{\partial a_0} = 0 \quad \text{and} \quad \frac{\partial S}{\partial a_1} = 0$$

Thus

$$\frac{\partial S}{\partial a_0} = \sum_{i=1}^{n} 2(y_i - a_1 x_i - a_0)(-1) = 0 \tag{5.16}$$

$$\frac{\partial S}{\partial a_1} = \sum_{i=1}^{n} 2(y_i - a_1 x_i - a_0)(-x_i) = 0 \tag{5.17}$$

Equation (5.16) becomes

$$\sum y_i - a_1 \sum x_i - a_0 \sum 1 = 0$$

$$na_0 + a_1 \sum x_i = \sum y_i \tag{5.18}$$

and (5.17) becomes

$$\sum y_i x_i + a_1 \sum x_i^2 + a_0 \sum x_i = 0$$

$$a_0 \sum x_i + a_1 \sum x_i^2 = \sum x_i y_i \tag{5.19}$$

Solving (5.18) and (5.19) for the two unknowns a_0, a_1, we get

$$a_0 = \frac{\sum y_i \sum x_i^2 - \sum x_i \sum x_i y_i}{n \sum x_i^2 - (\sum x_i)^2} \tag{5.20}$$

$$a_1 = \frac{n \sum x_i y_i - \sum x_i \sum y_i}{n \sum x_i^2 - (\sum x_i)^2} \tag{5.21}$$

These two equations are called "normal equations" and give the coefficients of the linear least squares fit line a_0, a_1. The coefficients a_0 and a_1 are called "regression coefficients." It can be noticed that the denominator is same in both (5.20) and (5.21); the summation is always over $i = 1$ to n.

Let us consider an example that demonstrates the usage of this formulation.

Example 3

Fit a straight line for the data given below:

x	1	2	4	5	6	8	9
y	2	5	7	10	12	15	19

Solution: Here $n = 7$. As nothing else is mentioned, let us consider x as the independent variable and y as the dependent variable.

Note: If it is given that y is the independent variable and x is the dependent variable, then the corresponding equation of straight line would be of $x = b_i y + b_0$ form, where b_0, b_1 have similar meanings and expressions as those for a_0, a_1 with x and y interchanged. Else re-designate and name the independent variable as x and dependent variable as y in whatever form the data is given; however, this aspect must be clearly explained before you start applying formulas for a_0, a_1.

Formulate the following table, based on the given x_i and y_i:

x_i *(Independent variable)*	y_i *(Dependent variable)*	x_i^2	x_iy_i
1	2	1	2
2	5	4	10
4	7	16	28
5	10	25	50
6	12	36	72
8	15	64	90
9	19	81	171
$\sum x_i = 35$	$\sum y_i = 70$	$\sum x_i^2 = 227$	$\sum x_iy_i = 453$

From (5.20) and (5.21), we have

$$a_0 = \frac{(70)(227)-(35)(453)}{(7)(227)-(35)^2} = \frac{3}{364} = 0.096$$

$$a_1 = \frac{(7)(453)-(35)(70)}{(7)(227)-(35)^2} = \frac{721}{364} = 1.98$$

Thus the approximate, best possible, straight line fit is

$$y = 1.98x + 0.096 \tag{5.22}$$

Further, another interesting fact can be described in terms of the averages of given x_i and y_i $(i = 1, \ldots, n)$

$$\langle x \rangle_{\text{av}} = \frac{\sum x_i}{n} = \text{Mean of } x = \frac{35}{7} = 5$$

$$\langle y \rangle_{\text{av}} = \frac{\sum y_i}{n} = \text{Mean of } y = \frac{70}{7} = 10$$

and the straight line fit obtained passes through this $x = 5$, $y = 10$ point. That is the linear least squares fit line always passes through the mean of the tabulated values of x and y.

Note: If y is the independent variable and x is the dependent variable, then the equation would be

$$x = b_1 y + b_0$$

This is the regression line of x on y. Equations (5.20) and (5.21) with x and y interchanged will yield b_0 and b_1, respectively. [This needs another column formation y_i^2 in the above table in Example 3.] Thus

$$b_0 = \frac{\sum x_i \sum y_i^2 - \sum y_i x_i}{n\sum y_i^2 - (\sum y_i)^2} = \frac{(35)(908)-(70)(453)}{(7)(908)-(70)^2} = \frac{70}{1456} = 0.048 \tag{5.23}$$

$$b_1 = \frac{n\sum x_i y_i - \sum y_i \sum x_i}{n\sum y_i^2 - (\sum y_i)^2} = \frac{(7)(458)-(35)(70)}{(7)(908)-(70)^2} = \frac{721}{1456} = 0.495 \tag{5.24}$$

Then, the straight line equation is

$$x = 0.495y + 0.048 \tag{5.25}$$

As you can see, (5.22) and (5.25) are not identical. However, both of them always pass through the means of x and y, namely point (5, 10) here.

ALGORITHM LINEAR REGRESSION $y = a_1x + a_0$

INPUT: n = Number of data points
(x_i, y_i), $i = 1$ to n: data points

OUTPUT: a_1, a_0 (taking x as the independent variable)

Step 1: Read n
Step 2: For $i = 1$ to n, in steps of 1, do step 3
Step 3: Read x_i, y_i
End of FOR loop of step 2
Step 4: Sum$x \leftarrow 0$
Step 5: Sum$xsq \leftarrow 0$
Step 6: Sum$y \leftarrow 0$
Step 7: Sum$xy \leftarrow 0$
Step 8: For $i = 1$ to n, in steps of 1, do upto step 12
Step 9: Sum$x \leftarrow$ Sum$x + x$
Step 10: Sum$xsq \leftarrow$ Sum$xsq + (x * x)$
Step 11: Sum$y \leftarrow$ Sum$y + y$
Step 12: Sum$xy \leftarrow$ Sum$xy + (x * y)$
End of FOR loop of step 8
Step 13: denom $\leftarrow (n *$ Sum$xsq) - ($Sum$x *$ Sum$x)$
Step 14: $a_0 \leftarrow ($Sum$y +$ Sum$xsq -$ Sum$x *$ Sum$xy)$/denom
Step 15: $a_1 \leftarrow (n *$ Sum$xy -$ Sum$x *$ Sum$y)$/denom
Step 16: Write a_1, a_0
Step 17: Stop

Polynomial Regression

It may not always be possible to say that the straight line fit is the best fit. Sometimes it may be necessary to fit a higher degree polynomial rather than a straight line. This can be seen easily from Figure 4. As it can be seen in Figure 4a, a straight line fit is reasonable, but the points in Figure 4b seem to fall on a parabola, which is a second degree polynomial. The resulting regression is a polynomial regression, which is in general nonlinear, the linear case being a specific one (when the order of the polynomial is one).

Let us now discuss how a polynomial of special degree can be fitted through a given set of points satisfying the minimum mean square error criterion. The actual degree or order of polynomial has to be given based on physical considerations.

A Quadratic (Polynomial) Fit

Let a quadratic (second degree) curve of the form

$$f(x) = \overline{y} = a_2x^2 + a_1x + a_0 \tag{5.26}$$

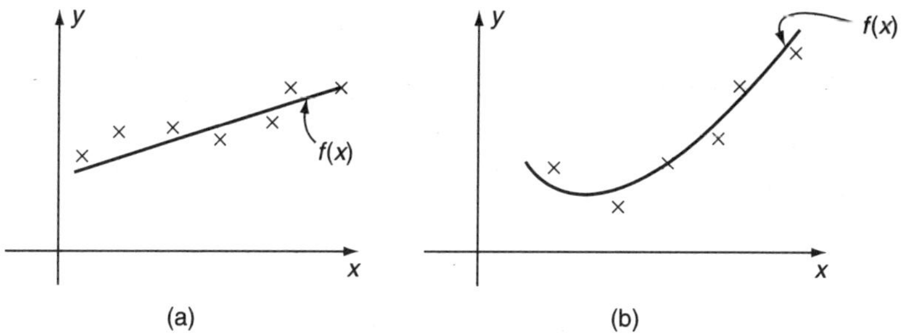

Figure 4 Least squares curve fitting: (a) Straight line fit; (b) polynomial fit.

be the fitting curve for the given set of points (x_i, y_i). Using (5.14), the sum S of the squares of the deviations is obtained as

$$S = \sum_{i=1}^{n} (y_i - \bar{y}_i)^2 = \Sigma(y_i - a_2x_i^2 - a_1x_i - a_0)^2 \tag{5.27}$$

To minimize S, partial differentiate S with respect to the three unknowns a_0, a_1, a_2 and set them equal to zero. As there are three unknowns and three partial derivatives, this will yield three simultaneous equations in terms of a_0, a_1, a_2, that is,

$$\frac{\partial S}{\partial a_0} = 0 \Rightarrow \sum_{i=1}^{n} 2(y_i - a_2x_i^2 - a_1x_i - a_0)(-1) = 0$$

$$\Rightarrow na_0 + (\Sigma x_i)a_1 + (\Sigma x_i^2)a_2 = \Sigma y_i \tag{5.28}$$

$$\frac{\partial S}{\partial a_1} = 0 \Rightarrow \sum_{i=1}^{n} 2(y_i - a_2x_i^2 - a_1x_i - a_0)(-x_i) = 0$$

$$\Rightarrow (\Sigma x_i)a_0 + (\Sigma x_i^2)a_1 + \Sigma x_i^3 a_2 = \Sigma x_i y_i \tag{5.29}$$

$$\frac{\partial S}{\partial a_2} = 0 \Rightarrow \sum_{i=1}^{n} 2(y_i - a_2x_i^2 - a_1x_i - a_0)(-x_i^2) = 0$$

$$\Rightarrow (\Sigma x_i^2)a_0 + (\Sigma x_i^3)a_1 + (\Sigma x_i^4)a_2 = \Sigma x_i^2 y_i \tag{5.30}$$

Equations (5.28)–(5.30) are three normal equations in three unknowns a_0, a_1, a_2. They are three linear simultaneous equations, which can be solved in our conventional way (or Gauss elimination method). Thus a second-order polynomial leads to a set of three equations in three unknowns. If we want to fit any nth degree polynomial, it will have $(n + 1)$ unknowns $(a_0, a_1, a_2, \ldots, a_n)$, and $(n + 1)$ simultaneous equations are to be solved (say, by Gauss elimination method). Some ill-conditioned equations (refer to Chapter 1) may also come into picture, and therefore there may be some numerical difficulties. Usually polynomial curve fit with lower order of n is attempted.

Even a second-degree $(n = 2)$ fit needs solution of three equations, which involve formation of table that has x_i^2, x_i^3, x_i^4, x_iy_i, columns and their summations. One such example is given below. However, algorithm and program are not being attempted as the process is complicated and we would have to involve Gauss elimination method as well.

Example 4

For the tabulated x_i, y_i points given below, find the best possible second-degree curve fit.

x	−4	−3	−2	−1	0	1	2	3	4	5
y	21	12	4	1	2	7	15	30	45	67

Solution: Formulate the following table:

x_i	y_i	x_i^2	x_i^3	x_i^4	x_iy_i	$x_i^2y_i$
−4	21	16	−64	256	−84	336
−3	12	9	−27	81	−36	108
−2	4	4	−8	16	−8	16
−1	1	1	−1	1	−1	1
0	2	0	0	0	0	0
1	7	1	1	1	7	7
2	15	4	8	16	30	60
3	30	9	27	81	90	270
4	45	16	64	256	180	720
5	67	25	125	625	335	1675
$\sum x_i = 5$	$\sum y_i = 204$	$\sum x_i^2 = 85$	$\sum x_i^3 = 125$	$\sum x_i^4 = 1333$	$\sum x_iy_i = 51$	$\sum x_i^2y_i = 3193$

Here $n = 10$. Substituting these sums in (5.28)–(5.30), we get the following equations:

$$\begin{aligned} 10a_0 + 5a_1 + 85a_2 &= 204 \\ 5a_0 + 85a_1 + 125a_2 &= 513 \\ 85a_0 + 125a_1 + 1333a_2 &= 3193 \end{aligned}$$

The first equation is kept as it is and Gauss elimination is used for solution. Eliminating the term a_0 from the second equation, terms a_0 and a_1 from the third equation (using triangularization process), we get

$$\begin{aligned} 10a_0 + 5a_1 + 85a_2 &= 204 \\ 0 + 82.5a_1 + 82.5a_2 &= 411 \\ 0 + 0 + 528a_2 &= 1048 \end{aligned}$$

Using back substitution, we have

$$a_2 = 1.98,\ a_1 = 3.0,\ a_0 = 2.07$$

Thus, the fitting curve would be

$$y = 1.98x^2 + 3x + 2.07$$

Nonlinear Curve Fitting

So far we have discussed linear and polynomial curve fitting using least squares method. There are various other problems, which are functions of time or some other variable, in which exponential, trigonometric, or geometric function would be the best possible fits. All these are again examples for nonlinear curve fitting.

Curve Fitting by a Sum of Exponential

Let us consider an exponential curve fit by least squares principle. Let

$$y = f(x) = ae^{-bx} \Rightarrow \ln y = \ln a + (-\,bx)\ln e \tag{5.31}$$

be the curve to be fitted, where a and b are unknown constants to be determined. We have to use a mathematical transformation appropriately, so that the nonlinear exponential function of (5.31) can be converted into some linear function $z(x)$. [*Note*: This sort of transformation may not be possible always. We shall consider those functions which are transferable to linear form.] Then the given data, which is likely to fit a particular nonlinear function (which, after suitable transformation, becomes a linear function) is utilized to formulate the table. As was done earlier, the least square error criterion is applied to determine the unknown coefficients that represent the straight line equation. It must be remembered here that the basic assumption in picking the least squares error method for curve fitting was that for linear functions it leads to "optimum fit" – the assumption being not valid for nonlinear curves. Still, the same method can be used because of its ease in implementation.

For fitting an exponential curve of the type given in (5.31), we use a logarithmic transformation:

$$z = \ln y \tag{5.32}$$

Thus

$$z = \ln y = \ln(ae^{-bx}) = \ln a + (-b)x \ln e \tag{5.33}$$

Comparing this with our linear relation $a_1x + a_0$, we have

$$a_1 = -b \tag{5.34}$$

$$a_0 = \ln a \quad \text{or} \quad a = e^{a_0} \tag{5.35}$$

Thus the new set of (transformed) equations is

$$z = (-b)x + \ln a = a_1x + a_0 \tag{5.36}$$

where $z = \ln y$. We have already solved (5.36) using the least squares principle, and the resulting a_1, a_0 are given by (5.20) and (5.21). [The situations are exactly identical, the only change is that (5.36) has Z on the left-hand side in place of y.] Thus, the normal equations are

$$na_0 + \left(\sum_{i=1}^{n} x_i\right)a_1 = \sum_{i=1}^{n} z_i = \sum_{i=1}^{n} \ln y_i \tag{5.37}$$

$$\left(\sum_{i=1}^{n} x_i\right)a_0 + \left(\sum_{i=1}^{n} x_i^2\right)a_1 = \sum_{i=1}^{n} x_i z_i = \sum_{i=1}^{n} (x_i \ln y_i) \tag{5.38}$$

and the coefficients are

$$a_0 = \frac{\sum \ln y_i \sum x_i^2 - \sum x_i \sum (x_i \ln y_i)}{n\sum x_i^2 - (\sum x_i)^2} \tag{5.39}$$

$$a_1 = \frac{n\sum x_i \ln y_i}{n\sum x_i^2 - (\sum x_i)^2} \tag{5.40}$$

The required unknowns a, b are given by

$$a = e^{a_0}, \quad b = -a_1 \tag{5.41}$$

and the fitted curve is

$$y = ae^{-bx} \tag{5.42}$$

Let us write an algorithm for this case.

ALGORITHM EXPONENTIAL FIT $y = ae^{-bx}$

In this algorithm we take x as the independent variable.

INPUT: n = Number of data points
(x_i, y_i), $i = 1$ to n: data points

OUTPUT: a_1, a_0, a, b

Step 1: Read n
Step 2: For $i = 1$ to n, in steps of 1, do step 3
Step 3: Read x_i, y_i
End of the FOR loop of step 2
Step 4: Sum$x \leftarrow 0.0$
Step 5: Sum$xsq \leftarrow 0.0$
Step 6: Sum$y \leftarrow 0.0$
Step 7: Sum$xy \leftarrow 0.0$
Step 8: For $i = 1$ to n, in steps of 1, do upto step 13
Step 9: Sum$x \leftarrow$ Sum$x + x$
Step 10: Sum$xsq \leftarrow$ Sum$xsq + (x * x)$
Step 11: $z \leftarrow \ln y$
Step 12: Sum$y \leftarrow$ Sum$y + z$
Step 13: Sum$xy \leftarrow$ Sum$xy + (x * z)$
End of FOR loop of step 8
Step 14: denom $\leftarrow n *$ Sumxsq – (Sum$x$ * Sum$x$)
Step 15: $a_0 \leftarrow$ (Sum$y$ * Sumxsq – Sum$x$ * Sumxy)/denom
Step 16: $a_1 \leftarrow$ (n * Sumxy – Sum$x$ * Sum$y$)/denom
Step 17: $a \leftarrow \exp(a_0)$
Step 18: b $\leftarrow -a_1$
Step 19: Write a, b
Step 20: Stop

Example 5

Fit the given data into an exponential curve of the form $y = ae^{bx}$

x	1.0	1.2	1.4	1.6
y	40.170	73.196	133.372	243.02

Solution: Here

$$y = ae^{bx}$$

$$z = \ln y = \ln a + bx = a_1 x + a_0$$

Therefore $a_0 = \ln a$ or $a = e^{a_0}$ and $b = a_1$. That is, a_0 and a_1 are to be determined as before, and then a, b. This calculation needs the following tabulation.

	x_i	y_i	$z_i = \ln y_i$	x_i^2	$x_i \ln y_i$
	1	40.17	3.6931	1	3.6931
	1.2	73.196	4.2931	1.44	5.1517
	1.4	133.372	4.8931	1.96	6.8503
	1.6	243.02	5.4931	2.56	8.7890
SUM	5.2	489.758	18.3724	6.96	24.4841

Here $n=4$, $\sum x_i = 5.2$, $\sum y_i = 489.758$, $\sum z_i = \sum \ln y_i = 18.3724$, $\sum x_i^2 = 6.9$, $\sum x_i \ln y_i = 24.4840$. Substituting these values in (5.39), (5.40), we get a_0, a_1. Thus

$$a_0 = \frac{(18.3724)(6.96) - (5.2)(24.4841)}{(4)(6.96) - (5.2)^2} = \frac{0.5446}{0.8} = 0.6932$$

$$a_1 = \frac{(4)(24.4841) - (5.2)(18.3724)}{(4)(6.96) - (5.2)^2} = \frac{2.3999}{0.8} = 2.9999$$

As defined earlier

$$a = \exp(a_0) = e^{0.6932} = 2.0$$

$$b = a_1 = 2.9999$$

and the equation for the curve fitted is

$$y = 2e^{2.999x}$$

Note: Had we started with an exponential fit of the form $y = ae^{-bx}$, the solution would have been $a = 2$, $b = -2.9999$.

Geometric Regression

This is also a nonlinear curve fit, which uses power function of x. Let us therefore consider an expression of the type

$$y = ax^b + c \tag{5.43}$$

where b is a noninteger. If $c = 0$, then we can very easily transform (5.43) into a linear equation using $\ln y$ transformation, as

$$z = \ln y = \ln a + b \ln x$$

which is again in $a_1 t + a_0$ form. Then $z = a_1 t + a_0$, where $z = \ln y$, $t = \ln x$, $a_1 = b$, $a_0 = \ln a$, and the unknowns can be easily evaluated, as explained in the previous subsection. However, when $c \neq 0$, with three unknowns a, b, c, there is no transformation which will linearize (5.43). Let us therefore assume (or fix) c (say $c = 1.0$), and then we can write

$$y - c = ax^b \tag{5.44}$$

Taking natural logs we get

$$\ln(y - c) = \ln a + b \ln x \tag{5.45}$$

This is in the form

$$z = a_1 t + a_0 \tag{5.46}$$

where $z = \ln(y-c)$, $a_0 = \ln a$, $a_1 = b$, $t = \ln x$. This linear expression in t for z is solved by the least squares method, as done before, and a_1, a_0 are evaluated. From a_1, a_0, we can get a, b values. Using (5.46) in (5.18) and (5.19), the normal equations for a geometric curve fit are obtained as

$$n \ln a + (\sum \ln x_i) b = \sum \ln(y_i - c) \tag{5.47}$$

$$(\sum \ln x_i) \ln a + \sum (\ln x_i)^2 \times b = \sum \ln x_i \times \ln(y_i - c) \tag{5.48}$$

From (5.20) and (5.21) the constants are evaluated as

$$a_0 = \ln a = \frac{\sum \ln(y_i - c) \times \sum (\ln x_i)^2 - \sum \ln x_i \sum \ln x_i \times \ln(y_i - c)}{n \sum (\ln x_i)^2 - (\sum \ln x_i)^2} \tag{5.49}$$

$$a_1 = b = \frac{n \sum \ln x_i \times \ln(y_i - c) - \sum \ln x_i \sum \ln(y_i - c)}{n \sum (\ln x_i)^2 - (\sum \ln x_i)^2} \tag{5.50}$$

Example 6

Compute a geometric curve fit for the following data points

x	1	2	3	4	5
y	1.7	3.894	6.735	10.1	13.916

Solution: Here $n = 5$. Let us choose $c = 0.5$.

	x_i	y_i	$\ln x_i$	$z_i = \ln(y_i - c)$	$(\ln x_i)^2$	$\ln(x_i)\ln(y_i - c)$
	1	1.7	0	0.1823	0	0
	2	3.894	0.6932	1.2220	0.4805	0.8471
	3	6.735	1.0986	1.8302	1.2069	2.0107
	4	10.1	1.3863	2.2618	1.9218	3.1355
	5	13.916	1.6094	2.5965	2.5902	4.1788
SUM	15	36.345	4.7875	8.0928	6.1994	10.1721

Using (5.49) and (5.50), the constants are obtained as

$$a_0 = \frac{(8.0928)(6.1994) - (4.7875)(10.1721)}{(5)(6.1994) - (4.7875)^2} = \frac{1.4716}{8.0768} = 0.1822$$

$$a_1 = \frac{(5)(10.1721) - (4.7875)(8.0928)}{8.0768} = 1.5001$$

Thus

$$a = \exp(a_0) = e^{0.1822} = 1.1999$$

$$b = a_1 = 1.5001$$

and the fitting curve is

$$y = 1.1999x^{1.5001} + 0.5$$

Fitting a Trigonometric Function

This is best suited for oscillating type of functions. A variation which involves trigonometric function is assumed first and the unknowns of the equation are determined by the method of least squares. Let us consider an equation

$$y = A\sin(\omega x + \phi) \tag{5.51}$$

to fit the given n data points, x_i, y_i. Assuming that ω is known, the other two unknown parameters, namely A and ϕ, are calculated as follows:

$$\begin{aligned} y &= A\sin(\omega x + \phi) \qquad [\phi \text{ is in radians}] \\ &= A\sin\omega x\cos\phi + A\cos\omega x\sin\phi \\ &= a_1\sin\omega x + a_2\cos\omega x \end{aligned} \tag{5.52}$$

where $a_1 = A\cos\phi$ and $a_2 = A\sin\phi$. Using (5.19), we can write the metric d and therefore S as

$$S = \sum_{i=1}^{n} d_i^2 = \sum_i (y_i - a_1\sin\omega x_i - a_2\cos\omega x_i)^2 \tag{5.53}$$

To minimize S, equate the partial derivatives of S with respect to a_1 and a_2 to zero and solve for a_1, a_2. Thus we have

$$\frac{\partial S}{\partial a_1} = 0$$

$$\Sigma 2(y_i - a_1\sin\omega x_i - a_2\cos\omega x_i)(-\sin\omega x_i) = 0$$

$$(\Sigma\sin^2\omega x_i)\,a_1 + (\Sigma\sin\omega x_i \times \cos\omega x_i)a_2 = \Sigma y_i\sin\omega x_i \tag{5.54}$$

Similarly,

$$\frac{\partial S}{\partial a_2} = 0$$

$$\Sigma 2(y_i - a_1\sin\omega x_i - a_2\cos\omega x_i)(-\cos\omega x_i) = 0$$

$$(\Sigma\sin\omega x_i \times \cos\omega x_i)a_1 + (\Sigma\cos^2\omega x_i)a_2 = \Sigma y_i\cos\omega x_i \tag{5.55}$$

Equations (5.54) and (5.55) should be solved to get a_1, a_2. These are two simultaneous linear equations in terms of two unknowns a_1, a_2 and can be solved easily, as was done earlier. From a_1, a_2, we obtain A and ϕ as shown below:

$$a_1 = A\cos\phi \quad \text{and} \quad a_2 = A\sin\phi$$

Therefore

$$A = \sqrt{a_1^2 + a_1^2}, \quad \phi = \tan^{-1}\left(\frac{a_2}{a_1}\right)$$

Hence we get the fitting curve as

$$y = A\sin(\omega x + \phi)$$

The algorithm and the program are a bit involved exercises, as the solution of the simultaneous equations is also to be brought into picture. These are therefore not attempted.

Other Functions for Curve Fitting

We have so far considered some important linear and nonlinear types of curve fitting, employing the method of least squares. In fact, the types of functions considered are not the only ones; there are various other types of functions available, in terms of which the data could be fitted. Some of them are: cubic splines, Chebyshev polynomial approximations and other orthogonal polynomials, trigonometric polynomials, power series expansions, etc. A detailed study is beyond the scope of the present coverage.

CONCEPT CHECK

1. Fit a straight line to the given points (x, y) by least squares. Show the details. Check your result by sketching the points and the line. Judge the goodness of fit.
   (a) (2, 0), (3, 4), (4, 10), (5, 16)
   (b) How does the line in part(a) change if you add a point far above it, say, (3, 20)?
   (c) (2.5, 8.0), (5.0, 6.9), (7.5, 6.2), (10.0, 5.0)
   (d) **(Ohm's law $U = Ri$)** Estimate the resistance R from the least squares line that fits $(i, U) = (2.0, 104), (4.0, 206), (6.0, 314), (10.0, 530)$.
   (e) **(Average speed)** Estimate the average speed v_{av} of a car traveling according to $s = v \times t$ [km] (s = distance traveled, t [h] = time) from $(t, s) = (9, 140), (10, 220), (11, 310), (12, 410)$.
   (f) **(Hooke's law $F = ks$)** Estimate the spring modulus k from the force F [lb] and extension s [cm], where $(F, s) = (1, 0.50), (2, 1.02), (4, 1.99), (6, 3.01), (10, 4.98), (20, 10.03)$.
2. Derive the normal equations (5.12).
3. Fit a parabola (5.11) to the given points (x, y) by least squares. Check by sketching.
   (a) (−1, 3), (0, 0), (1, 2), (2, 8)
   (b) (0, 4), (2, 2), (4, −1), (6, −5)
   (c)

Worker's time on duty x [h]	1	2			
Worker's reaction time [sec]	1.50	1.28	1.40	1.85	2.20

   (d) Fit (5.7) and (5.11) by least squares to (−1.0, 5.4), (−0.5, 4.1), (0, 3.9), (0.5, 4.8), (1.0, 6.3), (1.5, 9.3). Graph the data and the curves on common axes and comment.
   (e) **(Cubic parabola)** Derive the formula for the normal equations of a cubic least squares parabola.
   (f) Fit curves (5.7) and (5.11) and a cubic parabola by least squares to (−2, −35), (−1, −9), (0, −1), (1, −1), (2, 17), (3, 63). Graph the three curves and the points on common axes. Comment on the goodness of fit.
4. **TEAM PROJECT. The least squares approximation of a function $f(x)$ on an interval $a \leqq x \leqq b$ by a function**

$$F_m(x) = a_0 y_0(x) + a_1 y_1(x) + \cdots + a_m y_m(x)$$

where $y_0(x), \ldots, y_m(x)$ are given functions, requires the determination of the coefficients $a_0, \ldots, a_m$ such that

$$\int_a^b [f(x) - F_m(x)]^2\, dx \qquad (5.56)$$

becomes minimum. This integral is denoted by $\|f - F_m\|^2$, and $\|f - F_m\|$ is called the L_2-norm of $f - F_m$ (L suggesting Lebesgue). A necessary condition for that minimum is given by $\partial\|f - F_m\|^2/\partial a_j = 0$, $j = 0, \ldots, m$ [the analog of (5.10)].

   (a) Show that this leads to $m + 1$ normal equations $(j = 0, \ldots, m)$

$$\sum_{k=0}^{m} h_{jk} a_k = b_j \qquad (5.57)$$

where

$$h_{jk} = \int_a^b y_j(x) y_k(x) dx$$

$$b_j = \int_a^b f(x) y_j(x)\, dx$$

(b) **Polynomial.** What form does (5.57) take if $F_m(x) = a_0 + a_1x + \cdots + a_mx^m$? What is the coefficient matrix of (5.57) in this case when the interval is $0 \leqq x \leqq 1$?

(c) **Orthogonal functions.** What are the solutions of (5.57) if $y_0(x), \ldots, y_m(x)$ are orthogonal on the interval $a \leqq x \leqq b$?

5. Obtain an appropriate linear least squares fit for the data given below:

x	1	2	3	4	6	8
y	2.4	3.1	3.5	4.2	5	6

6. Fit a second degree polynomial for the data given below:

x	−2	−1	0	1	2
y	15	1	1	3	19

7. Fit a function of the form $y = ae^{bx}$ to the following data:

x	0	1/2	1	3/2	2	5/2
y	0.1	0.45	2.15	9.15	40.35	180.75

8. Derive the necessary equations, and fit the following data to a curve of the type:

$$y = \frac{1}{a + bx}$$

x	0	2	4	6	8
y	11	2.20	1	0.82	0.56

9. Derive the normal equations to fit the parabola $y = a + bx + cx^2$. [JNTU May 2006]

10. Fit a straight line for the following data

x	6	7	7	8	8	8	9	9	10
y	5	5	4	5	4	3	4	3	3

[JNTU Sep. 2008]

11. A chemical company, wishing to study the effect of extraction time (x) on the efficiency (y) of an extraction operation, obtained the data shown in the following table.

x	27	45	41	19	3	39	19	49	15	31
y	57	64	80	46	62	72	52	77	57	68

Fit a straight line to the given data by the method of least squares. [JNTU Sep. 2004]

12. Fit a straight line to the form $y = a + bx$ for the following data.

x	0	5	10	15	20	25
y	12	15	17	22	24	30

[JNTU May 2007]

13. By the method of least squares fit a parabola of the form $y = a + bx + cx^2$ for the following data:

x	2	4	6	8	10
y	3.07	12.85	31.47	57.38	91.29

[JNTU Sep. 2008]

14. Fit the curve of the form $y = ae^{bx}$.

x	77	100	185	239	285
y	2.4	3.4	7.0	11.1	19.6

[JNTU Sep. 2006]

5.3 Weighted Least Squares Approximation

It is a more general approach to minimize the weighted sum of the squares of the errors taken over all data points. If this sum is denoted by S, we have

$$S = W_1[y_1 - f(x_1)]^2 + W_2[y_2 - f(x_2)]^2 + \cdots + W_n[y_n - f(x_n)]^2$$

$$S = W_1e_1^{\,2} + W_2e_2^{\,2} + \cdots + W_ne_n^{\,2}$$

where W_i, $i = 1, 2, 3, \ldots, N$, are positive numbers and are known as weights. A weight is prescribed according to the relative accuracy of a data point. If all the data points are accurate, we set $W_i = 1$, for all i.

We consider the linear and nonlinear cases in the following subsections.

Linear Weighted Least Squares Approximation

Let

$$Y = a + bx \tag{5.58}$$

be the straight line to be fitted to the given data points $(x_1, y_1), (x_2, y_2), \ldots, (x_n, y_n)$. Then

$$s = \sum_{i=1}^{n} W_i [y_i - (a + bx_i)]^2 \tag{5.59}$$

For maxima and minima, we have

$$\frac{\partial s}{\partial a} = \frac{\partial s}{\partial b} = 0$$

$$\frac{\partial s}{\partial a} = -2\sum_{i=1}^{n} W_i [y_i - (a + bx_i)] = 0 \tag{5.60}$$

$$\frac{\partial s}{\partial b} = -2\sum_{i=1}^{n} W_i [y_i - (a + bx_i)] x_i = 0 \tag{5.61}$$

Simplifying (5.60) and (5.61), we get

$$\sum_{i=1}^{n} W_i y_i = a\sum_{i=1}^{n} W_i + b\sum_{i=1}^{n} W_i x_i \tag{5.62}$$

and

$$\sum_{i=1}^{n} W_i x_i y_i = a\sum_{i=1}^{n} W_i x_i + b\sum_{i=1}^{n} W_i x_i^2 \tag{5.63}$$

Equations (5.62) and (5.63) are known as normal equations and solving these normal equations, we get the values of a and b. Substituting these values in (5.58), we get the required least squares approximation.

Nonlinear Weighted Least Squares Approximation

Consider the polynomial

$$y = a_0 + a_1 x + a_2 x^2 + \cdots + a_k x^k$$

which satisfies the data points (x_i, y_i), $i = 1, 2, \ldots, n$. Then we have

$$S = \sum_{i=1}^{n} W_i [y_i - (a_0 + a_1 x + \cdots + a_k x^k)]$$

For the error S to be minimum, the following condition should be satisfied:

$$\frac{\partial s}{\partial a_0} = \frac{\partial s}{\partial a_1} = \cdots = \frac{\partial s}{\partial a_k} = 0$$

When $\partial s / \partial a_0 = 0$, we have

$$a_0\sum_{i=1}^{n} W_i + a_1\sum_{i=1}^{n} W_i x_i + \cdots + a_k\sum_{i=1}^{n} W_i x_i^k = \sum_{i=1}^{n} W_i y_i$$

When $\partial s/\partial a_1 = 0$, we have

$$a_0 \sum_{i=1}^{n} W_i x_i + a_1 \sum_{i=1}^{n} W_i x_i^2 + \cdots + a_k \sum_{i=1}^{n} W_i x_i^{k+1} = \sum_{i=1}^{n} W_i x_i y_i$$

Proceeding this way we have that when $\partial s/\partial a_k = 0$,

$$a_0 \sum_{i=1}^{n} W_i x_i^k + a_1 \sum_{i=1}^{n} W_i x_i^{k+1} + \cdots + a_k \sum_{i=1}^{n} W_i x_i^{2k} = \sum_{i=1}^{n} W_i x_i^k y_i$$

Solving these equations, the constants, $a_0, a_1, \ldots, a_k$ can be obtained.

Example 7

Fit a straight line to the following data giving weights to x as 2, 2, 1, 1, 2 by the method of least squares.

Solution: Let the straight line be

$$y = a + bx \tag{5.64}$$

For the linear weighted least square approximation, the normal equations are:

$$\sum wy = a\sum w + b\sum wx \tag{5.65}$$

$$\sum wxy = a\sum wx + b\sum wx^2 \tag{5.66}$$

The calculations are tabulated as follows:

	x	y	w	wx	wxy	wx^2	wy
	0	1	2	0	0	0	2
	1	1.8	2	2	3.6	2	3.6
	2	3.3	1	2	6.6	4	3.3
	3	4.5	1	3	9.0	9	4.5
	4	6.5	2	8	52.0	32	13.0
SUM	10	16.9	8	15	71.2	47	26.4

Substituting these values in (5.65) and (5.66) respectively we have

$$26.4 = 8a + 15b$$
$$71.2 = 15a + 47b$$

Solving for a and b we get $a = 1.14$ and $b = 1.15$. Substituting these values in (5.64) we get

$$y = 1.14 + 1.15x$$

which is required equation.

Example 8

Fit a second-degree parabola to the following data giving weights 2, 1, 2, 1, 2 to x by the method of least squares.

Solution: Let the equation of parabola be

$$y = a + bx + cx^2 \tag{5.67}$$

For a nonlinear weighted least square approximation, the normal equations are:

$$\sum wy = a\sum w + b\sum wx + c\sum wx^2 \tag{5.68}$$

$$\sum wxy = a\sum wx + b\sum wx^2 + c\sum wx^3 \tag{5.69}$$

$$\sum wx^2y = a\sum wx^2 + b\sum wx^3 + c\sum wx^4 \tag{5.70}$$

The calculations are tabulated as follows:

x	y	w	$X = x - 2$	wx	wy	wx^2	wxy	wx^2y	wx^3	wx^4
0	1.9	2	−2	−4	3.8	8	−7.6	15.2	−168	32
1	2.8	1	−1	−1	2.8	1	−2.8	2.8	−1	1
2	3.7	2	0	0	7.4	0	7.4	0	0	0
3	4.5	1	1	1	4.5	1	4.5	4.5	1	1
4	6.6	2	2	4	13.2	8	52.8	25.2	16	32
SUM		8	0	0	14.7	11	20.5	75.3	0	66

Substituting these values in (5.68)–(5.70), we have

$$\begin{aligned} 31.7 &= 8a + 0 + 18c \\ 20.5 &= 0 + 18b + 0 \\ 75.3 &= 18a + 0 + 66c \end{aligned}$$

Solving these equations, we get

$$a = 3.6118,\ b = 1.1389,\ c = 0.1559$$

The required parabola is

$$\begin{aligned} y &= 3.6118 + 1.1389X + 0.1559X^2 \\ &= 3.6118 + 1.1389\,(x - 2) + 0.1559\,(x - 2)^2 \\ &= 3.6118 + 1.1389x - 2.2764 + 0.1559(x^2 - 4x + 4) \\ &= 3.6118 + 1.1389x - 2.2764 + 0.1559x^2 - 0.6236x + 0.6236 \\ &= 1.959 + 0.5153x + 0.1559x^2 \end{aligned}$$

Therefore the required equation is

$$y = 1.959 + 0.5153x + 0.1559x^2$$

5.4 Numeric Integration

Numeric integration means the numeric evaluation of integrals

$$J = \int_a^b f(x)dx$$

where a and b are given and f is a function given analytically by a formula or empirically by a table of values. Geometrically, J is the area under the curve of f between a and b (Figure 5).

We know that if f is such that we can find a differentiable function F whose derivative is f, then we can evaluate J by applying the familiar formula

$$J = \int_a^b f(x)dx = F(b) - F(a) \quad [\text{where } F'(x) = f(x)]$$

However, applications often lead to integrals whose analytic evaluation would be very difficult or even impossible, or whose integrand is an empirical function given by recorded numeric values. Then we may obtain approximate numeric values of the integral by a numeric integration method.

Rectangular Rule and Trapezoidal Rule

Numeric integration methods are obtained by approximating the integrand f by functions that can easily be integrated. The simplest formula, the **rectangular rule**, is obtained if we subdivide the interval of integration $a \leqq x \leqq b$ into n subintervals of equal length $h = (b-a)/n$ and in each subinterval approximate f by the constant $f(x_j^*)$, the value of f at the midpoint (x_j^*) of the jth subinterval (Figure 6). Then f is approximated by a **step function** (piecewise constant function), the n rectangles in Figure 6 have the areas $f(x_1^*)h, \ldots, f(x_n^*)h$, and the **rectangular rule** is

$$J = \int_a^b f(x)dx \approx h[f(x_1^*) + f(x_2^*) + \cdots + f(x_n^*)] \qquad \left(h = \frac{b-a}{n}\right) \tag{5.71}$$

The **trapezoidal rule** is generally more accurate. We obtain it if we take the same subdivision as before and approximate f by a broken line of segments (chords) with endpoints $[a, f(a)]$, $[x_1, f(x_1)]$, ..., $[b, f(b)]$ on the curve of f (Figure 7). Then the area under the curve of f between a and b is approximated by n trapezoids of areas

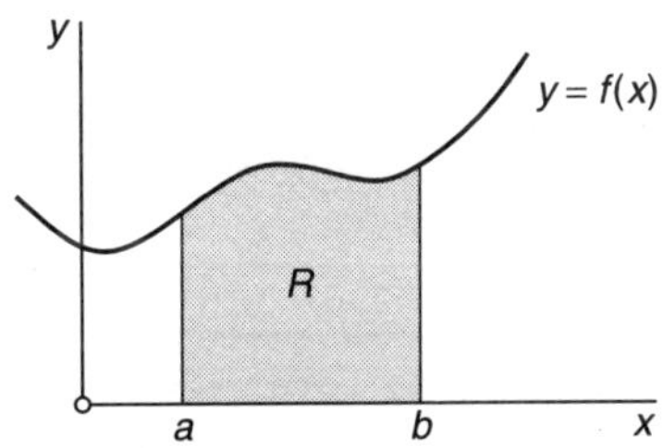

Figure 5 Geometric interpretation of a definite integral.

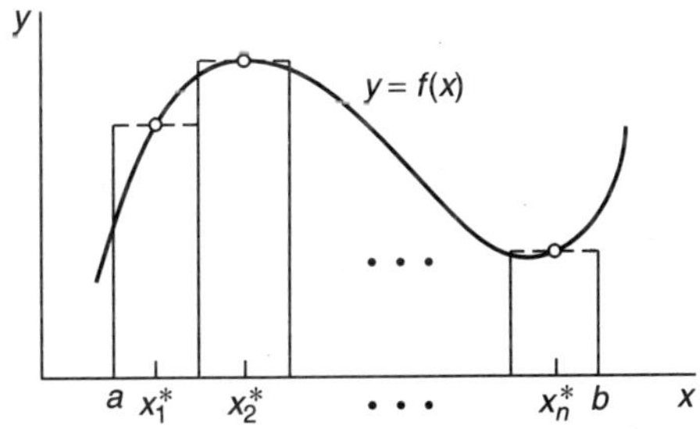

Figure 6 Rectangular rules.

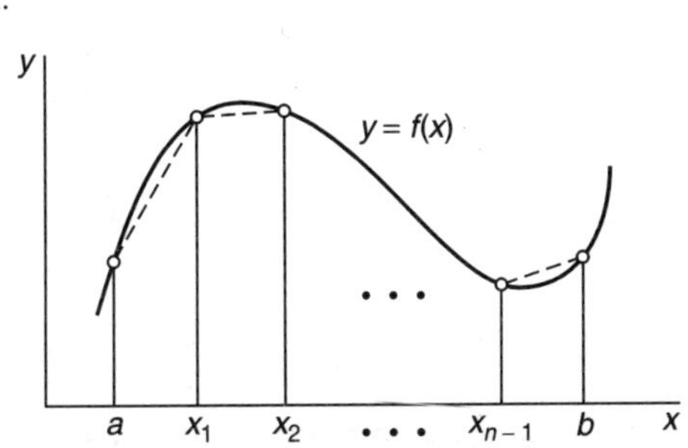

Figure 7 Trapezoidal rule.

$$\frac{1}{2}[f(a)+f(x_1)]h, \quad \frac{1}{2}[f(x_1)+f(x_2)]h, \quad \ldots, \quad \frac{1}{2}[f(x_{n-1})+f(b)]h$$

By taking their sum we obtain the **trapezoidal rule**

$$\int_a^b f(x)dx \approx h\left[\frac{1}{2}f(a)+f(x_1)+f(x_2)+\cdots+f(x_{n-1})+\frac{1}{2}f(b)\right] \tag{5.72}$$

where $h=(b-a)/n$, as in (5.71). The x_j's and a and b are called **nodes**.

Example 9 Trapezoidal Rule

Evaluate $J=\int_0^1 e^{-x^2}\,dx$ by means of (5.72) with $n=10$.

Solution: $J \approx 0.1(0.5\times 1.367879+6.778167)=0.746211$ from Table 1.

Table 1 Computations in Example 9

j	x_j	$x_j^{\,2}$	$e^{-x_j^2}$	
0	0	0	1.000000	
1	0.1	0.01		0.990050
2	0.2	0.04		0.960789
3	0.3	0.09		0.913931
4	0.4	0.16		0.852144
5	0.5	0.25		0.778801
6	0.6	0.36		0.697676
7	0.7	0.49		0.612626
8	0.8	0.64		0.527292
9	0.9	0.81		0.444858
10	1.0	1.00	0.367879	
SUM			1.367879	6.778167

Error Bounds and Estimate for the Trapezoidal Rule

An error estimate for the trapezoidal rule can be derived from the error estimate of "General Lagrange Interpolating Polynomial" explained in Chapter 4 with $n=1$ by integration as follows. For a single subinterval we have

$$f(x)-p_1(x)=(x-x_0)(x-x_1)\frac{f''(t)}{2}$$

with a suitable t depending on x, between x_0 and x_1. Integration over x from $a=x_0$ to $x_1=x_0+h$ gives

$$\int_{x_0}^{x_0+h} f(x)dx-\frac{h}{2}[f(x_0)+f(x_1)]=\int_{x_0}^{x_0+h}(x-x_0)\left((x-x_0-h)\frac{f''(t(x))}{2}dx\right)$$

Setting $x-x_0=v$ and applying the mean value theorem of integral calculus, which we can use because $(x-x_0)(x-x_0-h)$ does not change sign, we find that the right side equals

$$\int_0^h v(v-h)\,dv\,\frac{f''(\tilde{t})}{2} = \left(\frac{h^3}{3} - \frac{h^3}{2}\right)\frac{f''(\tilde{t})}{2} = -\frac{h^3}{12} f''(\tilde{t}) \tag{5.73}$$

where $\tilde{t}$ is a (suitable, unknown) value between x_0 and x_1. This is the error for the trapezoidal rule with $n = 1$, often called the **local error**. Hence the **error** ϵ of (5.72) with any n is the sum of such contributions from the n subintervals; since $h = (b - a)/n$, $nh^3 = n(b - a)^3/n^3$, and $(b - a)^2 = n^2h^2$, we obtain

$$\epsilon = -\frac{(b-a)^3}{12n^2} f''(\hat{t}) = -\frac{(b-a)}{12} h^2 f''(\hat{t}) \tag{5.74}$$

with (suitable, unknown) $\hat{t}$ between a and b. Because of (5.74) the trapezoidal rule (5.72) is also written as

$$J = \int_a^b f(x)dx = h\left[\frac{1}{2}f(a) + f(x_1) + \cdots + f(x_{n-1}) + \frac{1}{2}f(b)\right] - \frac{b-a}{12}h^2 f''(\hat{t}) \tag{5.75}$$

Error Bounds are now obtained by taking the largest value for f'', say, M_2, and the smallest value, M_2^*, in the interval of integration. Then (5.74) gives (note that K is negative)

$$KM_2 \leqq \epsilon \leqq KM_2^* \quad \text{where} \quad K = -\frac{(b-a)^3}{12n^2} = -\frac{b-a}{12}h^2 \tag{5.76}$$

Error Estimation by Halving h is advisable if h'' is very complicated or unknown, for instance, in the case of experimental data. Now we calculate by (5.72), first with h, obtaining, say, $J = J_h + \epsilon_h$, and then with $h/2$, obtaining $J = J_{h/2} + \epsilon_{h/2}$. Now if we replace h^2 in (5.74) with $(h/2)^2$, the error is multiplied by 1/4. Hence $\epsilon_{h/2} \approx \epsilon_h/4$ (not exactly because $\hat{t}$ may differ). Together, $J_{h/2} + \epsilon_{h/2} = J_h + \epsilon_h \approx J_h + 4\epsilon_{h/2}$. Thus $J_{h/2} - J_h = (4 - 1)\epsilon_{h/2}$. Division by 3 gives the error formula for $J_{h/2}$

$$\epsilon_{h/2} \approx \frac{1}{3}(J_{h/2} - J_h) \tag{5.77}$$

Example 10 Error Estimation for the Trapezoidal Rule by (5.76) and (5.77)

Estimate the error of the approximate value in Example 9 by (5.76) and (5.77).

Solution:

(a) ***Error bounds by*** **(5.76).** By differentiation, $f''(x) = 2(2x^2 - 1)e^{-x^2}$. Also, $f'''(x) > 0$ if $0 < x < 1$, so that the minimum and maximum occur at the ends of the interval. We compute $M_2 = f''(1) = 0.735759$ and $M_2^* = f''(0) = -2$. Furthermore, $K = -1/1200$, and (5.76) gives

$$-0.000614 \leqq \epsilon \leqq 0.001667$$

Hence the exact value of J must lie between

$$0.746211 - 0.000614 = 0.745597 \quad \text{and} \quad 0.746211 + 0.001667 = 0.747878$$

Actually, $J = 0.746\,824$, exact to 6D.

(b) ***Error estimate by*** **(5.77).** $J_h = 0.746211$ in Example 9. Also,

$$J_{h/2} = 0.05\left[\sum_{j=1}^{19} e^{-(j/20)^2} + \frac{1}{2}(1 + 0.367879)\right] = 0.746671$$

Hence $\epsilon_{h/2} = (J_{h/2} - J_h)/3 = 0.00153$ and $J_{h/2} + \epsilon_{h/2} = 0.746824$, exact to 6D.

Simpson's Rule of Integration

Piecewise constant approximation of f led to the rectangular rule (5.71), piecewise linear approximation to the trapezoidal rule (5.72), and piecewise quadratic approximation will lead to Simpson's rule, which is of great practical importance because it is sufficiently accurate for most problems, but still sufficiently simple.

To derive Simpson's rule, we divide the interval of integration $a \leqq x \leqq b$ into an ***even number*** of equal subintervals, say, into $n = 2m$ subintervals of length $h = (b - a)/(2m)$, with endpoints x_0 $(= a)$, $x_1, \ldots, x_{2m-1}, x_{2m}$ $(= b)$; see Figure 8. We now take the first two subintervals and approximate $f(x)$ in the interval $x_0 \leqq x \leqq x_2 = x_0 + 2h$ by the Lagrange polynomial $p_2(x)$ through (x_0, f_0), (x_1, f_1), (x_2, f_2), where $f_j = f(x_j)$. From (5.74) we obtain

$$p_2(x) = \frac{(x-x_1)(x-x_2)}{(x_0-x_1)(x_0-x_2)} f_0 + \frac{(x-x_0)(x-x_2)}{(x_1-x_0)(x_1-x_2)} f_1 + \frac{(x-x_0)(x-x_1)}{(x_2-x_0)(x_2-x_1)} f_2 \tag{5.78}$$

The denominators in (5.78) are $2h^2$, $-h^2$, and $2h^2$, respectively. Setting $s = (x - x_1)/h$, we have

$$x - x_1 = sh, \qquad x - x_0 = x - (x_1 - h) = (s+1)h$$

$$x - x_2 = x - (x_1 + h) = (s-1)h$$

and we obtain

$$p_2(x) = \frac{1}{2} s(s-1) f_0 - (s+1)(s-1) f_1 + \frac{1}{2}(s+1) s f_2$$

We now integrate with respect to x from x_0 to x_2. This corresponds to integrating with respect to s from -1 to 1. Since $dx = h\,ds$, the result is

$$\int_{x_0}^{x_2} f(x)dx \approx \int_{x_0}^{x_2} p_2(x)dx = h\left(\frac{1}{3} f_0 + \frac{4}{3} f_1 + \frac{1}{3} f_2\right) \tag{5.79}$$

A similar formula holds for the next two subintervals from x_2 to x_4, and so on. By summing all these m formulas we obtain **Simpson's rule**

$$\int_a^b f(x)dx \approx \frac{h}{3}(f_0 + 4f_1 + 2f_2 + 4f_3 + \cdots + 2f_{2m-2} + 4f_{2m-1} + f_{2m}) \tag{5.80}$$

where $h = (b-a)/(2m)$ and $f_j = f(x_j)$. We now show an algorithm for Simpson's rule.

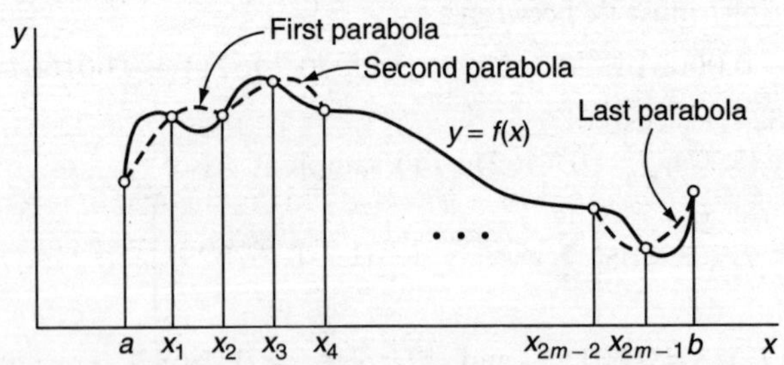

Figure 8 Simpson's rule.

ALGORITHM SIMPSON $(a, b, m, f_0, f_1, \ldots, f_{2m})$

This algorithm computes the integral $J = \int_a^b f(x)\,dx$ from given values $f_j = f(x_j)$ at equidistant $x_0 = a$, $x_1 = x_0 + h$, ..., $x_{2m} = x_0 + 2mh = b$ by Simpson's rule (5.80), where $h = (b - a)/(2m)$.

INPUT: $a, b, m, f_0, \ldots, f_{2m}$

OUTPUT: Approximate value $\tilde{J}$ of J

Compute

$$s_0 = f_0 + f_{2m}$$
$$s_1 = f_1 + f_3 + \cdots + f_{2m-1}$$
$$s_2 = f_2 + f_4 + \cdots + f_{2m-2}$$
$$h = (b-a)/2m$$
$$\tilde{J} = \frac{h}{3}(s_0 + 4s_1 + 2s_2)$$

OUTPUT $\tilde{J}$ Stop.

End SIMPSON

Error of Simpson's Rule (5.80)

If the fourth derivative $f^{(4)}$ exists and is continuous on $a \leqq x \leqq b$, the **error** of (5.80), call it ϵ_s, is

$$\epsilon_s = -\frac{(b-a)^5}{180(2m)^4} f^{(4)}(\hat{t}) = -\frac{(b-a)}{180} h^4 f^{(4)}(\hat{t}) \tag{5.81}$$

here $\hat{t}$ is a suitable unknown value between a and b. This is obtained similarly to (5.74). With this we may also write Simpson's rule (5.80) as

$$\int_a^b f(x)dx = \frac{h}{3}(f_0 + 4f_1 + \cdots + f_{2m}) - \frac{(b-a)}{180} h^4 f^{(4)}(\hat{t}) \tag{5.82}$$

Error Bounds

By taking for $f^{(4)}$ in (5.81) the maximum M_4 and minimum M_4^* on the interval of integration we obtain from (5.81) the error bounds (note that C is negative)

$$CM_4 \leqq \epsilon_s \leqq CM_4^* \quad \text{where} \quad C = -\frac{(b-a)^5}{180(2m)^4} = -\frac{(b-a)}{180} h^4 \tag{5.83}$$

Degree of Precision (DP) *of an integration formula.* This is the maximum degree of arbitrary polynomials for which the formula gives exact values of integrals over any intervals. Hence for the trapezoidal rule,

$$\text{DP} = 1$$

because we approximate the curve of f by portions of straight lines (linear polynomials).

For Simpson's rule we might expect DP = 2 (why?). Actually,

$$\text{DP} = 3$$

by (5.83) because $f^{(4)}$ is identically zero for a cubic polynomial. This makes Simpson's rule sufficiently accurate for most practical problems and accounts for its popularity.

Numeric Stability *with respect to rounding* is another important property of Simpson's rule. Indeed, for the sum of the roundoff errors ϵ_j of the $2m + 1$ values f_j in (5.80) we obtain, since $h = (b - a)/2m$,

$$\frac{h}{3}|\epsilon_0 + 4\epsilon_1 + \cdots + \epsilon_{2m}| \leqq \frac{(b-a)}{3 \times 2m} 6mu = (b-a)u$$

where u is the rounding unit [$u = (1/2) \times 10^{-6}$ if we round off to 6D]. Also $6 = 1 + 4 + 1$ is the sum of the coefficients for a pair of intervals in (5.80); take $m = 1$ in (5.80) to see this. The bound $(b - a)u$ is independent of m, so that it cannot increase with increasing m, that is, with decreasing h. This proves stability.

Newton–Cotes Formulas (General Quadrature)

We mention that the trapezoidal and Simpson rules are special *closed Newton–Cotes formulas,* that is, integration formulas in which $f(x)$ is interpolated at equally spaced nodes by a polynomial of degree n ($n = 1$ for trapezoidal, $n = 2$ for Simpson), and **closed** means that a and b are nodes ($a = x_0$, $b = x_n$). $n = 3$ (the three-eighths rule) and a higher n are used occasionally. From $n = 8$ on, some of the coefficients become negative, so that a positive f_j could make a negative contribution to an integral, which is absurd.

Example 11 Simpson's Rule. Error Estimate

Evaluate

$$J = \int_0^1 e^{-x^2}\, dx$$

by Simpson's rule with $2m = 10$ and estimate the error.

Solution: Since $h = 0.1$, Table 2 gives

$$J \approx \frac{0.1}{3}(1.367879 + 4 \times 3.740266 + 2 \times 3.037901) = 0.746825$$

Estimate of error. Differentiation gives $f^{(4)}(x) = 4(4x^4 - 12x^2 + 3)e^{-x^2}$. By considering the derivative $f^{(5)}$ of $f^{(4)}$ we find that the largest value of $f^{(4)}$ in the interval of integration occurs at 0 and the smallest value at $x^* = (2.5 - 0.5\sqrt{10})^{1/2}$. Computation gives the values $M_4 = f^{(4)}(0) = 12$ and $M_4^* = f^{(4)}(x^*) = -7.419$. Since $2m = 10$ and $b - a = 1$, we obtain $C = -1/1800000 = -0.00000056$. Therefore, from (5.83),

$$-0.000007 \leqq \epsilon_s \leqq 0.000005$$

Hence J must lie between $0.746825 - 0.000007 = 0.746818$ and $0.746825 + 0.000005 = 0.746830$, so that at least four digits of our approximate value are exact. Actually, the value 0.746825 is exact to 5D because $J = 0.746824$ (exact to 6D). Thus our result is much better than that in Example 9 obtained by the trapezoidal rule, whereas the number of operations is nearly the same in both cases.

Table 2 Computations in Example 11

	j	x_j	x_j^2	$e^{-x_j^2}$		
	0	0	0	1.000000		
	1	0.1	0.01		0.990050	
	2	0.2	0.04			0.960789
	3	0.3	0.09		0.913931	
	4	0.4	0.16			0.852144
	5	0.5	0.25		0.778801	
	6	0.6	0.36			0.697676
	7	0.7	0.49		0.612626	
	8	0.8	0.64			0.527292
	9	0.9	0.81		0.444858	
	10	1.0	1.00	0.367879		
SUM				1.367879	3.740266	3.037901

Instead of picking an $n = 2m$ and then estimating the error by (5.83), as in Example 11, it is better to require an accuracy (e.g., 6D) and then determine $n = 2m$ from (5.83).

Example 12 Determination of $n = 2m$ in Simpson's Rule from the Required Accuracy

What n should we choose in Example 11 to get 6D-accuracy?

Solution: Using $M_4 = 12$ (which is bigger in absolute value than M_4^*), we get from (5.83), with $b - a = 1$ and the required accuracy,

$$|CM_4| = \frac{12}{180(2m)^4} = \frac{1}{2}\times 10^{-6}, \quad \text{thus} \quad m = \left[\frac{2\times 10^6 \times 12}{180\times 2^4}\right]^{1/4} = 9.55$$

Hence we should choose $n = 2m = 20$. Do the computation, which parallels that in Example 11.

Note that the error bounds in (5.76) or (5.83) may sometimes be loose, so that in such a case a smaller $n = 2m$ may already suffice.

Error Estimation for Simpson's Rule by Halving h

The idea is the same as in (5.77) and gives

$$\epsilon_{h/2} \approx \frac{1}{15}(J_{h/2} - J_h) \tag{5.84}$$

J_h is obtained by using h and $J_{h/2}$ by using $h/2$, and $\epsilon_{h/2}$ is the error of $J_{h/2}$.

Derivation. In (5.77) we had 1/3 as the reciprocal of 3 = 4 − 1 and $1/4 = (1/2)^2$ resulted from h^2 in (5.74) by replacing h with $h/2$. In (5.84) we have 1/15 as the reciprocal of 15 = 16 − 1 and $1/16 = (1/2)^4$ results from h^4 in (5.81) by replacing h with $h/2$.

Example 13 Error Estimation for Simpson's Rule by Halving

Integrate

$$f(x) = \frac{1}{4}\pi x^4 \cos\frac{1}{4}\pi x$$

from 0 to 2 with $h = 1$ and apply (5.84).

Solution: The exact 5D-value of the integral is $J = 1.25953$. Simpson's rule gives

$$J_h = \frac{1}{3}[f(0) + 4f(1) + f(2)] = \frac{1}{3}(0 + 4 \times 0.555360 + 0) = 0.740480$$

$$J_{h/2} = \frac{1}{6}\left[f(0) + 4f\left(\frac{1}{2}\right) + 2f(1) + 4f\left(\frac{3}{2}\right) + f(2)\right]$$

$$= \frac{1}{6}[0 + 4 \times 0.045351 + 2 \times 0.555361 + 4 \times 1.521579 + 0] = 1.22974$$

Hence (5.84) gives $\epsilon_{h/2} = (1.22974 - 0.74048)/15 = 0.032617$ and thus $J \approx J_{h/2} + \epsilon_{h/2} = 1.26236$, with an error -0.00283, which is less in absolute value than 1/10 of the error 0.02979 of $J_{h/2}$. Hence the use of (5.84) was well worthwhile.

Adaptive Integration

The idea is to adapt step h to the variability of $f(x)$. That is, where f varies but little, we can proceed in large steps without causing a substantial error in the integral, but where f varies rapidly, we have to take small steps in order to stay everywhere close enough to the curve of f.

Changing h is done systematically, usually by halving h, and automatically (not "by hand") depending on the size of the (estimated) error over a subinterval. The subinterval is halved if the corresponding error is still too large, that is, larger than a given **tolerance** TOL (maximum admissible absolute error), or is not halved if the error is less than or equal to TOL.

Adapting is one of the techniques typical of modern software. In connection with integration it can be applied to various methods. We explain it here for Simpson's rule. In Table 3 a star means that for that subinterval, TOL has been reached.

Example 14 Adaptive Integration with Simpson's Rule

Integrate

$$f(x) = \frac{1}{4}\pi x^4 \cos\frac{1}{4}\pi x$$

from $x = 0$ to 2 by adaptive integration and with Simpson's rule and TOL [0, 2] = 0.0002.

Solution: Table 3 shows the calculations. Figure 9 shows the integrand $f(x)$ and the adapted intervals used. The first two intervals ([0, 0.5], [0.5, 1.0]) have length 0.5, hence $h = 0.25$ [because we use $2m = 2$ subintervals in Simpson's rule (5.82)]. The next two intervals ([1.00, 1.25], [1.25, 1.50]) have length 0.25 (hence $h = 0.125$) and the last four intervals have length 0.125.

Sample computations. For 0.740480 see Example 13. Formula (5.84) gives $(0.123716 - 0.122794)/15 = 0.000061$. Note that 0.123716 refers to [0, 0.5] and [0.5, 1], so that we must subtract the value corresponding to [0, 1] in the line before. TOL [0, 2] = 0.0002 gives 0.0001 for subintervals of length 1, 0.00005 for length 0.5, etc. The value of the integral obtained is the sum of the values marked by an asterisk (for which the error estimate has become less than TOL). This gives

$$J \approx 0.123716 + 0.528895 + 0.388263 + 0.218486 = 1.25936$$

The exact 5D-value is $J = 1.25953$. Hence the error is 0.00017. This is about 1/200 of the absolute value of that in Example 13. Our more extensive computation has produced a much better result.

Table 3 Computations in Example 14

Interval	*Integral*	*Error* (5.84)	*TOL*	*Comment*
[0, 2]	0.740480		0.0002	
[0, 1]	0.122794			
[1, 2]	1.10695			
	Sum = 1.22974	0.032617	0.0002	Divide further
[0.0, 0.5]	0.004782			
[0.5, 1.0]	0.118934			
	Sum = 0.123716*	0.000061	0.0001	TOL reached
[1.0, 1.5]	0.528176			
[1.5, 2.0]	0.605821			
	Sum = 1.13300	0.001803	0.0001	Divide further
[1.00, 1.25]	0.200544			
[1.25, 1.50]	0.328351			
	Sum = 0.528895*	0.000048	0.00005	TOL reached
[1.50, 1.75]	0.388235			
[1.75, 2.00]	0.218457			
	Sum = 0.606692	0.000058	0.00005	Divide further
[1.500, 1.625]	0.196244			
[1.625, 1.750]	0.192019			
	Sum = 0.388263*	0.000002	0.000025	TOL reached
[1.750, 1.875]	0.153405			
[1.875, 2.000]	0.065078			
	Sum = 0.218483*	0.000002	0.000025	TOL reached

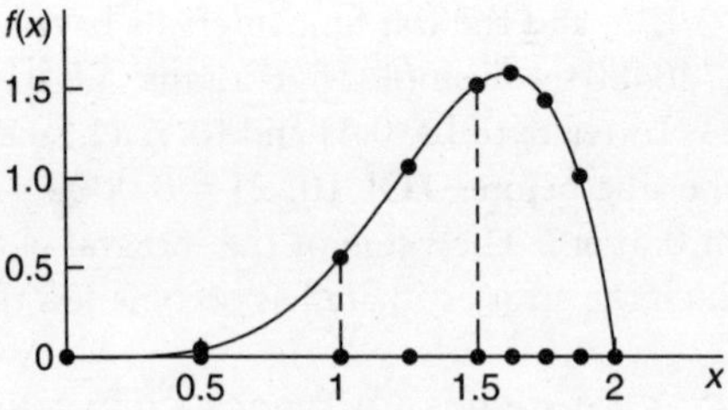

Figure 9 Adaptive integration in Example 14.

Gauss Integration Formulas

Maximum Degree of Precision

Our integration formulas discussed so far use function values at *predetermined* (equidistant) x values (nodes) and give exact results for polynomials not exceeding a certain degree [called the *degree of precision*]. But we can get much more accurate integration formulas as follows. We set

$$\int_{-1}^{1} f(t)dt \approx \sum_{j=1}^{n} A_j f_j \; [f_j = f(t_j)] \tag{5.85}$$

with fixed n, and $t = \pm 1$ obtained from $x = a, b$ by setting $x = [a(t-1) + b(t+1)]/2$. Then we determine the n coefficients $A_1, \ldots, A_n$ and n nodes $t_1, \ldots, t_n$ so that (5.85) gives exact results for polynomials of degree k as high as possible. Since $n + n = 2n$ is the number of coefficients of a polynomial of degree $2n - 1$, it follows that $k \leqq 2n - 1$.

Gauss has shown that exactness for polynomials of degree not exceeding $2n - 1$ (instead of $n - 1$ for predetermined nodes) can be attained, and he has given the location of the t_j (= the jth zero of the Legendre polynomial P_n) and the coefficients A_j which depend on n but not on $f(t)$, and are obtained by using Lagrange's interpolation polynomial. With these t_j and A_j, formula (5.85) is called a **Gauss integration formula** or *Gauss quadrature formula.* Its degree of precision is $2n - 1$, as just explained. Table 4 gives the values needed for $n = 2, \ldots, 5$.

Table 4 Gauss Integration: Nodes t_j and Coefficients A_j

n	*Nodes* t_j	*Coefficients* A_j	*Degree of precision*
	−0.5773502692	1	
2			3
	0.5773502692	1	
	−0.7745966692	0.5555555556	
3	0	0.8888888889	5
	0.7745966692	0.5555555556	

Table 4 (Continued)

n	*Nodes* t_j	*Coefficients* A_j	*Degree of precision*
	−0.8611363116	0.3478548451	
	−0.3399810436	0.6521451549	
4			7
	0.3399810436	0.6521451549	
	0.8611363116	0.3478548451	
	−0.9061798459	0.2369268851	
	−0.5384693101	0.4786286705	
5	0	0.5688888889	9
	0.5384693101	0.4786286705	
	0.9061798459	0.2369268851	

Example 15 Gauss Integration Formula with $n = 3$

Evaluate the integral in Example 11 by the Gauss integration formula (5.85) with $n = 3$.

Solution: We have to convert our integral from 0 to 1 into an integral from −1 to 1. We set $x = (t + 1)/2$ Then $dx = dt/2$, and (5.85) with $n = 3$ and the above values of the nodes and the coefficients yields

$$\int_0^1 \exp(-x^2)\,dx = \frac{1}{2}\int_{-1}^{1} \exp\left(-\frac{1}{4}(t+1)^2\right)dt$$

$$\approx \frac{1}{2}\left[\frac{5}{9}\exp\left\{-\frac{1}{4}\left(1\ \sqrt{\frac{3}{5}}\right)^2\right\} + \frac{8}{9}\exp\left(-\frac{1}{4}\right) + \frac{5}{9}\exp\left\{-\frac{1}{4}\left(1+\sqrt{\frac{3}{5}}\right)^2\right\}\right] = 0.746815$$

(exact to 6D: 0.746825), which is almost as accurate as the Simpson result obtained in Example 11 with a much larger number of arithmetic operations. With 3 function values (as in this example) and Simpson's rule we would get $1/6(1 + 4e^{-0.25} + e^{-1}) = 0.747180$, with an error over 30 times that of the Gauss integration.

Example 16 Gauss Integration Formula with $n = 4$ and 5

Integrate

$$f(x) = \frac{1}{4}\pi x^4 \cos\frac{1}{4}\pi x$$

from $x = 0$ to 2 by Gauss. Compare with the adaptive integration in Example 14 and comment.

Solution: Here $x = t + 1$ gives

$$f(t) = \frac{1}{4}\pi(t+1)^4 \cos\left[\frac{1}{4}\pi(t+1)\right]$$

as needed in (5.85). For $n = 4$, we calculate (6S)

$$J \approx A_1 f_1 + \cdots + A_4 f_4 = A_1(f_1 + f_4) + A_2(f_2 + f_3)$$
$$= 0.347855(0.000290309 + 1.02570) + 0.652145(0.129464 + 1.25459) = 1.25950$$

The error is 0.00003 because $J = 1.25953$ (6S). Calculating with 10S and $n = 4$ gives the same result; so the error is due to the formula, not rounding. For $n = 5$ and 10S we get $J \approx 1.259526185$, too large by the amount 0.000000250 because $J = 1.259525935$ (10S). The accuracy is impressive, particularly if we compare the amount of work with that in Example 14.

Gauss integration is of considerable practical importance. Whenever the integrand f is given by a formula (not just by a table of numbers) or when experimental measurements can be set at times t_j (or whatever t represents) shown in Table 3, then the great accuracy of Gauss integration outweighs the disadvantage of the complicated t_j and A_j (which may have to be stored). Also, Gauss coefficients A_j are positive for all n, in contrast with some of the Newton–Cotes coefficients for larger n.

Of course, there are frequent applications with equally spaced nodes, so that Gauss integration does not apply (or has no great advantage if one first has to get the t_j in (5.85) by interpolation).

Since the endpoints −1 and 1 of the interval of integration in (5.85) are not zeros of P_n, they do not occur among $t_0, \dots, t_n$, and the Gauss formula (5.85) is called, therefore, an **open formula**, in contrast with a **closed formula**, in which the endpoints of the interval of integration are t_0 and t_n.

CONCEPT CHECK

1. **(Rectangular rule)** Evaluate the integral in Example 9 by the rectangular rule (5.71) with a subinterval of length 0.1.
2. Derive a formula for lower and upper bounds for the rectangular rule and apply it to Prob. 1.
3. Evaluate the integrals numerically as indicated and determine the error by using an integration formula known from calculus.

$$F(x)=\int_1^x \frac{dx^*}{x^*},\quad G(x)=\int_0^x \frac{dx^*}{\cos^2 x^*},$$

$$H(x)=\int_0^x x^* e^{-x^*} dx^*$$

   (a) $F(2)$ by (5.72), $n = 10$
   (b) $F(2)$ by (5.80), $n = 10$
   (c) $G(1)$ by (5.72), $n = 10$
   (d) $G(1)$ by (5.80), $n = 10$
   (e) $H(4)$ by (5.72), $n = 10$
   (f) $H(4)$ by (5.80), $n = 10$
4. Estimate the error by halving.
   (a) In Prob. 3(c)
   (b) In Prob. 3(d)
   (c) In Prob. 3(e)
   (d) In Prob. 3(f)
5. The following integrals cannot be evaluated by the usual methods of calculus. Evaluate them as indicated.

$$\mathrm{Si}(x) = \int_0^x \frac{\sin x^*}{x^*}\, dx$$

$$\mathrm{S}(x) = \int_0^x \sin (x^{*2})\, dx^{2*}$$

$$\mathrm{C}(x) = \int_0^x \cos (x^{*2})\, dx^*$$

   Si(x) is the **sine integral**. S(x) and C(x) are the **Fresnel integrals**.
   (a) Si(1) by (5.72), $n = 5$, $n = 10$
   (b) Using the values in Prob. 5(a), obtain a better value for Si(1). *Hint.* Use (5.77).
   (c) Si(1) by (5.80), $2m = 2$, $2m = 4$
   (d) Obtain a better value in Prob. 5(c). *Hint.* Use (5.84).
   (e) Si(1) by (5.80), $2m = 10$
   (f) S(1.25) by (5.80), $2m = 10$
   (g) C(1.25) by (5.80), $2m = 10$
6. **(Stability)** Prove that the trapezoidal rule is stable with respect to rounding.

7. Integrate by (5.85) with $n = 5$:
   (a) $1/x$ from 1 to 3
   (b) $\cos x$ from 0 to $\pi/2$
   (c) e^{-x^2} from 0 to 1
   (d) $\sin(x^2)$ from 0 to 1.25
8. **(Given TOL)** Find the smallest n in computing the integral of $1/x$ from 1 to 2 for which 5D-accuracy is guaranteed **(a)** by (5.76) in the use of (5.72), **(b)** by (5.83) in the use of (5.80). Compare and comment.
9. **TEAM PROJECT. Romberg Integration** (W. Romberg, *Norske Videnskab. Trondheim, Førh.* 28, Nr. 7, 1955). This method uses the trapezoidal rule and gains precision stepwise by halving h and adding an error estimate. Do this for the integral of $f(x) = e^{-x}$ from $x = 0$ to $x = 2$ with TOL $= 10^{-3}$, as follows.
   Step 1. Apply the trapezoidal rule (5.72) with $h = 2$ (hence $n = 1$) to get an approximation J_{11}. Halve h and use (5.72) to get J_{21} and an error estimate
$$\epsilon_{21} = \frac{1}{2^2 - 1}(J_{21} - J_{11})$$
   If $|\epsilon_{21}| \leqq$ TOL, stop. The result is $J_{22} = J_{21} + \epsilon_{21}$.
   Step 2. Show that $\epsilon_{21} = -0.066596$, hence $|\epsilon_{21}| >$ TOL and go on. Use (5.72) with $h/4$ to get J_{31} and add to it the error estimate $\epsilon_{31} = 1/3\,(J_{31} - J_{21})$ to get the better $J_{32} = J_{31} + \epsilon_{31}$. Calculate
$$\epsilon_{32} = \frac{1}{2^4 - 1}(J_{32} - J_{22}) = \frac{1}{15}(J_{32} - J_{22}$$
   If $|\epsilon_{32}| \leqq$ TOL, stop. The result is $J_{33} = J_{32} + \epsilon_{32}$. (Why does $2^4 = 16$ come in?) Show that we obtain $\epsilon_{32} = -0.000266$, so that we can stop. Arrange your J- and ϵ-values in a kind of "difference table."

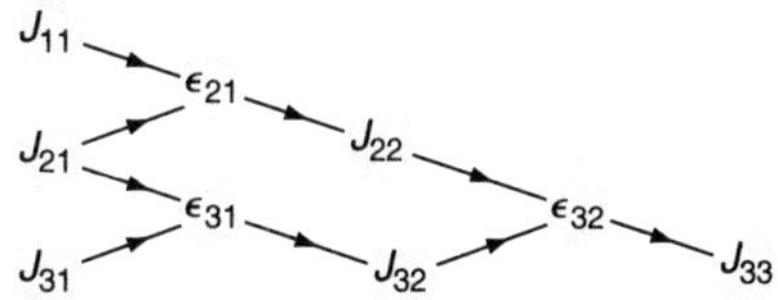

   If $|\epsilon_{32}|$ were greater than TOL, you would have to go on and calculate in the next step J_{41} from (5.72) with $h = 1/4$; then
$$J_{42} = J_{41} + \epsilon_{41} \quad \text{with} \quad \epsilon_{41} = 1/3(J_{41} - J_{31})$$
$$J_{43} = J_{42} + \epsilon_{42} \quad \text{with} \quad \epsilon_{42} = 1/15(J_{42} - J_{32})$$
$$J_{44} = J_{43} + \epsilon_{43} \quad \text{with} \quad \epsilon_{43} = 1/63(J_{43} - J_{33})$$
   where $63 = 2^6 - 1$. (How does this come in?) Apply the Romberg method to the integral of $f(x) = (\pi x^4/4)\cos(\pi x/4)$ from $x = 0$ to 2 with TOL $= 10^{-4}$.

5.5 Numeric Differentiation

Numeric differentiation is the computation of values of the derivative of a function f from given values of f. Numeric differentiation should be avoided whenever possible, because, whereas *integration* is a *smoothing* process and is not affected much by small inaccuracies in function values, *differentiation* tends to make matters *rough* and generally gives values of f' much less accurate than those of f – remember that the derivative is the limit of the difference quotient, and in the latter you usually have a small difference of large quantities that you then divide by a small quantity. However, the formulas to be obtained will be basic in the numeric solution of differential equations.

We use the notations $f_j' = f'(x_j)$, $f_j'' = f''(x_j)$, etc., and may obtain rough approximation formulas for derivatives by remembering that
$$f'(x) = \operatorname*{Lt}_{h \to 0} \frac{f(x+h) - f(x)}{h}$$
This suggests
$$f_{1/2}' \approx \frac{\delta f_{1/2}}{h} = \frac{f_1 - f_0}{h} \tag{5.86}$$

Similarly, for the second derivative we obtain

$$f_1'' \approx \frac{\delta^2 f_1}{h^2} = \frac{f_2 - 2f_1 + f_0}{h^2} \text{ etc.} \tag{5.87}$$

More accurate approximations are obtained by differentiating suitable Lagrange polynomials. Differentiating (5.78) and remembering that the denominators in (5.78) are $2h^2, -h^2, 2h^2$, we have

$$f'(x) \approx p_2'(x) = \frac{2x - x_1 - x_2}{2h^2} f_0 - \frac{2x - x_0 - x_2}{h^2} f_1 + \frac{2x - x_0 - x_1}{2h^2} f_2$$

Evaluating this at x_0, x_1, x_2, we obtain the "three-point formulas"

$$\begin{aligned}
&\text{(a)}\ f_0'' \approx \frac{1}{2h}(-3f_0 + 4f_1 - f_2) \\
&\text{(b)}\ f_1'' \approx \frac{1}{2h}(-f_0 + f_2) \\
&\text{(c)}\ f_2'' \approx \frac{1}{2h}(f_0 - 4f_1 + 3f_2)
\end{aligned} \tag{5.88}$$

Applying the same idea to the Lagrange polynomial $p_4(x)$, we obtain similar formulas, in particular,

$$f_2'' \approx \frac{1}{12h}(f_0 - 8f_1 + 8f_3 - f_4) \tag{5.89}$$

CONCEPT CHECK

1. Consider $f(x) = x^4$ for $x_0 = 0, x_1 = 0.2, x_2 = 0.4, x_3 = 0.6, x_4 = 0.8$. Calculate f_2' from (5.88a), (5.88b), (5.88c), (5.89). Determine the errors. Compare and comment.
2. A **"four-point formula"** for the derivative is

$$f_2' \approx \frac{1}{6h}(-2f_1 - 3f_2 + 6f_3 - f_4)$$

Apply it to $f(x) = x^4$ with $x_1, \ldots, x_4$ determine the error, and compare it with that in the case of (5.89)

3. The derivative $f'(x)$ can also be approximated in terms of first-order and higher order differences:

$$f'(x_0) \approx \frac{1}{h}\left(\Delta f_0 - \frac{1}{2}\Delta^2 f_0 + \frac{1}{3}\Delta^3 f_0 - \frac{1}{4}\Delta^4 f_0 + - \cdots\right)$$

Compute $f'(0,4)$ from this formula, using differences up to and including first order, second order, third order, fourth order.

Solved Problems

Q 1. Derive normal equations to fit the straight line $y = a + bx$.

[JNTU April/May 2007, May/June 2006, Nov./Dec. 2005]

Answer: The equation of a straight line is

$$y = a + bx \tag{5.90}$$

To find the curve of best fit for data (x_i, y_i), where $i = 1, 2, \ldots, n$. Finding the normal equations by method of least squares we minimize the error, that is multiply (5.90) with x, we get

$$xy = ax + bx^2 \tag{5.91}$$

Applying summation on both sides of (5.90), we get

$$\sum_{i=1}^{n} y_i = a\sum_{i=1}^{n} 1 + b\sum_{i=1}^{n} x_i$$

Since $\sum_{i=1}^{n} 1 = 1+1+1+\cdots+1\ (n \text{ times}) = n$, we have

$$\sum_{i=1}^{n} y_i = na + b\sum_{i=1}^{n} x_i \tag{5.92}$$

where, $i = 1$ to n is the size of the data. Multiplying with Σx_i on both sides of (5.91), we get

$$\sum_{i=1}^{n} x_i\, y_i = a\sum_{i=1}^{n} x_i + b\sum_{i=1}^{n} x_i^2 \tag{5.93}$$

Equations (5.92) and (5.93) are the normal equations for the best fit of straight line $y = a + bx$. On solving (5.92) and (5.92) we get the values of a and b. Therefore, on substituting values of a and b in (5.90), we get the line of best fit.

Q 2. Derive normal equations to fit the parabola $y = a + bx + cx^2$. [JNTU May/June 2006]

Answer: The equation of parabola is

$$y = a + bx + cx^2 \tag{5.94}$$

Let (5.94) pass through the points $(x_1, y_1), (x_2, y_2), \ldots, (x_n, y_n)$. Therefore

$$y_i = a + bx_i + x_i^2 \tag{5.95}$$

where, $i = 1, 2, \ldots, n$. Let E_i be the error found between the expected and the observed values of $y = y_i$, that is

$$E_i = y_i - (a + bx_i + x_i^2) \qquad i = 1, 2, \ldots, n$$

Therefore, the sum of the squares of these errors is

$$E = \sum_{i=1}^{n} E_i^2 = \sum_{i=1}^{n} (y_i - a - bx_i - x_i^2)^2$$

To find the minimum E, we use

$$\frac{\partial E}{\partial a} = 0, \quad \frac{\partial E}{\partial b} = 0, \quad \frac{\partial E}{\partial c} = 0$$

Therefore, the normal equations are

$$\Sigma y_i = na + b\Sigma x_i + c\Sigma x_i^2$$
$$\Sigma x_i y_i = a\Sigma x_i + b\Sigma x_i^2 + c\Sigma x_i^3$$
$$\Sigma x_i^2 y_i = a\Sigma x_i^2 + b\Sigma x_i^3 + c\Sigma x_i^4$$

Generalizing the above equations, we get

$$\Sigma y = na + b\Sigma x + c\Sigma x^2$$
$$\Sigma xy = a\Sigma x + b\Sigma x^2 + c\Sigma x^3$$
$$\Sigma x^2 y = a\Sigma x^2 + b\Sigma x^3 + c\Sigma x^4$$

On solving the above normal equations we obtain the values of a, b, c. The values of a, b, c are substituted in (5.94) to obtain the parabola of best fit.

Q 3. Fit a parabola of the form $y = a + bx + cx^2$ to the following data.

x	1	2	3	4	5	6	7
y	2.3	5.2	9.7	16.5	29.4	35.5	54.4

[JNTU April/May 2007, Aug./Sep. 2006]

Answer: The equation of a parabola is of the form

$$y = a + bx + cx^2$$

The normal equations are

$$\sum y = na + b\sum x + c\sum x^2 \tag{5.96}$$

$$\sum xy = a\sum x + b\sum x^2 + c\sum x^3 \tag{5.97}$$

$$\sum x^2 y = a\sum x^2 + b\sum x^3 + c\sum x^4 \tag{5.98}$$

where $n = 7$

The above three equations are solved to get the equations of a parabola.

Let $X = x - 4$. Therefore, normal equations change as

$$\sum y = na + b\sum X + c\sum X^2 \tag{5.99}$$

$$\sum Xy = a\sum X + b\sum X^2 + c\sum X^3 \tag{5.100}$$

$$\sum X^2 y = a\sum X^2 + b\sum X^3 + c\sum X^4 \tag{5.101}$$

Tabulating them we get

X	y	X	X^2	X^3	X^4	Xy	X^2y
1	2.3	–3	9	–27	81	–69	20.7
2	5.2	–2	4	–8	16	–10.4	20.8
3	9.7	–1	1	–1	1	–9.7	9.7
4	16.5	0	0	0	0	0	0
5	29.4	1	1	1	1	29.4	29.4
6	35.5	2	4	8	16	71	14.2
7	54.4	3	9	27	81	163.2	489.6
SUM	$\sum y = 153$	$\sum X = 0$	$\sum X^2 = 28$	$\sum X^3 = 0$	$\sum X^4 = 196$	$\sum Xy = 236.6$	$\sum X^2 y = 712.2$

Therefore, substituting the above values in (5.99)–(5.101), we get

$$153 = 7a + b \times 0 + c \times 28 \tag{5.102}$$

$$236.6 = a \times 0 + b \times 28 + c \times 0 \tag{5.103}$$

$$712.2 = a \times 28 + b \times 0 + c \times 196 \tag{5.104}$$

From (5.104) we get

$$236.6 = 28 \times b \Rightarrow b = 8.45$$

Solving (5.102) and (5.104) for a and c we get

$$7a + \ \ 28c = 153 \qquad (5.105)$$

$$28a + 196c = 712.2 \qquad (5.106)$$

Multiplying (5.105) with 4 and subtracting from (5.106) we get

$$-84c = -100.2 \Rightarrow c = 1.1928$$

Substituting the value of c in (5.105) we get

$$7a + 28 \times 1.1928 = 153$$

$$7a = 153 - 28 \times 1.1928$$

$$a = 17.0857$$

Therefore, substituting the values of a, b, c in the parabola of the form $y = a + bX + cX^2$ and using $X = x - 4$ we get

$$\begin{aligned} y &= a + b\,(x-4) + c(x-4)^2 \\ &= a + bx - 4b + c(x^2 - 8x + 16) \\ &= a + bx - 4b + cx^2 - 8cx + 16c \\ &= cx^2 + (b - 8c)x + (a - 4b + 16c) \\ &= 1.1928x^2 + (8.45 - 8 \times 1.1928)x + (17.0857 - 4 \times 8.45 + 16 \times 1.1928) \\ &= 1.1928x^2 + (8.45 - 9.5424)x + (17.0857 - 33.80 + 19.0848) \\ &= 1.1928x^2 - 1.0924x + 2.3705 \end{aligned}$$

Therefore, $y = 1.1928x^2 - 1.0294x + 2.3705$ is the required equation of the parabola.

Q 4. The table below shows the temperature $f(t)$ as a function of time

t	1	2	3	4	5	6	7
$f(t)$	81	75	80	83	78	70	60

Use Simpson's one-third method to estimate $\int_1^7 f(t)\,dt$. **[JNTU April/May 2007, Aug./Sep. 2006]**

Answer: From Simpson's one-third rule, we get

$$\int_a^b f(x)\,dx = \frac{h}{3}[(y_0 + y_n) + 4(y_1 + y_3 + y_5 + \cdots + y_{n-1}) + 2(y_2 + y_4 + y_6 + \cdots + y_{n-2})] \qquad (5.107)$$

where $a = 1$, $b = 7$ and $n = 6$

The given data can be tabulated as

t	1	2	3	4	5	6	7
$f(t)$	$81 = y_0$	$75 = y_1$	$80 = y_2$	$83 = y_3$	$78 = y_4$	$70 = y_5$	$60 = y_6$

Here, $n = 6$. Therefore

$$h = \frac{b-a}{n} = \frac{7-1}{6} = \frac{6}{6} = 1$$

Subtituting all values in (5.107) we get

$$\int_1^7 f(t)\,dt = \frac{1}{3}[(81+60)+4(75+83+70)+2(80+78)]$$

$$= \frac{1}{3}(141+912+316)$$

$$= \frac{1}{3}(1369) = 456.33$$

Q 5. Given that

x	4.0	4.2	4.4	4.6	4.8	5.0	5.2
log(x)	1.3863	1.4351	1.4816	1.5261	1.5686	1.6094	1.6487

Evaluate $\int_4^{5.4} \log x\, dx$ by Simpson's 3/8 rule. **[JNTU May/June 2006]**

Answer: Given that, $y_0 = 1.3863$, $y_1 = 1.4351$, $y_2 = 1.4816$, $y_3 = 1.5261$, $y_4 = 1.5686$, $y_5 = 1.6094$, $y_6 = 1.6487$. Now Simpson's 3/8 rule we have

$$f(x) = \int_4^{5.2} \log x\, dx = \frac{3h}{8}[(y_0+y_n)+3(y_1+y_2+y_4+y_5+\cdots+y_{n-1})+2(y_3+y_6+y_9+\cdots+y_{n-3})]$$

The common difference $h = 0.2$. Substituting all values and $n = 6$ we get

$$f(x) = \frac{3(0.2)}{8}[(y_0+y_6)+3(y_1+y_2+y_4+y_5)+2(y_3)]$$

$$= \frac{3(0.2)}{8}[(1.3863+1.6487)+3(1.4351+1.4816+1.586+1.6094)+2(1.5261)]$$

Therefore

$$\int_4^{5.2} \log x dx = 0.075(3.035+18.2841+3.0522) = 0.0075(24.3713) = 1.8278475$$

Summary

The fitting of a polynomial $p(x) = b_0 + b_1x + \cdots + b_mx^m$ through given data (points in the xy plane) $(x_1, y_1), \ldots, (x_n, y_n)$ by the method of **least squares.**

Polynomial interpolation means the determination of a polynomial $p_n(x)$ such that $p_n(x_j) = f_j$, where $j = 0, \ldots, n$ and $(x_0, f_0), \ldots, (x_n, f_n)$ are measured or observed values, values of a function, etc. $p_n(x)$ is called an *interpolation polynomial*.

Simpson's rule numeric integration is

$$\int_a^b f(x)dx \approx \frac{h}{3}(f_0+4f_1+2f_2+4f_3 +\cdots+2f_{2m-2}+4f_{2m-1}+f_{2m}) \qquad (5.108)$$

with equally spaced nodes $x_j = x_0 + j_h$, $j = 1, \ldots, 2m$, $h = (b-a)/(2m)$, and $f_j = f(x_j)$. It is simple but accurate enough for many applications. Its degree of precision is DP = 3 because the error (5.81), involves h^4. A more practical error estimate is (5.84)

$$\epsilon_{h/2} = \frac{1}{15}(J_{h/2} - J_h)$$

obtained by first computing with step h, then with step $h/2$, and then taking 1/15 of the difference of the results.

Simpson's rule is the most important of the **Newton–Cotes formulas**, which are obtained by integrating Lagrange interpolation polynomials, linear ones for the **trapezoidal rule** (5.72), quadratic for Simpson's rule, cubic for the *three-eights rule*, etc.

Adaptive integration is integration that adjusts ("*adapts*") the step (automatically) to the variability of $f(x)$. **Gauss integration**, is important because of its great accuracy (DP = $2n - 1$, compared to Newton–Cotes's DP = $n - 1$ or n). This is achieved by an optimal choice of the nodes, which are not equally spaced. **Numeric differentiation** is also discussed.

Review Questions and Problems

1. What is least squares approximation? What are the normal equations?
2. What is tridiagonalization? When would you apply it?
3. Fit straight line to (–2, 0.1), (0, 1.9), (2, 3.8), (4, 6.1), (6, 7.8).
4. Fit quadratic parabola to (1, 9), (2, 5), (3, 4), (4, 5), (5, 7).
5. List and compare numeric integration methods. When would you apply them?
6. What does adaptive integration mean? Why is it useful?
7. Why is numeric differentiation generally more delicate than numeric integration?
8. Compute the integral of x^3 from 0 to 1 by the trapezoidal rule with $n = 5$. What error bounds are obtained from (5.76)? What is the actual error of the result? Why is this result larger than the exact value?
9. Compute the integral of $\cos(x^2)$ from 0 to 1 by Simpson's rule with $2m = 2$ and $2m = 4$ and estimate the error by (5.84).
10. Compute the integral of $\cos x$ from 0 to $\pi/2$ by the three-eighths rule

$$\int_a^b f(x)dx = \frac{3}{8}h(f_0 + 3f_1 + 3f_2 + f_3) - \frac{1}{80}(b-a)h^4 f^{(\text{iv})}(\hat{t})$$

and give error bounds; here $a \leqq \hat{t} \leqq b$ and $x_j = a + (b-a)j/3$, $j = 0, \ldots, 3$.

6 Numerical Solutions of Initial Value Problems in ODEs

LEARNING OBJECTIVES

After reading this chapter, you will be able to understand the following:

- Taylor series method.
- Picard's method.
- Euler's method.
- Runge–Kutta method.
- Adams–Bashforth–Moulton method.
- Predictor–corrector method.

Numeric methods for differential equations are of great practical importance to the engineer and physicist because practical problems often lead to differential equations that cannot be solved by one of the methods discussed in the previous chapters. Also, sometimes an ODE does have a solution formula, which, however, in some specific cases may become so complicated that one prefers to apply a numeric method instead. Sections 6.2–6.5 discuss single-step methods whereas Secs. 6.6–6.8 discuss multistep methods.

6.1 Numerical Solutions of First-Order ODEs

We know that an ODE of the first order is of the form $F(x, y, y') = 0$ and can often be written in the explicit form $y' = f(x, y)$. An **initial value problem** for this equation is of the form

$$y' = f(x, y), \qquad y(x_0) = y_0 \tag{6.1}$$

where x_0 and y_0 are given and we assume that the problem has a unique solution on some open interval $a < x < b$ containing x_0.

In this section we shall discuss methods of computing approximate numeric values of the solution $y(x)$ of (6.1) at the equidistant points on the x axis

$$x_1 = x_0 + h, \qquad x_2 = x_0 + 2h, \qquad x_3 = x_0 + 3h, \ldots$$

where the **step size** h is a fixed number, for instance, 0.2 or 0.1 or 0.01, whose choice we discuss later in this section. Those methods are **step-by-step methods**, using the same formula in each step. Such formulas are suggested by the Taylor series

$$y(x + h) = y(x) + hy'(x) + \frac{h^2}{2}y''(x) + \cdots \tag{6.2}$$

For a small h the higher powers $h^2, h^3, \ldots$ are very small. This suggests the crude approximation

$$\begin{aligned} y(x + h) &\approx y(x) + hy'(x) \\ &= y(x) + hf(x, y) \end{aligned}$$

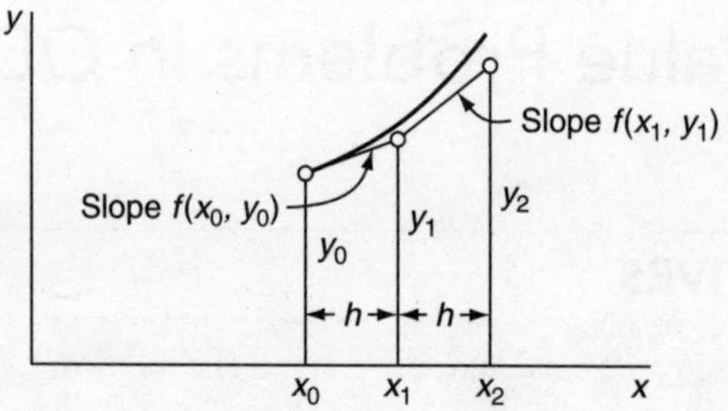

Figure 1 Euler method.

(with the second line obtained from the given ODE) and the following iteration process. In the first step we compute

$$y_1 = y_0 + hf(x_0, y_0)$$

which approximates $y(x_1) = y(x_0 + h)$. In the second step we compute

$$y_2 = y_1 + hf(x_1, y_1)$$

which approximates $y(x_2) = y(x_0 + 2h)$, etc., and in general

$$y_{n+1} = y_n + hf(x_n, y_n) \qquad (n = 0, 1, \ldots) \tag{6.3}$$

This is called the **Euler method** or the **Euler–Cauchy method**. Geometrically it is an approximation of the curve of $y(x)$ by a polygon whose first side is tangent to this curve at x_0 (see Figure 1). This crude method is hardly ever used in practice, but since it is simple, it nicely explains the principle of methods based on the Taylor series.

Taylor's formula with remainder has the form

$$y(x+h) = y(x) + hy'(x) + \frac{1}{2}h^2 y''(\xi)$$

(where $x \leqq \xi \leqq x + h$). It shows that in the Euler method the *truncation error in each step* or **local truncation error** is proportional to h^2, written $O(h^2)$, where O suggests *order.* Now over a fixed x interval in which we want to solve an ODE the number of steps proportional to $1/h$. Hence the *total error* or **global error** is proportional to $h^2(1/h) = h$. For this reason, the Euler method is called a **first-order method.** In addition, there are **roundoff errors** in this and other methods, which may affect the accuracy of the values $y_1, y_2, \ldots$ more and more as n increases, as we shall see.

6.2 Euler's Method

Euler was the first to devise an approximate method to solve the ordinary differential equation. It uses the simplest extrapolation techniques for obtaining a solution. We shall explain this method by solving a hypothetical example.

Example 1

Solve $dy/dx = f(x, y)$ given that y at $x_0 = y_0$.

Solution: We have to obtain a set of values of y at different x_i by solving the given equation. Here

$$x_i = x_0 + ih$$

where $i = 1, 2, 3, \ldots$, and h is the step size. We obtain the solution by integrating $dy/dx = f(x, y)$. At $x = x_1 = x_0 + h$, $y = y_1$, we have

$$y = y_0 + \int_{x_0}^{x} f(x, y)\, dx \tag{6.4}$$

Assuming that $f(x, y) = f(x_0, y_0)$ (i.e., a constant) in the range $x_0 \leqq x \leqq x_1$, we have

$$y_1 \approx y_0 + f(x_0, y_0)(x_1 - x_0) = y_0 + hf(x_0, y_0) \tag{6.5}$$

This implies that $y(x_1) = y(x_0 + h) = y_0 + hf(x_0, y_0)$. We started with the point (x_0, y_0) and extrapolated the point (x_1, y_1). Now these two points are to be joined by a straight line, as we have assumed above that (x, y), which is nothing but dy/dx (the slope of the curve), is constant between $x = x_0$ and $x = x_1$.

The next solution point y_2 at $x = x_2 = x_0 + 2h = x_1 + h$ is also obtained assuming $f(x, y) = f(x_1, y_1) =$ a constant in $x_1 \leqq x \leqq x_2$ in an identical fashion as

$$y_2 = y_1 + \int_{x_1}^{x_2} f(x, y)dx \approx y_1 + hf(x_1, y_1) = y(x_2) = y(x_0 + 2h) \tag{6.6}$$

The extrapolation curve from (x_1, y_1) to (x_2, y_2) is again a straight line (or linear relation). We can similarly proceed to get other point y_3, from y_3 to y_4 and so on. In general, the $(i + 1)$th point may be obtained from the ith point using the relation

$$y_{i+1} = y_i + hf(x_i, y_i) \tag{6.7}$$

This is the Euler's formula. This is illustrated in Figure 2. It can be seen that this process is very slow. The approximate solution is assumed to follow a straight line corresponding to the tangent at the starting point (as it is determined from the slope). Thus Euler's method is basically a technique

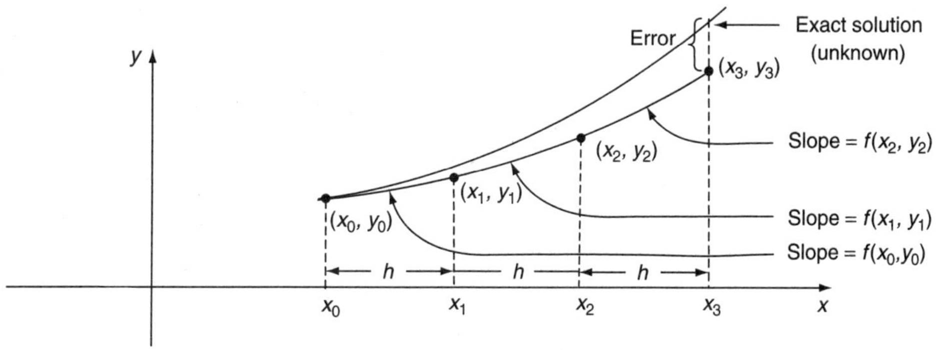

Figure 2 Illustrating Euler's method (three steps).

of developing a **piecewise linear** approximation to the solution. It is evident that error will be there, unless the actual solution itself is linear. Error tends to accumulate at each step, as shown in the figure. The local error at each step can be reduced by making h smaller, but there is a limit to how small h can be made from efficiency and accuracy considerations.

[As a result, this method is unsuitable for practical use and so a modification is provided for it, which resulted in a modified Euler method. This gives better results, however this is not in our purview.] To demonstrate this method, let us consider an example.

Example 2

Solve

$$y' = \frac{x+y}{x}$$

given that $y(2) = 2$. Estimate y at $x = 2.5$ using Euler's method. [Here $x_0 = 2$, $y_0 = 2$, and $y(2.5)$ is to be estimated.]

Solution:

(a) Let us choose $h = 0.25$ here. This implies 2 steps, that is

$$x_0 = 2, \quad x_1 = 2.25, \quad x_2 = 2.5 = x_0 + 2h$$

We first get y_1 and then y_2:

$$y_1 \cong y_0 + hf(x_0, y_0)$$

Therefore

$$y_1 = y(2.25) = 2 + 0.25 \times \left(\frac{2+2}{2}\right) = 2.5$$

Now

$$y_2 \cong y_1 + hf(x_1, y_1)$$

$$y_2 = y(2.5) = 2.5 + 0.25 \times \left(\frac{2.25+2.5}{2.25}\right) = 3.028$$

(b) We can improve the accuracy by decreasing h, as this choice is in our hands (no h is prescribed). But this will increase the number of steps. As $x_0 = 2$, x_i (required) $= 2.5$

$$x_i = 2.5 = x_0 + ih = 2 + ih$$

Therefore

$$h = \frac{0.5}{i}$$

Choosing 5 steps, $i = 5$, we have $h = 0.1$. With $h = 0.1$, we have

$$y_1 = y(2.1) = 2 + 0.1\left(\frac{2+2}{2}\right) = 2.2$$

$$y_2 = y(2.2) = 2.2 + 0.1\left(\frac{2.1+2.2}{2.1}\right) = 2.405$$

$$y_3 = y(2.3) = 2.405 + 0.1\left(\frac{2.2 + 2.405}{2.2}\right) = 2.614$$

$$y_4 = y(2.4) = 2.614 + 0.1\left(\frac{2.3 + 2.614}{2.3}\right) = 2.828$$

$$y_5 = y(2.5) = 2.828 + 0.1\left(\frac{2.4 + 2.828}{2.4}\right) = 3.046$$

[The exact solution is given by $y = x\,(1 + \ln x/2)$, and thus $y(2.5) = 3.058$]. The error is considerable even with $h = 0.1$.

Remarks: As error build-up in Euler's method is substantial, the method is rarely used.

Example 3 Euler Method

Apply the Euler method to the following initial value problem, choosing $h = 0.2$ and computing $y_1, \ldots, y_5$:

$$y' = x + y, \qquad y(0) = 0 \tag{6.8}$$

Solution: Here $f(x, y) = x + y$; hence $f(x_n, y_n) = x_n + y_n$, and we see that (6.8) becomes

$$y_{n+1} = y_n + 0.2(x_n + y_n)$$

Table 1 Euler Method Applied to (6.7) in Example 3 and Error

n	x_n	y_n	$0.2(x_n + y_n)$	*Exact values*	*Error* ϵ_n
0	0.0	0.000	0.000	0.000	0.000
1	0.2	0.000	0.040	0.021	0.021
2	0.4	0.040	0.088	0.092	0.052
3	0.6	0.128	0.146	0.222	0.094
4	0.6	0.274	0.215	0.426	0.152
5	1.0	0.489		0.718	0.229

Table 1 shows the computations, the values of the exact solution

$$y(x) = e^x - x - 1$$

obtained from (6.8) and the error. In practice the exact solution is unknown, but an indication of the accuracy of the values can be obtained by applying the Euler method once more with step $2h = 0.4$, letting $y_n{}^*$ denote the approximation now obtained, and comparing corresponding approximations. This computation is:

x_n	$y_n{}^*$	$0.4(x_n + y_n)$	y_n *in Table* 1	*Difference* $y_n - y_n{}^*$
0.0	0.000	0.000	0.000	0.000
0.4	0.000	0.160	0.040	0.040
0.8	0.160		0.274	0.114

Let ϵ_n and $\epsilon_n{}^*$ be the errors of the computations with h and $2h$, respectively. Since the error is of order h^2, in a switch from h to $2h$ it is multiplied by $2^2 = 4$, but since we need only half as many steps as before, it will be multiplied only by $4/2 = 2$. Hence $\epsilon_n{}^* \approx 2\epsilon_n$ so that the difference is $\epsilon_n{}^* - \epsilon_n \approx 2\epsilon_n - \epsilon_n = \epsilon_n$. Now $y = y_n + \epsilon_n = y_n{}^* + \epsilon_n{}^*$ by the definition of error; hence $\epsilon_n{}^* - \epsilon_n = y_n - y_n{}^*$ indicates ϵ_n qualitatively. In our computations, $y_2 - y_2{}^* = 0.04 - 0 = 0.04$ (actual error 0.052, see Table 1) and $y_4 - y_4{}^* = 0.274 - 0.160 = 0.114$ (actually 0.152).

Example 4 Euler Method for a Nonlinear ODE

Figure 3 concerns the initial value problem

$$y' = (y - 0.01x^2)^2 \sin(x^2) + 0.02x, \quad y(0) = 0.4$$

and shows the curve of the solution $y = 1/[2.5 - S(x)] + 0.01x^2$ where $S(x)$ is the Fresnel integral. It also shows 80 approximate values for $0 \leqq x \leqq 4$ obtained by the Euler method from (6.3),

$$y_{n+1} = y_n + 0.05\,[(y_n - 0.01x_n{}^2)^2 \sin(x_n{}^2) + 0.02x_n]$$

Although $h = 0.05$ is smaller than h in Example 3, the accuracy is still not good. It is interesting that the error is not monotonically increasing, obviously since the solution is not monotone. We shall return to this ODE in the problem set.

Automatic Variable Step Size Selection in Modern Numeric Software

The idea of adaptive integration as motivated and explained applies equally well to the numeric solution of ODEs. It now concerns automatically changing the step size h depending on the variability of $y' = f$ determined by

$$y'' = f' = f_x + f_y y' = f_x + f_y f \tag{6.9}$$

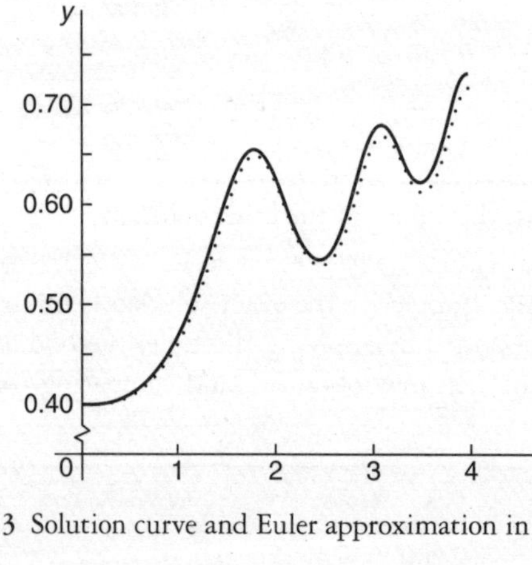

Figure 3 Solution curve and Euler approximation in Example 4.

Accordingly, modern software automatically selects variable step sizes h_n so that the error of the solution will not exceed a given maximum size TOL (suggesting *tolerance*). Now for the Euler method, when the step size is $h = h_n$, the local error at x_n is about $h_n^{\,2}|y''(x_n)|/2$. We require that this be equal to a given tolerance TOL,

$$\text{(a)}\quad \frac{1}{2}h_n^{\,2}|y''(x_n)| = \text{TOL}, \quad \text{thus} \quad \text{(b)}\quad h_n = \sqrt{\frac{2\,\text{TOL}}{|y''(x_n)|}} \tag{6.10}$$

$y''(x)$ must not be zero on the interval J: $x_0 \leqq x = x_N$ on which the solution is wanted. Let K be the minimum of $|y''(x)|$ on J and assume that $K > 0$. Minimum $|y''(x)|$ corresponds to maximum $h = H = \sqrt{2\,\text{TOL}/K}$ by (6.10). Thus, $\sqrt{2\,\text{TOL}} = H\sqrt{K}$. We can insert this into (6.10b), obtaining by straightforward algebra

$$h_n = \varphi(x_n)H \quad \text{where} \quad \varphi(x_n) = \sqrt{\frac{K}{|y''(x_n)|}} \tag{6.11}$$

For other methods, automatic step size selection is based on the same principle.

Improved Euler Method

By taking more terms in (6.2) into account we obtain numeric methods of higher order and precision. But there is a practical problem. If we substitute $y' = f(x, y(x))$ into (6.2), we have

$$y(x+h) = y(x) + hf + \frac{1}{2}h^2 f' + \frac{1}{6}h^3 f'' + \cdots \tag{6.12}$$

Now y in f depends on x, so that we have f' as shown in (6.9) and f'', f''' even much more cumbersome. The ***general strategy*** now is to avoid the computation of these derivatives and to replace it by computing f for one or several suitably chosen auxiliary values of (x, y). "Suitably" means that these values are chosen to make the order of the method as high as possible (to have high accuracy). Let us discuss two such methods that are of practical importance, namely, the improved Euler method and the (classical) Runge–Kutta method.

In the **improved Euler method** or **improved Euler–Cauchy method** (sometimes also called **Heun method**), in each step we compute first the auxiliary value

$$y_{n+1}^* = y_n + hf(x_n, y_n) \tag{6.13a}$$

and then the new value

$$y_{n+1} = y_n + \frac{1}{2}h\,[f(x_n, y_n) + f(x_{n+1}, y_{n+1}^*)] \tag{6.13b}$$

This method has a simple geometric interpretation. In fact, we may say that in the interval from x_n to $x_n + (1/2)h$ we approximate the solution y by the straight line through (x_n, y_n) with slope $f(x_n, y_n)$, and then we continue along the straight line with slope $f(x_{n+1}, y_{n+1}^*)$ until x reaches x_{n+1}.

The improved Euler–Cauchy method is a **predictor–corrector method** because in each step we first ***predict*** a value by (6.13a) and then ***correct*** it by (6.13b). In algorithmic form, using the notations $k_1 = hf(x_n, y_n)$ in (6.13a) and $k_2 = hf(x_{n+1}, y_{n+1}^*)$ in (6.13b), we can write this method as shown in the following algorithm.

ALGORITHM EULER (f, x_0, y_0, h, N)

This algorithm computes the solution of the initial value problem $y' = f(x, y), y(x_0) = y_0$ at equidistant points $x_1 = x_0 + h, x_2 = x_0 + 2h, \ldots, x_N = x_0 + Nh$; here f is such that this problem has a unique solution on the interval $[x_0, x_N]$.

INPUT: Initial values x_0, y_0, step size h, number of steps N

OUTPUT: Approximation y_{n+1} to the solution $y(x_{n+1})$ at $x_{n+1} = x_0 + (n + 1)h$, where $n = 0, \ldots, N - 1$

For $n = 0, 1, \ldots, N - 1$ do:

$x_{n+1} = x_n + h$

$k_1 = hf(x_n, y_n)$

$k_2 = hf(x_{n+1}, y_n + k_1)$

$y_{n+1} = y_n + 1/2\,(k_1 + k_2)$

OUTPUT x_{n+1}, y_{n+1}

End

Stop

End EULER

Example 5 Improved Euler Method

Apply the improved Euler method to the initial value problem (6.8), choosing $h = 0.2$, as before.

Solution: For the present problem we have in Table 2

$$k_1 = 0.2(x_n + y_n)$$

$$k_2 = 0.2[x_n + 0.2 + y_n + 0.2(x_n + y_n)]$$

$$y_{n+1} = y_n + \frac{0.2}{2}(2.2x_n + 2.2y_n + 0.2) = y_n + 0.22(x_n + y_n) + 0.02$$

Table 2 shows that our present results are more accurate than those in Example 3.

Table 2 Improved Euler Method Applied to (6.7) and Error

n	x_n	y_n	$0.22(x_n + y_n) + 0.02$	*Exact values* (4D)	*Error*
0	0.0	0.0000		0.0000	0.0000
1	0.2	0.0200	0.0200	0.0214	0.0014
2	0.4	0.0884	0.0684	0.0918	0.0034
3	0.6	0.2158	0.1274	0.2221	0.0063
4	0.8	0.4153	0.1995	0.4255	0.0102
5	1.0	0.7027	0.2874	0.7183	0.0156

Error of the Improved Euler Method

The local error is of order h^3 and the global error of order h^2, so that the method is a second-order method. We now prove this.

PROOF: Setting $\tilde{f}_n = f(x_n, y(x_n))$ and using (6.12), we have

$$y(x_n + h) - y(x_n) = h\tilde{f}_n + \frac{1}{2}h^2\tilde{f}_n' + \frac{1}{6}h^3\tilde{f}_n'' + \cdots \tag{6.14a}$$

Approximating the expression in the brackets in (6.13b) by $\tilde{f}_n + \tilde{f}_{n+1}$ and again using the Taylor expansion, we obtain from (6.13a)

$$\begin{aligned} y_{n+1} - y_n &\approx \frac{1}{2}h[\tilde{f}_n + \tilde{f}_{n+1}] \\ &= \frac{1}{2}h[\tilde{f}_n + (\tilde{f}_n' + h\tilde{f}_n' + \frac{1}{2}h^2\tilde{f}_n'' + \cdots)] \\ &= h\tilde{f}_n + \frac{1}{2}h^2\tilde{f}_n' + \frac{1}{4}h^3\tilde{f}_n'' + \cdots \end{aligned} \tag{6.14b}$$

where ′ (prime) implies d/dx_n, etc. Subtraction of (6.14b) from (6.14a) gives the local error

$$\frac{h^3}{6}\tilde{f}_n'' - \frac{h^3}{4}\tilde{f}_n'' + \cdots = -\frac{h^3}{12}\tilde{f}_n'' + \cdots$$

Since the number of steps over a fixed x-interval is proportional to $1/h$, the global error is of order $h^3/h = h^2$, so that the method is of second order. ■

Backward Euler Method and Stiff ODEs

The **backward Euler formula** for numerically solving (6.1) is

$$y_{n+1} = y_n + hf(x_{n+1}, y_{n+1}) \quad (n = 0, 1, \ldots) \tag{6.15}$$

This formula is obtained by evaluating the right side at the *new* location (x_{n+1}, y_{n+1}); this is called the **backward Euler scheme**. For known y_n it gives y_{n+1} *implicitly*, so it defines an **implicit method,** in contrast to the Euler method (6.3), which gives y_{n+1} explicitly. Hence (6.15) must be solved for y_{n+1}. How difficult this is depends on f in (6.1). For a linear ODE this provides no problem, as Example 6 (below) illustrates. The method is particularly useful for "stiff" ODEs, as they occur quite frequently in the study of vibrations, electric circuits, chemical reactions, etc. The situation of stiffness is roughly as follows:

Error terms of the methods considered so far involve a higher derivative. And we ask what happens if we let *h increase.* Now if the error (the derivative) grows fast but the desired solution also grows fast, nothing will happen. However, if that solution does not grow fast, then with growing h the error term can take over to an extent that the numeric result becomes completely nonsensical, as in Figure 4. Such an ODE for which h must thus be restricted to small values, and the physical system the ODE models, are called **stiff.** This term is suggested by a mass–spring system with a stiff spring (spring with a large k). Example 6 illustrates that implicit methods remove the difficulty of increasing h in the case of stiffness: it can be shown that in the application of an implicit method the solution remains stable under any increase of h, although the accuracy decreases with increasing h.

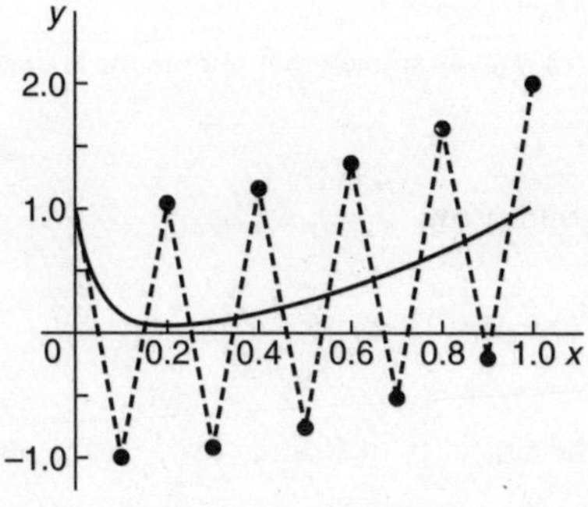

Figure 4 Euler method with $h = 0.1$ for the stiff ODE in Example 6 and exact solution.

Example 6 Backward Euler Method and Stiff ODE

The initial value problem

$$y' = f(x, y) = -20y + 20x^2 + 2x, \qquad y(0) = 1$$

has the solution (verify!)

$$y = e^{-20x} + x^2$$

Solution: The backward Euler formula (6.15) is

$$y_{n+1} = y_n + hf(x_{n+1}, y_{n+1}) = y_n + h(-20y_{n+1} + 20x_{n+1}^2 + 2x_{n+1})$$

Noting that $x_{n+1} = x_n + h$, taking the term $-20hy_{n+1}$ to the left and dividing we obtain

$$y_{n+1} = \frac{y_n + h[20(x_n + h)^2 + 2(x_n + h)]}{1 + 20h} \tag{6.16}$$

The numeric results in Table 3 show the following. Stability of the backward Euler method for $h = 0.05$ and also for $h = 0.2$ with an error increase by about a factor 4 for $h = 0.2$, stability of the Euler method for $h = 0.05$ but instability for $h = 0.1$ (Figure 4), stability of RK for $h = 0.1$ but instability for $h = 0.2$. This illustrates that the ODE is stiff. Note that even in the case of stability the approximation of the solution near $x = 0$ is poor.

Table 3 Backward Euler Method (BEM) for Example 6 and Comparison with Euler and RK Methods

x	*BEM* $h = 0.05$	*BEM* $h = 0.2$	*Euler* $h = 0.05$	*Euler* $h = 0.1$	*RK* $h = 0.1$	*RK* $h = 0.2$	*Exact*
0.0	1.00000	1.00000	1.00000	1.00000	1.00000	1.000	1.00000
0.1	0.26188		0.00750	−1.00000	0.34500		0.14534
0.2	0.10484	0.24800	0.03750	1.04000	0.15333	5.093	0.05832
0.3	0.10809		0.08750	−0.92000	0.12944		0.05832
0.4	0.16640	0.20960	0.15750	1.16000	0.17482	25.48	0.16034

Table 3 (Continued)

x	*BEM* $h=0.05$	*BEM* $h=0.2$	*Euler* $h=0.05$	*Euler* $h=0.1$	*RK* $h=0.1$	*RK* $h=0.2$	*Exact*
0.5	0.25347		0.24750	−0.76000	0.25660		0.25004
0.6	0.36274	0.37792	0.35750	1.36000	0.36387	127.0	0.36001
0.7	0.49256		0.48750	−0.52000	0.49296		0.49001
0.8	0.64252	0.65158	0.63750	1.64000	0.64265	634.0	0.64000
0.9	0.81250		0.80750	−0.20000	0.81255		0.81000
1.0	1.00250	1.01032	0.99750	2.00000	1.00252	3168	1.00000

CONCEPT CHECK

1. Do 10 steps. Solve the problem exactly. Compute the error. (Show the details.)
   (a) $y' = y$, $y(0) = 1$, $h = 0.1$
   (b) $y' = y$, $y(0) = 1$, $h = 0.01$
   (c) $y' = (y - x)^2$, $y(0) = 0$, $h = 0.1$
   (d) $y' = (y + x)^2$, $y(0) = 0$, $h = 0.1$
2. Do 10 steps. Solve exactly. Compute the error. (Show the details.)
   (a) $y' = y$, $y(0) = 1$, $h = 0.1$. Compare with Prob. 1(a) and comment.
   (b) **(Logistic population)** $y' = y - y^2$, $y(0) = 0.2$, $h = 0.1$
   (c) $y' - xy^2 = 0$, $y(0) = 1$, $h = 0.1$
   (d) $y' + y \tan x = \sin 2x$, $y(0) = 1$, $h = 0.1$
   (e) Do Prob. 2(c) using the Euler method with $h = 0.1$ and compare the accuracy.
   (f) $y' = 1 - (1/4)\,y^2$, $y(0) = 0$, $h = 0.1$

6.3 Picard's Method

This method is also known as "Picard's Method of successive approximations." Consider

$$\frac{dy}{dx} = f(x, y) \quad \text{and} \quad y = y_0 \text{ at } x = x_0 \tag{6.17}$$

Integrating the given equation, we get

$$y = \int_{x_0}^{x} f(x, y)dx + C$$

As $y = y_0$ at $x = x_0$, we have $y_0 = C$. Therefore

$$y = y_0 + \int_{x_0}^{x} f(x, y)\,dx \tag{6.18}$$

As y is some function of x, (6.18) can also be expressed as

$$y(x) = y_0 + \int_{x_0}^{x} f(x, y)\,dx$$

Equation (6.18), in which the unknown function y appears in the integrand, is called an "integral equation." This can be solved by successive approximation method (also known as the method of iteration).

The first approximation to y is obtained by putting y_0 for y on the right-hand side of (6.18). Then $y = y^{(1)}$ is the value of y and is given by

$$y^{(1)} = y_0 + \int_{x_0}^{x} f(x, y_0)\, dx \tag{6.19}$$

As y_0 is a constant, the integral on the right of (6.19) can be solved, which yields $y^{(1)}$. This is again substituted on right-hand side of (6.18) to get the second approximation to y, which is $y^{(2)}$:

$$y^{(2)} = y_0 + \int_{x_0}^{x} f(x, y^{(1)})\, dx$$

Proceeding in this way, we get $y^{(3)}, y^{(4)}, \ldots, y^{(n-1)}$ and $y^{(n)}$, where

$$y^{(n)} = y_0 + \int_{x_0}^{x} f(x, y^{(n-1)})\, dx$$

with $n = 1, 2, 3$, etc., and when $x = x_0$, we have $y = y_0$ as the initial condition, or $y^{(0)} = y_0$ becomes the initial or starting value for this iterative method. We thus proceed from $y^{(0)}$ to $y^{(1)}$ to $y^{(2)}$ to $y^{(n)}$ till two successive approximations are sufficiently close. This method is not of much practical use.

Let us workout an example, for completeness sake.

Example 7

Solve the equation

$$y' = x + y^2$$

given $y = 1$ at $x = 0$.

Solution: Here $x_0 = 0$, $y_0 = 1$. From (6.18), we have

$$y = y_0 + \int_{x_0}^{x} f(x, y)\, dx$$

Using the method of successive approximations, we have

$$y^{(0)} = y_0 = 1$$

$$y^{(1)} = y_0 + \int_{x_0}^{x} f(x, y_0)\, dx = 1 + \int_{0}^{x} (x + 1^2)\, dx = 1 + \frac{x^2}{2} + x$$

$$y^{(2)} = 1 + \int_{0}^{x} f(x, y^{(1)})\, dx = 1 + \int_{0}^{x} \left[x + \left(1 + x + \frac{x^2}{2}\right)^2 \right] dx = 1 + x + \frac{3}{2}x^2 + \frac{2}{3}x^3 + \frac{1}{4}x^4 + \frac{1}{20}x^5$$

The integrations thus become more and more involved and the resulting expressions are too big, as we proceed to higher approximations. The method therefore has very limited applications.

Note: This is not at all a practical method and its algorithm/program development is quite complicated. We therefore do not attempt to write an algorithm on the same.

CONCEPT CHECK

1. Solve $y = x^2 + y^2$ given $y(0) = 0$ using Picard's method correct to three places of decimals.
2. Obtain $y(0.2)$ by Picard's method given

$$y' = \frac{y-x}{y+x}, \ y(0) = 1$$

3. Solve $y' = y - x^2$, $y(0) = 1$ by Picard's method upto the third approximation. Hence, find the value of $y(0.1)$, $y(0.2)$.

6.4 Taylor's Series Method

The methods that have been or will be discussed in this chapter make use of Taylor series truncated after first order, either directly or in the evaluation of the mean slope on an interval. It is often desirable to avoid this truncation, so that we are free to use higher order terms from the Taylor series.

We can illustrate the Taylor method by considering the differential equation

$$y' = \frac{dy}{dx} = f(x, y)$$

given $y = y_0$ at $x = x_0$. Let $y(x)$ be the exact solution of the above differential equation. Expanding $y(x)$ in Taylor's series around the initial point $x = x_0$, we get

$$y(x) = y_0 + (x - x_0)\frac{dy}{dx} + \frac{(x-x_0)^2}{2!}\frac{d^2y}{dx^2} + \frac{(x-x_0)^3}{3!}\frac{d^3y}{dx^3} + \cdots \tag{6.20}$$

Then extrapolation for y at any $x = x_0 + h$ is given by

$$y(x_0 + h) = y_0 + h \times \left.\frac{dy}{dx}\right|_{x=x_0} + \frac{h^2}{2!} \times \left.\frac{d^2y}{dx^2}\right|_{x=x_0} + \cdots$$

We know that y is a function of x and y' or dy/dx is a function of x, y. Then

$$y'' = \frac{d^2y}{dx^2} = \frac{d}{dx}\left(\frac{dy}{dx}\right) = \frac{d}{dx} f(x, y) = \frac{\partial}{\partial x} f(x, y) + \frac{\partial}{\partial y} f(x, y) \ \frac{dy}{dx}$$

Let

$$\frac{\partial}{\partial x} f(x, y) = f_x, \ \frac{\partial}{\partial y} f(x, y) = f_y, \ \frac{dy}{dx} = f(x, y) = f$$

for convenience. Then

$$y'' = f_x + f_y \times f_x + f \times f_y$$

$$y''' = \frac{d}{dx}(y'') = f_{xx} + 2f \times f_{xy} + f^2 f_{yy} + f_x f_y + f f_y^{\,2}$$

Therefore (6.20) can be written as

$$y(x_0+h) = y_0 + hf(x_0, y_0) + \frac{h^2}{2!}(f_x + ff_y) + \frac{h^3}{3!}(\cdots) \quad (y = y_0 \quad \text{at} \quad x = x_0) \tag{6.21}$$

Limiting the summation up to three terms and neglecting h^3 term and other higher powers of h, we get

$$y(x_0+h) \approx y_0 + hf(x_0, y_0) + \frac{h^2}{2!}[f_x(x_0, y_0) + f(x_0, y_0) \times f_y(x_0, y_0)] \tag{6.22}$$

This will definitely have truncation error which will predominantly be the h^3 term.

In general, depending on the accuracy required, as many number of terms are to be considered for summation in (6.21). Instead of going for expressions for y', y''', etc., in terms of f, f_x, f_y, it would be easier to derive y'', y''', etc., straight away from the given $f(x, y)$ expression.

Taylor series method is not applicable in many cases, because the derivatives of $f(x, y)$ are not always easy to obtain; even more, this method is particularly unsuitable if $f(x, y)$ is given in a tabular form.

Example 8

From the Taylor series for $y(x)$, find $y(0.1)$ correct to 4 decimal places if $y' = x - y^2$ and $y(0) = 1$.

Solution: Given

$$y' = \frac{dy}{dx} = f(x, y) = x - y^2$$

Also $y(x_0 = 0) = y_0 = 1$. The Taylor series for $y(x)$ is given by

$$y(x) = 1 + xy' + \frac{x^2}{2!}y''\ (0, 1) + \frac{x^3}{3!}\ y'''\ (0, 1) + \frac{x^4}{4!}y^{\text{iv}}\ (0,\ 1) + \cdots$$

The derivatives are

$$y' = x - y^2 \Rightarrow y'(0, 1) = -1$$

$$y'' = \frac{d}{dx}(y') = 1 - 2yy' \Rightarrow y''(0, 1) = 3$$

$$y''' = \frac{d}{dx}(y'') = -2yy'' - 2(y')^2 \Rightarrow y'''(0, 1) = -8$$

$$y^{\text{iv}} = \frac{d}{dx}(y''') = -2y'y'' - 2yy''' - 4y'y'' = -6y'y'' - 2yy''' \Rightarrow y^{\text{iv}}(0, 1) = 34$$

Using these values, the Taylor series becomes

$$y(x) = 1 - x + \frac{3}{2}x^2 - \frac{4}{3}x^3 + \frac{17}{12}x^4 - \cdots$$

To obtain $y(0, 1)$ correct to 4 decimals, it is necessary that terms upto x^4 should be considered, and then

$$y(0.1) = 0.9138$$

Note: If we consider only the first two terms in the Taylor series, we see that the expression is similar to that of Euler's method. That is, Euler's method is a special case of Taylor series method.

CONCEPT CHECK

1. Find $y(0.1)$, $y(0.2)$, $z(0.1)$, $z(0.2)$ given

$$\frac{dy}{dx} = x + z, \quad \frac{dz}{dx} = x - y^2$$

and $y(0) = 2$, $z(0) = 1$ by using Taylor's series method.

2. Solve $y' = x + y$ given $y(1) = 0$. Find $y(1.1)$ and $y(1.2)$ by Taylor's series method.

3. Solve the equation

$$\frac{dy}{dx} = x - y^2$$

with the conditions $y(0) = 1$ and $y'(0) = 1$. Find $y(0.2)$ and $y(0.4)$ using Taylor's series method.

6.5 Runge–Kutta Methods (RK Methods)

A method of great practical importance and much greater accuracy than that of the improved Euler method is the *classical Runge–Kutta method of fourth order,* which we call briefly the **Runge–Kutta method**. The algorithm is shown as for better understanding. We see that in each step we first compute four auxiliary quantities k_1, k_2, k_3, k_4 and then the new value y_{n+1}. The method is well suited to the computer because it needs no special starting procedure, makes light demand on storage, and repeatedly uses the same straightforward computational procedure. It is numerically stable.

Note that if f depends only on x, this method reduces to Simpson's rule of integration (explained in Chapter 5). Note further that $k_1, \ldots, k_4$ depend on n and generally change from step to step.

Example 9 Classical Runge–Kutta Method

Apply the Runge–Kutta method to the initial value problem (6.8) in Example 3, choosing $h = 0.2$, as before and computing five steps.

Solution: For the present problem we have $f(x, y) = x + y$. Hence

$$k_1 = 0.2(x_n + y_n)$$
$$k_2 = 0.2(x_n + 0.1 + y_n + 0.5k_1)$$
$$k_3 = 0.2(x_n + 0.1 + y_n + 0.5k_2)$$
$$k_4 = 0.2(x_n + 0.2 + y_n + k_3)$$

Table 4 shows the results and their errors, which are smaller by factors 10^3 and 10^4 than those for the two Euler methods. See also Table 5. We mention in passing that since the present $k_1, \ldots, k_4$ are simple, operations were saved by substituting k_1 into k_2, then k_2 into k_3, etc.; the resulting formula is shown in Column 4 of Table 4.

Table 4 Runge–Kutta Method Applied to (6.7)

n	x_n	y_n	$0.2214(x_n+y_n)+0.0214$	*Exact values* (6D) $y=e^x-x-1$	$10^6\times$ *Error of* y_n
0	0.0	0	0.021400	0.000000	0
1	0.2	0.021400	0.070418	0.021403	3
2	0.4	0.091818	0.130289	0.091825	7
3	0.6	0.222107	0.203414	0.222119	12
4	0.8	0.425521	0.292730	0.425541	20
5	1.0	0.718251		0.718282	31

Table 5 Comparison of the Accuracy of the Three Methods Under Consideration in the Case of the Initial Value Problem (6.7), with $h = 0.2$

x	$y=e^x-x-1$	*Error*		
		Euler	*Error improved error*	*Runge–Kutta*
0.2	0.021403	0.021	0.0014	0.000003
0.4	0.091825	0.052	0.0034	0.000007
0.6	0.222119	0.094	0.0063	0.000011
0.8	0.425541	0.152	0.0102	0.000020
1.0	0.718282	0.229	0.0156	0.000031

ALGORITHM RUNGE–KUTTA (f, x_0, y_0, h, N)

This algorithm computes the solution of the initial value problem $y' = f(x, y)$, $y(x_0) = y_0$ at equidistant points

$$x_1 = x_0 + h,\ x_2 = x_0 + 2h,\ \ldots,\ x_N = x_0 + Nh;$$

here f is such that this problem has a unique solution on the interval $[x_0, x_N]$.

INPUT: Function f, initial values x_0, y_0, step size h, number of steps N

OUTPUT: Approximation y_{n+1} to the solution $y(x_{n+1})$ at $x_{n+1} = x_0 + (n + 1)h$, where $n = 0, 1, \ldots, N - 1$

For $n = 0, 1, \ldots, N-1$ do:

$k_1 = hf(x_n, y_n)$

$k_2 = hf(x_n + h/2, y_n + k_1/2)$

$k_3 = hf(x_n + h/2, y_n + k_2/2)$

$k_4 = hf(x_n + h, y_n + k_3)$

$x_{n+1} = x_n + h$

$y_{n+1} = y_n + (1/6)(k_1 + 2k_2 + 2k_3 + k_4)$

OUTPUT x_{n+1}, y_{n+1}

End

Stop

End RUNGE–KUTTA

Error and Step Size Control

The idea of adaptive integration has analogs for Runge–Kutta (and other) methods. For RK (Runge–Kutta), if we compute in each step approximations $\tilde{y}$ and $\tilde{\tilde{y}}$ with step sizes h and $2h$, respectively, the latter has error per step equal to $2^5 = 32$ times that of the former; however, since we have only half as many steps for $2h$, the actual factor is $2^5/2 = 16$, so that, say,

$$\epsilon^{(2h)} \approx 16\epsilon^{(h)}$$

and thus

$$y^{(h)} - y^{(2h)} = \epsilon^{(2h)} - \epsilon^{(h)} \approx (16-1)\epsilon^{(h)}$$

Hence the error $\epsilon = \epsilon^{(h)}$ for step size h is about

$$\epsilon = \frac{1}{15}(\tilde{y} - \tilde{\tilde{y}}) \tag{6.23}$$

where $\tilde{y} - \tilde{\tilde{y}} = y^{(h)} - y^{(2h)}$, as said before. Table 6 illustrates (6.23) for the initial value problem

$$y' = (y - x - 1)^2 + 2, \qquad y(0) = 1 \tag{6.24}$$

the step size $h = 0.1$ and $0 \leqq x \leqq 0.4$. We see that the estimate is close to the actual error. This method of error estimation is simple but may be unstable.

Table 6 Runge–Kutta Method Applied to the Initial Value Problem (6.24) and Error Estimate (6.23) (Exact Solution $y = \tan x + x + 1$)

x	$\tilde{y}$ (Step size h)	$\tilde{\tilde{y}}$ (Step size $2h$)	Error estimate (6.23)	Actual error	Exact solution (9D)
0.0	1.000000000	1.000000000	0.000000000	0.000000000	1.000000000
0.1	1.200334589			0.000000083	1.200334672
0.2	1.402709878	1.402707408	0.000000165	0.000000157	1.402710036
0.3	1.609336039			0.000000210	1.609336250
0.4	1.822792993	1.822788993	0.000000267	0.000000226	1.822793219

Runge–Kutta–Fehlberg (RKF)

E. Fehlberg proposed and developed error control by using two RK methods of different orders to go from (x_n, y_n) to (x_{n+1}, y_{n+1}). The difference of the computed y values at x_{n+1} gives an error estimate to be used for step size control. Fehlberg discovered two RK formulas that together need only 6 function evaluations per step. We present these formulas here because RKF has become quite popular. For instance, Maple uses it (also for systems of ODEs).

Fehlberg's fifth-order RK method is

$$y_{n+1} = y_n + \gamma_1 k_1 + \cdots + \gamma_6 k_6 \tag{6.25a}$$

with coefficient vector $\gamma = [\gamma_1 \cdots \gamma_6]$,

$$\gamma = \begin{bmatrix} \frac{16}{135} & 0 & \frac{6656}{12825} & \frac{28561}{56430} & -\frac{9}{50} & \frac{2}{55} \end{bmatrix} \tag{6.25b}$$

His **fourth-order RK method** is

$$y_{n+1}^* = y_n + \gamma_1^* k_1 + \ldots + \gamma_5^* k_5 \tag{6.26a}$$

with coefficient vector

$$\gamma^* = \begin{bmatrix} \frac{25}{216} & 0 & \frac{1408}{2565} & \frac{2197}{4104} & -\frac{1}{5} \end{bmatrix} \tag{6.26b}$$

In both formulas we use only six different function evaluations altogether, namely,

$$\begin{aligned}
k_1 &= hf(x_n, y_n) \\
k_2 &= hf\left(x_n + \frac{1}{4}h, y_n + \frac{1}{4}k_1\right) \\
k_3 &= hf\left(x_n + \frac{3}{8}h, y_n + \frac{3}{32}k_1 + \frac{9}{32}k_2\right) \\
k_4 &= hf\left(x_n + \frac{12}{13}h, y_n + \frac{1932}{2197}k_1 - \frac{7200}{2197}k_2 - \frac{7296}{2197}k_3\right) \\
k_5 &= hf\left(x_n + h, y_n + \frac{439}{216}k_1 - 8k_2 + \frac{3680}{513}k_3 - \frac{845}{4104}k_4\right) \\
k_6 &= hf\left(x_n + \frac{1}{2}h, y_n - \frac{8}{27}k_1 + 2k_2 - \frac{3544}{2565}k_3 + \frac{1859}{4104}k_4 - \frac{11}{40}k_5\right)
\end{aligned} \tag{6.27}$$

The difference of (6.25a) and (6.26a) gives the **error estimate**

$$\epsilon_{n+1} \approx y_{n+1} - y_{n+1}^* = \frac{1}{360}k_1 - \frac{128}{4275}k_3 - \frac{2197}{75240}k_4 + \frac{1}{50}k_5 + \frac{2}{55}k_6 \tag{6.28}$$

Example 10 Runge–Kutta–Fehlberg

For the initial value problem (6.24) we obtain from (6.25)–(6.27) with $h = 0.1$ in the first step the 12S-values

$$k_1 = 0.200000000000, \qquad k_2 = 0.200062500000$$
$$k_3 = 0.200140756867, \qquad k_4 = 0.200856926154$$
$$k_5 = 0.201006676700, \qquad k_6 = 0.200250418651$$

$$y_1^* = 1.20033466949$$

$$y_1 = 1.20033467253$$

and the error estimate

$$\epsilon_1 \approx y_1 - y_1^* = 0.00000000304$$

The exact 12S-value is $y(0.1) = 1.20033467209$. Hence the actual error of y_1 is -4.4×10^{-10}, smaller than that in Table 6 by a factor 200.

Table 7 Methods Considered and Their Order (= their Global Error)

Method	*Function evaluation per step*	*Global error*	*Local error*
Euler	1	$O(h)$	$O(h^2)$
Improved Euler	2	$O(h^2)$	$O(h^3)$
RK (fourth order)	4	$O(h^4)$	$O(h^5)$
RKF	6	$O(h^5)$	$O(h^6)$

Table 7 summarizes essential features of the methods in this section. It can be shown that these methods are *numerically stable.* They are **one-step methods** because in each step we use the data of just *one* preceding step, in contrast to **multistep methods** where in each step we use data from *several* preceding steps, as we shall see later.

CONCEPT CHECK

1. Do 10 steps. Compare as indicated. Comment. (Show the details.)
   (a) $y' - xy^2 = 0$, $y(0) = 1$, $h = 0.1$. Compare with $y' - xy^2 = 0$, $y(0) = 1$, $h = 0.1$. Apply (6.22) to y_{10}.
   (b) $y' = y - y^2$, $y(0) = 0.2$, $h = 0.1$. Compare with $y' - y - y^2$, $y(0) = 0.2$, $h = 0.1$. Apply (6.22) to y_{10}.
   (c) $y' = (1 + x^{-1})y$, $y(1) = e$, $h = 0.2$
   (d) $y' = (1/2)(y/x - x/y)$, $y(2) = 2$, $h = 0.2$
   (e) $y' + y \tan x = \sin 2x$, $y(0) = 1$, $h = 0.1$
   (f) In Prob. 1(e) use $h = 0.2$ (5 steps) and compare the error.
   (g) $y' + 5x^4y^2 = 0$, $y(0) = 1$, $h = 0.2$
2. **Kutta's third-order method** is defined by $y_{n+1} = y_n + (1/6)(k_1 + 4k_2 + k_3^*)$ with k_1 and k_2 as in RK and $k_3^* = hf(x_{n+1}, y_n - k_1 + 2k_2)$. Apply this method to (6.7) in Example 3. Choose $h = 0.2$ and do 5 steps.

6.6 Adams–Bashforth–Moulton (ABM) Methods

Adams–Bashforth Methods

We consider an initial value problem

$$y' = f(x, y), \quad y(x_0) = y_0 \tag{6.29}$$

as before, with f such that the problem has a unique solution on some open interval containing x_0. We integrate $y' = f(x, y)$ from x_n to $x_{n+1} = x_n + h$. This gives

$$\int_{x_n}^{x_{n+1}} y'(x)dx = y(x_{n+1}) - y(x_n) = \int_{x_n}^{x_{n+1}} f(x, y(x))dx$$

Now comes the main idea. We replace $f(x, y(x))$ by an interpolation polynomial $p(x)$, so that we can later integrate. This gives approximations y_{n+1} of $y(x_{n+1})$ and y_n of $y(x_n)$ as

$$y_{n+1} = y_n + \int_{x_n}^{x_{n+1}} p(x)dx \tag{6.30}$$

Different choices of $p(x)$ will now produce different methods. We explain the principle by taking a cubic polynomial, namely, the polynomial $p_3(x)$ that at (equidistant) $x_n, x_{n-1}, x_{n-2}, x_{n-3}$, has the respective values

$$\begin{aligned} f_n &= f(x_n, y_n) \\ f_{n-1} &= f(x_{n-1}, y_{n-1}) \\ f_{n-2} &= f(x_{n-2}, y_{n-2}) \\ f_{n-3} &= f(x_{n-3}, y_{n-3}) \end{aligned} \tag{6.31}$$

This will lead to a practically useful formula. We can obtain $p_3(x)$ from Newton's backward difference formula (discussed in Chapter 4):

$$p_3(x) = f_n + r\nabla f_n + \frac{1}{2}r(r+1)\nabla^2 f_n + \frac{1}{6}r(r+1)(r+2)\nabla^3 f_n$$

where

$$r = \frac{x - x_n}{h}$$

We integrate $p_3(x)$ over x from x_n to $x_{n+1} = x_n + h$, thus over r from 0 to 1. Since $x = x_n + hr$, we have $dx = h\,dr$. The integral of $r(r + 1)/2$ is 5/12 and that of $r(r + 1)(r + 2)/6$ is 3/8. We thus obtain

$$\int_{x_n}^{x_{n+1}} p_3\, dx = h\int_0^1 p_3\, dr = h\left(f_n + \frac{1}{2}\nabla f_n + \frac{5}{12}\nabla^2 f_n + \frac{3}{8}\nabla^3 f_n\right) \tag{6.32}$$

It is practical to replace these differences by their expressions in terms of f:

$$\begin{aligned} \nabla f_n &= f_n - f_{n-1} \\ \nabla^2 f_n &= f_n - 2f_{n-1} + f_{n-2} \\ \nabla^3 f_n &= f_n - 3f_{n-1} + 3f_{n-2} - f_{n-3} \end{aligned}$$

We substitute this into (6.32) and collect terms. This gives the multistep formula of the **Adams–Bashforth method** *of fourth order*

$$y_{n+1} = y_n + \frac{h}{24}(55f_n - 59f_{n-1} + 37f_{n-2} - 9f_{n-3}) \tag{6.33}$$

It expresses the new value y_{n+1} [approximation of the solution y of (6.29) at x_{n+1}] in terms of 4 values of f computed from the y values obtained in the preceding 4 steps. The local truncation error is of order h^5, as can be shown, so that the global error is of order h^4; hence (6.33) does define a fourth-order method.

Adams–Moulton Methods

Adams–Moulton methods are obtained if for $p(x)$ in (6.30) we choose a polynomial that interpolates $f(x, y(x))$ at $x_{n+1}, x_n, x_{n-1}, \ldots$ (as opposed to $x_n, x_{n-1}, \ldots$ used before; this is the main point). We explain the principle for the cubic polynomial $\tilde{p}_3(x)$ that interpolates at $x_{n+1}, x_n, x_{n-1}, x_{n-2}$. (Before we had $x_n, x_{n-1}, x_{n-2}, x_{n-3}$.) Again using Newton's backward difference formula but now setting $r = (x - x_{n+1})/h$, we have

$$\tilde{p}_3(x) = f_{n+1} + r\nabla f_{n+1} + \frac{1}{2}r(r+1)\nabla^2 f_{n+1} + \frac{1}{6}r(r+1)(r+2)\nabla^3 f_{n+1}$$

We now integrate over x from x_n to x_{n+1} as before. This corresponds to integrating over r from −1 to 0. We obtain

$$\int_{x_n}^{x_{n+1}} \tilde{p}_3(x)\,dx = h\left(f_{n+1} - \frac{1}{2}\nabla f_{n+1} - \frac{1}{12}\nabla^2 f_{n+1} - \frac{1}{24}\nabla^3 f_{n+1}\right)$$

Replacing the differences as before gives

$$y_{n+1} = y_n + \int_{x_n}^{x_{n+1}} \tilde{p}_3(x)\,dx = y_n + \frac{h}{24}(9f_{n+1} + 19f_n - 5f_{n-1} + f_{n-2}) \tag{6.34}$$

This is usually called an **Adams–Moulton formula**. It is an **implicit formula** because $f_{n+1} = f(x_{n+1}, y_{n+1})$ appears on the right, so that it defines y_{n+1} only *implicitly*, in contrast to (6.33), which is an **explicit formula**, not involving y_{n+1} on the right. To use (6.34) we must ***predict*** a value $y_{n+1}{}^*$, for instance, by using (6.33), that is,

$$y_{n+1}{}^* = y_n + \frac{h}{24}(55f_n - 59f_{n-1} + 37f_{n-2} - 9f_{n-3}) \tag{6.35a}$$

The ***corrected*** new value y_{n+1} is then obtained from (6.34) with f_{n+1} replaced by $f_{n+1}{}^* = f(x_{n+1}, y_{n+1}{}^*)$ and the other f's as in (6.34); thus,

$$y_{n+1} = y_n + \frac{h}{24}(9f_{n+1}{}^* + 19f_n - 5f_{n-1} + f_{n-2}) \tag{6.35b}$$

This **predictor–corrector method** [(6.35a) and (6.35b)] is usually called the **Adams–Moulton method** *of fourth order* discussed in next section. It has the advantage over RK that (6.35) gives the error estimate

$$\epsilon_{n+1} \approx \frac{1}{15}(y_{n+1} - y_{n+1}{}^*)$$

as can be shown. Sometimes the name "Adams–Moulton method" is reserved for the method with *several* corrections per step by (6.35b) until a specific accuracy is reached. Popular codes exist for both versions of the method.

Getting Started

In (6.33) we need f_0, f_1, f_2, f_3. Hence from (6.31) we see that we must first compute y_1, y_2, y_3 by some other method of comparable accuracy, for instance, by RK or by RKF.

Example 11 Adams–Bashforth Prediction (6.35a), Adams–Moulton Correction (6.35b)

Solve the initial value problem

$$y' = x + y, \qquad y(0) = 0 \tag{6.36}$$

by (6.35a), (6.35b) on the interval $0 \leqq x \leqq 2$, choosing $h = 0.2$.

Solution: We compute starting values y_1, y_2, y_3 by the classical Runge–Kutta method. Then in each step we predict by (6.35a) and make one correction by (6.35b) before we execute the next step. The results are shown and compared with the exact values in Table 8. We see that the corrections improve the accuracy considerably. This is typical.

Table 8 Adams–Moulton Method Applied to the Initial Value Problem (6.36); Predicted Values Computed by (6.35a) and Corrected Values by (6.35b)

n	x_n	*Starting* y_n	*Predicted* y_n^*	*Corrected* y_n	*Exact values*	$10^6 \times$ *error of* y_n
0	0.0	0.000000			0.000000	0
1	0.2	0.021400			0.021403	3
2	0.4	0.091818			0.091825	7
3	0.6	0.222107			0.222119	12
4	0.8		0.425361	0.425529	0.425541	12
5	1.0		0.718066	0.718270	0.718282	12
6	1.2		1.119855	1.120106	1.120117	11
7	1.4		1.654885	1.655191	1.655200	9
8	1.6		2.352653	2.353026	2.353032	6
9	1.8		3.249190	3.249646	3.249647	1
10	2.0		4.388505	4.389062	4.389056	–6

Comments on Comparison of Methods

An Adams–Moulton formula is generally much more accurate than an Adams–Bashforth formula of the same order. This justifies the greater complication and expense in using the former. The method [(6.35a) and (6.35b)] is ***numerically stable***, whereas the exclusive use of (6.35a) might cause instability. Step size control is relatively simple. If |Corrector–Predictor| > TOL, use interpolation to generate "old" results at half the current step size and then try $h/2$ as the new step.

Whereas the Adams–Moulton formula [(6.35a) and (6.35b)] needs only 2 evaluations per step, Runge–Kutta needs 4; however, with Runge–Kutta one may be able to take a step size more than twice as large, so that a comparison of this kind (widespread in the literature) is meaningless.

CONCEPT CHECK

1. Carry out and show the details of the calculations leading to (6.32)–(6.35) in the text.
2. Solve the initial value problems by Adams–Moulton, 10 steps with 1 correction per step. Solve exactly and compute the error. (Use RK where no starting values are given.)
   (a) $y' = y$, $y(0) = 1$, $h = 0.1$ (1.105171, 1.221403, 1.349859)
   (b) $y' = -0.2xy$, $y(0) = 1$, $h = 0.2$
   (c) $y' = 2xy$, $y(0) = 1$, $h = 0.1$
   (d) $y' = 1 + y^2$, $y(0) = 0$, $h = 0.1$
   (e) $y' = x/y$, $y(1) = 3$, $h = 0.2$
   (f) $y' = (x + y - 4)^2$, $y(0) = 4$, $h = 0.2$, only 7 steps (why?)
   (g) $y' = 1 - 4y^2$, $y(0) = 0$, $h = 0.1$
   (h) $y' = x + y$, $y(0) = 0$, $h = 0.1$ (0.00517083, 0.0214026, 0.0498585)
   (i) Show that by applying the method in the text to a polynomial of second degree we obtain

$$y_{n+1}^* = y_n + \frac{h}{12}(23f_n - 16f_{n-1} + 5f_{n-2})$$

$$y_{n+1} = y_n + \frac{h}{12}(5f_{n+1} + 8f_n - f_{n-1})$$

3. Use Prob. 2(i) to solve $y' = 2xy$, $y(0) = 1$ (10 steps, $h = 0.1$, RK starting values). Compare with the exact solution and comment.

4. How much can you reduce the error in Prob. 3 by halving h (20 steps, $h = 0.05$)? First guess, then compute.

6.7 Predictor–Corrector Methods

In Runge–Kutta type of method, any point y_{i+1} is estimated using only the previous point y_i at x_i. None of the previous points like $y_{i-1}, y_{i-2}, \ldots$, on the solution curve are necessary. Thus all past information regarding the behavior of the solution is ignored and each step starts afresh. The process may be easier, but this is inefficient and it would be desirable to use number of previous points (not just one alone) in evolving a solution. The predictor–corrector methods do this precisely.

The predictor–corrector methods use information from several previous known points to compute the next one, as indicated in Figure 5.

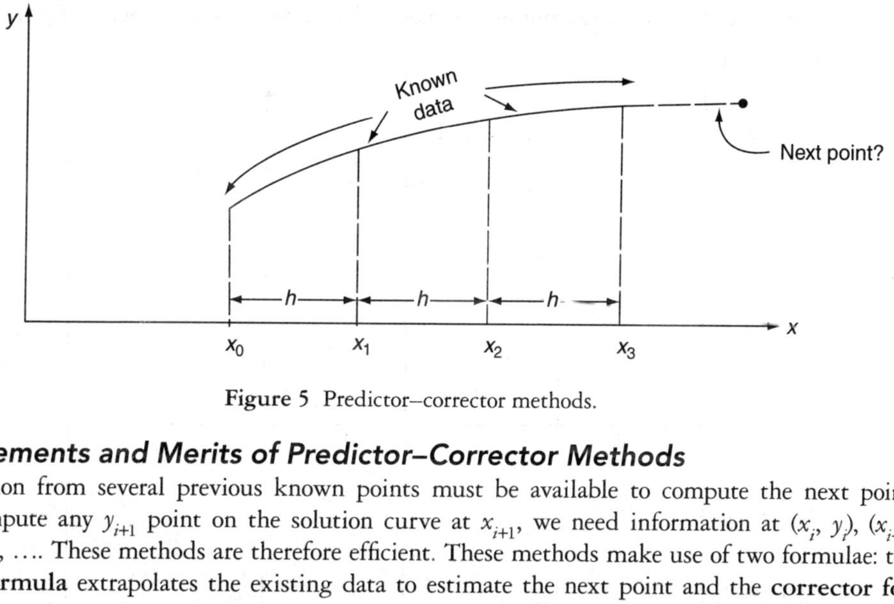

Figure 5 Predictor–corrector methods.

Requirements and Merits of Predictor–Corrector Methods

Information from several previous known points must be available to compute the next point, that is to compute any y_{i+1} point on the solution curve at x_{i+1}, we need information at (x_i, y_i), (x_{i-1}, y_{i-1}), (x_{i-2}, y_{i-2}), ... These methods are therefore efficient. These methods make use of two formulae: the **predictor formula** extrapolates the existing data to estimate the next point and the **corrector formula** improves on this estimate. (In some cases, the corrector formula can be applied repeatedly until some convergence criterion is satisfied.) Some predictor–corrector methods use backward differences and some (like Milne's method, which we shall discuss) use forward differences. Note that backward and forward differences were explained in detail in Chapter 4.

Illustration

Assume that by Runge–Kutta method (or similar method) the solutions at any x_i and x_{i+1} are found to be y_i and y_{i+1}. The solution for y_{i+2} at x_{i+2} may be extrapolated from y_i and y_{i+1} by fitting a quadratic curve through them. Let us assume that this quadratic is given by

$$y = a + b\,(x - x_i) + c(x - x_i)\,(x - x_i - h) \tag{6.37}$$

The unknown coefficients a, b, c may be obtained from

$$y = y_i \quad \text{at} \quad x = x_i \tag{6.38}$$

$$y = y_{i+1} \quad \text{at} \quad x = x_{i+1} \tag{6.39}$$

$$\frac{dy}{dx} \quad \text{at} \quad (x_{i+1},\ y_{i+1}) = f(x_{i+1},\ y_{i+1}) \tag{6.40}$$

Using (6.38)–(6.40) in (6.37), we get

$$a = y_i$$

$$b = \frac{y_{i+1} - y_i}{h} \tag{6.41}$$

$$c = \frac{1}{h} f(x_{i+1}, y_{i+1}) - \frac{y_{i+1} - y_i}{h^2}$$

Using a, b, c in (6.37), we can get y_{i+2} as

$$y_{i+2} = y_i + 2hf(x_{i+1}, y_{i+1}) = y_{i+2}{}^p \tag{6.42}$$

The superscript p indicates that $y_{i+2}{}^p$ is an extrapolated or predicted value. Whether this prediction is reasonable or not may be checked by examining the plot of derivative dy/dx, as shown in Figure 6.

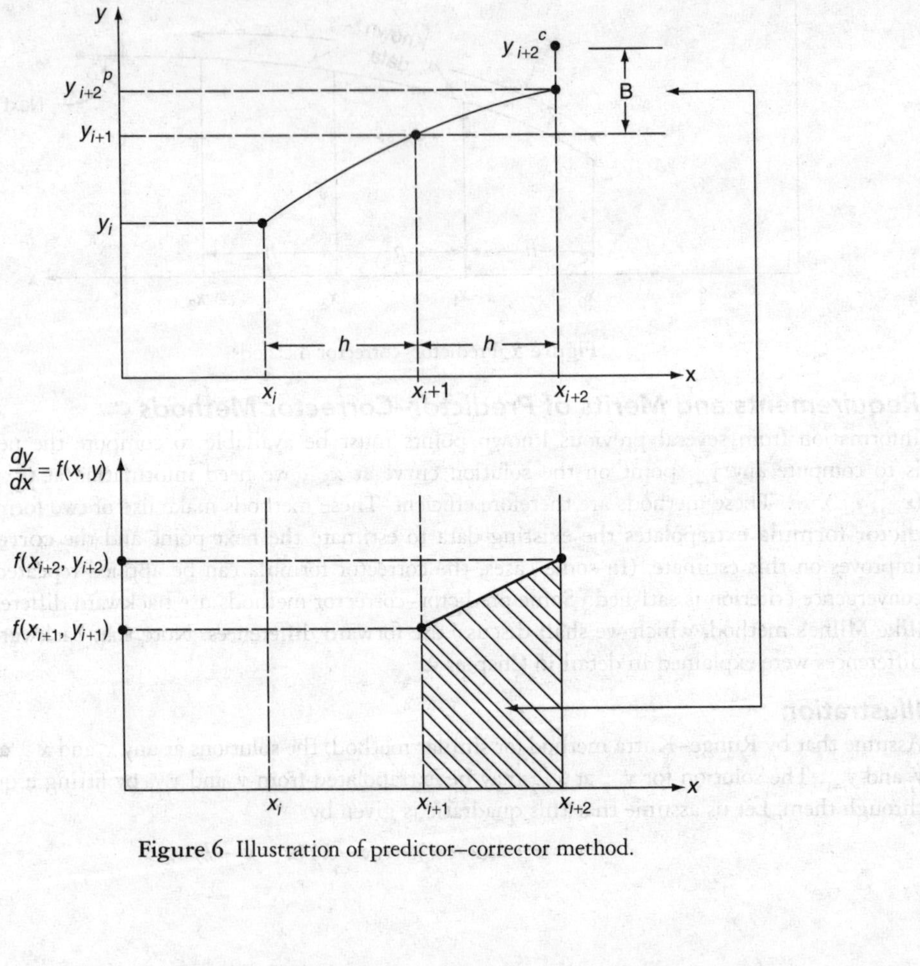

Figure 6 Illustration of predictor–corrector method.

As $y(x)$ is assumed to be a quadratic in (6.37), dy/dx would be linear. The value of dy/dx at x_{i+1} is known and its value at x_{i+2} may be estimated as $f(x_{i+2}, y_{i+2})$ by using the predicted value $y_{i+2}{}^p$ from (6.42). The area under the dy/dx curve between x_{i+1} and x_{i+2} may thus be evaluated using trapezoidal rule (Simpson's rule could also be used with appropriate number of divisions). This area is given by

$$B = \frac{h}{2}\,[f(x_{i+1},\ y_{i+1}) + f(x_{i+2},\ y_{i+2}{}^p)] \tag{6.43}$$

Another estimate of y_{i+2} may be obtained by adding this area B to y_{i+1}. This new estimate is called the corrected value of y_{i+2} and is designated as y_{i+2}^c. It is given by

$$y_{i+2}{}^c = y_{i+1} + \frac{h}{2}[f(x_{i+1},\ y_{i+1}) + f(x_{i+2}, y_{i+2}{}^p)] \tag{6.44}$$

Equations (6.42) and (6.44) constitute a predictor–corrector method of developing the solution of a differential equation from points (x_i, y_i) and (x_{i+1}, y_{i+1}) to (x_{i+2}, y_{i+2}). This method thus illustrates the basic principle of predictor–corrector methods which utilize information on previous two points.

Milne's Method

This is one of the popular predictor–corrector methods. This method uses a formula deduced by Milne as a predictor and the familiar Simpson's rule (for integration/area calculations) as a corrector. The method is of fourth order, that is, the dominant error term in both the predictor and corrector includes h^5. The method requires four initial values to get started. That is the past four points of the solution are used to extrapolate the next point using a fourth degree polynomial curve fit for $y(x)$. Therefore corrector uses Simpson's rule for integration (true for cubic fit, dy/dx being cubic). The formulas are given by

$$y_{i+1}{}^p = y_{i-3} + \frac{4h}{3}[2y_{i-1}{}' + 2\ y_{i-2}{}'] \tag{6.45}$$

Milne's predictor formula

$$y_{i+1}{}^c = y_{i-1} + \frac{h}{3}[y_{i-1}{}' + 4y_i{}' + y_{i-1}{}'] \tag{6.46}$$

Simpson's Rule (Corrector Formula)

The truncation errors of the predictor and corrector formulas are, respectively,

$$\frac{28}{90}h^5 y^{\mathrm{v}}(\xi_1) \quad \text{and} \quad \frac{-1}{90}h^5 y^{\mathrm{v}}(\xi_2)$$

where ξ_1, ξ_2 are the values of x at which the fifth derivatives (in each formula) have an upper bound. Thus, it can be seen that the error term associated with the corrector formula is smaller compared to that associated with the predictor fomula.

Example 12

Solve $y' = 1 + y^2$ given $y(0) = 0$ and compute $y(0.8)$ using Milne's method.

Solution: For Milne's method, we need the information on four previous points. As the initial condition is specified at $x = 0$, and we have to calculate y at $x = 0.8$, the interval 0–0.8 can be conveniently divided into four equal divisions, each of size 0.2. There are thus five points (four steps). To estimate $y(0.8)$, we should also have $y(0)$, $y(0.2)$, $y(0.4)$, $y(0.6)$. Note that $y(0)$ is given, and therefore we have to use some method (Euler's, Runge–Kutta, etc.) to calculate $y(0.2)$, $y(0.4)$, and $y(0.6)$. We shall use Runge–Kutta fourth-order formula to estimate $y(0.2)$, $y(0.4)$, $y(0.6)$. Thus

$$h = 0.2,\ x_0 = 0,\ y_0 = 0,\quad f(x, y) = y' = 1 + y^2$$

(a) We calculate $y(0.2)$

$$k_1 = 0.2 \times (1 + 0) = 0.2$$
$$k_2 = 0.2 \times f(0.1, 0.1) = 0.2 \times 1.01 = 0.202$$
$$k_3 = 0.2 \times f(0.1, 0.101) = 0.2 \times 1.010201 = 0.20204$$
$$k_4 = 0.2 \times f(0.2, 0.20204) = 0.2 \times 1.04082 = 0.20816$$

Therefore

$$y(0.2) = 0 + \frac{1}{6}(k_1 + 2k_2 + 2k_3 + k_4) = 0.2027 \text{ (upto 4 decimals)}$$

(b) Now we calculate $y(0.4)$. [The initial point now is $y(0.2) = 0.2027$.] Therefore for $x_0 = 0.2$, $y_0(0.2) = 0.2027$, $h = 0.2$, $f(x, y) = 1 + y^2$, we have

$$k_1 = 0.2[1 + (0.2027)^2] = 0.2082$$
$$k_2 = 0.2[1 + (0.3068)^2] = 0.2188$$
$$k_3 = 0.2[1 + (0.3121)^2] = 0.2195$$
$$k_4 = 0.2[1 + (0.4222)^2] = 0.2356$$

Therefore

$$y(0.4) = 0.2027 + 0.2201 = 0.4228$$

(c) The value $y(0.6)$ is calculated. We know that $y(0.4) = 0.4228$. Therefore for $h = 0.2$, $x_0 = 0.4$, $y_0 = 0.4228$, $f(x, y) = 1 + y^2$, proceeding in a similar way as above, we get

$$y(0.6) = 0.6841$$

Now we have to compute $y(0.8)$ using predictor–corrector (Milne's) method. Here, we have to use the following: $y(0) = 0$, $y(0.2) = 0.2027$, $y(0.4) = 0.4228$, $y(0.6) = 0.6841$. We are given $f(x, y) = y' = 1 + y^2$ and $h = 0.2$. We know that

$$y_4^{\,p} = y_0 + \frac{4h}{3}(2y_3' - y_2' + 2y_1')$$

Now

(a) $y_0 = y(0) = 0$

(b) $y_1 = y(0.2) = 0.2027$, $x_1 = 0.2$. Therefore

$$y_1' = f(x_1, y_1) = 1 + (0.2027)^2 = 1.0411$$

(c) $y_2 = y(0.4) = 0.4228$, $x_2 = 0.4$. Therefore

$$y_2' = f(x_2, y_2) = 1 + (0.4228)^2 = 1.1787$$

(d) $y_3 = y(0.6) = 0.6841$, $x_3 = 0.6$. Therefore

$$y_3' = f(x_3, y_3) = 1 + (0.6841)^2 = 1.4681$$

Thus

$$y_4^p = 0 = \frac{0.2\times 4}{3}(2\times 1.4681 + 2\times 1.0411 - 1.1787) = 1.0239$$

Now $y(x_4) = y(0.8) = y_4 = 1.0239$. Therefore

$$y_4' = f(x_4, y_4) = f(0.8, 1.0239) = 1 + (1.0239)^2 = 2.0483$$

We have got the predicted value now. The correct value of $y(0.8)$ is

$$y_4^c = y_2 \frac{h}{2}[y_2' + 4y_3' + y_4'] = 0.4228 + (1.1787 + 4\times 1.4681 + 2.0483) = 1.0294$$

This is the corrector output.

ALGORITHM MILNE'S (FOURTH-ORDER) METHOD

INPUT: x_0, y_0: initial condition
$f(x, y)$: functional relation for slope y'
$x_4 = x_f$: at which y is to be computed
Integer n_x: Number of steps (choose in such a way that $h = x_f - x_0/n_3$ is a finite fraction, but not an unending fraction. That is $x_0 + n_s h$ should be exactly equal to x_f.
Typical value of $n_s = 4$

OUTPUT: $y(x_f)$ or $y(x_4)$

Requirements: As only one initial point $y(x_0) = y_0$ is being specified, the other three starting points required for Milne's method are to be obtained from Runge–Kutta fourth-order method.

6.8 Methods for Systems and Higher Order ODEs

Initial value problems for first-order systems of ODEs are of the form

$$y' = f(x, y), \qquad y(x_0) = y_0 \tag{6.47}$$

in components

$$\begin{aligned} y_1' &= f_1(x, y_1, \ldots, y_m), & y_1(x_0) &= y_{10} \\ y_2' &= f_2(x, y_1, \ldots, y_m), & y_2(x_0) &= y_{20} \\ &\cdots\cdots & &\cdots\cdots \\ y_m' &= f_m(x, y_1, \ldots, y_m), & y_m(x_0) &= y_{m0} \end{aligned}$$

f is assumed to be such that the problem has a unique solution $y(x)$ on some open x interval containing x_0.

Before explaining solution methods it is important to note that (6.47) includes initial value problems for single mth-order ODEs,

$$y^{(m)} = f(x, y, y', y'', \ldots, y^{(m-1)}) \tag{6.48}$$

and initial conditions $y(x_0) = K_1, y'(x_0) = K_2, \cdots, y^{(m-1)}(x_0) = K_m$ as special cases. Indeed, the connection is achieved by setting

$$y_1 = y, \qquad y_2 = y', \qquad y_3 = y'', \qquad \ldots, \qquad y_m = y^{(m-1)} \tag{6.49}$$

Then we obtain the system

$$\begin{aligned} y_1' &= y_2 \\ y_2' &= y_3 \\ &\vdots \\ y_{m-1}' &= y_m \\ y_m' &= f(x, y_1, \ldots, y_m) \end{aligned} \tag{6.50}$$

and the initial conditions $y_1(x_0) = K_1, y_2(x_0) = K_2, \ldots, y_m(x_0) = K_m$.

Euler Method for Systems

Methods for single first-order ODEs can be extended to systems (6.47) simply by writing vector functions $\mathbf{y}$ and $\mathbf{f}$ instead of scalar functions y and f, whereas x remains a scalar variable.

We begin with the Euler method. Just as for a single ODE this method will not be accurate enough for practical purposes, but it nicely illustrates the extension principle.

Example 13 Euler Method for a Second-Order ODE. Mass–Spring System

Solve the initial value problem for a damped mass–spring system

$$y'' + 2y' + 0.75y = 0, \qquad y(0) = 3, \qquad y'(0) = -2.5$$

by the Euler method for systems with step $h = 0.2$ for x from 0 to 1 (where x is time).

Solution: The Euler method (6.3) generalizes to systems of the form

$$\mathbf{y}_{n+1} = \mathbf{y}_n + h\mathbf{f}(x_n, \mathbf{y}_n) \tag{6.51}$$

in components

$$y_{1,\,n+1} = y_{1,\,n} + hf_1(x_n, y_{1,\,n}, y_{2,\,n})$$

$$y_{2,\,n+1} = y_{2,\,n} + hf_2(x_n, y_{1,\,n}, y_{2,\,n})$$

and similarly for systems of more than two equations. By (6.49) the given ODE converts to the system

$$y_1{}' = f_1(x, y_1, y_2) = y_2$$
$$y_2{}' = f_2(x, y_1, y_2) = -2y_2 - 0.75y_1$$

Hence (6.51) becomes

$$y_{1,\,n+1} = y_{1,\,n} + 0.2y_{2,\,n}$$
$$y_{2,\,n+1} = y_{2,\,n} + 0.2(-2y_{2,\,n} - 0.75y_{1,\,n})$$

The initial conditions are $y(0) = y_1(0) = 3$, $y'(0) = y_2(0) = -2.5$. The calculations are shown in Table 9. As for single ODEs, the results would not be accurate enough for practical purposes. The example merely serves to illustrate the method because the problem can be readily solved exactly,

$$y = y_1 = 2e^{-0.5x} + e^{-1.5x}, \qquad \text{thus} \qquad y' = y_2 = -e^{-0.5x} - 1.5e^{-1.5x}$$

Table 9 Euler Method for Systems in Example 13 (Mass–Spring System)

n	x_n	$y_{1,n}$	y_1 *exact* (5D)	*Error* ($\epsilon_1 = y_1 - y_{1,n}$)	$y_{2,n}$	y_2 *exact* (5D)	*Exact* ($\epsilon_2 = y_2 - y_{2,n}$)
0	0.0	3.00000	3.00000	0.00000	−2.50000	−2.50000	0.00000
1	0.2	2.50000	2.55049	0.05049	−1.95000	−2.01606	−0.06606
2	0.4	2.11000	2.18627	0.76270	−1.54500	−1.64195	−0.09695
3	0.6	1.80100	1.88821	0.08721	−1.24350	−1.35067	−0.10717
4	0.8	1.55230	1.64183	0.08953	−1.01625	−1.12211	−0.10586
5	1.0	1.34905	1.43619	0.08714	−0.84260	−0.94123	−0.09863

Runge–Kutta Methods for Systems

As for Euler methods, we obtain RK methods for an initial value problem (6.47) simply by writing vector formulas for vectors with m components, which for $m = 1$ reduce to the previous scalar formulas. Thus for the *classical* **RK method** *of fourth order*. We obtain

$$\mathbf{y}(x_0) = \mathbf{y}_0 \qquad \text{(Initial values)} \tag{6.52}$$

and for each step $n = 0, 1, \ldots, N - 1$ we obtain the 4 auxiliary quantities

$$\begin{aligned} \mathbf{k}_1 &= h\mathbf{f}(x_n, \mathbf{y}_n) \\ \mathbf{k}_2 &= h\mathbf{f}\left(x_n + \frac{1}{2}h,\ \mathbf{y}_n + \frac{1}{2}\mathbf{k}_1\right) \\ \mathbf{k}_3 &= h\mathbf{f}\left(x_n + \frac{1}{2}h,\ \mathbf{y}_n + \frac{1}{2}\mathbf{k}_2\right) \\ \mathbf{k}_4 &= h\mathbf{f}(x_n + h,\ \mathbf{y}_n + \mathbf{k}_3) \end{aligned} \tag{6.53}$$

and the new value [approximation of the solution $y(x)$ at $x_{n+1} = x_0 + (n + 1)h$]

$$\mathbf{y}_{n+1} = \mathbf{y}_n + \frac{1}{6}(\mathbf{k}_1 + 2\mathbf{k}_2 + 2\mathbf{k}_3 + \mathbf{k}_4) \tag{6.54}$$

Example 14 RK Method for Systems. Airy's Equation. Airy Function Ai(*x*)

Solve the initial value problem

$$y'' = xy, \quad y(0) = 1/[3^{2/3} \times \Gamma(2/3)] = 0.35502805, \quad y'(0) = -1/[3^{1/3} \times \Gamma(1/3)] = -0.25881940$$

by the Runge–Kutta method for systems with $h = 0.2$; do 5 steps. This is **Airy's equation**, which arose in optics. Γ is the gamma function. The initial conditions are such that we obtain a standard solution, the **Airy function** Ai(x), a special function that has been thoroughly investigated.

Solution: For $y'' = xy$, setting $y_1 = y$, $y_2 = y_1' = y'$ we obtain the system (6.50)

$$y_1' = y_2$$
$$y_2' = xy_1$$

Hence $\mathbf{f} = [f_1 \quad f_2]^T$ in (6.47) has the components $f_1(x, y) = y_2, f_2(x, y) = xy_1$. We now write (6.52)–(6.54) in components. The initial conditions (6.52) are $y_{1,0} = 0.35502805$, $y_{2,0} = -0.25881940$. In (6.53) we have fewer subscripts by simply writing $\mathbf{k}_1 = \mathbf{a}$, $\mathbf{k}_2 = \mathbf{b}$, $\mathbf{k}_3 = \mathbf{c}$, $\mathbf{k}_4 = \mathbf{d}$, so that $\mathbf{a} = [a_1 \quad a_2]^T$, etc. Then (6.53) takes the form

$$\mathbf{a} = h\begin{bmatrix} y_{2,n} \\ x_n y_{1,n} \end{bmatrix}$$

$$\mathbf{b} = h\begin{bmatrix} y_{2,n} + \frac{1}{2}a_2 \\ \left(x_n + \frac{1}{2}h\right)\left(y_{1,n} + \frac{1}{2}a_1\right) \end{bmatrix}$$

$$\mathbf{c} = h\begin{bmatrix} y_{2,n} + \frac{1}{2}b_2 \\ \left(x_n + \frac{1}{2}h\right)\left(y_{1,n} + \frac{1}{2}b_1\right) \end{bmatrix} \tag{6.55}$$

$$\mathbf{d} = h\begin{bmatrix} y_{2,n} + c_2 \\ (x_n + h)(y_{1,n} + c_1) \end{bmatrix}$$

For example, the second component of $\mathbf{b}$ is obtained as follows. $\mathbf{f}(x, \mathbf{y})$ has the second component $f_2(x, \mathbf{y}) = xy_1$. Now in $\mathbf{b}$ ($= \mathbf{k}_2$) the first argument is

$$x = x_n + \frac{1}{2}h$$

The second argument in $\mathbf{b}$ is

$$\mathbf{y} = \mathbf{y}_n + \frac{1}{2}\mathbf{a}$$

and the first component of this is

$$y_1 = y_{1,n} + \frac{1}{2}a_1$$

Together,

$$xy_1 = \left(x_n + \frac{1}{2}h\right)\left(y_{1,n} + \frac{1}{2}a_1\right)$$

Similarly for the other components in (6.55). Finally,

$$\mathbf{y}_{n+1} = \mathbf{y}_n + \frac{1}{6}(\mathbf{a} + 2\mathbf{b} + 2\mathbf{c} + \mathbf{d})$$

Table 10 shows the values $y(x) = y_1(x)$ of the Airy function Ai(x) and of its derivative $y'(x) = y_2(x)$ as well as of the (rather small!) error of $y(x)$.

Table 10 RK Method for Systems: Values $y_{1,n}(x_n)$ of the Airy Function Ai(x) in Example 14

n	x_n	$y_{1,n}(x_n)$	$y_1(x_n)$ *exact* (8D)	$10^8 \times$ *error of* y_1	$y_{2,n}(x_n)$
0	0.0	0.35502805	0.35502805	0	−0.25881940
1	0.2	0.30370303	0.30370315	12	−0.25240464
2	0.4	0.25474211	0.25474235	24	−0.23583073
3	0.6	0.20979973	0.20980006	33	−0.21279185
4	0.8	0.16984596	0.16984632	36	−0.18641171
5	1.0	0.13529207	0.13529242	35	−0.15914687

Runge–Kutta–Nyström Methods (RKN Methods)

RKN methods are direct extensions of RK methods (Runge–Kutta methods) to second-order ODEs $y'' = f(x, y, y')$, as given by the Finnish mathematician E. J. Nyström. The best known of these uses the following formulas, where $n = 0, 1, \ldots, N-1$ (N the number of steps):

$$k_1 = \frac{1}{2}hf(x_n, y_n, y_n')$$

$$k_2 = \frac{1}{2}hf\left(x_n + \frac{1}{2}h, y_n + K, y_n' + k_1\right) \quad \text{where} \quad K = \frac{1}{2}h\left(y_n' + \frac{1}{2}k_1\right) \tag{6.56a}$$

$$k_3 = \frac{1}{2}hf\left(x_n + \frac{1}{2}h, y_n + K, y_n' + k_2\right)$$

$$k_4 = \frac{1}{2}hf(x_n + h, y_n + L, y_n' + 2k_3) \quad \text{where} \quad L = h(y_n' + k_3)$$

From this we compute the approximation y_{n+1} of $y(x_{n+1})$ at $x_{n+1} = x_0 + (n+1)h$:

$$y_{n+1} = y_n + h\left(y_n' + \frac{1}{3}(k_1 + k_2 + k_3)\right) \tag{6.56b}$$

and the approximation y_{n+1}' of the derivative $y'(x_{n+1})$ needed in the next step,

$$y_{n+1}' = y_n' + \frac{1}{3}(k_1 + 2k_2 + 2k_3 + k_4) \tag{6.56c}$$

RKN for ODEs y″ = f(x, y) Not Containing y′

Then $k_2 = k_3$ in (6.56), which makes the method particularly advantageous and reduces (6.56) to

$$\begin{aligned}
k_1 &= \frac{1}{2}hf(x_n, y_n) \\
k_2 &= \frac{1}{2}hf\left(x_n + \frac{1}{2}h, y_n + \frac{1}{2}h(y_n' + \frac{1}{2}k_1)\right) = k_3 \\
k_4 &= \frac{1}{2}hf(x_n + h, y_n + h(y_n' + k_2)) \\
y_{n+1} &= y_n + h\left(y_n' + \frac{1}{3}(k_1 + 2k_2)\right) \\
y_{n+1}' &= y_n' + \frac{1}{3}(k_1 + 4k_2 + k_4)
\end{aligned} \tag{6.57}$$

Our work in Examples 14 and 15 also illustrates that usefulness of methods for ODEs in the computation of values of "higher transcendental functions."

Backward Euler Method for Systems. Stiff Systems

The backward Euler formula (6.15) generalizes to systems in the form

$$\mathbf{y}_{n+1} = \mathbf{y}_n + h\mathbf{f}(x_{n+1}, \mathbf{y}_{n+1}) \qquad (n = 0, 1, \ldots) \tag{6.58}$$

Example 15 RKN Method. Airy's Equation. Airy Function Ai(*x*)

For the problem in Example 14 and $h = 0.2$ as before we obtain from (6.57) simply $k_1 = 0.1x_ny_n$ and

$$k_2 = k_3 = 0.1(x_n + 0.1)(y_n + 0.1y_n' + 0.05k_1), \qquad k_4 = 0.1(x_n + 0.2)(y_n + 0.2y_n' + 0.2k_2)$$

Table 11 shows the results. The accuracy is the same as in Example 14, but the work was much less.

Table 11 Runge–Kutta–Nyström Method Applied to Airy's Equation, Computation of the Airy Function y = Ai(x)

x_n	y_n	y_n'	$y(x)$ *exact* (8D)	$10^8 \times$ *error of* y_n
0.0	0.35502805	−0.25881940	0.35502805	0
0.2	0.30370304	−0.25240464	0.30370315	11
0.4	0.25474211	−0.23583070	0.30370315	24
0.6	0.20979974	−0.21279172	0.25474235	32
0.8	0.16984599	−0.18641134	0.16984632	33
1.0	0.13529218	−0.15914609	0.13529242	24

This is again an implicit method, giving $\mathbf{y}_{n+1}$ implicitly for given $\mathbf{y}_n$. Hence (6.58) must be solved for $\mathbf{y}_{n+1}$. For a linear system this is shown in the next example. This example also illustrates that, similar to the case of a single ODE, the method is very useful for **stiff systems.** These are systems of ODEs whose matrix has eigenvalues λ of very different magnitudes, having the effect that the step in direct methods, RK for example, cannot be increased beyond a certain threshold without losing stability. ($\lambda = -1$ and -10 in Example 16, but larger differences do occur in applications.

Example 16 Backward Euler Method for Systems of ODEs. Stiff Systems

Compare the backward Euler method (6.58) with the Euler and the RK methods for numerically solving the initial value problem

$$y'' + 11y' + 10y = 10x + 11, \qquad y(0) = 2, \qquad y'(0) = -10$$

converted to a system of first-order ODEs.

Solution: The given problem can easily be solved, obtaining

$$y = e^{-x} + e^{-10x} + x$$

so that we can compute errors. Conversion to a system by setting $y = y_1, y' = y_2$ [see (6.50)] gives

$$\begin{aligned} y_1' &= y_2, & y_1(0) &= 2 \\ y_2' &= -10y_1 - 11y_2 + 10x + 11, & y_2(0) &= -10 \end{aligned}$$

The coefficient matrix

$$\mathbf{A} = \begin{bmatrix} 0 & 1 \\ -10 & -11 \end{bmatrix}$$

has the characteristic determinant

$$\begin{vmatrix} -\lambda & 1 \\ -10 & -\lambda - 11 \end{vmatrix}$$

whose value is $\lambda^2 + 11\lambda + 10 = (\lambda + 1)(\lambda + 10)$. Hence the eigenvalues are -1 and -10. The backward Euler formula is

$$\mathbf{y}_{n+1} = \begin{bmatrix} y_{1,n+1} \\ y_{2,n+1} \end{bmatrix} = \begin{bmatrix} y_{1,n} \\ y_{2,n} \end{bmatrix} + h\begin{bmatrix} y_{2,n+1} \\ -10y_{1,n+1} - 11y_{2,n+1} + 10x_{n+1} + 11 \end{bmatrix}$$

Reordering terms gives the linear system in the unknowns $y_{1,n+1}$ and $y_{2,n+1}$

$$\begin{aligned} y_{1,n+1} - hy_{2,n+1} &= y_{1,n} \\ 10hy_{1,n+1} + (1 + 11h)y_{2,n+1} &= y_{2,n} + 10h(x_n + h) + 11h \end{aligned}$$

The coefficient determinant is $D = 1 + 11h + 10h^2$, and Cramer's rule gives the solution

$$\mathbf{y}_{n+1} = \frac{1}{D}\begin{bmatrix} (1 - 11h)y_{1,n} + hy_{2,n} + 10h^2x_n + 11h^2 + 10h^3 \\ -10hy_{1,n} + y_{2,n} + 10hx_n + 11h + 10h^2 \end{bmatrix}$$

Table 12 Backward Euler Method (BEM) for Example 16; Comparison with Euler and RK

x_n	*BEM* $h = 0.2$	*BEM* $h = 0.4$	*Euler* $h = 0.1$	*Euler* $h = 0.2$	*RK* $h = 0.2$	*RK* $h = 0.3$	*Exact*
0.0	2.00000	2.00000	2.00000	2.00000	2.00000	2.00000	2.00000
0.2	1.36667		1.01000	0.00000	1.35207		1.15407
0.4	1.20556	1.31429	1.56100	2.04000	1.18144		1.08864
0.6	1.21574		1.13144	0.11200	1.18585	3.03947	1.15129
0.8	1.29460	1.35020	1.23047	2.20960	1.26168		1.24966
1.0	1.40599		1.34868	0.32768	1.37200		1.36792
1.2	1.53627	1.57243	1.48243	2.46214	1.50257	5.07569	1.50120
1.4	1.67954		1.62877	0.60972	1.64706		1.64660
1.6	1.83272	1.86191	1.78530	2.76777	1.80205		1.80190
1.8	1.99386		1.95009	0.93422	1.96535	8.72329	1.96530
2.0	2.16152	2.18625	2.12158	3.10737	2.13536		2.13534

Table 12 shows the following:

(a) Stability of the backward Euler method for $h = 0.2$ and 0.4 (and in fact for any h; try $h = 5.0$) with decreasing accuracy for increasing h.

(b) Stability of the Euler method for $h = 0.1$ but instability for $h = 0.2$.

(c) Stability of RK for $h = 0.2$ but instability for $h = 0.3$.

Figure 7 shows the Euler method for $h = 0.18$, an interesting case with initial jumping (for about $x < 3$) but later monotone following the solution curve of $y = y_1$.

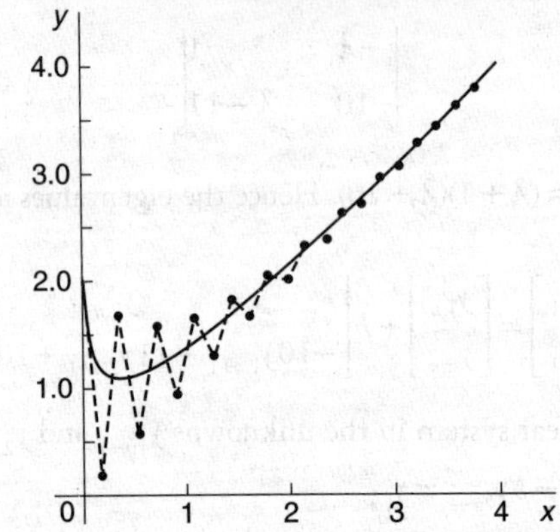

Figure 7 Euler method with $h = 0.18$ in Example 16.

CONCEPT CHECK

1. Verify the calculations in Example 13.
2. Solve by the Euler method:
   (a) $y_1' = -3y_1 + y_2$, $y_2' = y_1 - 3y_2$, $y_1(0) = 2$, $y_2(0) = 0$, $h = 0.1$, 5 steps
   (b) $y_1' = y_1$, $y_2' = y_2$, $y_1(0) = 1$, $y_2(0) = -1$, $h = 0.2$, 5 steps
   (c) $y_1' = y_1$, $y_2' = -y_2$, $y_1(0) = 2$, $y_2(0) = 2$, $h = 0.1$, 10 steps
   (d) $y'' + 4y = 0$, $y(0) = 1$, $y'(0) = 0$, $h = 0.2$, 5 steps
   (e) $y'' - y = x$, $y(0) = 1$, $y'(0) = -2$, $h = 0.1$, 5 steps
   (f) $y_1' = -y_1 + y_2$, $y_2' = -y_1 - y_2$, $y_1(0) = 0$, $y_2(0) = 4$, $h = 0.1$, 10 steps
   (g) Verify the formulas and calculations for the Airy equation in Example 14.
3. Solve by the classical RK:
   (a) The system in Prob. 2(f). How much smaller is the error?
   (b) The ODE in Prob. 2(c). By what factor did the error decrease?
   (c) **Undamped Pendulum.** $y'' + \sin y = 0$, $y(\pi) = 0$, $y'(\pi) = 1$, $h = 0.2$, 5 steps.
   (d) **Bessel Function J_0.** $xy'' + y' + xy = 0$, $y(1) = 0.765198$, $y'(1) = -0.440051$, $h = 0.5$, 5 steps.
   (e) $y_1' = -4y_1 + y_2$, $y_2' = y_1 - 4y_2$, $y_1(0) = 0$, $y_2(0) = 2$, $h = 0.1$, 5 steps.
   (f) The system in Prob. 2(a). How much smaller is the error?
4. Verify the calculations for the Airy equation in Example 15.
5. **Do by RKN:** Prob. 3(d) (Bessel function J_0). Compare the results.
   (a) $y'' - xy' + 4y = 0$, $y(0) = 3$, $y'(0) = 0$, $h = 0.2$, 5 steps (Exact: $y = x^4 - 6x^2 + 3$.)
   (b) $(x^2 - x)y'' - xy' + y = 0$, $y(1/2) = 1 - (\ln 2/2)$, $y'(1/2) = 1 - \ln 2$, $h = 0.1$, 4 steps
   (c) Prob. 3(c). Compare the results.

Solved Problems

Q 1. Using Taylor's series method solve

$$y' = xy + y^2, \quad y(0) = 1 \quad \text{at} \quad x = 0.1,\ 0.2,\ 0.3$$

[JNTU Aug./Sep. 2008]

Answer: Here, $x_0 = 0$, $y_0 = 1$. Therefore

$$y_0' = x_0 y_0 + y_0^2 = (0)1 + 1 = 1 \tag{6.59}$$

$$y_0'' = x_0 y_0' + y_0 + 2y_0 y_0' = 0(1) + 1 + 2(1)(1) = 3 \tag{6.60}$$

$$y_0''' = x_0 y_0'' + 2y_0' + 2y_0'^2 + 2y_0 y_0'' = 0(3) + 2(1) + 2(1) + 2(1)(3) = 10 \tag{6.61}$$

$$y_0^{(iv)} = x_0 y_0''' + y_0'' + 2y_0'' + 4y_0' y_0'' + 2y_0' y_0'' + 2y_0' y_0''' = 0(10) + 3 + 6 + 12 + 6 + 20 = 47 \tag{6.62}$$

(a) For $x = 0.1$: By Taylor's series, we have,

$$y(0.1) = y_1 = y_0 + hy_0' + \frac{h^2}{2!} y_0'' + \frac{h^3}{3!} y_0''' + \frac{h^4}{4!} y_0^{iv} + \cdots$$

where $h = 0.1$. Substituting the values we get

$$y(0.1) = 1 + (0.1)1 + \frac{0.01}{2}(3) + \frac{0.001}{6}(10) + \frac{0.00001}{2}(47)$$

$$= 1.1 + 0.015 + 0.00166 + 0.0001958 = 1.11686$$

(b) For $x = 0.2$: By Taylor's series, we have

$$y(0.2) = y_2 = y_1 + hy_1{}' + \frac{h^2}{2!}y_1{}'' + \frac{h^3}{3!}y_1{}''' + \frac{h^4}{4!}y_1{}^{\text{iv}} + \cdots$$

Here $x_1 = 0.1$ and $y_1 = 1.11686$. We calculate the other values. Now

$$y_1{}' = x_1 y_1 + y_1{}^2 = (0.1)(1.11686) + (1.11686)^2$$
$$= 0.111686 + 1.2474 = 1.359086$$

$$y_1{}'' = x_1 y_1{}' + y_1 + 2y_1\, y_1{}' = (0.1)(1.359086) + (1.11686) + 2(1.11686)(1.359086) = 4.28859$$

Similarly, $y_1{}''' = 16.42026$ and $y_1{}^{\text{iv}} = 86.15743$ (verify!)

Now, substituting the values we get

$$y(0.2) = 1.11686 + (0.1)(1.359086) + \frac{(0.01)}{2}(4.28859) + \frac{(0.001)}{6}(16.42026) + \frac{(0.0001)}{24}(86.15743)$$

$$= 1.2773$$

(iii) For $x = 0.3$: By Taylor's series, we have

$$y(0.3) = y_2 + hy_2{}' + \frac{h^2}{2!}y_2{}'' + \frac{h^3}{3!}y_2{}''' + \frac{h^4}{4!}y_2{}^{\text{iv}} + \cdots$$

Here $x_2 = 0.2$ and $y_2 = 1.2773$. We now calculate the other values:

$$y_2{}' = (0.2)(1.2773) + (1.2773)^3 = 0.25546 + 1.63146 = 1.88692$$

$$y_2{}'' = (0.2)(1.88692) + (1.2773) + 2(1.2773)(1.88692) = 0.3774 + 1.2773 + 4.8203 = 6.475$$

Similarly, $y_2{}''' = 28.7308$ and $y_2{}^{\text{iv}} = 171.8737$ (verify!). Substituting we get

$$y(0.3) = 1.2773 + (0.1)(1.88692) + \frac{(0.01)}{2}(6.475) + \frac{(0.001)}{6} + (28.7308) + \frac{(0.0001)}{24}(171.8737)$$

$$= 1.503872$$

Q 2. Find $y(0.1)$ and $y(0.2)$ using Taylor's series method given that, $y' = y^2 + x$ and $y(0) = 1$.

[JNTU Aug./Sep. 2006]

OR

Tabulate $y(0.1)$, $y(0.2)$, and $y(0.3)$ using Taylor's series method given that $y' = y^2 + x$ and $y(0) = 1$.

[JNTU Aug./Sep. 2006, May/June 2006]

Answer: The given equation is $y' = y^2 + x$. At $x = 0$, we get

$$y'(0) = 1 + 0 = 1$$

Differentiating the given equation successively with respect to x, we get

$$y'' = 2yy' + 1 \Rightarrow y''(0) = 2\times1\times1+1 = 3$$

$$y''' = 2yy'' + 2(y'')^2 \Rightarrow y'''(0) = 2\times1\times3+2\times1^2 = 8$$

$$y^{\text{iv}} = 2yy''' + 2y''y' + 4y'y'' \Rightarrow y^{\text{iv}}(0) = 2\times1\times8+2\times3\times1+4\times1\times3 = 34$$

By Taylor's series method, we have,

$$f(x) = y_0 + xy'(0) + \frac{x_2}{2!}y''(0) + \frac{x_3}{3!}y'''(0) + \frac{x_4}{4!}y^{\text{iv}}(0) + \cdots$$

$$= 1 + 1\times x + \frac{x_2}{2}\times3 + \frac{x_3}{6}\times8 + \frac{x_4}{24}\times34 \qquad \text{(Neglecting higher powers)}$$

Substituting $x = 0.1$, $x = 0.2$, $x = 0.3$ in the above equation, we get

$$y(0.1) = 1 + 0.1 + \frac{(0.1)^2}{2}\times3 + \frac{(0.1)^3\times8}{6} + \frac{(0.1)^4}{24}\times34$$

$$= 1 + 0.1 + \frac{0.03}{2} + \frac{0.004}{3} + \frac{0.0017}{12}$$

$$= 1 + 0.1 + 0.015 + 0.0013 + 0.0001 = 1.1164$$

$$y(0.2) = 1 + 0.2 + \frac{(0.2)^2}{2}\times3 + \frac{(0.2)^3}{6}\times8 + \frac{(0.2)^4}{24}\times34$$

$$= 1 + 0.2 + \frac{0.04}{2}\times3 + \frac{0.008}{6}\times8 + \frac{0.0016}{24}\times34$$

$$= 1 + 0.2 + 0.06 + 0.0106 + 0.002267 = 1.2728$$

$$y(0.3) = 1 + (0.3)\times1 + \frac{(0.3)^2}{2}\times3 + \frac{(0.3)^3}{6}\times8 + \frac{(0.3)^4}{24}\times34$$

$$= 1 + 0.3 + 0.135 + 0.36 + 0.011475 = 1.482475$$

Q 3. Solve the equation $dy/dx = x - y^2$ with the conditions $y(0) = 1$ and $y'(0) = 1$. Find $y(0.2)$ and $y(0.4)$ using Taylor's series method. **[JNTU Aug./Sep. 2008, May/June 2008]**

Answer: We have that

$$\frac{dy}{dx} = y' = x - y^2 \qquad (6.63)$$

Differentiating (6.63) with respect to x, we get

$$y'' = 1 - 2yy' \tag{6.64}$$

$$y''' = -2(y')^2 - 2yy'' \tag{6.65}$$

$$y^{\text{iv}} = -4y'y'' - 2y'y'' - 2yy''' \tag{6.66}$$

It is given that $y_0 = 1$ at $x_0 = 0$ and $y_0' = 1$ at $x_0 = 0$. According to Taylor's series expansion,

$$y_1 = y_0 + hy_0' + \frac{h^2}{2!}y_0'' + \frac{h^3}{3!}y_0''' + \frac{h^4}{4!}y_0^{\text{iv}} + \cdots \tag{6.67}$$

Substituting the given conditions in (6.63)–(6.66), we get

$$y_0' = x_0 - (y_0)^2 = 0 - (1)^2 = -1$$
$$y_0'' = 1 - 2y_0y_0' = 1 - 2(1)(-1) = 3$$
$$y_0''' = -2(y_0')^2 - 2y_0y_0'' = -2(-1)^2 - 2(1)(3) = -2 - 6 = -8$$
$$\begin{aligned} y_0^{\text{iv}} &= -4(y_0')y_0'' - 2y_0'y_0'' - 2y_0y_0''' \\ &= -4(-1)3 - 2(-1)3 - 2(1)(-8) \\ &= 12 + 6 + 16 = 34 \end{aligned}$$

Substituting these and using $h = 0.2$ in (6.67) we get

$$\begin{aligned} y(0.2) &= 1 + (0.2)(-1) + \frac{0.04}{2}\times 3 + \frac{0.008}{6}(-8) + \frac{0.0016}{24}\times 34 \\ &= 1 - 0.2 + 0.06 - 0.010667 + 0.002267 = 0.8516 \end{aligned}$$

Therefore, $y(0.2) = 0.8516$, similarly

$$\begin{aligned} y(0.4) &= 1 + \frac{0.4}{1}\times y_0' + \frac{(0.4)^2}{2}\times y_0'' + \frac{(0.4)^3}{6}\times y_0''' + \frac{(0.4)^4}{24}\times y_0^{\text{iv}} \\ &= 1 + (0.4)\times(-1) + \frac{(0.4)^2}{2}\times 3 + \frac{(0.4)^3}{6}\times(-8) + \frac{(0.4)^4}{24}\times 34 \\ &= 1 - 0.4 + 0.24 - 0.085333 + 0.036267 = 0.790934 \end{aligned}$$

Q 4. Obtain $y(0.1)$ given

$$y' = \frac{y - x}{y + x}, \; y(0) = 1$$

by Picard's method. **[JNTU Aug./Sep. 2008]**

Answer: The given function is

$$\frac{dy}{dx} = \frac{y - x}{y + x} = f(x, y)$$

Also $y(0) = 1$ that is, at $x_0 = 0$, $y_0 = 1$. By Picard's method of successive approximation we have

$$y = y_0 + \int_0^x f(x, y)dx$$

$$y = 1 + \int_0^x \frac{y-x}{y+x} dx \qquad (6.68)$$

First approximation: Put $y = 1$ is RHS of (6.68) and adding +1, −1 in the numerator we get

$$y^{(1)} = 1 + \int_0^x \frac{1-x+1-1}{1+x} dx = 1 + \int_0^x \frac{(1+1)-(1+x)}{1+x} dx$$

$$= 1 + \int_0^x \left(\frac{2}{1+x} - \frac{1+x}{1+x}\right) dx = 1 + \int_0^x \left(\frac{2}{1+x} - 1\right) dx$$

$$= 1 - [x - 2\log(1+x)]_0^x$$

$$= 1 - x + 2\log(1+x) \qquad (6.69)$$

Second approximation: Put RHS of (6.69) in (6.68), we get

$$y^{(2)} = 1 + \int_0^x \frac{y^{(1)} - x}{y^{(1)} + x} dx$$

$$= 1 + \int_0^x \frac{1 - x + 2\log(1+x) - x}{1 - x + 2\log(1+x) + x} dx$$

$$= 1 + \int_0^x \frac{1 - 2x + 2\log(1+x)}{1 + 2\log(1+x)} dx$$

$$= 1 + \int_0^x \left[1 - \frac{2x}{1 + 2\log(1+x)}\right] dx$$

$$= 1 + x\Big|_0^x - 2\int_0^x \frac{x}{1 + 2\log(1+x)} dx$$

As the integration is tedious in second approximation, we stop at first approximation. We get

$$y(0.1) = 1 - 0.1 + 2\log(1 + 0.1)$$

$$= 0.9 + 2\log(1.1)$$

$$= 0.9 + 0.08278 = 0.98278$$

Q 5. Given $y' = x^2 - y$, $y(0) = 1$, find correct to four decimal places the value of $y(0.1)$, by using Euler's method. **[JNTU May/June 2008]**

OR

Find $y(0.1)$ and $y(0.2)$ using Euler's modified formula given that $dy/dx = x^2 - y$, $y(0) = 1$.

[JNTU May/June 2006]

Answer: To find $y(0.1)$ and $y(0.2)$ using Euler's modified formula. We have

$$\frac{dy}{dx} = x^2 - y,\ y(0) = 1$$

Now the function is $f(x, y) = x^2 - y$. For $x_0 = 0$ and $y_0 = 1$ we get

$$f(x_0, y_0) = 0^2 - 1 = -1$$

Modified Euler's method is

$$y_{k+1}^{(i)} = y_0 + \frac{h}{2}[f(x_k, y_k) + f(x_{k+1}, y_{k+1}^{i-1})]$$

Substituting $h = 0.1$, $x_0 = 0, y_0 = 1$, we have

$$\begin{aligned} y_1^{(0)} &= y_0 + hf(x_0, y_0) \\ &= 1 + (0.1)(0^2 - 1) \\ &= 1 + (0.1)(-1) \\ &= 1 - 0.1 = 0.9 \end{aligned}$$

Using Euler's modified method, for $i = 1$, $k = 0$, $x = 0$, we have

$$y_1^{(1)} = y_0 + \frac{h}{2}[f(x_0, y_0) + f(x_1, y_1^{(0)})]$$

Here $y_0 = 1$, $h = 0.1$, $f(x_0, y_0) = -1$, $x_1 = 0.1$, $y_1^{(0)} = 0.9$. Therefore

$$f(x_1, y_1^{(0)}) = x_1^2 - y_1^{(0)} = (0.1)^2 - 0.9 = -0.895$$

Substituting these we get

$$\begin{aligned} y_1^{(1)} &= 1 + \frac{0.1}{2}(-1 - 0.895) \\ &= 1 + \frac{0.1}{2}(-1.895) = 1 - 0.0945 = 0.9055 \end{aligned}$$

Therefore $y_1^{(1)} = 0.905$. By Eulers method,

$$y_2^{(0)} = y_1 + hf(x_1, y_1) = 0.905 + (0.1)(-0.895) = 0.8155$$

Applying Euler's modified method, $i = 1$, $k = 1$, we get

$$\begin{aligned} y_2^{(1)} &= y_1^{(1)} + \frac{h}{2}[f(x_1, y_1) + f(x_2, y_2^{(0)})] \\ &= 0.905 + \frac{0.1}{2}[-0.895 + [(0.2)^2 - 0.8155]] \\ &= 0.905 + (-0.083525) \\ &= 0.821475 \end{aligned}$$

$$y_2^{(2)} = y_1 + \frac{h}{2}[f(x_1, y_1) + f(x_2, y_2^{(1)})]$$
$$= 0.905 + \frac{0.1}{2}[-0.895 + [(0.2)^2 - 0.821475]]$$
$$= 0.821176$$

$$y_2^{(3)} = y_1 + \frac{h}{2}[f(x_1, y_1) + f(x_2, y_2^{(2)})]$$
$$= 0.905 + \frac{0.1}{2}[-0.895 + [(0.2)^2 - 0.821176]]$$
$$= 0.905 - 0.08381 = 0.82119$$

Therefore $y_2 = 0.8211$. Hence, $y(0.1) = 0.905$ and $y(0.2) = 0.8211$.

Q 6. Solve $y' = x/y$, $y(0) = 20$ with $h = 0.2$ using Euler's modified method.

[JNTU Nov./Dec. 2005, May/June 2004]

Answer: The given differential equation is

$$\frac{dy}{dx} = y' = \frac{x}{y} \quad \text{with} \quad h = 0.2,\ x_0 = 0,\ y_0 = 20$$

Now

$$f(x_0, y_0) = \frac{x_0}{y_0} = \frac{0}{20} = 0$$

Finding the initial value using Euler's method that is

$$y_1^{(0)} = y_0 + hf(x_0, y_0)$$
$$y_1^{(0)} = 20 + 0 = 20$$

Now $x_1 = 0.2$, $y_1^{(0)} = 20$. Therefore

$$f(x_1, y_1^{(0)}) = f(0.2,\ 20) = \frac{0.2}{20} = 0.01$$

Applying Euler's modified method we get

$$y_1^{(1)} = y_0 + \frac{h}{2}[f(x_0,\ y_0) + f(x_1,\ y_1^{(0)})]$$
$$= 20 + \frac{0.2}{2}(0 + 0.01) = 20 + 0.01 = 20.001$$

Again

$$y_1^{(2)} = y_0 + \frac{h}{2}[f(x_0, y_0) + f(x_1, y_1^{(1)})]$$

We now calculate $f(x_1, y_1^{(1)})$:

$$f(x_1, y_1^{(1)}) = \frac{0.2}{20.001} = 0.009995$$

Again

$$y_1^{(2)} = 20 + \frac{0.2}{2}(0 + 0.009995) = 20.0009995$$

Finally

$$f(x_1, y_1^{(2)}) = \frac{0.2}{20.0009995} = 0.0099995$$

Therefore

$$y_1^{(3)} = y_0 + \frac{h}{2}[f(x_0, y_0) + f(x_1, y_1^{(2)})]$$

$$= 20 + \frac{0.2}{2}(0 + 0.0099995) = 20.00099995 \text{ correct to 6 decimals.}$$

Q 7. Solve $dy/dx = xy$ using RK method for $x = 0.2$ given $y(0) = 1$, $y'(0) = 0$ taking $h = 0.2$.

[JNTU Aug./Sep. 2008, May/June 2008]

Answer: It is given that

$$\frac{dy}{dx} = xy$$

The given conditions are $y(0) = 1$, $y'(0) = 0$ and $h = 0.2$. Applying Runge–Kutta method, we have for $x_0 = 0, y_0 = 1$.

$$k_1 = h \times f(x_0, y_0) = 0.2 \times f(0,1) = 0.2 \times (0 \times 1) = 0$$

$$k_2 = h \times f\left(x_0 + \frac{h}{2},\ y_0 + \frac{k_1}{2}\right)$$

$$= 0.2 \times f\left(0 + \frac{0.2}{2},\ 1 + 0\right) = 0.2(0.1 \times 1) = 0.02$$

$$k_3 = h \times f\left(x_0 + \frac{h}{2},\ y_0 + \frac{k_2}{2}\right)$$

$$= 0.2 \times f\left(0 + \frac{0.2}{2},\ 1 + \frac{0.02}{2}\right) = 0.2(0.1 \times 1.01) = 0.0202$$

$$k_4 = h \times f(x_0 + h,\ y_0 + k_3)$$

$$= 0.2 \times f(0 + 0.2,\ 1 + 0.0202)$$

$$= 0.2(0.2 \times 1.0202) = 0.0408$$

Now

$$y_1 = y(0.2) = y_0 + \frac{1}{6}[k_1 + 2(k_2 + k_3) + k_4]$$

Therefore

$$y(0.1) = y_1 = 1 + \frac{1}{6}[0 + 2(0.02 + 0.0202) + 0.0408]$$

$$= 1 + \frac{1}{6}[0 + 0.04 + 0.0404 + 0.0408] = 1.0202$$

Hence, $y_1 = 1.0202$.

Summary

In this chapter we discussed numerics for ODEs. Methods for initial value problems

$$y' = f(x, y), \qquad y(x_0) = y_0 \tag{6.70}$$

involving a first-order ODE are obtained by truncating the Taylor series

$$y(x+h) = y(x) + hy(x) + \frac{h^2}{2} y''(x) + \cdots$$

where, by (1), $y' = f$, $y'' = f' = \partial f/\partial x + (\partial f/\partial y)y'$, etc. Truncating after the term hy', we get the *Euler method,* in which we compute step by step

$$y_{n+1} = y_n + hf(x_n, y_n) \quad (n = 0, 1, \ldots) \tag{6.71}$$

Taking one more term into account, we obtain the *improved Euler method.* Both methods show the basic idea but are too inaccurate in most cases.

Truncating after the term in h^4, we get the important classical **Runge–Kutta (RK) method** of fourth order. The crucial idea in this method is the replacement of the cumbersome evaluation of derivatives by the evaluation of $f(x, y)$ at suitable points (x, y); thus in each step we first compute four auxiliary quantities.

$$\begin{aligned} k_1 &= hf(x_n, y_n) \\ k_2 &= hf\left(x_n + \frac{1}{2}h, y_n + \frac{1}{2}k_1\right) \\ k_3 &= hf\left(x_n + \frac{1}{2}h, y_n + \frac{1}{2}k_2\right) \\ k_4 &= hf(x_n + h, y_n + k_3) \ldots \end{aligned} \tag{6.72}$$

and then the new value

$$y_{n+1} = y_n + \frac{1}{6}(k_1 + 2k_2 + 2k_3 + k_4) \tag{6.73}$$

Error and step size control are possible by step halving or by **RKF** (Runge–Kutta–Fehlberg).

The methods in Secs. 6.1–6.5 are **one-step methods** since they get y_{n+1} from the result y_n of a single step. A **multistep method** uses the values of $y_n, y_{n-1}, \ldots$ of several steps for computing y_{n+1}. Integrating cubic interpolation polynomials gives the **Adams–Bashforth predictor**.

$$y_{n+1}^* = y_n + \frac{1}{24}h(55f_n - 59f_{n-1} + 37f_{n-2} - 9f_{n-3}) \tag{6.74}$$

where $f_j = f(x_j, y_j)$, and an **Adams–Moulton corrector** (the actual new value)

$$y_{n+1} = y_n + \frac{1}{24}h(9f_{n+1}^* + 19f_n - 5f_{n-1} + f_{n-2}) \tag{6.75}$$

where $f_{n+1}^* = f(x_{n+1}, y_{n+1}^*)$. Here, to get started, y_1, y_2, y_3 must be computed by the Runge–Kutta method or by some other accurate method.

Section 6.8 concerned the extension of Euler and RK methods to systems:

$$\mathbf{y}' = \mathbf{f}(x, \mathbf{y})$$

thus $y_j' = f_j(x, y_1, \ldots, y_m)$, $j = 1, \ldots, m$.
This includes single mth order ODEs, which are reduced to systems. Second-order equations can also be solved by **RKN** (Runge–Kutta–Nyström) **methods.** These are particularly advantageous for $y'' = f(x, y)$ with f not containing y'.

Review Questions and Problems

1. Explain the Euler and improved Euler methods in geometrical terms.
2. How did we obtain numeric methods by using the Taylor series?
3. In each Runge–Kutta step we computed auxiliary values. How many? Why?
4. What are one-step and multistep methods? Give examples.
5. What is the idea of a predictor–corrector method? Mention some of these methods.
6. What is the idea of the Runge–Kutta–Fehlberg method?
7. How can Runge–Kutta be generalized to systems of ODEs?
8. Solve $y' = 2xy$, $y(0) = 1$, by the Euler method with $h = 0.1$, 10 steps. Compute the error.
9. Solve $y' = 1 + y^2$, $y(0) = 0$, by the improved Euler method with $h = 0.1$, 5 steps. Compute the error.
10. Solve $y' = (x + y - 4)^2$, $y(0) = 4$, by RK with $h = 0.2$, 7 steps.
11. Solve Prob. 9 by RK with $h = 0.1$, 5 steps. Compute the error. Compare with Prob. 9.
12. **(Fair comparison)** Solve $y' = 2x^{-1}\sqrt{y - \ln x} + x^{-1}$, $y(1) = 0$ for $1 \leqq x \leqq 1.8$ (a) by the Euler method with $h = 0.1$, (b) by the improved Euler method with $h = 0.2$, (c) by RK with $h = 0.4$. Verify that the exact solution is $y = (\ln x)^2 + \ln x$. Compute and compare the errors. Why is the comparison fair?
13. Compute e^x for $x = 0, 0.1, \ldots, 1.0$ by applying RK to $y' = y$, $y(0) = 1$, $h = 0.1$. Show that the result is 5D-exact.
14. Solve $y' = (x + y)^2$, $y(0) = 0$ by RK with $h = 0.2$, 5 steps.
15. Show that by applying the ABM method to a polynomial of first degree we obtain the multistep predictor and corrector formulas

$$y_{n+1}^* = y_n + \frac{h}{2}(3f_n - f_{n-1})$$

$$y_{n+1}^* = y_n + \frac{h}{2}(f_{n+1}^* + f_n)$$

where $f_{n+1}^* = f(x_{n+1}, y_{n+1}^*)$.

16. Apply the multistep method in Prob. 15 to the initial value problem $y' = x + y$, $y(0) = 0$, choosing $h = 0.2$ and doing 5 steps. Compare with the exact values.
17. Solve $y' = (y - x - 1)^2 + 2$, $y(0) = 1$ for $0 \leqq x \leqq 1$ by Adams–Moulton with $h = 0.1$ and starting values 1, 1.200334589, 1.402709878, 1.609336039.
18. Solve $y'' + y = 0$, $y(0) = 0$, $y'(0) = 1$ by RKN with $h = 0.2$, 5 steps. Find the error.
19. Solve $y_1' = -4y_1 + 3y_2$, $y_2' = 5y_1 - 6y_2$, $y_1(0) = 3$, $y_2(0) = -5$, by RK for systems, $h = 0.1$, 5 steps.
20. Solve $y_1' = -5y_1 + 3y_2$, $y_2' = -3y_1 - 5y_2$, $y_1(0) = 2$, $y_2(0) = 2$ by RK for systems, $h = 0.1$, 5 steps.

7 Fourier Series

LEARNING OBJECTIVES

After reading this chapter, you will be able to understand the following:

- Fourier series.
- Convergence of Fourier series.
- Periodic functions.
- Dirichlet's conditions.
- Even and odd functions.
- Half-range expansions.

Fourier series are infinite series designed to represent general periodic functions in terms of simple ones, namely, cosines and sines. They constitute a very important tool, in particular in solving problems that involve ODEs and PDEs.

In this chapter we discuss Fourier series and their engineering use from a practical point of view, in connection with ODEs and with the approximation of periodic functions. The *theory* of Fourier series is complicated, but we shall see that the *application* of these series is rather simple. Fourier series are in a certain sense more universal than the familiar Taylor series in calculus because many *discontinuous* periodic functions of practical interest can be developed in Fourier series but, of course, do not have Taylor series representations.

7.1 An Introduction

Fourier series are the basic tool for representing periodic functions, which play an important role in applications. A function $f(x)$ is called a **periodic function** if $f(x)$ is defined for all real x (perhaps except at some points, such as $x = \pm\pi/2, \pm 3\pi/2, \ldots$ for $\tan x$) and if there is some positive number p, called a **period** of $f(x)$, such that

$$f(x+p) = f(x) \quad \text{for all } x \tag{7.1}$$

The graph of such a function is obtained by periodic repetition of its graph in any interval of length p (Figure 1).

Familiar periodic functions are the cosine and sine functions. Examples of functions that are not periodic are x, x^2, x^3, e^x, $\cosh x$, and $\ln x$, to mention just a few.

If $f(x)$ has period p, it also has the period $2p$ because (7.1) implies

$$f(x+2p) = f([x+p]+p) = f(x+p) = f(x), \ldots$$

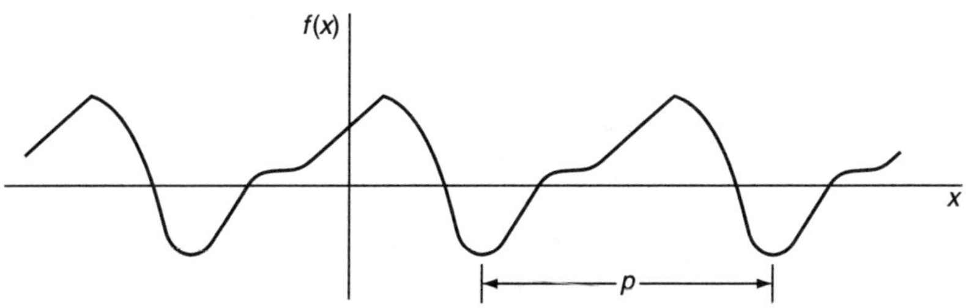

Figure 1 Periodic function.

Thus for any integer $n = 1, 2, 3, \ldots,$

$$f(x+np) = f(x) \quad \text{for all } x \tag{7.2}$$

Furthermore if $f(x)$ and $g(x)$ have period p, then $af(x) + bg(x)$ with any constants a and b also has the period p. Our problem in the first few sections of this chapter will be the representation of various ***functions $f(x)$ of period 2π*** in terms of the simple functions

$$1, \ \cos x, \quad \sin x, \ \cos 2x, \quad \sin 2x, \ \ldots, \ \cos nx, \quad \sin nx, \ \ldots \tag{7.3}$$

All these functions have the period 2π. They form the so-called **trigonometric system**. Figure 2 shows the first few of them (except for the constant 1, which is periodic with any period).

The series to be obtained will be a **trigonometric series,** that is, a series of the form

$$a_0 + a_1 \cos x + b_1 \sin x + a_2 \cos 2x + b_2 \sin 2x + \cdots = a_0 + \sum_{n=1}^{\infty} (a_n \cos nx + b_n \sin nx) \tag{7.4}$$

where $a_0, a_1, b_1, a_2, b_2, \ldots$ are constants, called the **coefficients** of the series. We see that each term has the period 2π. Hence *if the coefficients are such that the series converges, its sum will be a function of period 2π.*

It can be shown that if the series on the left side of (7.4) converges, then inserting parentheses on the right gives a series that converges and has the same sum as the series on the left. This justifies the equality in (7.4).

Now suppose that $f(x)$ is a given function of period 2π and is such that it can be **represented** by a series (7.4), that is, (7.4) converges and, moreover, has the sum $f(x)$. Then, using the equality sign, we write

$$f(x) = a_0 + \sum_{n=1}^{\infty} (a_n \cos nx + b_n \sin nx) \tag{7.5}$$

and call (7.5) the **Fourier series** of $f(x)$. We shall prove that in this case the coefficients of (7.5) are the so-called **Fourier coefficients** of $f(x)$, given by the **Euler formulas**

$$\begin{aligned} &\text{(a)} \qquad a_0 = \frac{1}{2\pi} \int_{-\pi}^{\pi} f(x)\,dx \\ &\text{(b)} \qquad a_n = \frac{1}{\pi} \int_{-\pi}^{\pi} f(x) \cos nx \, dx \qquad (n = 1, 2, \ldots) \\ &\text{(c)} \qquad b_n = \frac{1}{\pi} \int_{-\pi}^{\pi} f(x) \sin nx \, dx \qquad (n = 1, 2, \ldots) \end{aligned} \tag{7.6}$$

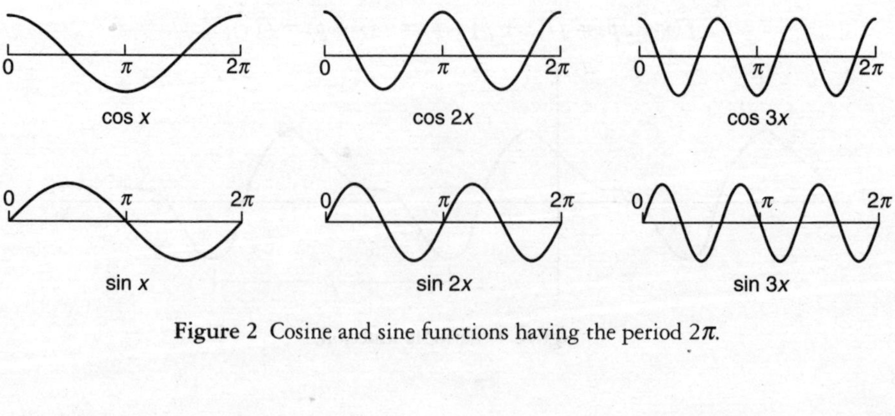

Figure 2 Cosine and sine functions having the period 2π.

The name "Fourier series" is sometimes also used in the exceptional case that (7.5) with coefficients (7.6) does not converge or does not have the sum $f(x)$ – this may happen but is merely of theoretical interest.

A Basic Example

Before we derive the Euler formulas (7.6), let us become familiar with the application of (7.5) and (7.6) in the case of an important example. Since your work for other functions will be quite similar, try to fully understand every detail of the integrations, which because of the n involved differ somewhat from what you have practiced in calculus. Do not just routinely use your software, but make observations: How are continuous functions (cosines and sines) able to represent a given discontinuous function? How does the quality of the approximation increase if you take more and more terms of the series? Why are the approximating functions, called the **partial sums** of the series, always zero at 0 and π? Why is the factor $1/n$ (obtained in the integration) important?

Example 1 Periodic Rectangular Wave (Figure 3a)

Find the Fourier coefficients of the periodic function $f(x)$ in Figure 3a. The formula is

$$f(x)=\begin{cases}-k & \text{if} \quad -\pi<x<0 \\ k & \text{if} \quad 0<x<\pi\end{cases} \quad \text{and} \quad f(x+2\pi)=f(x) \tag{7.7}$$

Functions of this kind occur as external forces acting on mechanical systems, electromotive forces in electric circuits, etc. (The value of $f(x)$ at a single point does not affect the integral; hence we can leave $f(x)$ undefined at $x=0$ and $x=\pm\pi$.)

Solution: From (7.6a) we obtain $a_0=0$. This can also be seen without integration, since the area under the curve of $f(x)$ between $-\pi$ and π is zero. From (7.6b),

$$a_n=\frac{1}{\pi}\int_{-\pi}^{\pi} f(x)\cos nx\,dx=\frac{1}{\pi}\left[\int_{-\pi}^{0}(-k)\cos nx\,dx+\int_{0}^{\pi}k\cos nx\,dx\right]$$

$$=\frac{1}{\pi}\left[-k\frac{\sin nx}{n}\bigg|_{-\pi}^{0}+k\frac{\sin nx}{n}\bigg|_{0}^{\pi}\right]=0$$

because $\sin nx=0$ at $-\pi$, 0, and π for all $n=1, 2, \ldots$. Similarly, from (7.6c) we obtain

$$b_n=\frac{1}{\pi}\int_{-\pi}^{\pi} f(x)\sin nx\,dx=\frac{1}{\pi}\left[\int_{-\pi}^{0}(-k)\sin nx\,dx+\int_{0}^{\pi}k\sin nx\,dx\right]$$

$$=\frac{1}{\pi}\left[k\left(\frac{\cos nx}{n}\right)\bigg|_{-\pi}^{0}-k\left(\frac{\cos nx}{n}\right)\bigg|_{0}^{\pi}\right]$$

Since $\cos(-\alpha)=\cos\alpha$ and $\cos 0=1$, this yields

$$b_n=\frac{k}{n\pi}[\cos 0-\cos(-n\pi)-\cos n\pi+\cos 0]=\frac{2k}{n\pi}(1-\cos n\pi)$$

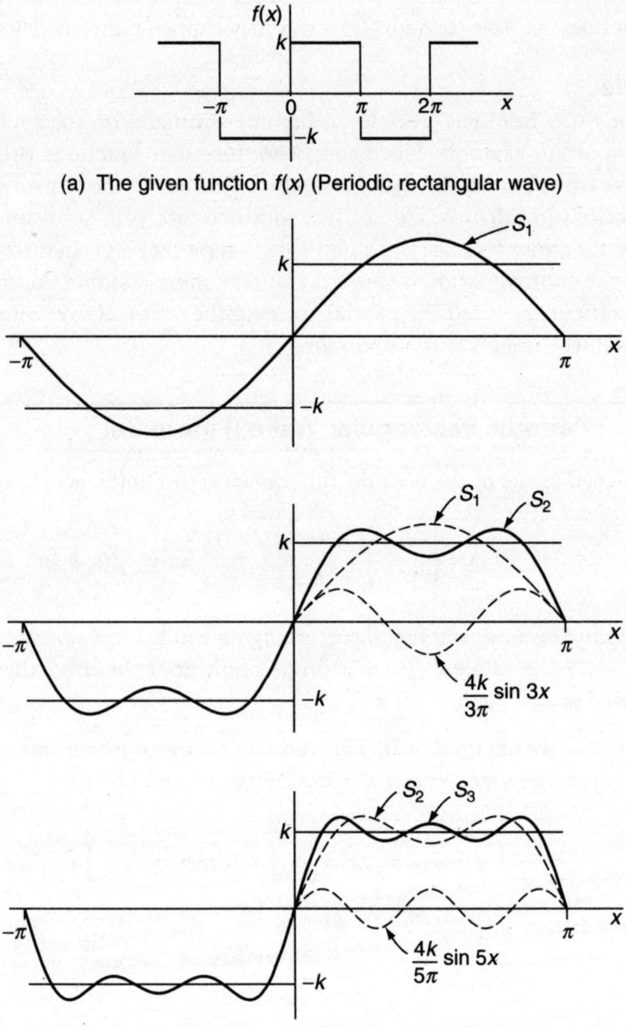

(a) The given function $f(x)$ (Periodic rectangular wave)

(b) The first three partial sums of the corresponding Fourier series

Figure 3 Example 1.

Now, $\cos \pi = -1$, $\cos 2\pi = 1$, $\cos 3\pi = -1$, etc. In general,

$$\cos n\pi = \begin{cases} -1 & \text{for odd } n \\ 1 & \text{for even } n \end{cases}$$

and thus

$$1 - \cos n\pi = \begin{cases} 2 & \text{for odd } n \\ 0 & \text{for even } n \end{cases}$$

Hence the Fourier coefficients b_n of our function are

$$b_1 = \frac{4k}{\pi},\quad b_2 = 0,\quad b_3 = \frac{4k}{3\pi},\quad b_4 = 0,\quad b_5 = \frac{4k}{5\pi},\ \cdots$$

Since the a_n are zero, the Fourier series of $f(x)$ is

$$\frac{4k}{\pi}\left(\sin x + \frac{1}{3}\sin 3x + \frac{1}{5}\sin 5x + \cdots\right) \tag{7.8}$$

The partial sums are

$$S_1 = \frac{4k}{\pi}\sin x,\qquad S_2 = \frac{4k}{\pi}\left(\sin x + \frac{1}{3}\sin 3x\right),\ \text{etc.,}$$

Their graphs in Figure 3 seem to indicate that the series is convergent and has the sum $f(x)$, the given function. We notice that at $x = 0$ and $x = \pi$, the points of discontinuity of $f(x)$, all partial sums have the value zero, the arithmetic mean of the limits $-k$ and k of our function, at these points.

Furthermore, assuming that $f(x)$ is the sum of the series and setting $x = \pi/2$, we have

$$f\left(\frac{\pi}{2}\right) = k = \frac{4k}{\pi}\left(1 - \frac{1}{3} + \frac{1}{5} - + \cdots\right)$$

Thus

$$1 - \frac{1}{3} + \frac{1}{5} - \frac{1}{7} + - \cdots = \frac{\pi}{4}$$

This is a famous result obtained by Leibniz in 1673 from geometric considerations. It illustrates that the values of various series with constant terms can be obtained by evaluating Fourier series at specific points.

Derivation of the Euler Formulas (7.6)

The key to the Euler formulas (7.6) is the **orthogonality** of (7.3), a concept of basic importance, as follows.

THEOREM 1 [Orthogonality of the trigonometric system (7.3)]: *The trigonometric system* (7.3) *is orthogonal on the interval* $-\pi \leqq x \leqq \pi$ *(hence also on* $0 \leqq x \leqq 2\pi$ *or any other interval of length* 2π *because of periodicity); that is, the integral of the product of any two functions in* (7.3) *over that interval is* 0, *so that for any integers n and m,*

$$\begin{aligned}
&\text{(a)}\quad \int_{-\pi}^{\pi} \cos nx \cos mx\, dx = 0, \qquad n \neq m\\
&\text{(b)}\quad \int_{-\pi}^{\pi} \sin nx \sin mx\, dx = 0, \qquad n \neq m\\
&\text{(c)}\quad \int_{-\pi}^{\pi} \sin nx \cos mx\, dx = 0, \qquad n \neq m \text{ or } n = m
\end{aligned} \tag{7.9}$$

PROOF: This follows simply by transforming the integrands trigonometrically from products into sums. In (7.9a) and (7.9b), we have

$$\int_{-\pi}^{\pi} \cos nx \cos mx \, dx = \frac{1}{2}\int_{-\pi}^{\pi} \cos(n+m)x \, dx + \frac{1}{2}\int_{-\pi}^{\pi} \cos(n-m)x \, dx$$

$$\int_{-\pi}^{\pi} \sin nx \sin mx \, dx = \frac{1}{2}\int_{-\pi}^{\pi} \cos(n-m)x \, dx - \frac{1}{2}\int_{-\pi}^{\pi} \cos(n+m)x \, dx$$

Since $m \neq n$ (integer!), the integrals on the right are all 0. Similarly, in (7.9c), for all integer m and n (without exception; do you see why?)

$$\int_{-\pi}^{\pi} \sin nx \cos mx \, dx = \frac{1}{2}\int_{-\pi}^{\pi} \sin(n+m)x \, dx + \frac{1}{2}\int_{-\pi}^{\pi} \sin(n-m)x \, dx = 0 + 0$$ ■

Application of Theorem 1 to the Fourier Series (7.5)

We prove (7.6a). Integrating on both sides of (7.5) from $-\pi$ to π, we get

$$\int_{-\pi}^{\pi} f(x)\, dx = \int_{-\pi}^{\pi}\left[a_0 + \sum_{n=1}^{\infty}(a_n \cos nx + b_n \sin nx)\right] dx$$

We now assume that termwise integration is allowed. (We shall say in the proof of Theorem 2 when this is true.) Then we obtain

$$\int_{-\pi}^{\pi} f(x)\, dx = a_0 \int_{-\pi}^{\pi} dx + \sum_{n=1}^{\infty}\left(a_n \int_{-\pi}^{\pi} \cos nx \, dx + b_n \int_{-\pi}^{\pi} \sin nx \, dx\right)$$

The first term on the right equals $2\pi a_0$. Integration shows that all the other integrals are 0. Hence division by 2π gives (7.6a).

We prove (7.6b). Multiplying (7.5) on both sides by $\cos mx$ with any ***fixed*** positive integer m and integrating from $-\pi$ to π, we have

$$\int_{-\pi}^{\pi} f(x) \cos mx \, dx = \int_{-\pi}^{\pi}\left[a_0 + \sum_{n=1}^{\infty}(a_n \cos nx + b_n \sin nx)\right] \cos mx \, dx \tag{7.10}$$

We now integrate term by term. Then on the right we obtain an integral of $a_0 \cos mx$, which is 0; an integral of $a_n \cos nx \cos mx$, which is $a_m\pi$ for $n = m$ and 0 for $n \neq m$ by (7.9a); and an integral of $b_n \sin nx \cos mx$, which is 0 for all n and m by (7.9c). Hence the right side of (7.10) equals $a_m\pi$. Division by π gives (7.6b) (with m instead of n).

We finally prove (7.6c). Multiplying (7.5) on both sides by $\sin mx$ with any ***fixed*** positive integer m and integrating from $-\pi$ to π, we get

$$\int_{-\pi}^{\pi} f(x) \sin mx \, dx = \int_{-\pi}^{\pi}\left[a_0 + \sum_{n=1}^{\infty}(a_n \cos nx + b_n \sin nx)\right] \sin mx \, dx \tag{7.11}$$

Integrating term by term, we obtain on the right an integral of $a_0 \sin mx$, which is 0; an integral of $a_n \cos nx \sin mx$, which is 0 by (7.9c); and an integral of $b_n \sin nx \sin mx$, which is $b_m\pi$ if $n = m$ and 0 if $n \neq m$ by (7.9b). This implies (7.6c) (with n denoted by m). This completes the proof of the Euler formulas (7.6) for the Fourier coefficients. ■

Convergence and Sum of a Fourier Series

The class of functions that can be represented by Fourier series is surprisingly large and general. Sufficient conditions valid in most applications are as follows.

THEOREM 2 (Representation by a Fourier Series): *Let $f(x)$ be periodic with period 2π and piecewise continuous in the interval $-\pi \leqq x \leqq \pi$. Furthermore, let $f(x)$ have a left-hand derivative and a right-hand derivative at each point of that interval. Then the Fourier series (7.5) of $f(x)$ [with coefficients (7.6)] converges. Its sum is $f(x)$, except at points x_0 where $f(x)$ is discontinuous. There the sum of the series is the average of the left- and right-hand limits of $f(x)$ at x_0.*

PROOF: We prove convergence in Theorem 2. We prove convergence for a continuous function $f(x)$ having continuous first and second derivatives. Integrating (7.6b) by parts, we obtain.

$$a_n = \frac{1}{\pi}\int_{-\pi}^{\pi} f(x)\cos nx\, dx = \left.\frac{f(x)\sin nx}{n\pi}\right|_{-\pi}^{\pi} - \frac{1}{n\pi}\int_{-\pi}^{\pi} f'(x)\sin nx\, dx$$

The first term on the right is zero. Another integration by parts gives

$$a_n = \left.\frac{f'(x)\cos nx}{n^2\pi}\right|_{-\pi}^{\pi} - \frac{1}{n^2\pi}\int_{-\pi}^{\pi} f''(x)\cos nx\, dx$$

The first term on the right is zero because of the periodicity and continuity of $f'(x)$. Since f'' is continuous in the interval of integration, we have

$$|f''(x)| < M$$

for an appropriate constant M. Furthermore, $|\cos nx| \leqq 1$. It follows that

$$|a_n| = \frac{1}{n^2\pi}\left|\int_{-\pi}^{\pi} f''(x)\cos nx\, dx\right| < \frac{1}{n^2\pi}\int_{-\pi}^{\pi} M\, dx = \frac{2M}{n^2}$$

Similarly, $|b_n| < 2\,M/n^2$ for all n. Hence the absolute value of each term of the Fourier series of $f(x)$ is at most equal to the corresponding term of the series

$$|a_0| + 2M\left(1 + 1 + \frac{1}{2^2} + \frac{1}{2^2} + \frac{1}{3^2} + \frac{1}{3^2} + \cdots\right)$$

which is convergent. Hence that Fourier series converges and the proof is complete. The proof of convergence in the case of a piecewise continuous function $f(x)$ and the proof that under the assumptions in the theorem the Fourier series (7.5) with coefficients (7.6) represents $f(x)$ are substantially more complicated. ■

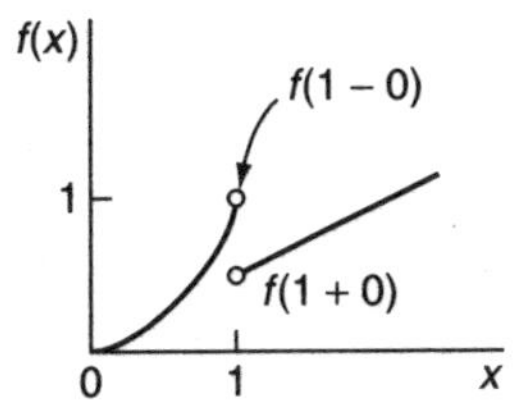

Figure 4 Left- and right-hand limits $f(1-0)=1$, $f(1+0)=1/2$ of the function $f(x)=x^2$ if $x<1$ and $f(x)=x/2$ otherwise.

Example 2 **Convergence at a Jump as Indicated in Theorem 2**

The rectangular wave in Example 1 has a jump at $x = 0$. Its left-hand limit there is $-k$ and its right-hand limit is k (Figure 3). Hence the average of these limits is 0. The Fourier series (7.8) of the wave does indeed converge to this value when $x = 0$ because then all its terms are 0. Similarly for the other jumps. This is in agreement with Theorem 2.

Summary

A Fourier series of a given function $f(x)$ of period 2π is a series of the form (7.5) with coefficients given by the Euler formulas (7.6). Theorem 2 gives conditions that are sufficient for this series to converge and at each x to have the value $f(x)$, except at discontinuities of $f(x)$, where the series equals the arithmetic mean of the left-hand and right-hand limits of $f(x)$ at that point.

CONCEPT CHECK

1. **(Calculus review)** Review integration techniques for integrals as they are likely to arise from the Euler formulas, for instance, definite integrals of $x \cos nx$, $x^2 \sin nx$, $e^{-2x} \cos nx$, etc.
2. The *fundamental period* is the smallest positive period. Find it for
   (a) $\cos x$, $\sin x$, $\cos 2x$, $\sin 2x$, $\cos \pi x$, $\sin \pi x$, $\cos 2\pi x$, $\sin 2\pi x$
   (b) $\cos nx$, $\sin nx$, $\cos (2\pi x/k)$, $\sin (2\pi x/k)$, $\cos (2\pi nx/k)$, $\sin (2\pi nx/k)$
3. Show that $f = const$ is periodic with any period but has no fundamental period.
4. If fx) and $g(x)$ have period p, show that $h(x) = af(x) + bg(x)$ (a, b, constant) has the period p. Thus all functions of period p form a **vector space.**
5. **(Change of scale)** If $f(x)$ has period p, show that $f(ax)$, $a \neq 0$, and $f(x/b)$, $b \neq 0$, are periodic functions of x of periods p/a and bp, respectively. Give examples.
6. Sketch or graph $f(x)$, of period 2π, which for 2π, $-\pi < x < \pi$ is given as follows.
   (a) $f(x) = x$ (b) $f(x) = e^{-|x|}$
   (c) $f(x) = \pi - |x|$ (d) $f(x) = |\sin 2x|$
   (e) $f(x) = \begin{cases} -x^3 & \text{if } -\pi < x < 0 \\ x^3 & \text{if } 0 < x < \pi \end{cases}$
   (f) $f(x) = \begin{cases} 1 & \text{if } -\pi < x < 0 \\ \cos \frac{1}{2} x & \text{if } 0 < x < \pi \end{cases}$
7. Showing the details of your work, find the Fourier series of the given $f(x)$, which is assumed to have the period 2π. Sketch or graph the partial sums up to that including $\cos 5x$ and $\sin 5x$.

(a)

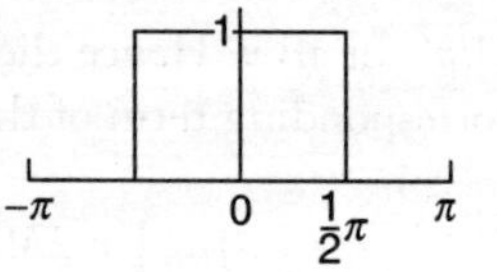

(b)

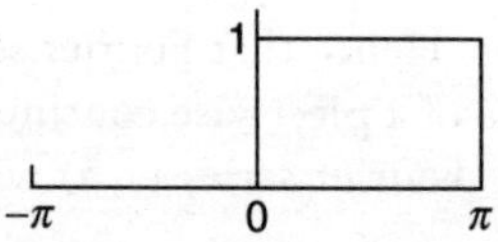

(c)

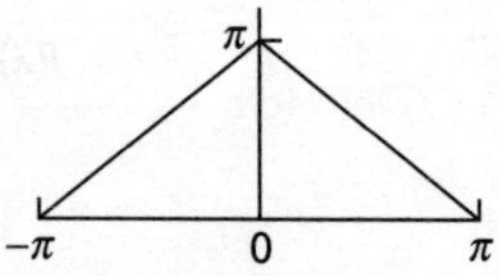

(d)

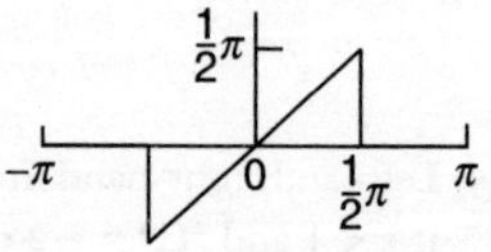

(e)

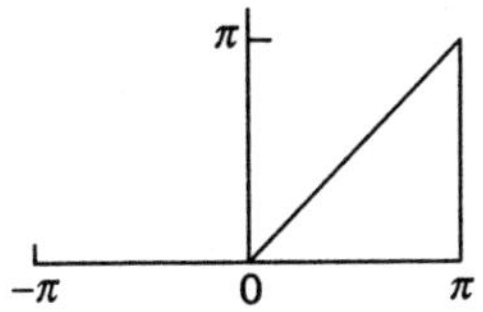

(f)

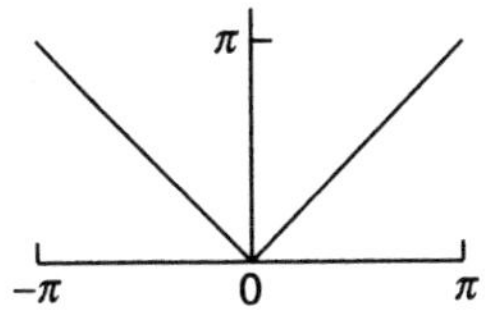

(g)

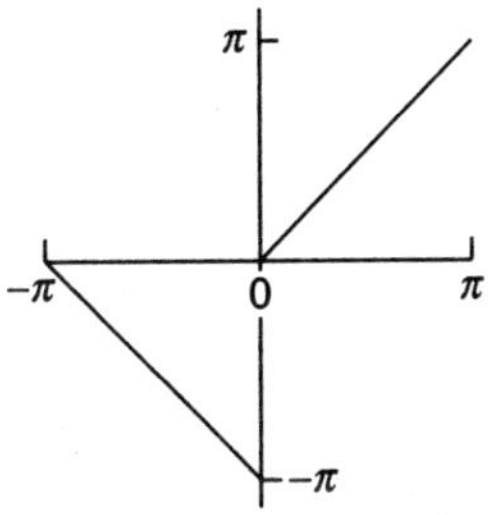

(h)

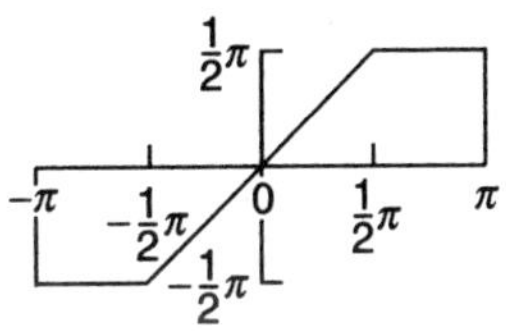

(i) $f(x) = x^2 (-\pi < x < \pi)$

(j) $f(x) = x^2 (0 < x < 2\pi)$

(k) $f(x) = \begin{cases} x^2 & \text{if } -(1/2)\pi < x < (1/2)\pi \\ (1/4)\pi^2 & \text{if } (1/2)\pi < x < (3/2)\pi \end{cases}$

(l) $f(x) = \begin{cases} -4x & \text{if } -\pi < x < 0 \\ 4x & \text{if } 0 < x < \pi \end{cases}$

8. **(Discontinuities)** Verify the last statement in Theorem 2 for the discontinuities of $f(x)$ in Prob. 7(a).
9. **PROJECT. Euler Formulas in Terms of Jumps Without Integration.** Show that for a function whose third derivative is identically zero,

$$a_n = \frac{1}{n\pi}\left[-\sum j_s \sin nx_s - \frac{1}{n}\sum j_s' \cos nx_s + \frac{1}{n^2}\sum j_s'' \sin nx_s\right]$$

$$b_n = \frac{1}{n\pi}\left[\sum j_s \cos nx_s - \frac{1}{n}\sum j_s' \sin nx_s - \frac{1}{n^2}\sum j_s'' \cos nx_s\right]$$

where $n = 1, 2, \ldots$ and we sum over all the jumps j_s, j_s', j_s'' of f, f', f'', respectively, located at x_s.
10. Apply the formulas in Project 9 to the function in Prob. 7(h) and compare the results.

7.2 Periodic Functions

The functions considered so far had period 2π, for the simplicity of the formulas. Of course, periodic functions in applications will generally have other periods. However, we now show that the transition from period $p = 2\pi$ to a period $2L$ is quite simple. The notation $p = 2L$ is practical because L will be the length of a violin string or the length of a rod in heat conduction, and so on.

The idea is simply to find and use a ***change of scale*** that gives from a function $g(v)$ of period 2π a function of period $2L$. Now from (7.5) and (7.6) in the last section with $g(v)$ instead of $f(x)$ we have the Fourier series

$$g(v) = a_0 + \sum_{n=1}^{\infty}(a_n \cos nv + b_n \sin nv) \tag{7.12}$$

with coefficients

$$a_0 = \frac{1}{2\pi}\int_{-\pi}^{\pi} g(v)\,dv$$

$$a_n = \frac{1}{\pi}\int_{-\pi}^{\pi} g(v)\cos nv\, dv$$
$$b_n = \frac{1}{\pi}\int_{-\pi}^{\pi} g(v)\sin nv\, dv \qquad (7.13)$$

We can now write the change of scale as $v = kx$ with k such that the old period $v = 2\pi$ gives for the new variable x the new period $x = 2L$. Thus, $2\pi = k2L$. Hence $k = \pi/L$ and

$$v = kx = \pi x/L \qquad (7.14)$$

This implies $dv = (\pi/L)\, dx$, which upon substitution into (7.13) cancels $1/2\pi$ and $1/\pi$ and gives instead the factors $1/2L$ and $1/L$. Writing

$$g(v) = f(x) \qquad (7.15)$$

we thus obtain from (7.12) the **Fourier series** of the function $f(x)$ of period $2L$

$$f(x) = a_0 + \sum_{n=1}^{\infty}\left(a_n \cos\frac{n\pi}{L}x + b_n \sin\frac{n\pi}{L}x\right) \qquad (7.16)$$

with the **Fourier coefficients** of $f(x)$ given by the **Euler formulas**

$$\text{(a)}\quad a_0 = \frac{1}{2L}\int_{-L}^{L} f(x)dx$$
$$\text{(b)}\quad a_n = \frac{1}{L}\int_{-L}^{L} f(x)\cos\frac{n\pi x}{L}\, dx \qquad (n = 1, 2, \ldots) \qquad (7.17)$$
$$\text{(c)}\quad b_n = \frac{1}{L}\int_{-L}^{L} f(x)\sin\frac{n\pi x}{L}\, dx \qquad (n = 1, 2, \ldots)$$

We continue to call (7.5) with any coefficients a **trigonometric series.** We can integrate from 0 to $2L$ or over any other interval of length $p = 2L$.

Example 3 Periodic Rectangular Wave

Find the Fourier series of the function (Figure 5)

$$f(x) = \begin{cases} 0 & \text{if} \quad -2 < x < -1 \\ k & \text{if} \quad -1 < x < 1 \\ 0 & \text{if} \quad 1 < x < 2 \end{cases}$$

$p = 2L = 4$, $L = 2$.

Solution: From (7.17a) we obtain $a_0 = k/2$ (verify!). From (7.17b) we obtain

$$a_n = \frac{1}{2}\int_{-2}^{2} f(x)\cos\frac{n\pi x}{2}\, dx = \frac{1}{2}\int_{-1}^{1} k\cos\frac{n\pi x}{2}\, dx = \frac{2k}{n\pi}\sin\frac{n\pi}{2}$$

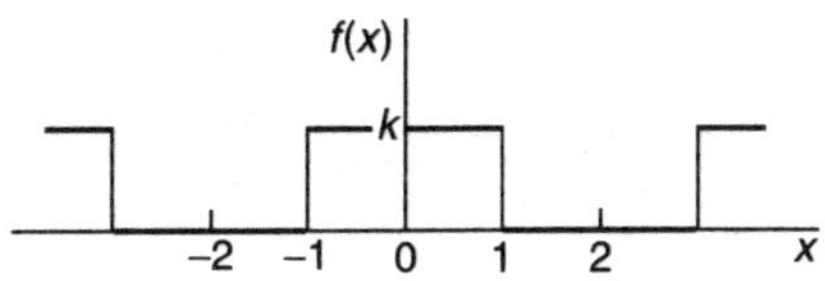

Figure 5 Example 3.

Thus $a_n = 0$ if n is even, and $a_n = 2k/n\pi$ if $n = 1, 5, 9, \ldots$, $a_n = -2k/n\pi$ if $n = 3, 7, 11, \ldots$. From (7.17c) we find that $b_n = 0$ for $n = 1, 2, \ldots$. Hence the Fourier series is

$$f(x) = \frac{k}{2} + \frac{2k}{\pi}\left(\cos\frac{\pi}{2}x - \frac{1}{3}\cos\frac{3\pi}{2}x + \frac{1}{5}\cos\frac{5\pi}{2}x - + \cdots\right)$$

Example 4 Periodic Rectangular Wave

Find the Fourier series of the function (Figure 6)

$$f(x) = \begin{cases} -k & \text{if} \quad -2 < x < 0 \\ k & \text{if} \quad 0 < x < 2 \end{cases}$$

$p = 2L = 4$, $L = 2$.

Solution: From (7.17a) $a_0 = 0$. From (7.17b), with $1/L = 1/2$,

$$\begin{aligned} a_n &= \frac{1}{2}\left[\int_{-2}^{0} (-k)\cos\frac{n\pi x}{2}\,dx + \int_{0}^{2} k\cos\frac{n\pi x}{2}\,dx\right] \\ &= \frac{1}{2}\left[-\frac{2k}{n\pi}\sin\frac{n\pi x}{2}\bigg|_{-2}^{0} + \frac{2k}{n\pi}\sin\frac{n\pi x}{2}\bigg|_{0}^{2}\right] = 0 \end{aligned}$$

so that the Fourier series has no cosine terms. From (7.17c),

$$\begin{aligned} b_n &= \frac{1}{2}\left[\frac{2k}{n\pi}\cos\frac{n\pi x}{2}\bigg|_{-2}^{0} - \frac{2k}{n\pi}\cos\frac{n\pi x}{2}\bigg|_{0}^{2}\right] \\ &= \frac{k}{n\pi}(1 - \cos n\pi - \cos n\pi + 1) \\ &= \begin{cases} 4k/n\pi & \text{if } n = 1, 3, \ldots \\ 0 & \text{if } n = 2, 4, \ldots \end{cases} \end{aligned}$$

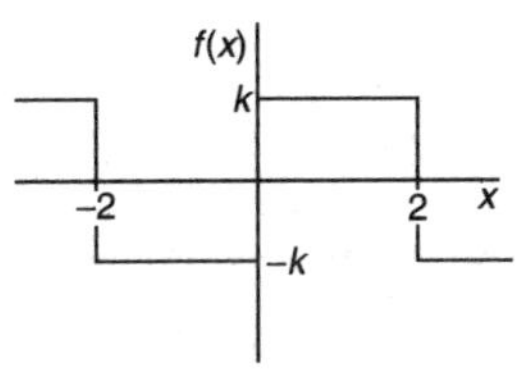

Figure 6 Example 4.

Hence the Fourier series of $f(x)$ is

$$f(x)=\frac{4k}{\pi}\left(\sin\frac{\pi}{2}+\frac{1}{3}\sin\frac{3\pi}{2}x+\frac{1}{5}\sin\frac{5\pi}{2}x+\cdots\right)$$

It is interesting that we could have derived this from (7.8) namely, by the scale change (7.14). Indeed, writing v instead of x, we have in (7.8),

$$\frac{4k}{\pi}\left(\sin v+\frac{1}{3}\sin 3v+\frac{1}{5}\sin 5v+\cdots\right)$$

Since the period 2π in v corresponds to $2L = 4$, we have $k = \pi/L = \pi/2$ and $v = kx = \pi x/2$ in (7.14); hence we obtain the Fourier series of $f(x)$, as before.

Example 5 Half-Wave Rectifier

A sinusoidal voltage $E \sin\omega t$, where t is time, is passed through a half-wave rectifier that clips the negative portion of the wave (Figure 7). Find the Fourier series of the resulting periodic function

$$u(t)=\begin{cases}0 & \text{if } -L<t<0\\ E\sin\omega t & \text{if } 0<t<L\end{cases}$$

$p = 2L = 2\pi/\omega,\ L = \pi/\omega.$

Solution: Since $u = 0$ when $-L < t < 0$, we obtain from (7.17a), with t instead of x,

$$a_0=\frac{\omega}{2\pi}\int_0^{\pi/\omega} E\sin\omega t\,dt=\frac{E}{\pi}$$

and from (7.17b) we have with $x = \omega t$ and $y = n\omega t$

$$a_n=\frac{\omega}{\pi}\int_0^{\pi/\omega} E\sin\omega t\cos n\omega t\,dt=\frac{\omega E}{2\pi}\int_0^{\pi/\omega}[\sin(1+n)\,\omega t+\sin(1-n)\,\omega t]\,dt$$

If $n = 1$, the integral on the right is zero, and if $n = 2, 3, \ldots$, we readily obtain

$$a_n=\frac{\omega E}{2\pi}\left[-\frac{\cos(1+n)\omega t}{(1+n)\omega}-\frac{\cos(1-n)\omega t}{(1-n)\omega}\right]_0^{\pi/\omega}$$

$$=\frac{E}{2\pi}\left(\frac{-\cos(1+n)\pi+1}{1+n}+\frac{-\cos(1-n)\pi+1}{1-n}\right)$$

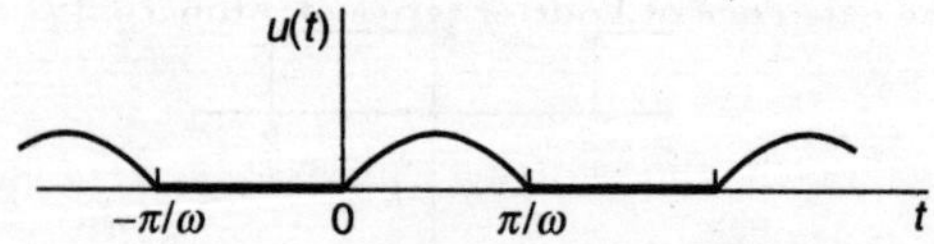

Figure 7 Half-wave rectifier.

If n is odd, this is equal to zero, and for even n we have

$$a_n = \frac{E}{2\pi}\left(\frac{2}{1+n} + \frac{2}{1-n}\right) = -\frac{2E}{(n-1)(n+1)\pi} \quad (n = 2, 4, \ldots)$$

In a similar fashion we find from (7.17c) that $b_1 = E/2$ and $b_n = 0$ for $n = 2, 3, \ldots$. Consequently,

$$u(t) = \frac{E}{\pi} + \frac{E}{2}\sin\omega t - \frac{2E}{\pi}\left(\frac{1}{1\times 3}\cos 2\omega t + \frac{1}{3\times 5}\cos 4\omega t + \cdots\right)$$

7.3 CONCEPT CHECK

1. Find the Fourier series of the function $f(x)$, of period $p = 2L$, and sketch or graph the first three partial sums. (Show the details of your work.)
   (a) $f(x) = -1\ (-2 < x < 0), f(x) = 1\ (0 < x < 2),$ $p = 4$
   (b) $f(x) = 0\ (-2 < x < 0), f(x) = 4\ (0 < x < 2),$ $p = 4$
   (c) $f(x) = x^2\ (-1 < x < 1), p = 2$
   (d) $f(x) = \pi x^3/2\ (-1 < x < 1), p = 2$
   (e) $f(x) = \sin\pi x\ (0 < x < 1), p = 1$
   (f) $f(x) = \cos\pi x\ (-1/2 < x < 1/2), p = 1$
   (g) $f(x) = |x|\ (-1 < x < 1), p = 2$
   (h) $f(x) = \begin{cases} 1+x & \text{if } -1 < x < 0 \\ 1-x & \text{if } \ \ 0 < x < 1 \end{cases} \quad (p = 2)$
   (i) $f(x) = 1 - x^2\ (-1 < x < 1), p = 2$
   (j) $f(x) = 0\ (-2 < x < 0), f(x) = x\ (0 < x < 2),$ $p = 4$
   (k) $f(x) = -x\ (-1 < x < 0), f(x) = x\ (0 < x < 1),$ $f(x) = 1\ (1 < x < 3), p = 4$
2. **(Rectifier)** Find the Fourier series of the function obtained by passing the voltage $v(t) = V_0 \cos 100\pi t$ through a half-wave rectifier.
3. Show that the familiar identities

$$\cos^3 x = \frac{3}{4}\cos x + \frac{1}{4}\cos 3x$$

$$\sin^3 x = \frac{3}{4}\sin x - \frac{1}{4}\sin 3x$$

   can be interpreted as Fourier series expansions. Develop $\cos^4 x$.
4. Obtain the series in Prob. 1(g) from that in Prob. 1(h).
5. Obtain the series in Prob. 1(f) from that in Prob. 1(e).
6. Using Prob. 1(c), show that

$$1 - \frac{1}{4} + \frac{1}{9} - \frac{1}{16} + - \cdots = \frac{1}{12}\pi^2$$

7. Show that

$$1 + \frac{1}{4} + \frac{1}{9} + \frac{1}{16} + \cdots = \frac{1}{6}\pi^2$$

7.3 Dirichlet's Conditions

Dirichlet's conditions for the existence of Fourier series of a function $f(x)$ in the interval $(\pi, \pi + 2\pi)$ are as follows:

1. $f(x)$ is uniformly bounded on the fundamental interval $(2\pi, \pi)$ of period 2π, that is, $|f(x)| \le M$ for all $2\pi < x < \pi$, where M is a constant.
2. $f(x)$ has no more than a finite number of finite discontinuities in $(2\pi, \pi)$.
3. $f(x)$ has a finite number of strict maxima or minima in $(2\pi, \pi)$.

These conditions are sufficient conditions for a real-valued periodic function $f(x)$ to be equal to the sum of its Fourier series at each point where f is continuous. Note that these conditions are sufficient but not necessary.

7.4 Change of Interval

Sometimes we require an expansion of a given function over an interval different from the intervals $(0, 2\pi)$, $(-\pi, \pi)$, or $(0, \pi)$. Suppose that the given function $f(x)$ is periodic and is defined for the interval $-c < x < c$, where c is any positive real number.

If we put

$$x = \frac{cz}{\pi}$$

then

$$f(x) = f\left(\frac{cz}{\pi}\right)$$

Also as x varies from $-c$ to c, z varies from $-\pi$ to π. Now since $f(x) = f(cz/\pi)$ is regarded as a function of z, it can be represented by a Fourier series for the interval

$$-\pi \le z \le \pi$$

so that

$$a_0 = \int_{-\pi}^{\pi} f\left(\frac{cz}{\pi}\right) dz$$

Changing the variable we get

$$a_0 = \frac{1}{c}\int_{-c}^{c} f(x)\, dx$$

$$a_n = \int_{-\pi}^{\pi} f\left(\frac{cz}{\pi}\right) \cos nz\, dz$$

Now changing the variable z to x we get

$$a_n = \frac{1}{c}\int_{-c}^{c} f(x) \cos\left(\frac{n\pi x}{c}\right) dx$$

Similarly

$$b_n = \frac{1}{\pi}\int_{-\pi}^{\pi} f\left(\frac{cz}{\pi}\right) \sin nz\, dz$$

$$= \frac{1}{c}\int_{-c}^{c} f(x)\left(\frac{n\pi x}{c}\right) dx$$

The series is

$$f(x) = \frac{a_0}{2} + a_1 \cos\frac{\pi x}{c} + a_2 \cos\frac{2\pi x}{c} + a_3 \cos\frac{3\pi x}{c} + \cdots$$

$$+ b_1 \sin\frac{\pi x}{c} + b_2 \sin\frac{2\pi x}{c} + b_3 \sin\frac{3\pi x}{c} + \cdots$$

where

$$a_n = \frac{1}{c}\int_{-c}^{c} f(x)\cos\frac{n\pi x}{c}\,dx \text{ and } b_n = \frac{1}{c}\int_{-c}^{c} f(x)\sin\frac{n\pi x}{c}\,dx$$

Example 6

Obtain the Fourier series for the function

$$f(x) = \begin{cases} -c & -a < x < 0 \\ c & 0 < x < a \end{cases}$$

Solution: Here

$$\begin{aligned} a_n &= \frac{1}{a}\int_{-a}^{a} f(x)\cos\left(\frac{n\pi x}{a}\right)dx \\ &= \frac{1}{a}\left[\int_{-a}^{0}(-c)\cos\left(\frac{n\pi x}{a}\right)dx + \int_{0}^{a} c\cos\left(\frac{n\pi x}{a}\right)dx\right] \\ &= \frac{1}{a}\left[-\frac{ac}{n\pi}\left[\sin\frac{n\pi x}{a}\right]_{-a}^{0} + \frac{ac}{n\pi}\left[\sin\frac{n\pi x}{a}\right]_{0}^{a}\right] \\ &= 0 \end{aligned}$$

$$\begin{aligned} b_n &= \frac{1}{a}\int_{-a}^{0}(-c)\sin\left(\frac{n\pi x}{a}\right)dx + \int_{-a}^{0} c\sin\left(\frac{n\pi x}{a}\right)dx \\ &= \frac{1}{a}\left[-\frac{ac}{n\pi}\left[-\cos\left(\frac{n\pi x}{a}\right)\right]_{-a}^{0} + \frac{ac}{n\pi}\left[-\cos\left(\frac{n\pi x}{a}\right)\right]_{0}^{a}\right] \\ &= \frac{c}{n\pi}(1-\cos n\pi - \cos n\pi + 1) \\ &= \frac{2c}{n\pi}(1-\cos n\pi) \end{aligned}$$

Two cases arise: If n is even, $b_n = 0$; and if n is odd, $b_n = 4c/n\pi$. Hence the series is

$$f(x) = \frac{4c}{\pi}\left[\sin\frac{\pi x}{a} + \frac{1}{3}\sin\frac{3\pi x}{a} + \frac{1}{5}\sin\frac{5\pi x}{a} + \cdots\right]$$

Example 7

Find the Fourier expansion for the function

$$f(x) = \begin{cases} 2 & -2 < x < 0 \\ x & 0 < x < 2 \end{cases}$$

Solution: Here $c = 2$, therefore

$$a_n = \frac{1}{2}\int_{-2}^{2} f(x)\cos\left(\frac{n\pi x}{2}\right)dx$$

$$= \frac{1}{2}\int_{-2}^{0} 2\cos\left(\frac{n\pi x}{2}\right)dx + \frac{1}{2}\int_{0}^{2} x\cos\left(\frac{n\pi x}{2}\right)dx$$

$$= \left[\frac{2}{n\pi}\sin\left(\frac{n\pi x}{2}\right)\right]_{-2}^{0} + \frac{1}{2}\left[\frac{2x}{n\pi}\sin\left(\frac{n\pi x}{2}\right) + \frac{4}{n^2\pi^2}\cos\left(\frac{n\pi x}{2}\right)\right]_{0}^{2}$$

$$= \frac{1}{2}\left[\frac{4}{n^2\pi^2}\cos n\pi - \frac{4}{n^2\pi^2}\right]$$

Two cases arise: If n is even, $\cos n\pi = 1$ then $a_n = 0$ and if n is odd, $\cos n\pi = -1$ then $a_n = -4/n^2\pi^2$. Now

$$a_0 = \frac{1}{2}\int_{-2}^{0} 2dx + \frac{1}{2}\int_{0}^{2} x\,dx = 2 + 1 = 3$$

Also

$$b_n = \frac{1}{2}\int_{-2}^{0} 2\sin\left(\frac{n\pi x}{2}\right)dx + \frac{1}{2}\int_{0}^{2} x\sin\left(\frac{n\pi x}{2}\right)dx$$

$$= \left[\frac{-2}{n\pi}\cos\left(\frac{n\pi x}{2}\right)\right]_{-2}^{0} + \frac{1}{2}\left[\frac{-2x}{n\pi}\cos\left(\frac{n\pi x}{2}\right) + \frac{4}{n^2\pi^2}\sin\left(\frac{n\pi x}{2}\right)\right]_{0}^{2}$$

$$= \left(-\frac{2}{n\pi} + \frac{2}{n\pi}\cos n\pi\right) + \frac{1}{2}\left(-\frac{4}{n\pi}\cos n\pi\right)$$

$$= -\frac{2}{n\pi}$$

Therefore, the required series is

$$f(x) = \frac{3}{2} - \frac{4}{\pi^2}\left(\cos\frac{\pi x}{2} + \frac{1}{3^2}\cos\frac{3\pi x}{2} + \frac{1}{5^2}\cos\frac{5\pi x}{2} + \cdots\right)$$
$$- \frac{2}{\pi}\left(\sin\frac{\pi x}{2} + \frac{1}{2}\sin\frac{2\pi x}{2} + \frac{1}{3}\sin\frac{3\pi x}{2} + \cdots\right)$$

By substituting $x = 1$, we get

$$1 = \frac{3}{2} - \frac{2}{\pi}\left(1 - \frac{1}{3} + \frac{1}{5} - \frac{1}{7} + \cdots\right)$$

$$1 - \frac{1}{3} + \frac{1}{5} - \frac{1}{7} + \cdots = \frac{\pi}{4}$$

Putting $x = 2$, we get

$$2 = \frac{3}{2} - \frac{4}{\pi^2}\left(-1 - \frac{1}{3^2} - \frac{1}{5^2} + \cdots\right)$$

$$1 + \frac{1}{3^2} + \frac{1}{5^2} + \frac{1}{7^2} + \cdots = \frac{\pi^2}{8}$$

Example 8

Taking the period (0, 4) expand the following function in the Fourier series

$$f(x) = \begin{cases} x & 0 < x < 2 \\ 0 & 2 < x < 4 \end{cases}$$

Solution: We have

$$a_n = \frac{1}{2}\int_0^4 f(x)\cos\left(\frac{n\pi x}{2}\right)dx \quad [\text{here } 2c = 4,\ 0 < x < 2c]$$

$$= \frac{1}{2}\int_0^2 x\cos\left(\frac{n\pi x}{2}\right)dx$$

$$= \frac{1}{2}\left[\frac{2x}{n\pi}\sin\left(\frac{n\pi x}{2}\right) + \frac{4}{n^2\pi^2}\cos\left(\frac{n\pi x}{2}\right)\right]_0^2$$

$$= \frac{1}{2}\left(\frac{4}{n^2\pi^2}\cos n\pi - \frac{4}{n^2\pi^2}\right)$$

Now from the above equation we have

$$a_n = \begin{cases} -\dfrac{4}{n^2\pi^2} & \text{when } n \text{ is odd} \\ 0 & \text{when } n \text{ is even} \end{cases}$$

Therefore

$$a_0 = \frac{1}{2}\int_0^2 x\,dx = 1$$

$$b_n = \frac{1}{2}\int_0^2 x\sin\left(\frac{n\pi x}{2}\right)dx$$

$$= \frac{1}{2}\left[-\frac{2x}{n\pi}\cos\left(\frac{n\pi x}{2}\right) + \frac{4}{n^2\pi^2}\sin\left(\frac{n\pi x}{2}\right)\right]_0^2$$

$$= \frac{1}{2}\left[\frac{-4}{n\pi}\cos n\pi\right]$$

$$= \frac{2}{n\pi} \text{ or } \frac{-2}{n\pi} \quad (\text{according as } n \text{ is odd or even})$$

The required series, therefore, is

$$f(x)=\frac{1}{2}-\frac{4}{\pi^2}\left(\cos\frac{\pi x}{2}+\frac{1}{3^2}\cos\frac{3\pi x}{2}+\frac{1}{5^2}\cos\frac{5\pi x}{2}+\cdots\right)$$
$$+\frac{2}{\pi}\left(\sin\frac{\pi x}{2}-\frac{1}{2}\sin\frac{2\pi x}{2}+\frac{1}{3}\sin\frac{3\pi x}{2}-\frac{1}{4}\sin\frac{4\pi x}{2}+\cdots\right)$$

CONCEPT CHECK

1. Find a Fourier series to represent $f(x) = x^2$ in the interval $(0, 2\pi)$. OR Expand $f(x) = x^2$, $0 < x < \pi$ as a Fourier series. [JNTU 2008]
2. Find the Fourier series representing $f(x) = x$, $0 < x < 2\pi$. Sketch the graph of $f(x)$ from -4π to 4π. [JNTU 2007, 2005, 2003]
3. Obtain the Fourier series for $f(x) = x - x^2$ in the interval $[-\pi, \pi]$. Hence show that
$$\sum_{n=1}^{\infty}\frac{(-1)^{n-1}}{n^2}=\frac{\pi^2}{12}$$
(or) $$\frac{\pi^2}{12}=\frac{1}{1^2}-\frac{1}{2^2}+\frac{1}{3^2}-\frac{1}{4^2}+\cdots$$
[JNTU Dec. 2002]
4. Expand
$$f(x)=\left(\frac{\pi-x}{2}\right)^2$$
$0<x<2\pi$ in a Fourier series. [JNTU 2004]
5. Obtain in Fourier series expansion of $f(x)$ given that $f(x) = (\pi - x)^2$ in $0 < x < 2\pi$ and deduce the value of
$$\frac{1}{1^2}+\frac{1}{2^2}+\frac{1}{3^2}+\cdots+\frac{\pi^2}{6}$$ [JNTU 2008]
6. Find the Fourier series to represent the function e^{-ax} from $x = -\pi$ to $x = \pi$. Deduce from this that
$$\frac{\pi}{\sinh\pi}=2\left[\frac{1}{2^2+1}-\frac{1}{3^2+1}+\frac{1}{4^2+1}-\cdots\right]$$
7. Find the Fourier series to represent the function $f(x)$ given by
$$f(x)=\begin{cases}-k & \text{for } -\pi<x<0\\ k & \text{for } 0<x<\pi\end{cases}$$
Hence show that
$$1-\frac{1}{3}+\frac{1}{5}-\frac{1}{7}+\cdots=\frac{\pi}{4}$$
[JNTU Dec. 2005, 2002]
8. Find the Fourier series of the periodic function defined as
$$f(x)=\begin{cases}-\pi, & -\pi<x<0\\ x, & 0<x<\pi\end{cases}$$
Hence deduce that
$$\frac{1}{1^2}+\frac{1}{3^2}+\frac{1}{5^2}+\cdots=\frac{\pi^2}{8}$$
[JNTU 2007, 2006, 2004]
9. Obtain Fourier series for
$$f(x)=\begin{cases}x & \text{in } -\pi<x<0\\ 0 & \text{in } 0<x<\pi/2\\ x-\frac{\pi}{2} & \text{in } \frac{\pi}{2}<x<\pi\end{cases}$$ [JNTU 2006]

7.5 Even and Odd Functions

The function in Example 3 is *even,* and its Fourier series has only *cosine* terms. The function in Example 4 is *odd*, and its Fourier series has only *sine* terms.

Recall that g is **even** if $g(-x) = g(x)$, so that its graph is symmetric with respect to the vertical axis (Figure 8). A function h is **odd** if $h(-x) = -h(x)$ (Figure 9).

Now the cosine terms in the Fourier series (7.16) are even and the sine terms are odd. So it should not be a surprise that an even function is given by a series of cosine terms and an odd function by a series of sine terms. Indeed, the following holds.

THEOREM 3 (Fourier Cosine Series, Fourier Sine Series): *The Fourier series of an* ***even*** *function of period 2L is a* ***"Fourier cosine series"***

$$f(x) = a_0 + \sum_{n=1}^{\infty} a_n \cos\frac{n\pi}{L}x \quad (f \text{ even}) \tag{7.18}$$

with coefficients (note: integration from 0 to L only!)

$$a_0 = \frac{1}{L}\int_0^L f(x)dx, \quad a_n = \frac{2}{L}\int_0^L f(x)\cos\frac{n\pi x}{L}dx \quad (n = 1, 2, \ldots) \tag{7.19}$$

The Fourier series of an ***odd*** *function of period 2L is a* ***"Fourier sine series"***

$$f(x) = \sum_{n=1}^{\infty} b_n \sin\frac{n\pi}{L}x \quad (f \text{ odd}) \tag{7.20}$$

with coefficients

$$b_n = \frac{2}{L}\int_0^L f(x)\sin\frac{n\pi x}{L}dx \tag{7.21}$$

PROOF: Since the definite integral of a function gives the area under the curve of the function between the limits of integration, we have

$$\int_{-L}^{L} g(x)dx = 2\int_0^L g(x)dx \quad \text{(for even } g)$$

$$\int_{-L}^{L} h(x)dx = 0 \quad \text{(for odd } h)$$

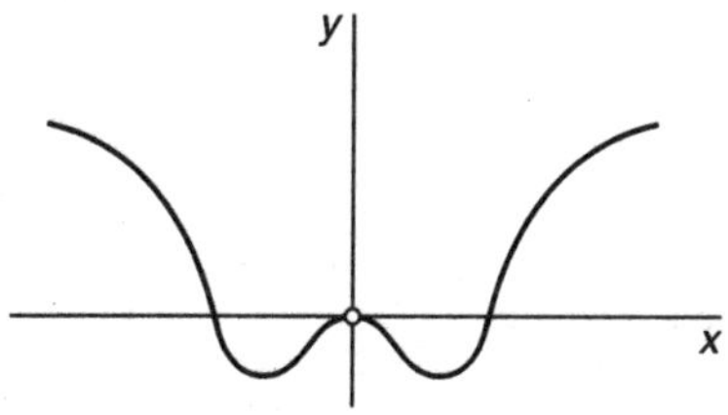

Figure 8 Even function.

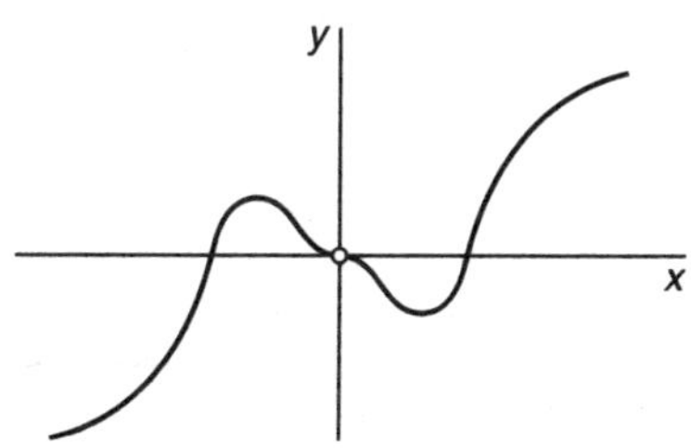

Figure 9 Odd function.

as is obvious from the graphs of g and h. (Give a formal proof.) Now let f be even. Then (7.17a) gives a_0 in (7.19). Also, the integrand in (7.17b) is even (a product of even functions is even), so that (7.17b) gives a_n in (7.19). Furthermore, the integrand in (7.17c) is the even f times the odd sine, so that the integrand (the product) is odd, the integral is zero, and there are no sine terms in (7.18).

Similarly, if f is odd, the integrals for a_0 and a_n in (7.17a) and (7.17b) are zero, f times the sine in (7.17c) is even, (7.17c) implies (7.21), and there are no cosine terms in (7.20). ■

The Case of Period 2π

If $L = \pi$, then

$$f(x) = a_0 + \sum_{n=1}^{\infty} a_n \cos nx \quad (f \text{ even})$$

with coefficients

$$a_0 = \frac{1}{\pi}\int_0^{\pi} f(x)\,dx, \quad a_n = \frac{2}{\pi}\int_0^{\pi} f(x)\cos nx\,dx \quad (n = 1, 2, \ldots) \tag{7.22}$$

and

$$f(x) = \sum_{n=1}^{\infty} b_n \sin nx \quad (f \text{ odd})$$

with coefficients

$$b_n = \frac{2}{\pi}\int_0^{\pi} f(x)\sin nx\,dx \quad (n = 1, 2, \ldots) \tag{7.23}$$

For instance, $f(x)$ in Example 1 is odd and is represented by a Fourier sine series. Further simplifications result from the following property, whose very simple proof is left to the student.

THEOREM 4 (Sum and Scalar Multiple): *The Fourier coefficients of a sum $f_1 + f_2$ are the sums of the corresponding Fourier coefficients of f_1 and f_2. The Fourier coefficients of cf are c times the corresponding Fourier coefficients of f.*

Example 9 Rectangular Pulse

The function $f^*(x)$ in Figure 10 is the sum of the function $f(x)$ in Example 1 and the constant k. Hence, from that example and Theorem 2 we conclude that

$$f^*(x) = k + \frac{4k}{\pi}\left(\sin x + \frac{1}{3}\sin 3x + \frac{1}{5}\sin 5x + \cdots\right)$$

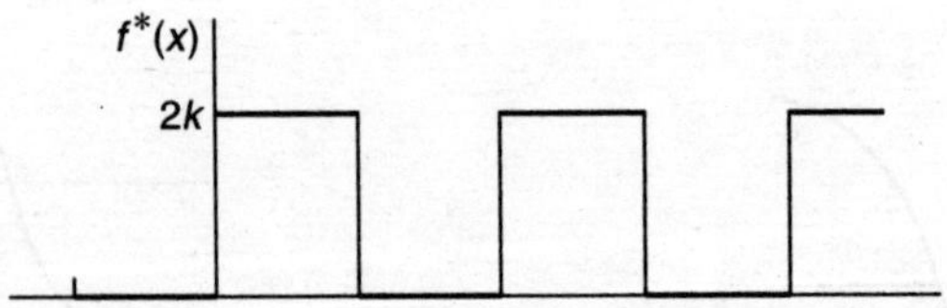

Figure 10 Example 9.

Example 10 Half-Wave Rectifier

The function $u(t)$ in Example 5 has a Fourier cosine series plus a single term $v(t) = (E/2)\sin \omega t$. We conclude from this and Theorem 2 that $u(t) - v(t)$ must be an even function. Verify this graphically. (See Figure 11.)

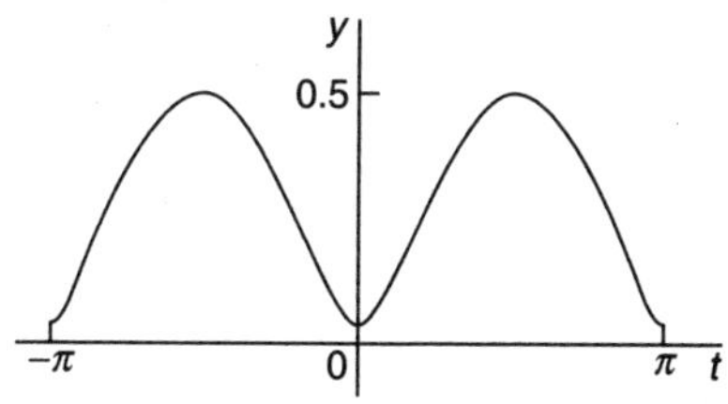

Figure 11 $u(t) - v(t)$ with $E = 1$, $\omega = 1$.

Example 11 Sawtooth Wave

Find the Fourier series of the function (Figure 12)

$$f(x) = x + \pi \quad \text{if} \quad -\pi < x < \pi \qquad \text{and} \qquad f(x + 2\pi) = f(x)$$

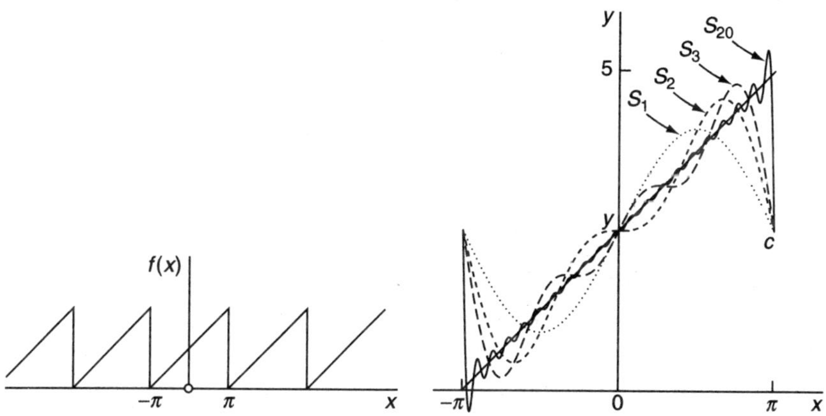

(a) The function $f(x)$ (b) Partial sums S_1, S_2, S_3, S_{20}

Figure 12 Example 11.

Solution: We have $f = f_1 + f_2$, where $f_1 = x$ and $f_2 = \pi$. The Fourier coefficients of f_2 are zero, except for the first one (the constant term), which is π. Hence, by Theorem 4, the Fourier coefficients a_n, b_n are those of f_1, except for a_0, which is π. Since f_1 is odd, $a_n = 0$ for $n = 1, 2, \ldots$, and

$$b_n = \frac{2}{\pi}\int_0^{\pi} f_1(x)\sin nx\,dx = \frac{2}{\pi}\int_0^{\pi} x\sin nx\,dx$$

Integrating by parts, we obtain

$$b_n = \frac{2}{\pi}\left[\frac{-x\cos nx}{n}\Bigg|_0^{\pi} + \frac{1}{n}\int_0^{\pi}\cos nx\,dx\right] = -\frac{2}{n}\cos n\pi$$

Hence, $b_1 = 2$, $b_2 = -2/2$, $b_3 = 2/3$, $b_4 = -2/4$, ..., and the Fourier series of $f(x)$ is

$$f(x) = \pi + 2\left(\sin x - \frac{1}{2}\sin 2x + \frac{1}{3}\sin 3x - \cdots\right)$$

CONCEPT CHECK

1. Are the following functions even, odd, or neither even nor odd?
   (a) $|x|$, $x^2 \sin nx$, $x + x^2$, $e^{-|x|}$, $\ln x$, $x \cosh x$
   (b) $\sin(x^2)$, $\sin^2 x$, $x \sinh x$, $|x^3|$, $e^{\pi x}$, xe^x, $\tan 2x$, $x/(1 + x^2)$
2. Are the following functions, which are assumed to be periodic of period 2π, even, odd, or neither even nor odd?
   (a) $f(x) = x^3\ (-\pi < x < \pi)$
   (b) $f(x) = x^2\ (-\pi/2 < x < 3\pi/2)$
   (c) $f(x) = e^{-4x}\ (-\pi < x < \pi)$
   (d) $f(x) = x^3 \sin x\ (-\pi < x < \pi)$
   (e) $f(x) = x|x| - x^3\ (-\pi < x < \pi)$
   (f) $f(x) = 1 - x + x^3 - x^5\ (-\pi < x < \pi)$
   (g) $f(x) = 1/(1 + x^2)$ if $-\pi < x < 0$, $f(x) = -1/(1 + x^2)$ if $0 < x < \pi$
3. **PROJECT. Even and Odd Functions.**
   (a) Are the following expressions even or odd? Sums and products of even functions and of odd functions. Products of even times odd functions. Absolute values of odd functions. $f(x) + f(-x)$ and $f(x) - f(-x)$ for arbitrary $f(x)$.
   (b) Write e^{kx}, $1/(1 - x)$, $\sin(x + k)$, $\cosh(x + k)$ as sums of an even and an odd function.
   (c) Find all functions that are both even and odd.
   (d) Is $\cos^3 x$ even or odd? $\sin^3 x$? Find the Fourier series of these functions. Do you recognize familiar identities?
4. Is the given function even or odd? Find its Fourier series. Sketch or graph the function and some partial sums. (Show the details of your work.)
   (a) $f(x) = \pi - |x|\quad (-\pi < x < \pi)$
   (b) $f(x) = 2x|x|\quad (-1 < x < 1)$
   (c) $f(x) = \begin{cases} x & \text{if} \quad -\pi/2 < x < \pi/2 \\ \pi - x & \text{if} \quad \pi/2 < x < 3\pi/2 \end{cases}$
   (d) $f(x) = \begin{cases} \pi e^{-x} & \text{if} \quad -\pi < x < 0 \\ \pi e^{x} & \text{if} \quad 0 < x < \pi \end{cases}$
   (e) $f(x) = \begin{cases} 2 & \text{if} \quad -2 < x < 0 \\ 0 & \text{if} \quad 0 < x < 2 \end{cases}$
   (f) $f(x) = \begin{cases} 1 - \frac{1}{2}|x| & \text{if} \quad -2 < x < 2 \\ 0 & \text{if} \quad 2 < x < 6 \end{cases} \quad (p = 8)$

7.6 Fourier Coefficients when f(x) is Even or Odd

Half-Range Expansions

Half-range expansions are Fourier series. The idea is simple and useful. Figure 13 explains it. We want to represent $f(x)$ in Figure 13a by a Fourier series, where $f(x)$ may be the shape of a distorted violin string or the temperature in a metal bar of length L, for example. Now comes the idea. We could extend $f(x)$ as a function of period L and develop the extended function into a Fourier series. But this series would in general contain *both* cosine *and* sine terms. We can do better and get simpler series. Indeed, for our given f we can calculate Fourier coefficients from (7.19) or from (7.21) in Theorem 3. And we have a choice and can take what seems more practical. If we use (7.19), we get (7.18). This is the **even periodic extension** f_1 of f in Figure 13b. If we choose (7.21) instead, we get (7.20), the **odd periodic extension** f_2 of f in Figure 13c.

Both extensions have period $2L$. This motivates the name **half-range expansions:** f is given (and of physical interest) only on half the range, half the interval of periodicity of length $2L$.

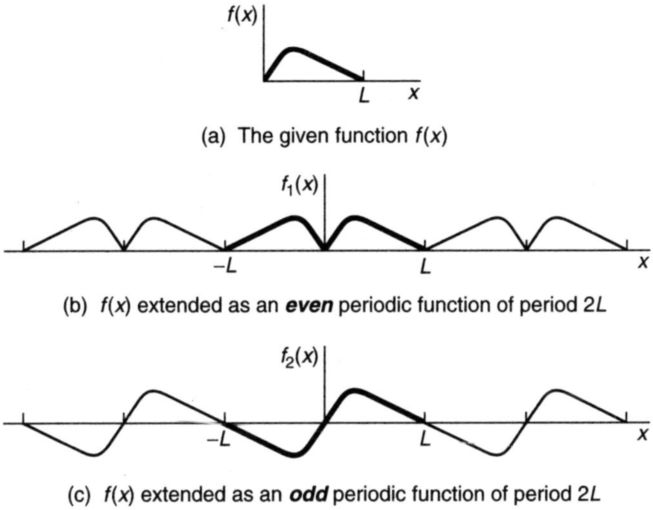

(a) The given function $f(x)$

(b) $f(x)$ extended as an ***even*** periodic function of period $2L$

(c) $f(x)$ extended as an ***odd*** periodic function of period $2L$

Figure 13 (a) Function $f(x)$ given on an interval $0 \leqq x \leqq L$; (b) **even extension** to the full "range" (interval) $-L \leqq x \leqq L$ (heavy curve) and the periodic extension of perod $2L$ to the x axis; (c) **odd extension** to $-L \leqq x \leqq L$ (heavy curve) and the periodic extension of period $2L$ to the x axis.

Example 12 "Triangle" and Its Half-Range Expansion

Find the two half-range expansions of the function (Figure 14)

$$f(x)=\begin{cases}\dfrac{2k}{L}x & \text{if} \quad 0<x<\dfrac{L}{2}\\[2mm] \dfrac{2k}{L}(L-x) & \text{if} \quad \dfrac{L}{2}<x<L\end{cases}$$

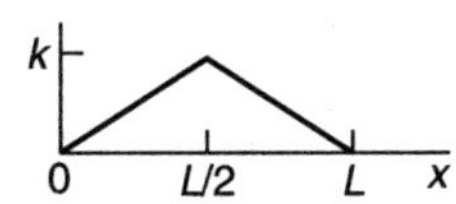

Figure 14 The given function in Example 12.

Solution: (a) ***Even periodic extension.*** From (7.19) we obtain

$$a_0=\frac{1}{L}\left[\frac{2k}{L}\int_0^{L/2} x\,dx+\frac{2k}{L}\int_{L/2}^{L}(L-x)dx\right]=\frac{k}{2}$$

$$a_n = \frac{2}{L}\left[\frac{2k}{L}\int_0^{L/2} x\cos\frac{n\pi}{L}x\,dx + \frac{2k}{L}\int_{L/2}^{L}(L-x)\cos\frac{n\pi}{L}x\,dx\right]$$

We consider a_n. For the first integral we obtain by integration by parts

$$\int_0^{L/2} x\cos\frac{n\pi}{L}x\,dx = \frac{Lx}{n\pi}\sin\frac{n\pi}{L}x\Big|_0^{L/2} - \frac{L}{n\pi}\int_0^{L/2}\sin\frac{n\pi}{L}x\,dx$$

$$= \frac{L^2}{2n\pi}\sin\frac{n\pi}{2} + \frac{L^2}{n^2\pi^2}\left(\cos\frac{n\pi}{2}-1\right)$$

Similarly, for the second integral we obtain

$$\int_{L/2}^{L}(L-x)\cos\frac{n\pi}{L}x\,dx = \frac{L}{n\pi}(L-x)\sin\frac{n\pi}{L}x\Big|_{L/2}^{L} + \frac{L}{n\pi}\int_{L/2}^{L}\sin\frac{n\pi}{L}x\,dx$$

$$= \left[0 - \frac{L}{n\pi}\left(L-\frac{L}{2}\right)\sin\frac{n\pi}{2}\right] - \frac{L^2}{n^2\pi^2}\left[\cos n\pi - \cos\frac{n\pi}{2}\right]$$

We insert these two results into the formula for a_n. The sine terms cancel and so does a factor L^2. This gives

$$a_n = \frac{4k}{n^2\pi^2}\left(2\cos\frac{n\pi}{2} - \cos n\pi - 1\right)$$

Thus,

$$a_2 = -16k/(2^2\pi^2), \qquad a_6 = -16k/(6^2\pi^2), \qquad a_{10} = -16k/(10^2\pi^2), \quad \ldots$$

and $a_n = 0$ if $n \neq 2, 6, 10, 14, \ldots$. Hence the first half-range expansion of $f(x)$ is (Figure 15a)

$$f(x) = \frac{k}{2} - \frac{16k}{\pi^2}\left(\frac{1}{2^2}\cos\frac{2\pi}{L}x + \frac{1}{6^2}\cos\frac{6\pi}{L}x + \cdots\right)$$

This Fourier cosine series represents the even periodic extension of the given function $f(x)$ of period $2L$.

(b) ***Odd periodic extension.*** Similarly, from (7.21) we obtain

$$b_n = \frac{8k}{n^2\pi^2}\sin\frac{n\pi}{2} \tag{7.24}$$

Hence the other half-range expansion of $f(x)$ is (Figure 15b)

$$f(x) = \frac{8k}{\pi^2}\left(\frac{1}{1^2}\sin\frac{\pi}{L}x - \frac{1}{3^2}\sin\frac{3\pi}{L}x + \frac{1}{5^2}\sin\frac{5\pi}{L}x - + \cdots\right)$$

This series represents the odd periodic extension of $f(x)$ of period $2L$.

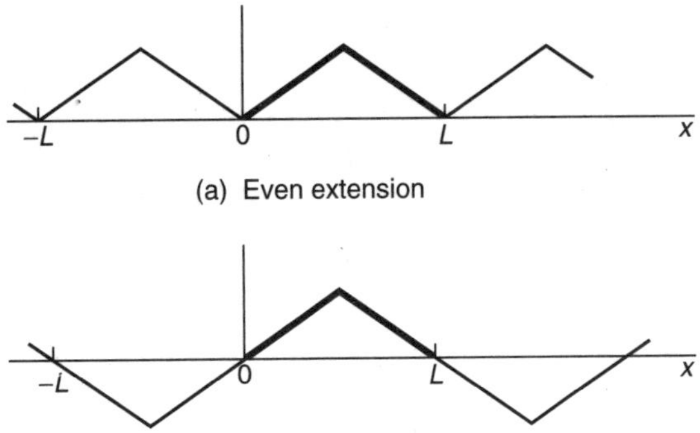

(a) Even extension

(b) Odd extension

Figure 15 Periodic extensions of $f(x)$ in Example 12.

CONCEPT CHECK

Find (a) the Fourier cosine series, (b) the Fourier sine series. Sketch $f(x)$ and its two periodic extensions. (Show the details of your work.)

1. $f(x) = 1 \quad (0 < x < 2)$
2. $f(x) = x(0 < x < 1/2)$
3. $f(x) = 2 - x \quad (0 < x < 2)$
4. $f(x) = \begin{cases} 0 & 0<x<2 \\ 1 & 2<x<4 \end{cases}$
5. $f(x) = \begin{cases} 1 & 0<x<1 \\ 2 & 1<x<2 \end{cases}$
6. $f(x) = \begin{cases} x & 0<x<\pi/2 \\ \pi/2 & \pi/2<x<\pi \end{cases}$
7. $f(x) = x \quad (0 < x < L)$
8. $f(x) = x^2 \quad (0 < x < L)$
9. $f(x) = \pi - x \quad (0 < x < \pi)$

Solved Problems

Q 1. Obtain the Fourier series expansion of $f(x)$ given that $f(x) = kx(\pi - x)$ in $0 < x < 2$ where k is a constant. **[JNTU Aug./Sep. 2008, May/June 2008]**

Answer: The given function is $f(x) = kx(\pi - x)$.

$$kx(\pi - x) = \frac{a_0}{2} + \sum_{n=1}^{\infty}(a_n \cos nx + b_n \sin nx) \tag{7.25}$$

We now calculate a_0, a_n, and b_n.

$$a_0 = \frac{1}{\pi}\int_0^{2\pi} f(x)dx = \frac{1}{\pi}\int_0^{2\pi} kx(\pi - x)dx = \frac{k}{\pi}\int_0^{2\pi}(\pi x - x^2)dx = \frac{k}{\pi}\left[\frac{\pi x^2}{2} - \frac{x^3}{3}\right]_0^{2\pi}$$

$$= \frac{k}{\pi}\left[\frac{\pi(2\pi)^2}{2} - \frac{(2\pi)^3}{3}\right] = \frac{k}{\pi}\left[\frac{4\pi^3}{2} - \frac{8\pi^3}{3}\right] = \frac{k}{\pi}\left[\frac{6\pi^3 - 8\pi^3}{3}\right] = \frac{k}{\pi}\left(\frac{-2\pi^3}{3}\right) = k\left(\frac{-2\pi^2}{3}\right)$$

$$a_n = \frac{1}{\pi}\int_0^{2\pi} f(x)\cos nx\,dx = \frac{1}{\pi}\int_0^{2\pi} k(\pi x - x^2)\cos nx\,dx$$

$$= \frac{k}{\pi}\left[(\pi x - x^2)\left(\frac{\sin nx}{n}\right) - (\pi - 2x)\left(\frac{\cos nx}{n^2}\right) + (-2)\left(\frac{-\sin nx}{n^3}\right)\right]_0^{2\pi}$$

$$= \frac{k}{\pi}\left[(2\pi^2 - 4\pi^2)(0) + -(\pi - 4\pi)\left(\frac{+1}{n^2}\right) + \left(\frac{\pi}{n^2}\right)\right] \text{ [since } \sin 2n\pi = 0 \text{ and } \cos 2n\pi = 1\text{]}$$

$$= \frac{k}{\pi}\left[(2\pi^2 - 4\pi^2)(0) - (\pi - 4\pi)\frac{1}{n^2} + \frac{\pi}{n^2}\right] = \frac{+4k}{n^2}$$

$$b_n = \frac{1}{\pi}\int_0^{2\pi} k(\pi x - x^2)\sin nx\,dx$$

$$= \frac{k}{\pi}\left[(\pi x - x^2)\left(\frac{-\cos nx}{n}\right) - (\pi - 2x)\left(\frac{-\sin nx}{n^2}\right) + (-2)\left(\frac{-\cos nx}{n^3}\right)\right]_0^{2x}$$

$$= \frac{k}{\pi}\left[(2\pi^2 - 4\pi^2)\left(\frac{-1}{n}\right) - 0 - 2\left(\frac{-1}{n^3}\right) + (-2)\left(\frac{-1}{n^3}\right)\right]$$

$$= \frac{k}{\pi}\left(\frac{2\pi^2}{n}\right) = \frac{2\pi k}{n}$$

Replacing a_n, b_n, and a_0 in (7.25) we get the Fourier series as

$$f(x) = \frac{1}{2}\left[k\left(\frac{-2\pi^2}{3}\right)\right] + \sum_{n=1}^{\infty}\left[\frac{-4k}{n^2}\cos nx + \frac{2\pi k}{n}\sin nx\right]$$

$$= \left[\frac{-k\pi^2}{3} + \sum_{n=1}^{\infty}\frac{2\pi k}{n}\sin nx - \frac{4k}{n^2}\cos nx\right]$$

Q 2. Show that for $-\pi < x < \pi$

$$\sin ax = \frac{2\sin\pi}{\pi}\left(\frac{\sin x}{1^2 - a^2} - \frac{2\sin 2x}{2^2 - a^2} + \frac{3\sin 3x}{3^2 - a^2}\cdots\right)$$

[JNTU May 2005, May/June 2004]

OR

Show that

$$\sin ax = \frac{2\sin a\pi}{\pi}\left[\frac{\sin x}{1^2 - a^2} - \frac{2\sin 2x}{2^2 - a^2} + \frac{3\sin 3x}{3^2 - a^2}\cdots\right]$$

in the interval $-\pi < x < \pi$.

Answer: Since $f(x) = \sin ax$ is known to be an odd function, the Fourier expansion of it will consist of sine terms only, which is given by

$$f(x) = \sum_{n=1}^{\infty} b_n \sin nx \tag{7.26}$$

where

$$b_n = \frac{1}{\pi}\int_a^b f(x)\sin\frac{n\pi x}{l}\,dx = \frac{2}{\pi}\int_0^{\pi} \sin ax \sin nx\,dx$$

Since $2\sin A \sin B = \cos(A - B) - \cos(A + B)$ we can write b_n as

$$b_n = \frac{1}{\pi}\int_0^{\pi} \cos(a-n)x - \cos(a+n)x\,dx$$

$$= \frac{1}{\pi}\left[\sin\frac{(a-n)x}{a-n} - \sin\frac{(a+n)x}{a+n}\right]_0^{\pi}$$

$$= \frac{1}{\pi}\left[\frac{\sin a\pi \cos n\pi - \cos a\pi \sin n\pi}{a-n} - \frac{\sin a\pi \cos n\pi + \cos a\pi \sin n\pi}{a+n}\right]$$

$$= \frac{1}{\pi}\left[\frac{\sin a\pi \cos n\pi}{a-n} - \frac{\sin a\pi \cos n\pi}{a+n}\right]$$

Since $\cos n\pi = (-1)^n$ and $\sin n\pi = 0$ we have

$$b_n = \frac{1}{\pi}\left[\sin a\pi \cos n\pi\left(\frac{1}{a-n} - \frac{1}{a+n}\right)\right]$$

$$= \frac{1}{\pi}\sin a\pi(-1)^n\left(\frac{a+n-a+n}{a^2-n^2}\right)$$

$$= \frac{(-1)^n 2n}{\pi(a^2-n^2)}\sin a\pi \tag{7.27}$$

Substituting (7.27) in (7.26), we get

$$f(x) = \frac{2\sin a\pi}{\pi}\sum_{n=1}^{\infty}\frac{n(-1)^n}{a^2-n^2}\sin nx = \frac{2\sin a\pi}{\pi}\sum_{n=1}^{\infty}\frac{n(-1)^{n+1}}{n^2-a^2}\sin nx$$

$$\sin ax = \frac{2\sin a\pi}{\pi}\left(\frac{\sin x}{1^2-a^2} - \frac{2\sin 2x}{2^2-a^2} + \frac{3\sin 3x}{3^2-a^2}\cdots\right)$$

Q 3. Find the Fourier series to represent the function $f(x) = |\sin x|, -\pi < x < \pi$.

[JNTU Aug./Sep. 2006, May/June 2004]

Answer: Since $f(x) = |\sin x|$ is an even function, its Fourier series consists of cosine terms only. Therefore

$$f(x) = \frac{a_0}{2} + \sum_{n=1}^{\infty} a_n \cos nx \tag{7.28}$$

We now find the values a_0 and a_n:

$$a_0 = \frac{1}{\pi}\int_{-\pi}^{\pi} f(x)dx = \frac{2}{\pi}\int_0^{\pi} f(x)\,dx = \frac{2}{\pi}\int_0^{\pi} \sin x\,dx = \frac{2}{\pi}[-\cos x]_0^{\pi} = \frac{2}{\pi}[-(-1)+1] = \frac{4}{\pi}$$

$$a_n = \frac{1}{\pi}\int_0^{\pi} f(x)\cos nx\,dx = \frac{2}{\pi}\int_0^{\pi} f(x)\cos nx\,dx = \frac{2}{\pi}\int_0^{\pi} \sin x \cos nx\,dx$$

$$= \frac{1}{\pi}\int_0^{\pi} [\sin(n+1)x - \sin(n-1)x]dx = \frac{1}{\pi}\left[\frac{-\cos(n+1)x}{n+1} + \frac{\cos(n-1)x}{n-1}\right]_0^{\pi}$$

$$= \frac{1}{\pi}\left[\frac{-\cos(n+1)\pi}{n+1} + \frac{\cos(n-1)\pi}{n-1} + \frac{1}{n+1} + \frac{1}{n-1}\right]$$

$$= \frac{1}{\pi}\left[\frac{-(-1)^{n+1}}{n+1} + \frac{(-1)^{n+1}}{1-n} + \frac{1}{n+1} + \frac{1}{n-1}\right]$$

$$= \frac{-1}{\pi}\left[\frac{-(-1)^{n+1}+1}{n+1}\right] + \frac{1}{\pi}\left[\frac{(-1)^{n-1}-1}{n-1}\right]$$

$$= \frac{-1}{\pi}\left[\frac{-(-1)^{n}(-1)+1}{n+1}\right] + \frac{1}{\pi}\left[\frac{(-1)^{n}(-1)^{-1}-1}{n-1}\right]$$

$$= \frac{-(-1)^n}{\pi(n+1)} - \frac{1}{\pi(n+1)} - \frac{(-1)^n}{\pi(n-1)} - \frac{1}{\pi(n-1)}$$

$$= \frac{-(-1)^n}{\pi} - \left[\frac{1}{n+1} - \frac{1}{n-1}\right] - \frac{1}{\pi}\left[\frac{1}{n+1} - \frac{1}{n-1}\right]$$

$$= \frac{(-1)^n}{\pi}\left[\frac{n-1-n-1}{n^2-1}\right] - \frac{1}{\pi}\left[\frac{n-1+n-1}{n^2-1}\right]$$

$$= \frac{-2(-1)^n}{\pi(n^2-1)} - \frac{2n}{\pi(n^2-1)} = \frac{-2[n+(-1)^n]}{\pi(n^2-1)}$$

If $n = 1$,

$$a_1 = \frac{1}{\pi}\int_{-\pi}^{\pi} |\sin x|\cos x\,dx = \frac{2}{\pi}\int_0^{\pi} \sin x \cos x\,dx$$

$$= \frac{1}{\pi}\int_0^{\pi} \sin 2x\,dx = \frac{1}{\pi}\left|-\frac{\cos 2x}{2}\right|_0^{\pi} = \frac{1}{2\pi}(1-\cos 2\pi) = 0$$

Hence the Fourier series can be written as

$$f(x)=\frac{2}{\pi}-\frac{4}{\pi}\sum_{n=1}^{\infty}\frac{\cos 2nx}{(2n)^2-1}$$

$$f(x)=\frac{2}{\pi}-\frac{4}{\pi}\sum_{n=1}^{\infty}\frac{\cos 2nx}{(2n-1)(2n+1)}$$

Q 4. An alternating current after passing through rectifier has the form

$$i=\begin{cases} I_0\sin x & \text{for } 0\le x\le\pi \\ 0 & \text{for } \pi\le x\le 2\pi \end{cases}$$

where I_0 is the maximum current and the period is 2π. Express i as a Fourier series.

{JNTU Aug./Sep. 2006, May 2005}

Answer: The Fourier series representation of $i(x)$ in $[0, 2\pi]$ is

$$i(x)=\frac{a_0}{2}+\sum_{n=1}^{\infty}a_n\cos nx+\sum_{n=1}^{\infty}b_n\sin nx$$

where

$$a_0=\frac{1}{\pi}\int_0^{2\pi}f(x)dx=\frac{1}{\pi}\left[\int_0^{\pi}f(x)dx+\int_{\pi}^{2\pi}f(x)dx\right]$$

$$=\frac{1}{\pi}+\int_0^{\pi}I_0\sin x\,dx+0=\frac{I_0}{\pi}\left[-\cos x\right]_0^{\pi}=\frac{I_0}{\pi}\{1+1\}=\frac{2I_0}{\pi}$$

Similarly,

$$a_n=\frac{1}{\pi}\left[\int_0^{\pi}I_0\sin x\cos nx\,dx+0\right]$$

$$=\frac{1}{2\pi}\left[\int_0^{\pi}\sin(1+n)x\,dx+\int_0^{\pi}\sin(1-n)x\,dx\right]$$

$$=\frac{I_0}{2\pi}\left[\frac{-\cos(1+n)x}{(1+n)}-\frac{-\cos(1-n)x}{(1-n)}\right]_0^{\pi}$$

$$=\frac{I_0}{2\pi}\left[\left\{\frac{-1}{1+n}\{\cos(1+n)p-1\}\right\}\left\{\frac{\cos(1-n)\pi-1}{1-n}\right\}\right]$$

$$=\frac{I_0}{2\pi}\left[\frac{-(-1)^{1+n}}{1+n}+\frac{(-1)^{1-n}}{1-n}\right]+\frac{2}{1-n^2}$$

$$=\frac{I_0}{2\pi}\{(-1)^n+1\}\left[\frac{2}{1-n^2}\right]$$

$$=\frac{I_0}{\pi}\{(-1)^n+1\}\frac{1}{(n+1)(1-n)}$$

Therefore

$$a_n = 2\frac{I_0}{\pi}\frac{(-1)^{n+1}}{(n+1)(1-n)} \quad (n \neq 1)$$

For $n = 1$,

$$a_1 = \frac{1}{\pi}\int_0^{\pi} I_0 \sin x \cos x \, dx = \frac{I_0}{2\pi}\left[\frac{-\cos 2x}{2}\right]_0^{\pi} = 0$$

Again,

$$b_n = \frac{1}{\pi}\int_0^{2\pi} I_0(x) \sin nx \, dx = \frac{1}{\pi}\left[\int_0^{\pi} f(x) \sin nx \, dx + \int_{\pi}^{2\pi} f(x) \sin nx \, dx\right]$$

$$= \frac{1}{\pi}\left[\int_0^{\pi} I_0 \sin x \sin nx \, dx\right]$$

Since $2 \sin A \sin B = \cos (A - B) - \cos (A + B)$, we can write b_n as

$$b_n = \frac{I_0}{2\pi}\int_0^{\pi} [\cos (n-1)x - \cos (n+1)x] \, dx$$

$$= \frac{I_0}{2\pi}\left[\frac{\sin (n-1)x}{n+1} - \frac{\sin (n+1)x}{n+1}\right]_0^{\pi} = 0$$

For all $n \neq 1$ and when $n = 1$, we have

$$b_1 = \frac{1}{\pi}\int_0^{\pi} I_0 \sin^2 x \, dx$$

Since $\sin^2 x = (1 - \cos 2x)/2$, we have

$$b_1 = \frac{I_0}{2\pi}\int_0^{\pi} (1 - \cos 2x) \, dx = \frac{I_0}{2\pi}\left[x - \frac{\sin 2x}{2}\right]_0^{\pi} = \frac{I_0}{2}$$

Hence with these values, the Fourier series becomes

$$i(x) = \frac{I_0}{\pi} - \frac{2I_0}{2\pi}\left\{\frac{\cos 2x}{1\times 3} + \frac{\cos 4x}{3\times 5} + \cdots\right\} + \frac{I_0}{2}\sin x$$

Q 5. Define a periodic function. Find the Fourier expansion for the function $f(x) = x + x^2, -1 < x < 1$.

[JNTU May/June 2006, Nov. 2005, Nov. 2004, Nov. 2003]

Answer: A function $f(x)$ defined for real numbers is said to be periodic if there is a real number $T > 0$ such that,

$$f(x + T) = f(x) \text{ for all } x \tag{7.29}$$

and the number T is called its period. Therefore from (7.29), we get

$$f(x \pm kT) = f(x) \text{ for all } x$$

for $k = 1, 2, 3, \ldots$; that is, $2T, 3T, 4T$, etc., are also periods of $f(x)$.

The Fourier series representation of $f(x)$, $-c < x < c$, is

$$f(x) = \frac{1}{2}a_0 + \sum_{n=1}^{\infty}\left[a_n\cos\left(\frac{n\pi x}{c}\right) + b_n\sin\left(\frac{n\pi x}{c}\right)\right] \quad (7.30)$$

where

$$a_0 = \frac{1}{c}\int_{-c}^{c} f(x)dx$$

$$a_n = \frac{1}{c}\int_{-c}^{c} f(x)\cos\left(\frac{n\pi x}{c}\right)dx \quad (7.31)$$

$$b_n = \frac{1}{c}\int_{-c}^{c} f(x)\sin\left(\frac{n\pi x}{c}\right)dx,\ n = 1, 2, \ldots$$

Taking $c = 1$ in (7.30) and (7.31), we get

$$f(x) = \frac{1}{2}a_0 + \sum_{n=1}^{\infty}\{a_n \cos(n\pi x) + b_n \sin(n\pi x)\} \quad (7.32)$$

where

$$a_0 = \int_{-1}^{1} f(x)dx = \int_{-1}^{1}(x - x^2)dx = \int_{-1}^{1} xdx - \int_{-1}^{1} x^2dx = 0 - 2\int_{0}^{1} x^2dx = -2\left|\frac{x^3}{3}\right|_0^1 = -\frac{2}{3}$$

$$a_n = \int_{-1}^{1} f(x)\cos(n\pi x)dx = \int_{-1}^{1}(x - x^2)\cos(n\pi x)dx = \int_{-1}^{1} x\cos(n\pi x)dx - \int_{-1}^{1} x^2\cos(n\pi x)dx$$

$$= 0 - 2\int_{0}^{1} x^2\cos(n\pi x)dx$$

Since the integrand of the first term is odd and integrand of second term is even, we get

$$a_n = -2\left[(x^2)\left(\frac{\sin n\pi x}{n\pi}\right) - (2x)\left(-\frac{\cos n\pi x}{n^2\pi^2}\right) + (2)\left(-\frac{\sin n\pi x}{n^3\pi^3}\right)\right]_{x=0}^{1}$$

$$= -2\left[\left\{\frac{\sin n\pi}{n\pi} + 2\frac{\cos n\pi}{n^2\pi^2} - \frac{2\sin(n\pi)}{n^3\pi^3}\right\} - \{0 + 0 + (-2)\times 0\}\right]$$

$$= -2\left[0 + \frac{2(-1)^n}{n^2\pi^2} - 2\times 0 - 0\right] \quad [\text{since } \sin n\pi = 0 \text{ and } \cos n\pi = (-1)^n \text{ for all } n]$$

$$= \frac{4(-1)^n}{n^2\pi^2} = \frac{4(-1)^{n+1}}{n^2\pi^2} \quad \text{for} \quad n = 1, 2, \ldots$$

Thus,

$$a_1 = \frac{4}{\pi^2};\ a_2 = \frac{-4}{4\pi^2} = \frac{-1}{\pi^2};\ a_3 = \frac{4}{9\pi^2};\ \ldots$$

Finally,

$$b_n = \int_{-1}^{1} f(x) \sin(n\pi x)dx = \int_{-1}^{1} (x - x^2) \sin(n\pi x)dx = \int_{-1}^{1} x \sin(n\pi x)dx - \int_{-1}^{1} (x^2) \sin(n\pi x)dx$$

$$= 2\int_{0}^{1} x \sin(n\pi x)dx - 0 = 2\left[(x)\left(-\frac{\cos n\pi x}{n\pi}\right) - (1)\left(-\frac{\sin n\pi x}{n^2\pi^2}\right)\right]_0^1$$

$$= 2\left[(1)\left(-\frac{\cos n\pi}{n\pi} + \frac{\sin n\pi}{n^2\pi^2}\right) - 0\right] = \frac{2\cos n\pi}{n\pi} = -\frac{2(-1)^n}{n\pi} = \frac{2(-1)^{n+1}}{n\pi}$$

$$b_n = \frac{2(-1)^{n-1}}{n\pi}$$

For $n = 1, 2, \ldots$ we get

$$b_1 = \frac{2}{\pi}$$

$$b_2 = \frac{2(-1)^n}{n\pi} = -\frac{1}{\pi}$$

$$b_3 = \frac{2}{3\pi}, \ldots$$

Hence, the Fourier series of $f(x) = x - x^2$ in $-1 < x < 1$ is

$$f(x) = -\frac{1}{3} + \frac{4}{\pi^2}\sum_{n=1}^{\infty}\frac{(-1)^{n+1}}{n^2} \cos(n\pi x) + \frac{2}{\pi}\sum_{n=1}^{\infty}\frac{(-1)^{n+1}}{n} \sin(n\pi x)$$

Q 6. Write the Dirichlet's conditions for the existence of Fourier series of a function $f(x)$ in the interval $(\alpha, \alpha + 2\pi)$. Find the Fourier series representing $f(x) = x, 0 < x < 2\pi$.

[JNTU Aug./Sep. 2007, Nov. 2005, Nov. 2004]

Answer:

Conditions

1. $f(x)$ is uniformly bounded on the fundamental interval $(-\pi, \pi)$ of period 2π, that is, $|f(x)| \le M$ for all $-\pi < x < \pi$ where M is a constant.
2. $f(x)$ has no more than a finite number of finite discontinuities in $(-\pi, \pi)$.
3. $f(x)$ has a finite number of strict maxima or minima in $(-\pi, \pi)$.

The above conditions are Dirichlet's conditions. We now find the Fourier series of $f(x)$ in the interval $(0, 2\pi)$ is

$$f(x) = \frac{a_0}{2} + \sum_{n=1}^{\infty} a_n \cos nx + \sum_{n=1}^{\infty} b_n \sin nx \tag{7.33}$$

$$a_0 = \frac{1}{\pi}\int_{0}^{2\pi} f(x)dx$$

$$= \frac{1}{\pi}\int_0^{2\pi} x\,dx = \frac{1}{\pi}\left[\frac{x^2}{2}\right]_0^{2\pi}$$

$$= \frac{1}{2\pi}[x^2]_0^{2\pi} = \frac{1}{2\pi}[4\pi^2 - 0] = \frac{4\pi^2}{2\pi} = 2\pi \tag{7.34}$$

$$a_n = \frac{1}{\pi}\int_0^{2\pi} f(x)\cos nx\,dx = \frac{1}{\pi}\int_0^{2\pi} x\cos nx\,dx$$

$$= \frac{1}{\pi}\left[x\frac{\sin nx}{n} - \frac{1}{n^2}\cos nx\right]_0^{2\pi}$$

$$= \frac{1}{\pi}\left[\left[2\pi\frac{\sin 2n\pi}{n} - \frac{2\pi}{n^2}\cos 2n\pi\right] - [0]\right] = \frac{1}{\pi}\left[0 - \frac{2\pi}{n^2}\right] = \frac{-2}{n^2} \tag{7.35}$$

$$b_n = \frac{1}{\pi}\int_0^{2\pi} f(x)\sin nx\,dx = \frac{1}{\pi}\int_0^{2\pi} x\sin nx\,dx$$

$$= \frac{1}{\pi}\left[-x\frac{\cos nx}{n} + \frac{\sin nx}{n^2}\right]_0^{2\pi}$$

$$= \frac{1}{\pi}\left[-2\pi\frac{\cos 2n\pi}{n} + \frac{\sin 2n\pi}{n^2}\right] = \frac{-2}{n} \tag{7.36}$$

Substituting the values of a_0, a_n, b_n from (7.34) – (7.36) in (7.33),

$$f(x) = \pi \pm m\sum_{n=1}^{\infty}\left(\frac{-1}{n^2}\right)\cos nx + \sum_{n=1}^{\infty}\left(\frac{-2}{n}\right)\sin nx$$

$$x = \pi - 2\sum_{n=1}^{\infty}\left(\frac{\cos nx}{n^2} + \frac{\sin nx}{n}\right)$$

Q 7. Find the Fourier series of periodicity 3 for $f(x) = 2x - x^2$ in $0 < x < 3$.

[JNTU Aug./Sep. 2008, May/June 2008]

Answer: The given function is

$$f(x) = 2x - x^2$$

$$f(x) = \frac{a_0}{2} + \sum_{n=1}^{\infty} a_n \cos\left(\frac{n\pi x}{l}\right) + b_n \sin\left(\frac{n\pi x}{l}\right)$$

Now $2l = 3$. This implies $l = 3/2$. We find values of a_0, a_n, b_n.

$$a_0 = \frac{1}{l}\int_0^3 f(x)\,dx = \frac{2}{3}\int_0^3 (2x - x^2)\,dx = \frac{2}{3}\left[2\left(\frac{x^2}{2}\right) - \left(\frac{x^3}{3}\right)\right]_0^3 = \frac{2}{3}\left[9 - \frac{27}{3}\right] = 0$$

$$a_n = \frac{1}{l}\int_0^3 f(x)\cos\frac{n\pi x}{l}\,dx = \frac{2}{3}\int_0^3 (2x - x^2)\cos\left(\frac{n\pi x}{l}\right)dx = \frac{2}{3}\int_0^3 (2x - x^2)\cos\left(\frac{2n\pi x}{3}\right)dx$$

Applying integration by parts, we get

$$a_n = \frac{2}{3}\left[(2x - x^2)\sin\frac{2n\pi x}{3}\times\left(\frac{3}{2n\pi}\right) + (2-2x)\cos\frac{2n\pi x}{3}\times\left(\frac{3}{2n\pi}\right)^2 + 2\sin\frac{2n\pi x}{3}\times\left(\frac{3}{2n\pi}\right)^3\right]_0^3$$

$$= \frac{2}{3}\left[\frac{-4(9)}{4\pi^2 n^2} + \frac{2(9)}{4\pi^2 n^2}\right] = \frac{2}{3}\left[\frac{-36+18}{4\pi^2 n^2}\right]$$

$$= \frac{2}{3}\left[\frac{-18}{4\pi^2 n^2}\right] = \frac{-3}{\pi^2 n^2}$$

Also

$$b_n = \frac{1}{l}\int_0^3 f(x)\sin\left(\frac{\pi n x}{l}\right)dx = \frac{2}{3}\int_0^3 (2x - x^2)\sin\left(\frac{2\pi n x}{3}\right)dx$$

Applying integration by parts, we get

$$b_n = \frac{2}{3}\left[(2x - x^2)\left[\frac{\cos 2\pi n x}{3}\right]\left(\frac{3}{2n\pi}\right) + (2-2x)\left(-\frac{\sin 2n\pi x}{3}\right)\left(\frac{3}{2n\pi}\right)^2 - 2\frac{\cos 2n\pi x}{3}\left(\frac{3}{2\pi x}\right)^3\right]_0^3$$

$$= \frac{2}{3}\left[3\left(\frac{3}{2n\pi}\right) - 2\frac{27}{8n^3\pi^3} + \frac{2}{3}\frac{27}{8n^3\pi^3}\right] = \frac{6}{2n\pi} = \frac{3}{n\pi}$$

Therefore the Fourier series for $f(x)$ is

$$(2x - x^2) = \sum_{n=1}^{\infty}\left[\frac{-3}{\pi^2 n^2}\cos\frac{2n\pi}{3}x + \frac{3}{n\pi}\frac{\sin 2n\pi}{3}x\right]$$

Q 8. Prove that the function $f(x) = x, 0 \leqq x \leqq \pi$ can be expanded in a series of sines as,

$$x = 2\left[\frac{\sin x}{1} - \frac{\sin 2x}{2} + \frac{\sin 3x}{3} - \cdots\right]$$

[JNTU Nov. 2005, Nov. 2004, Nov. 2003]

Answer: The problem is to represent $f(r) = x$ as a half-range sine series in the ranges $0 \le x < \pi$, which is given by

$$f(x) = \sum_{n=1}^{\infty} b_n \sin nx \tag{7.37}$$

where

$$b_n = \frac{2}{\pi}\int_0^{\pi} f(x)\sin nx\,dx \quad (n = 1, 2, \ldots) \tag{7.38}$$

$$b_n = \frac{2}{\pi}\int_0^{\pi} x \sin nx\, dx = \frac{2}{\pi}\left[-(x)\left(\frac{\cos nx}{n}\right)+(1)\left(\frac{\sin nx}{n^2}\right)\right]_0^{\pi}$$

$$= \frac{2}{\pi}\left[\left(-\frac{\pi\cos n\pi}{n}+\frac{\sin n\pi}{n^2}\right)+\left(0+\frac{\sin 0}{n}\right)\right] = +\frac{2}{\pi}\frac{\pi(-1)^n}{n} = \frac{2(-1)^{n+1}}{n}$$

Since $\cos n\pi = (-1)^n$, $\sin n\pi = 0$, for $n = 1, 2, \ldots$. Taking $n = 1, 2, 3, \ldots$ we get,

$$b_1 = 2,\ b_2 = -2\frac{1}{2},\ b_3 = 2\frac{1}{3},\ \text{etc.}$$

With all these values, (7.37) can be written as,

$$f(x) = x = 2\sum_{n=1}^{\infty}\frac{(-1)^{n+1}}{n}\sin nx$$

$$x = 2\left[\frac{\sin x}{1}-\frac{\sin 2x}{2}+\frac{\sin 3x}{3}-\cdots\right], \qquad 0 \le x \le \pi$$

Summary

Fourier series concern **periodic functions** $f(x)$ of period $p = 2L$, that is, by definition $f(x+p) = f(x)$ for all x and some fixed $p > 0$; thus, $f(x+np) = f(x)$ for any integer n. These series are of the form

$$f(x) = a_0 + \sum_{n=1}^{\infty}\left(a_n\cos\frac{n\pi}{L}x + b_n\sin\frac{n\pi}{L}x\right) \quad (7.39)$$

with coefficients, called the **Fourier coefficients** of $f(x)$, given by the Euler formulas

$$a_0 = \frac{1}{2L}\int_{-L}^{L} f(x)dx,\ a_n = \frac{1}{L}\int_{-L}^{L} f(x)\cos\frac{n\pi x}{L}dx$$

$$b_n = \frac{1}{L}\int_{-L}^{L} f(x)\sin\frac{n\pi x}{L}dx \quad (7.40)$$

where $n = 1, 2, \ldots$. For period 2π we simply have

$$f(x) = a_0 + \sum_{n=1}^{\infty}(a_n\cos nx + b_n\sin nx) \quad (7.41)$$

with the *Fourier coefficients* of $f(x)$

$$a_0 = \frac{1}{2\pi}\int_{-\pi}^{\pi} f(x)dx,\ a_n = \frac{1}{\pi}\int_{-\pi}^{\pi} f(x)\cos nx\, dx$$

$$b_n = \frac{1}{\pi}\int_{-\pi}^{\pi} f(x)\sin nx\, dx$$

Fourier series are fundamental in connection with periodic phenomena, particularly in models involving differential equations. If $f(x)$ is even $[f(-x) = f(x)]$ or odd $[f(-x) = -f(x)]$, they reduce to **Fourier cosine** or **Fourier sine series**, respectively. If $f(x)$ is given for $0 \leqq x \leqq L$ only, it has two half-range expansions of period $2L$, namely, a cosine and a sine series.

The set of cosine and sine functions in (1) is called the **trigonometric system**. Its most basic property is its **orthogonality** on an interval of length $2L$; that is, for all integers m and $n \ne m$ we have

$$\int_{-L}^{L}\cos\frac{m\pi x}{L}\cos\frac{n\pi x}{L}dx = 0$$

$$\int_{-L}^{L}\sin\frac{m\pi x}{L}\sin\frac{n\pi x}{L}dx = 0$$

and for all integers m and n,

$$\int_{-L}^{L}\cos\frac{m\pi x}{L}\sin\frac{n\pi x}{L}dx = 0$$

This orthogonality was crucial in deriving the Euler formulas (7.40).

Review Questions and Problems

1. What is a Fourier series? A Fourier sine series? A half-range expansion?
2. Can a discontinuous function have a Fourier series? A Taylor series? Explain.
3. Why do we start with period 2π? How do we proceed to functions of any period p?
4. What is the trigonometric system? Its main property by which we obtained the Euler formulas?
5. What do you know about the convergence of a Fourier series?
6. Find the Fourier series of $f(x)$ as given over one period. Sketch $f(x)$. (Show the details of your work.)

a. $f(x)=\begin{cases}-k \text{ if } & -1<x<0\\ k \text{ if } & 0<x<1\end{cases}$

b. $f(x)=\begin{cases}0 \text{ if } & -\pi/2<x<\pi/2\\ 1 \text{ if } & \pi/2<x<3\pi/2\end{cases}$

c. $f(x) = x\ (-2\pi < x < 2\pi)$

d. $f(x) = |x|\ (-2 < x < 2)$

e. $f(x)=\begin{cases}x & \text{if } -1<x<1\\ 2-x & \text{if } 1<x<3\end{cases}$

f. $f(x)=\begin{cases}-1-x & \text{if } -1<x<0\\ 1-x & \text{if } 0<x<1\end{cases}$

g. $f(x) = |\sin 8\pi x|\ (-1/8 < x < 1/8)$

h. $f(x) = e^x\ (-\pi < x < \pi)$

i. $f(x) = x^2\ (-\pi/2 < x < \pi/2)$

j. $f(x) = x\ (0 < x < 2\pi)$

8 Partial Differential Equations

LEARNING OBJECTIVES

After reading this chapter, you will be able to understand the following:

- Basic concepts of partial differential equation (PDE).
- How to form a PDE.
- How to solve both linear and nonlinear PDEs.
- Method of separation of variables.
- One-dimensional and two-dimensional wave equations.
- One-dimensional and two-dimensional heat equations.
- Steady-state conditions.

Partial differential equations (PDEs) are models of various physical and geometrical problems, arising when the unknown functions (the solutions) depend on two or more variables, usually on time t and one or several space variables. It is fair to say that only the simplest physical systems can be modeled by ODEs, whereas most problems in dynamics, elasticity, heat transfer, electromagnetic theory, and quantum mechanics require PDEs. Indeed, the range of applications of PDEs is enormous, compared to that of ODEs.

8.1 Basic Concepts

A **partial differential equation (PDE)** is an equation involving one or more partial derivatives of an (unknown) function that depends on two or more variables. The order of the highest derivative is called the **order** of the PDE. As for ODEs, second-order PDEs will be the most important ones in applications.

A PDE is of the form

$$F\left(x, y, z, \frac{\partial z}{\partial x}, \frac{\partial z}{\partial y}, \frac{\partial^2 z}{\partial x^2}, \ldots\right) = 0 \tag{8.1}$$

Some standard notations that would be used in this chapter are as follows:

$$\frac{\partial z}{\partial x} = z_x = p; \quad \frac{\partial z}{\partial y} = z_y = q$$

$$\frac{\partial^2 z}{\partial x \partial y} = z_{xy} = s; \quad \frac{\partial^2 z}{\partial y^2} = z_{yy} = t; \quad \frac{\partial^2 z}{\partial x^2} = z_{xx} = r$$

Example 1

The following equations

$$px - qy = (z - xy)^2$$

$$\frac{\partial^2 u}{\partial t^2} = c^2 \frac{\partial^2 u}{\partial x^2}$$

$$px + qy + pq = 0$$

are some examples of PDE.

8.2 Formation of PDE

Formation of PDF by Eliminating Arbitrary Constants

Let

$$F(x, y, z, a, b) = 0 \tag{8.2}$$

be the equation in which a and b are arbitrary constants. Differentiating this equation with respect to x and y we get

$$\frac{\partial F}{\partial x} + \frac{\partial F}{\partial z} \times \frac{\partial z}{\partial x} = 0 \tag{8.3}$$

$$\frac{\partial F}{\partial y} + \frac{\partial F}{\partial z} \times \frac{\partial z}{\partial y} = 0 \tag{8.4}$$

By eliminating a and b from (8.2)–(8.4) we get a PDE of the form

$$f(x, y, z, p, q) = 0$$

Example 2

Eliminate arbitrary constants and form a PDE for

$$z = (a + x)(b + y)$$

Solution: The given PDE is

$$z = (a + x)(b + y) \tag{8.5}$$

Differentiating with respect to x and y we get

$$p = b + y \tag{8.6}$$

$$q = a + x \tag{8.7}$$

Eliminating a and b from (8.5)–(8.7) we get

$$z = pq$$

Example 3

Eliminate arbitrary constants and form a PDE for

$$(x-a)^2+(y-b)^2=z^2\cot^2\alpha$$

where α is a parameter.

Solution: Differentiating the given equation partially with respect to x we get

$$2(x-a)+0=2zp\cot^2\alpha$$

$$x-a=zp\cot^2\alpha$$

Differentiating the given equation partially with respect to y we get

$$0+2(y-b)=2zq\cot^2\alpha$$

$$y-b=zq\cot^2\alpha$$

Substituting the values of $(x-a)$ and $(y-b)$ thus obtained in the given equation, we get

$$(zp\cot^2\alpha)^2+(zq\cot^2\alpha)^2=z^2\cot^2\alpha$$

$$p^2+q^2=\frac{1}{\cot^2\alpha}$$

$$p^2+q^2=\tan^2\alpha$$

Formation of PDE by Eliminating Arbitrary Functions

With One Arbitrary Function

Consider

$$z=f(u) \tag{8.8}$$

where f is function of u and u is function of x, y, z, that is $u(x, y, z)$. Here u is the known function. Differentiating (8.8) partially with respect to x by using chain rule we get

$$\frac{\partial z}{\partial x}=\frac{\partial f}{\partial u}\times\frac{\partial u}{\partial x}+\frac{\partial f}{\partial u}\times\frac{\partial u}{\partial z}\times\frac{\partial z}{\partial u} \tag{8.9}$$

$$\frac{\partial f}{\partial z}=\frac{\partial f}{\partial u}\times\frac{\partial u}{\partial y}+\frac{\partial f}{\partial u}\times\frac{\partial u}{\partial z}\times\frac{\partial z}{\partial y} \tag{8.10}$$

By eliminating arbitrary function f from (8.8)–(8.10) we get a PDE. [*Note*: (a) If we eliminate one arbitrary function, we get first-order PDE. (b) If we eliminate two or more (n) arbitrary functions, we get several PDEs. However we should take the least-ordered PDE. (c) Differentiating twice or more times in elimination process results in second and higher order PDEs.]

With More than One Arbitrary Function

Consider the equation

$$F(u,v)=0 \tag{8.11}$$

where $u=u(x,y,z)$, $v=v(x,y,z)$ are known functions of x, y, z. Differentiating (8.11) partially with respect to x, by chain rule, we get

$$\frac{\partial F}{\partial u}\left(\frac{\partial u}{\partial x}+\frac{\partial u}{\partial z}\frac{\partial z}{\partial x}\right)+\frac{\partial F}{\partial v}\left(\frac{\partial v}{\partial x}+\frac{\partial v}{\partial z}\frac{\partial z}{\partial x}\right)=0 \tag{8.12}$$

Differentiating (8.11) partially with respect to y by chain rule, we get

$$\frac{\partial F}{\partial u}\left(\frac{\partial u}{\partial y}+\frac{\partial u}{\partial z}\frac{\partial z}{\partial y}\right)+\frac{\partial F}{\partial v}\left(\frac{\partial v}{\partial y}+\frac{\partial v}{\partial z}\frac{\partial z}{\partial y}\right)=0 \tag{8.13}$$

Eliminating $\partial F/\partial u$ and $\partial F/\partial v$ from (8.12) and (8.13) we get

$$\begin{vmatrix} \dfrac{\partial u}{\partial x}+\dfrac{\partial u}{\partial z}\dfrac{\partial z}{\partial x} & \dfrac{\partial v}{\partial x}+\dfrac{\partial v}{\partial z}\dfrac{\partial z}{\partial x} \\ \dfrac{\partial u}{\partial y}+\dfrac{\partial u}{\partial z}\dfrac{\partial z}{\partial y} & \dfrac{\partial v}{\partial y}+\dfrac{\partial v}{\partial z}\dfrac{\partial z}{\partial y} \end{vmatrix}=0$$

By expanding this determinant we get PDE of the form

$$Pp+Qq=R$$

where $P=u_z v_y - u_y v_z$, $Q=u_x v_z - u_z v_x$, $R=u_y v_x - u_x v_y$ which is first-order linear PDE. Here

$$u_x=\frac{\partial u}{\partial x},\ u_y=\frac{\partial u}{\partial y},\ u_z=\frac{\partial u}{\partial z},\ v_x=\frac{\partial v}{\partial x},\ v_y=\frac{\partial v}{\partial y},\ v_z=\frac{\partial v}{\partial z}$$

Example 4

Eliminate arbitrary functions and form PDE for the following:

$$f(x^2+y^2+z^2,\ z^2-2xy)=0$$

Solution: Here

$$u=x^2+y^2+z^2,\quad v=z^2-2xy$$

Partially differentiating with respect to x we get

$$u_x=\frac{\partial u}{\partial x}=2x,\quad v_x=\frac{\partial v}{\partial x}=-2y$$

Partially differentiating with respect to y we get

$$u_y=\frac{\partial u}{\partial y}=2y,\quad v_y=\frac{\partial v}{\partial y}=-2x$$

Partially differentiating with respect to z we get

$$u_z = \frac{\partial u}{\partial z} = 2z, \qquad v_z = \frac{\partial v}{\partial z} = 2z$$

The PDE formed by eliminating f from given function is

$$Pp + Qq = R$$

where $P = u_z v_y - u_y v_z, Q = u_x v_z - u_z v_x, R = u_y v_x - u_x v_y, p = \partial z/\partial x, q = \partial z/\partial y$. Substituting the values, we get

$$P = 2z(-2x) - 2y(2z) = -4z(x+y)$$
$$Q = 2x(2z) - 2z(-2y) = 4z(x+y)$$
$$R = 2y(-2y) - 2x(-2x) = -4y^2 + 4x^2 = 4(x^2 - y^2)$$

Therefore the PDE is

$$-4z(x+y)p + 4z(x+y)q = 4(x^2 - y^2) = 4(x-y)(x+y)$$
$$-4zp + 4zq = 4(x-y)$$

Example 5

Eliminate arbitrary functions and form PDE for the following:

$$F(xy + z^2, x + y + z) = 0 \tag{8.14}$$

Solution: Here $u(x,y,z) = xy + z^2$, $v(x,y,z) = x + y + z$. Differentiating (8.14) partially with respect to x by chain rule, we get

$$\frac{\partial F}{\partial u}\left[\frac{\partial u}{\partial x} + \frac{\partial u}{\partial z}\frac{\partial z}{\partial x}\right] + \frac{\partial F}{\partial v}\left[\frac{\partial v}{\partial x} + \frac{\partial v}{\partial z} \times \frac{\partial z}{\partial x}\right] = 0 \tag{8.15}$$

$$F_u(y + 2zp) + F_v(1 + p) = 0$$

Differentiating (8.14) with respect to y we get

$$F_u(x + 2zq) + F_v(1 + q) = 0 \tag{8.16}$$

Eliminating F_u and F_v from (8.15) and (8.16) we get

$$\begin{pmatrix} y + 2zp & 1 + p \\ x + 2zq & 1 + q \end{pmatrix} \begin{pmatrix} F_u \\ F_v \end{pmatrix} = \begin{pmatrix} 0 \\ 0 \end{pmatrix}$$

Solving this equation we get

$$(2z - x)p + (y - 2z)q = x - y$$

which is the required equation.

CONCEPT CHECK

1. Eliminate h, k from $(x-h)^2+(y-k)^2+z^2=a^2$

   OR

   Eliminate a, b from $(x-a)^2+(y-b)^2=r^2$. [JNTU 2007, 2006]
2. Find the differential equation of all spheres of radius 5 having their centers in the xy plane.
3. Form the partial differential equation by eliminating the arbitrary constants

   $$z = ax^3 + by^3$$ [JNTU 2007]
4. Form the partial differential equation by eliminating the arbitrary constants a, b from

   $$z = a\log\left[\frac{b(y-1)}{1-x}\right]$$ [JNTU 2003, 2002]
5. Form the partial differential equation by eliminating the arbitrary constants from

   $$2z = \sqrt{x+a} + \sqrt{y+b}$$
6. Form the partial differential equation by eliminating the arbitrary functions from:

   (a) $z = f(x^2 - y^2)$ [JNTU 2006]

   (b) $z = f(x) + e^y g(x)$ [JNTU 2004, 2003]
7. Form a partial differential equation by eliminating the arbitrary functions $f(x)$ and $g(y)$ from $z = yf(x) + xg(y)$. [JNTU 2003]
8. Form a partial differential equation by eliminating the arbitrary functions f from $xyz = f(x^2+y^2+z^2)$. [JNTU 2006, 2003, 2001]
9. Form a partial differential equation by eliminating the arbitrary functions f from $z = xy + f(x^2+y^2)$. [JNTU 2005, 2002]
10. Form the partial differential equations by eliminating the arbitrary functions from $z = f(x-it) + g(x-it)$. [JNTU 2003]
11. Form the partial differential equations by eliminating the arbitrary function from

    $$z = y^2 + 2f\left(\frac{1}{x} + \log y\right)$$ [JNTU 2006, 2003]
12. Form the partial differential equations by eliminating the arbitrary function from $f(xy+z^2, x+y+z)=0$. [JNTU 2003]

8.3 Solution of a PDE

Classification of Integrals

The general form of first-order PDE is

$$F(x, y, z, p, q) = 0 \tag{8.17}$$

where $p = \partial z/\partial x$; $q = \partial z/\partial y$; x and y are independent variables; z is a dependent variable. The various integrals can be classified as:

1. ***Complete integral*** (CI): It is the solution of a PDE in which the number of independent variables equals the number of arbitrary constants a and b. $F(x, y, z, a, b) = 0$ is the CI of PDE (8.17).
2. ***Particular integral***: It is the solution of PDE which is obtained by giving particular values to the arbitrary constants a and b in the complete integral.
3. ***General solution***: General solution of a PDE is any arbitrary function F of given functions u, v, that is, $F(u, v) = 0$.
4. ***Singular integral***: Singular integral is the envelope of the surfaces (i.e., solutions of PDE) which is also satisfied by given PDE but it is not obtained from giving particular values to CI of PDE. It is only an envelope of all solutions (surfaces) of the given PDE.

Procedure to Obtain Singular Solution

Let the PDE be

$$F(x, y, z, p, q) = 0 \tag{8.18}$$

and its complete integral be

$$F(x, y, z, a, b) = 0 \tag{8.19}$$

Now partially differentiate (8.19) with respect to a and b, where a and b are arbitrary constants. We get

$$\frac{\partial F}{\partial a} = 0 \tag{8.20}$$

$$\frac{\partial F}{\partial b} = 0 \tag{8.21}$$

Eliminating a and b from (8.20) and (8.21) (if possible) we obtain the singular integral of the given PDE.
Note 1: The singular integral is obtained by eliminating "a" and "b" by solving the equations

$$F(x, y, z, a, b) = 0; \quad \frac{\partial F}{\partial a} = 0; \quad \frac{\partial F}{\partial b} = 0$$

The operation of elimination is equivalent to finding the envelope of all the surfaces included in the complete integral. The three equations give the point of contact of the particular surface, represented by complete integral, with the general envelope. Thus, singular integral represents the general envelope of all surfaces included in the complete integral.
Note 2: It may happen that the entire system of surfaces does not admit a general envelope. In such cases, the singular integral does not exist for the given partial differential equation.

Solution of a PDE by Direct Integration

If the given PDE is in a simple form, then we can integrate it directly to get the solution. The following cases may arise:

1. If the PDE is of the form

$$\frac{\partial^2 z}{\partial x \partial y} = 0$$

then the solution can be obtained by integrating the given PDE with respect to x. We get

$$\frac{\partial z}{\partial y} = C_1 \tag{8.22}$$

Here we are integrating the function with respect to x. So C_1 is either a constant with respect to x or it may be a function of y. Substituting this value in (8.22) we get

$$\frac{\partial z}{\partial y} = f_1(y) \text{ or a constant} \tag{8.23}$$

Now integrating (8.23) with respect to y we get

$$z = \int f_1(y)dy + C_2 = \int f_1(y)dy + g(x)$$

where $C_2 = g(x)$. Since we are integrating with respect to y, the function remains a constant. Therefore

$$z = f(x) + g(x)$$

2. When the PDE is of the form

$$\frac{\partial z}{\partial x} = 0$$

Integrating with respect to x, we get $z = f(y)$ as the solution.

3. When the PDE is of the form

$$\frac{\partial^2 z}{\partial y^2} = 0$$

Integrating with respect to y, we get

$$\frac{\partial z}{\partial y} = f(x)$$

Integrating again we get

$$z = [f(x)]y + g(x)$$

4. When the PDE is of the form

$$\frac{\partial^2 z}{\partial x \partial y} = \sin x \sin y$$

Integrating with respect to x, we get

$$\frac{\partial z}{\partial y} = -\cos x \sin y + f(y)$$

Again integrating with respect to y, we get

$$z = \cos x \cos y + [h(y) + g(x)]$$

5. When the PDE is of the form

$$\frac{\partial^2 z}{\partial x \partial y} = k$$

On integration with respect to x, we get

$$\frac{\partial z}{\partial y} = kx + f(y)$$

Second integration yields .

$$z = kxy + h(y) + g(x)$$

where

$$h(y) = \int f(y)dy$$

Solution of Linear PDEs of Order One

A PDE of order one is said to be **linear** if it is linear in p and q, that is if p and q are of first degree then the PDE is called a **linear PDE**. Otherwise it is called a **nonlinear PDE** (dealt with in the next subsection). Some examples of a linear PDE are:

1. $px - qy = z$
2. $q = 3x - 5y$

Lagrange Method

Consider two independent variables x and y and one dependent variable z. The general form of PDE (first order, first degree) in x, y, z is

$$Pp + Qq = R$$

where P, Q, and R are functions of x, y, and z; u and v are known functions in x, y, z. Consider the equation

$$\phi(u, v) = 0$$

where ϕ is an arbitrary function. In the previous section we have discussed the method to obtain a PDE after eliminating this arbitrary function. Let the PDE after eliminating the arbitrary constant be of the form

$$Pp + Qq = R \tag{8.24}$$

where

$$P = \frac{\partial u}{\partial y}\frac{\partial v}{\partial z} - \frac{\partial u}{\partial z}\frac{\partial v}{\partial y}$$

$$Q = \frac{\partial u}{\partial z}\frac{\partial v}{\partial x} - \frac{\partial u}{\partial x}\frac{\partial v}{\partial z}$$

$$R = \frac{\partial u}{\partial x}\frac{\partial v}{\partial y} - \frac{\partial u}{\partial y}\frac{\partial v}{\partial x} \tag{8.25}$$

Any function in u, v [i.e., $\phi(u,v) = 0$] which satisfies (8.25) is a solution to the PDE (8.24). Let $u(x, y, z) = a$ and $v(x, y, z) = b$, say. Then

$$u(x, y, z) = a$$

$$du = 0$$

$$\frac{\partial u}{\partial x}dx + \frac{\partial u}{\partial y}dy + \frac{\partial u}{\partial z}dz = 0 \tag{8.26}$$

Similarly

$$v(x, y, z) = b$$

$$dv = 0$$

$$\frac{\partial v}{\partial x}dx + \frac{\partial v}{\partial y}dy + \frac{\partial v}{\partial z}dz = 0 \tag{8.27}$$

From the above two equations we get

$$\frac{dx}{\dfrac{\partial u}{\partial y}\dfrac{\partial v}{\partial z} - \dfrac{\partial u}{\partial z}\dfrac{\partial v}{\partial y}} = \frac{dy}{\dfrac{\partial u}{\partial z}\dfrac{\partial v}{\partial x} - \dfrac{\partial u}{\partial y}\dfrac{\partial v}{\partial z}} = \frac{dz}{\dfrac{\partial u}{\partial x}\dfrac{\partial v}{\partial y} - \dfrac{\partial u}{\partial y}\dfrac{\partial v}{\partial x}} \tag{8.28}$$

Using (8.25) in (8.28) we get

$$\frac{dx}{P} = \frac{dy}{Q} = \frac{dz}{R}$$

These are called the auxiliary equations. The solutions of this auxiliary equations are $u = a$ and $u = b$. The relation between these, that is $\phi(u,v) = 0$, is called the general solution of the PDE.

Example 6

Solve $p\tan x + q\tan y = \tan z$. **[JNTU 2004]**

Solution: This is in the form of Lagrange PDE given by

$$Pp + Qq = R$$

The auxiliary equations are

$$\frac{dx}{P} = \frac{dy}{Q} = \frac{dz}{R}$$

Substituting the values P, Q, and R we get

$$\frac{dx}{\tan x} = \frac{dy}{\tan y} = \frac{dz}{\tan z} \tag{8.29}$$

Taking the first two equations of (8.29) we get

$$\frac{dx}{\tan x} = \frac{dy}{\tan y}$$

Integrating we get

$$\log(\sin x) = \log(\sin y) + \log\, c_1$$

$$c_1 = \frac{\sin x}{\sin y}$$

Now consider the last two equations of (8.29):

$$\frac{dy}{\tan y} = \frac{dz}{\tan z}$$

Integrating we get

$$\log(\sin y) = \log(\sin z) + \log c_2$$

$$c_2 = \frac{\sin y}{\sin z}$$

Therefore the general solution of Lagrange's PDE is

$$\phi\left(\frac{\sin x}{\sin y}, \frac{\sin y}{\sin z}\right) = 0$$

where ϕ is an arbitrary function.

Example 7

Solve $x^2p - y^2q - z^2 = 0$.

Solution: The given equation is in the form

$$Pp + Qq = R$$

where $P = x^2$, $Q = y^2$, $R = -z^2$. The auxiliary equations are

$$\frac{dx}{P} = \frac{dy}{Q} = \frac{dz}{R}$$

$$\frac{dx}{x^2} = \frac{dy}{y^2} = \frac{dz}{-z^2} \tag{8.30}$$

Consider the first two equations of (8.30):

$$\frac{dx}{x^2} = \frac{dy}{y^2}$$

Integrating both sides we get

$$\frac{1}{x} - \frac{1}{y} = c_1$$

Now consider the last two equations of (8.30):

$$\frac{dy}{y^2} = \frac{dz}{-z^2}$$

Integrating we get

$$\frac{1}{y} + \frac{1}{z} = c_2$$

The general solution of Lagrange's PDE is

$$\phi\left(\frac{1}{x} - \frac{1}{y}, \frac{1}{y} + \frac{1}{z}\right) = 0$$

or

$$\frac{1}{x} - \frac{1}{y} = f\left(\frac{1}{y} + \frac{1}{z}\right),$$

where ϕ or f are arbitrary functions.

Solution of Nonlinear Partial Differential Equations

A **nonlinear PDE** is a PDE that has partial derivatives with degree higher than one and also the products of these partial derivatives. A PDE is said to be fully solved when all the three types of integrals (complete integral, singular integral, and general integral) have been obtained.

In this section we will deal with various standard forms of equations that are divided into subsections. For easy understanding solved examples are also provided.

Standard Form I

Here the equations are of the form

$$f(p, q) = 0 \tag{8.31}$$

where x, y, z do not occur explicitly.

Complete Integral: The auxiliary equations are of the form

$$\frac{dp}{0} = \frac{dq}{0}$$

$$dp = 0$$

Integrating we get

$$p = \text{constant} = a \text{ (say)}$$

Substituting the value of p in (8.31) we get

$$f(a, q) = 0$$

$$q = f(a) = b \text{ (say)}$$

where b is such that $f(a, b) = 0$. Now

$$dz = pdx + qdy$$

$$dz = adx + bdy$$

$$z = ax + by + c \tag{8.32}$$

where c is an arbitrary constant. This is the *complete integral* of equation $f(p, q) = 0$.

General Solution: We now discuss the method to evaluate the *general solution*. Let $f(p, q) = 0$.

***Step* 1:** When solved for b we get $b = \phi(a)$.

***Step* 2:** Substituting in (8.32) we get

$$z = ax + \phi(a)y + c \tag{8.33}$$

To obtain general solution put $c = \psi(a)$, where ψ is an arbitrary function, in (8.33). Therefore

$$z = ax + \phi(a)y + \psi(a) \tag{8.34}$$

Then the general integral is obtained by differentiating (8.34) with respect to a. We get

$$0 = x + \phi'(a)y + \psi'(a)$$

which is the general solution.

Singular Integral: Consider (8.33). Differentiating it with respect to a gives

$$0 = x + \phi'(a)y$$

Differentiating (8.33) with respect to c we get

$$0 = 1$$

which is not possible. Therefore the singular integral does not exist in such a case.

Example 8

Solve $pq = k$.

Solution:

Complete solution: Let $p = a$. Then

$$qa = k \Rightarrow q = \frac{k}{a}$$

Therefore

$$z = ax + \frac{k}{a}y + c$$

This is the complete solution, where a, c are two arbitrary constants.

General solution: Let us take

$$c = f(a)$$

Then

$$z = ax + \frac{k}{a}y + f(a) \tag{8.35}$$

Differentiating with respect to a we get

$$0 = x - \frac{ky}{a^2} + f'(a) \tag{8.36}$$

Eliminating a from (8.35) and (8.36) we get general solution.

Singular solution: Consider complete solution

$$z = ax + \frac{k}{a}y + c \tag{8.37}$$

Differentiating (8.37) with respect to a we get

$$0 = x - \frac{ky}{a^2} + 0$$

Differentiating (8.38) with respect to c we get

$$0 = 1$$

which is absurd. Therefore singular solution is not possible.

Standard Form II

Here the equations are of the form

$$f(z,p,q)=0 \tag{8.38}$$

In this form the independent variables x, y do not occur explicitly. The subsidiary equations for (8.38) are

$$\frac{dp}{\partial f/\partial z}=\frac{dq}{\partial f/\partial z}$$

$$\frac{dp}{p}=\frac{dq}{q}$$

On integration we get $p=aq$, where a is a constant. Substituting this value of p in $f(z,p,q)=0$ we get

$$f(z,aq,q)=0$$

Therefore we have

$$dz = pdx + qdy$$

$$dz = aqdx + qdy$$

$$dz = q(adx + dy)$$

$$\frac{1}{q}dz = adx + dy$$

The complete solution is

$$\int \frac{1}{q}dz = ax + y + c$$

Example 9

Solve $z=p^2+q^2$.

Solution:

Complete integral: Substituting $p=aq$ in the given equation we get

$$z=(aq)^2+q^2=a^2q^2+q^2=q^2(a^2+1)$$

$$q^2=\frac{z}{a^2+1}\Rightarrow q=\sqrt{\frac{z}{a^2+1}} \tag{8.39}$$

Consider

$$dz = pdx + qdy = aqdx + qdy$$

$$\frac{1}{q}dz = adx + dy$$

Integrating we get

$$\int \frac{1}{q}dz = ax + y + c \tag{8.40}$$

Substituting the value of q from (8.39) to (8.40) we obtain

$$\int\sqrt{\frac{1+a^2}{z}}\,dz = ax - y - c$$

$$\int\frac{\sqrt{1+a^2}}{\sqrt{z}}\,dz = ax - y - c$$

$$\int\sqrt{1+a^2}\,z^{-1/2}dz = ax - y - c$$

$$\frac{\sqrt{1+a^2}}{1/2}\sqrt{z} = ax - y - c$$

$$2\sqrt{z(a^2+1)} = ax - y - c \quad (8.41)$$

Equation (8.41) is the complete integral of the given PDE.
General solution: Squaring both sides we get

$$4z(a^2+1) = (ax+y+c)^2 \quad (8.42)$$

Let us take $c = \phi(a)$. Therefore

$$4z(a^2+1) = [ax+y+\phi(a)]^2 \quad (8.43)$$

Differentiating (8.43) with respect to a we get

$$4z(2a) = [x+\phi'(a)](2)(ax+y+\phi)$$

Eliminating a from (8.43) we get the general solution of the given PDE.
Singular integral: Consider the complete solution

$$4z(a^2+1) = (ax+y+c)^2 \quad (8.44)$$

Differentiating (8.44) with respect to a we get

$$8az = 2a(ax+y+c) \quad (8.45)$$

Differentiating (8.44) with respect to c we get

$$0 = 2(ax+y+c) \quad (8.46)$$

Eliminating a and c from (8.42), (8.45), and (8.46) we get $z = 0$ which is the singular integral of the given PDE.

Example 10

Solve $z = px + qy + pq$.

Solution: The complete integral of given equation is $z = ax + by + ab$ where a and b are arbitrary constants.
General solution: Let us consider the complete integral

$$z = ax + by + ab \quad (8.47)$$

Substituting $b = f(a)$ in (8.47) we get

$$z = ax + f(a)y + af(a) \tag{8.48}$$

Differentiating with respect to a we get

$$0 = x + yf'(a) + f(a) + af'(a) \tag{8.49}$$

Eliminating a from (8.48) and (8.49) we get the general solution.

Singular integral: Consider the complete integral. Differentiating (8.47) with respect to a we get

$$0 = x + b$$

$$b = -x$$

Differentiating (8.47) with respect to b we get

$$0 = y + a$$

$$a = -y$$

Substitute values of a and b in $z = ax + by + ab$ we get

$$z = -xy - yx + (-x)(-y)$$

$$z = -xy - xy + xy$$

$$z = -xy$$

$$z + xy = 0$$

This is the singular integral.

Standard Form III

Here the equations are of the form

$$f(x, p) = g(y, q) \tag{8.50}$$

These equations do not contain the terms involving z and the terms containing x and p can be separated from those containing y and q. Hence this form is also called *variable separable form*.

Substituting $p = f_1(a, x),\ q = g_1(a, y)$ in equation $dz = pdx + qdy$ and solving we get

$$dz = f_1(a, x)dx + g_1(a, y)dy$$

Integrating both sides we have

$$z = \int f_1(a, x)dx + \int g_1(a, y)dy + c \tag{8.51}$$

where c is an arbitrary constant. Equation (8.51) is called the complete integral of the given PDE. General integral and singular integral, if they exist, they are obtained in the usual way.

Example 11

Solve $p^2 - q^2 = x - y$.

Solution: The given equation can be written as

$$p^2 - x = q^2 - y$$

Let

$$p^2 - x = q^2 - y = a$$

Then

$$p^2 = a + x \Rightarrow p = \sqrt{a+x}$$
$$q^2 = a + y \Rightarrow q^2 = y + a \Rightarrow q = \sqrt{y+a}$$

Substituting these values in

$$dz = pdx + qdy$$

We get

$$dz = \sqrt{a+x}\,dx + \sqrt{y+a}\,dy$$

Integrating both sides, we have

$$z = \frac{2}{3}(x+a)^{3/2} + \frac{2}{3}(y+a)^{3/2} + c$$

where c is an arbitrary constant. This equation is the complete solution of the given PDE. We can find general solution for this problem in the usual way and it is left for the students. For this problem singular integral does not exist.

Example 12

Solve

$$\left(\frac{p}{2}+x\right)^2 + \left(\frac{q}{2}+y\right)^2 = 1$$

Solution: The given equation can be written as

$$\left(\frac{p}{2}+x\right)^2 = 1 - \left(\frac{q}{2}+y\right)^2$$

Let

$$\left(\frac{p}{2}+x\right)^2 = 1 - \left(\frac{q}{2}+y\right)^2 = a^2 \text{ (constant)}$$

Solving we get

$$\left[\frac{p}{2}+x\right]^2 = a^2$$

$$\frac{p}{2} + x = a$$

$$p = 2(a-x) \qquad (8.52)$$

and

$$1-\left(\frac{q}{2}+y\right)^2=a^2$$

$$\frac{q}{2}+y=\sqrt{1-a^2}$$

$$q=2(\sqrt{1-a^2}-y) \tag{8.53}$$

We know that

$$dz=pdx+qdy$$

Substituting the values of p and q from (8.52) and (8.53) in the above equation, we get

$$dz=2(a-x)dx+2(\sqrt{1-a^2}-y)dy$$

Integrating, we get

$$\int dz=2\int(a-x)+2\int(\sqrt{1-a^2}-y)dy+c$$

$$z=2\left(ax-\frac{x^2}{2}\right)+2\left(\sqrt{1-a^2}\times y-\frac{y^2}{2}\right)+c$$

$$z=2ax-x^2+2\sqrt{1-a^2}\times y-y^2+c$$

which is the complete integral.

Standard Form IV

It is also called the Clairaut's form. The general equation is of the form

$$z=px+qy+f(p,q) \tag{8.54}$$

The auxiliary equations for this equation are

$$\frac{dp}{-p+p}=\frac{dq}{-q+q}$$

$$\frac{dp}{0}=\frac{dq}{0}$$

$$dp=0=dq$$

On integrating we get

$$p=a \quad \text{(say)}$$

$$q=b \quad \text{(say)}$$

where a and b are constants. Substituting these in the PDE (8.54) we get

$$z=ax+by+f(a,b) \tag{8.55}$$

Substituting the values of p and q in equation $dz = pdx + qdy$ we get

$$dz = adx + bdy$$

Integrating we get

$$z = ax + by + c \tag{8.56}$$

Equation (8.56) is the complete integral of the given PDE.

General Solution: To get the general solution, take $b = \phi(a)$ in (8.55). We have

$$z = ax + \phi(a)y + f(a, \phi(a)) \tag{8.57}$$

Differentiating (8.57) with respect to a we get

$$0 = x + \phi'(a)y + f'(a) \tag{8.58}$$

Eliminating a from (8.57) and (8.58) we get the general solution.

Singular Integral: Now we discuss the method to calculate the singular integral. Consider the complete integral

$$z = ax + by + f(a, b) = 0 \tag{8.59}$$

Differentiating (8.59) with respect to a we get

$$0 = x + \frac{\partial f}{\partial a} \tag{8.60}$$

Solving (8.60) we get the singular integral.

Example 13

Solve $pqz = p^2(qx + p^2) + q^2(py + q^2)$. [JNTU 2003, 2002]

Solution: The given equation is

$$pqz = p^2q\left(x + \frac{p^2}{q}\right) + q^2p\left(y + \frac{q^2}{p}\right)$$

$$z = p\left(x + \frac{p^2}{q}\right) + q\left(y + \frac{q^2}{p}\right)$$

$$= (px + qy) + \frac{p^3}{q} + \frac{q^3}{p}$$

Since it is in the form $z = px + qy + f(p, q)$, therefore the complete solution is obtained by writing $p = a$ and $q = b$. Hence

$$z = ax + by + \frac{a^3}{b} + \frac{b^3}{a}$$

which is the complete solution.

CONCEPT CHECK

1. Solve the partial differential equation
   $(y+z)p+(z+x)q=x+y$ [JNTU 2006]
2. Solve the partial differential equation
   $z(x-y)=px^2+qy^2$ [JNTU 2007]
3. Find the integral surface of
   $x(y^2+z)p-y(x^2+z)q=(x^2-y^2)z$
   which contains the straight line $x+y=0,\ z=1$.
   [JNTU 2006]
4. Solve the equation $p\sqrt{x}+q\sqrt{y}=\sqrt{z}$.
   [JNTU 2006]
5. Solve the equation $(y^2+z^2)p-xyq+zx=0$.
   [JNTU 2008]

8.4 Solution by Separating Variables

The model of a vibrating elastic string (a violin string, for instance) consists of the **one-dimensional wave equation**

$$\frac{\partial^2 u}{\partial t^2}=c^2\frac{\partial^2 u}{\partial x^2} \tag{8.61}$$

where $c^2=T/P$, for the unknown deflection $u(x, t)$ of the string, a PDE that we have just obtained, and some ***additional conditions***, which we shall now derive.

Since the string is fastened at the ends $x = 0$ and $x = L$, we have the two **boundary conditions**

$$\text{(a)}\quad u(0,t)=0,\quad \text{(b)}\quad u(L,t)=0\quad \text{for all } t \tag{8.62}$$

Furthermore, the form of the motion of the string will depend on its *initial deflection* (deflection at time $t = 0$), call it $f(x)$, and on its *initial velocity* (velocity at $t = 0$), call it $g(x)$. We thus have the two **initial conditions**

$$\text{(a)}\quad u(x,0)=f(x),\quad \text{(b)}\quad u_t(x,0)=g(x)\quad (0\leqq x\leqq L) \tag{8.63}$$

where $u_t = \partial u/\partial t$. We now have to find a solution of the PDE (8.61) satisfying the conditions (8.62) and (8.63). This will be the solution of our problem. We shall do this in three steps, as follows.

***Step* 1.** By the "**method of separating variables**" or *product method,* setting $u(x, t) = F(x)G(t)$, we obtain from (8.61) two ODEs, one for $F(x)$ and the other one for $G(t)$.

***Step* 2.** We determine solutions of these ODEs that satisfy the boundary conditions (8.62).

***Step* 3.** Finally, using **Fourier series**, we compose the solutions gained in Step 2 to obtain a solution of (8.61) satisfying both (8.62) and (8.63), that is, the solution of our model of the vibrating string.

We would use the same concept while deriving **one-dimensional heat equation**.

Step 1: Two ODEs from the Wave Equation (8.61)

In the **method of separating variables**, or *product method,* we determine solutions of the wave equation (8.61) of the form

$$u(x, t) = F(x)G(t) \tag{8.64}$$

which are a product of two functions, each depending only on one of the variables x and t. This is a powerful general method that has various applications in engineering mathematics, as we shall see in this chapter. Differentiating (8.64), we obtain

$$\frac{\partial^2 u}{\partial t^2} = F\ddot{G} \quad \text{and} \quad \frac{\partial^2 u}{\partial x^2} = F''G$$

where dots denote derivatives with respect to t and primes derivatives with respect to x. By inserting this into the wave equation (8.61) we have

$$F\ddot{G} = c^2 F''G$$

Dividing by c^2FG and simplifying gives

$$\frac{\ddot{G}}{c^2G} = \frac{F''}{F}$$

The variables are now separated, the left side depending only on t and the right side only on x. Hence both sides must be constant because if they were variable, then changing t or x would affect only one side, leaving the other unaltered. Thus, say,

$$\frac{\ddot{G}}{c^2G} = \frac{F''}{F} = k$$

Multiplying by the denominators gives immediately two ***ordinary*** DEs

$$F'' - kF = 0 \tag{8.65}$$

and

$$\ddot{G} - c^2 kG = 0 \tag{8.66}$$

Here, the **separation constant** k is still arbitrary.

Step 2: Satisfying the Boundary Conditions (8.62)

We now determine solutions F and G of (8.65) and (8.66) so that $u = FG$ satisfies the boundary conditions (8.62), that is

$$u(0,t) = F(0)G(t) = 0, \qquad u(L,t) = F(L)G(t) = 0 \qquad \text{for all } t \tag{8.67}$$

We first solve (8.65). If $G \equiv 0$, then $u = FG \equiv 0$, which is of no interest. Hence $G \not\equiv 0$ and then by (8.67)

$$\text{(a)} \quad F(0) = 0, \qquad \text{(b)} \quad F(L) = 0 \tag{8.68}$$

We show that k must be negative. For $k = 0$ the general solution of (8.65) is $F = ax + b$, and from (8.68) we obtain $a = b = 0$, so that $F \equiv 0$ and $u = FG \equiv 0$, which is of no interest. For positive $k = \mu^2$ a general solution of (8.65) is

$$F = Ae^{\mu x} + Be^{-\mu x}$$

and from (8.68) we obtain $F \equiv 0$ as before (verify!). Hence we are left with the possibility of choosing k negative, say, $k = -p^2$. Then (8.65) becomes $F'' + p^2F = 0$ and has as a general solution

$$F(x) = A\cos px + B\sin px$$

From this and (8.68) we have

$$F(0) = A = 0 \quad \text{and then} \quad F(L) = B\sin pL = 0$$

We must take $B \neq 0$ since otherwise $F \equiv 0$. Hence $\sin pL = 0$. Thus

$$pL = n\pi, \quad \text{so that} \quad p = \frac{n\pi}{L} \quad (n \text{ integer}) \tag{8.69}$$

Setting $B = 1$, we thus obtain infinitely many solutions $F(x) = F_n(x)$, where

$$F_n(x) = \sin\frac{n\pi}{L}x \quad (n = 1, 2, \ldots) \tag{8.70}$$

These solutions satisfy (8.68). [For negative integer n we obtain essentially the same solutions, except for a minus sign, because $\sin(-\alpha) = -\sin\alpha$.] We now solve (8.66) with $k = -p^2 = -(n\pi/L)^2$ resulting from (8.69), that is

$$\ddot{G} + \lambda_n{}^2 G = 0 \tag{8.71}$$

where

$$\lambda_n = cp = \frac{cn\pi}{L} \tag{8.72}$$

A general solution is

$$G_n(t) = B_n\cos\lambda_n t + B_n{}^*\sin\lambda_n t$$

Hence solutions of (8.61) satisfying (8.62) are $u_n(x, t) = F_n(x)G_n(t) = G_n(t)F_n(x)$, written as

$$u_n(x, t) = (B_n\cos\lambda_n t + B_n{}^*\sin\lambda_n t)\sin\frac{n\pi}{L}x \quad (n = 1, 2, \ldots) \tag{8.73}$$

These functions are called the **eigenfunctions**, or *characteristic functions,* and the values $\lambda_n = cn\pi/L$ are called the **eigenvalues**, or *characteristic values*, of the vibrating string. The set $\{\lambda_1, \lambda_2, \ldots\}$ is called the **spectrum**.

Discussion of Eigenfunctions

We see that each u_n represents a harmonic motion having the **frequency** $\lambda_n/2\pi = cn/2L$ cycles per unit time. This motion is called the nth **normal mode** of the string. The first normal mode is known as the *fundamental mode* ($n = 1$), and the others are known as *overtones*; musically they give the octave, octave plus fifth, etc. Since in (8.73)

$$\sin\frac{n\pi x}{L} = 0 \quad \text{at} \quad x = \frac{L}{n}, \frac{2L}{n}, \ldots, \frac{n-1}{n}L$$

the nth normal mode has $n - 1$ **nodes**, that is, points of the string that do not move (in addition to the fixed endpoints); see Figure 1.

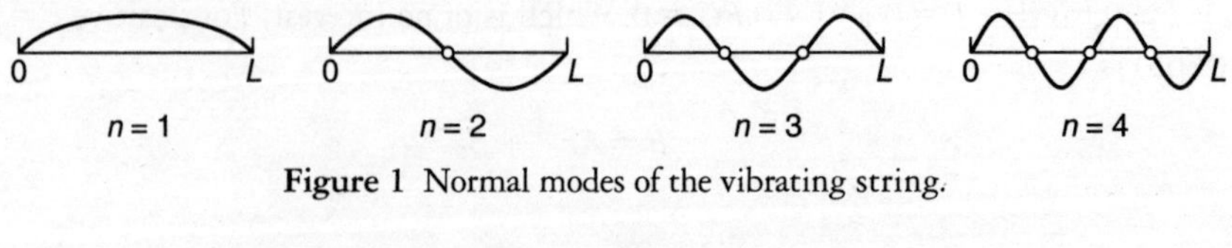

Figure 1 Normal modes of the vibrating string.

Figure 2 shows the second normal mode for various values of t. At any instant the string has the form of a sine wave. When the left part of the string is moving down, the other half is moving up, and conversely. For the other modes the situation is similar.

Tuning is done by changing the tension T. Our formula for the frequency $\lambda_n/2\pi = cn/2L$ of u_n with $c = \sqrt{T/\rho}$ confirms that effect because it shows that the frequency is proportional to the tension. T cannot be increased indefinitely, but can you see what to do to get a string with a high fundamental mode? (Think of both L and ρ.) Why is a violin smaller than a double-bass?

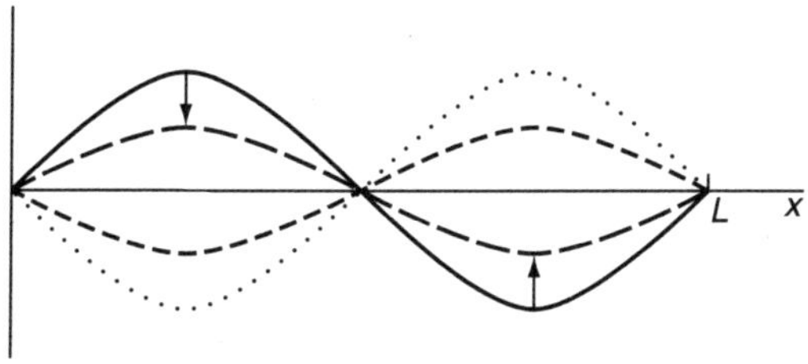

Figure 2 Second normal mode for various values of t.

Step 3: Solution of the Entire Problem. Fourier Series

The eigenfunctions (8.73) satisfy the wave equation (8.61) and the boundary conditions (8.62) (string fixed at the ends). A single u_n will generally not satisfy the initial conditions (8.63). But since the wave equation (8.61) is linear and homogeneous, it follows that the sum of finitely many solutions u_n is a solution of (8.61). To obtain a solution that also satisfies the initial conditions (8.63), we consider the infinite series (with $\lambda_n = cn\,\pi/L$ as before)

$$u(x,t)=\sum_{n=1}^{\infty} u_n(x,t)=\sum_{n=1}^{\infty}(B_n \cos\lambda_n t + B_n{}^* \sin\lambda_n t)\sin\frac{n\pi}{L}x \tag{8.74}$$

Satisfying Initial Condition (8.63a) (Given Initial Displacement)

From (8.74) and (8.63a) we obtain

$$u(x,0)=\sum_{n=1}^{\infty} B_n \sin\frac{n\pi}{L}x = f(x) \tag{8.75}$$

Hence we must choose the B_n's so that $u(x, 0)$ becomes the **Fourier sine series** of $f(x)$. Thus,

$$B_n = \frac{2}{L}\int_0^L f(x)\sin\frac{n\pi x}{L}dx, \quad n = 1, 2, \ldots \tag{8.76}$$

Satisfying Initial Condition (8.63b) (Given Initial Velocity)

Similarly, by differentiating (8.74) with respect to t and using (8.63b), we obtain

$$\left.\frac{\partial u}{\partial t}\right|_{t=0} = \left[\sum_{n=1}^{\infty}(-B_n\lambda_n \sin\lambda_n t + B_n{}^*\ \lambda_n \cos\lambda_n t)\sin\frac{n\pi x}{L}\right]_{t=0}$$

$$= \sum_{n=1}^{\infty} B_n{}^*\ \lambda_n \sin\frac{n\pi x}{L} = g(x)$$

Hence we must choose the B_n*'s so that for $t = 0$ the derivative $\partial u/\partial t$ becomes the Fourier sine series of $g(x)$. Thus, we have

$$B_n^* \lambda_n = \frac{2}{L}\int_0^L g(x)\sin\frac{n\pi x}{L}\,dx$$

Since $\lambda_n = cn\pi/L$, we obtain by division

$$B_n^* = \frac{2}{cn\pi}\int_0^L g(x)\sin\frac{n\pi x}{L}\,dx, \qquad n = 1, 2, \ldots \tag{8.77}$$

Result

Our discussion shows that $u(x, t)$ given by (8.74) with coefficients (8.76) and (8.77) is a solution of (8.61) that satisfies all the conditions in (8.62) and (8.63), provided the series (8.74) converges and so do the series obtained by differentiating (8.74) twice termwise with respect to x and t and have the sums $\partial^2 u/\partial x^2$ and $\partial^2 u/\partial t^2$, respectively, which are continuous.

Solution (8.74) Established

According to our derivation the solution (8.74) is at first a purely formal expression, but we shall now establish it. For the sake of simplicity we consider only the case when the initial velocity $g(x)$ is identically zero. Then the B_n^* are zero, and (8.74) reduces to

$$u(x, t) = \sum_{n=1}^{\infty} B_n \cos\lambda_n t \sin\frac{n\pi x}{L}, \qquad \lambda_n = \frac{cn\pi}{L} \tag{8.78}$$

It is possible to ***sum this series***, that is, to write the result in a closed or finite form. For this purpose we use the formula

$$\cos\frac{cn\pi}{L}t\sin\frac{n\pi}{L}x = \frac{1}{2}\left[\sin\left(\frac{n\pi}{L}(x-ct)\right) + \sin\left(\frac{n\pi}{L}(x+ct)\right)\right]$$

Consequently, we may write (8.78) in the form

$$u(x,t) = \frac{1}{2}\sum_{n=1}^{\infty} B_n \sin\left(\frac{n\pi}{L}(x-ct)\right) + \frac{1}{2}\sum_{n=1}^{\infty} B_n \sin\left(\frac{n\pi}{L}(x+ct)\right)$$

These two series are those obtained by substituting $x - ct$ and $x + ct$, respectively, for the variable x in the Fourier sine series (8.75) for $f(x)$. Thus

$$u(x, t) = \frac{1}{2}[f^*(x-ct) + f^*(x+ct)] \tag{8.79}$$

where f^* is the odd periodic extension of f with the period $2L$ (Figure 3). Since the initial deflection $f(x)$ is continuous on the interval $0 \leqq x \leqq L$ and zero at the endpoints, it follows from (8.79) that $u(x, t)$ is a continuous function of both variables x and t for all values of the variables. By differentiating (8.79) we see that $u(x, t)$ is a solution of (8.61), provided $f(x)$ is twice differentiable on the interval $0 < x < L$, and has one-sided second derivatives at $x = 0$ and $x = L$, which are zero. Under these conditions $u(x, t)$ is established as a solution of (8.61), satisfying (8.62) and (8.63) with $g(x) \equiv 0$.

Figure 3 Odd periodic extension of $f(x)$.

Generalized Solution

If $f'(x)$ and $f''(x)$ are merely piecewise continuous or if those one-sided derivatives are not zero, then for each t there will be finitely many values of x at which the second derivatives of u appearing in (8.61) do not exist. Except at these points the wave equation will still be satisfied. We may then regard $u(x, t)$ as a "**generalized solution,**" as it is called, that is, as a solution in a broader sense. For instance, a triangular initial deflection as in Example 13 (below) leads to a generalized solution.

Physical Interpretation of the Solution (8.79)

The graph of $f^*(x - ct)$ is obtained from the graph of $f^*(x)$ by shifting the latter ct units to the right (Figure 4). This means that $f^*(x - ct)$ $(c > 0)$ represents a wave that is traveling to the right as t increases. Similarly, $f^*(x + ct)$ represents a wave that is traveling to the left, and $u(x, t)$ is the superposition of these two waves.

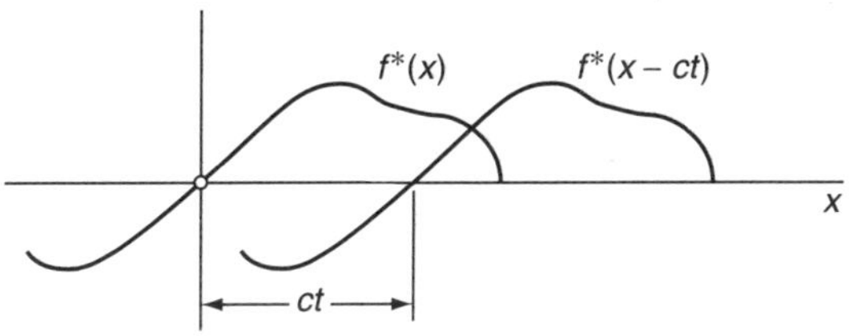

Figure 4 Interpretation of (8.79).

Example 14 Vibrating String if the Initial Deflection is Triangular

Find the solution of the wave equation (8.61) corresponding to the triangular initial deflection

$$f(x) = \begin{cases} \dfrac{2k}{L}x & \text{if } 0 < x < \dfrac{L}{2} \\ \dfrac{2k}{L}(L - x) & \text{if } \dfrac{L}{2} < x < L \end{cases}$$

and initial velocity zero. (Figure 5 shows $f(x) = u(x, 0)$ at the top.)

Solution: Since $g(x) \equiv 0$, we have $B_n^* = 0$ in (8.74). Thus (8.74) takes the form

$$u(x, t) = \frac{8k}{\pi^2}\left[\frac{1}{1^2}\sin\frac{\pi}{L}x\cos\frac{\pi c}{L}t - \frac{1}{3^2}\sin\frac{3\pi}{L}x\cos\frac{3\pi c}{L}t + - \cdots\right]$$

For graphing the solution we may use $u(x, 0) = f(x)$ and the above interpretation of the two functions in the representation (8.79). This leads to the graph shown in Figure 5.

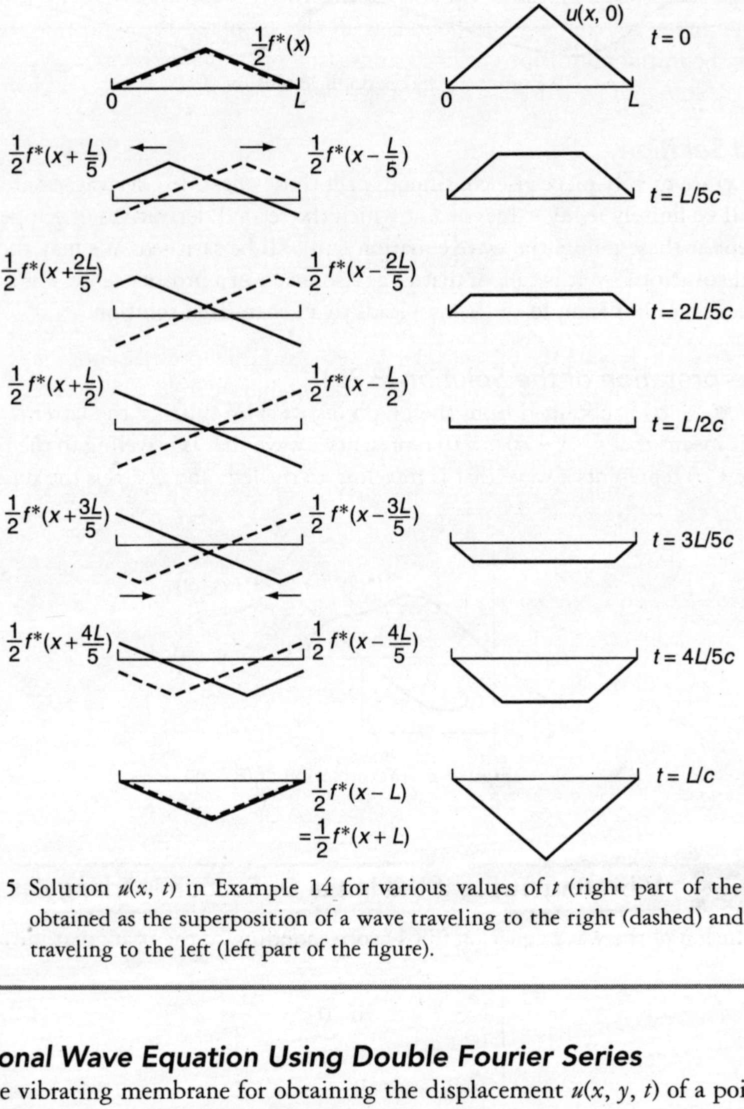

Figure 5 Solution $u(x, t)$ in Example 14 for various values of t (right part of the figure) obtained as the superposition of a wave traveling to the right (dashed) and a wave traveling to the left (left part of the figure).

Two-Dimensional Wave Equation Using Double Fourier Series

The model of the vibrating membrane for obtaining the displacement $u(x, y, t)$ of a point (x, y) of the membrane from rest ($u = 0$) at time t is

$$\frac{\partial^2 u}{\partial t^2} = c^2\left(\frac{\partial^2 u}{\partial x^2} + \frac{\partial^2 u}{\partial y^2}\right) \tag{8.80}$$

$$u = 0 \quad \text{on the boundary} \tag{8.81}$$

$$\begin{aligned} &\text{(a)} \quad u(x, y, 0) = f(x, y) \\ &\text{(b)} \quad u_t(x, y, 0) = g(x, y) \end{aligned} \tag{8.82}$$

Here (8.80) is the **two-dimensional wave equation** with $c^2 = T/\rho$ just derived, (8.81) is the **boundary condition** (membrane fixed along the boundary in the xy plane for all times $t \geqq 0$), and (8.82a) and (8.82b) are the **initial conditions** at $t = 0$, consisting of the given *initial displacement* (initial shape) $f(x, y)$ and the given *initial velocity* $g(x, y)$, where $u_t = \partial u/\partial t$.

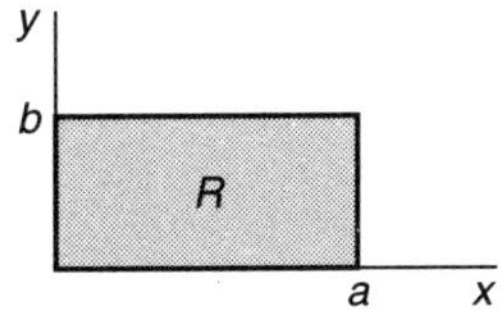

Figure 6 Rectangular membrane.

As a first important model, let us consider the **rectangular membrane** R in Figure 6, which is simpler than the circular drumhead to follow. Then the boundary in (8.81) is the rectangle in Figure 6. We shall solve this problem in three steps:

Step 1. By separating variables, setting $u(x, y, t) = F(x, y)G(t)$ and later $F(x, y) = H(x)Q(y)$ we obtain from (8.80) an ODE (8.83) for G and later from a PDE (8.84) for F two ODEs (8.85) and (8.86) for H and Q.

Step 2. From the solutions of those ODEs we determine solutions (8.92) of (8.80) (**"eigenfunctions"** u_{mn}) that satisfy the boundary condition (8.81).

Step 3. We compose the u_{mn} into a double series (8.93) solving the whole model (8.80)–(8.82).

Step 1: Three ODEs From the Wave Equation (8.60)

To obtain ODEs from (8.80), we apply two successive separations of variables. In the first separation we set $u(x, y, t) = F(x, y)G(t)$. Substitution into (8.80) gives

$$F\ddot{G} = c^2(F_{xx}G + F_{yy}G)$$

where subscripts denote partial derivatives and dots denote derivatives with respect to t. To separate the variables, we divide both sides by c^2FG:

$$\frac{\ddot{G}}{c_2G} = \frac{1}{F}(F_{xx} + F_{yy})$$

Since the left side depends only on t, whereas the right side is independent of t, both sides must equal a constant. By a simple investigation we see that only negative values of that constant will lead to solutions that satisfy (8.81) without being identically zero. Denoting that negative constant by $-v^2$, we have

$$\frac{\ddot{G}}{c^2G} = \frac{1}{F}(F_{xx} + F_{yy}) = -v^2$$

This gives two equations: for the **"time function"** $G(t)$ we have the ODE

$$\ddot{G} + \lambda^2 G = 0 \qquad \text{where} \quad \lambda = cv \tag{8.83}$$

and for the **"amplitude function"** $F(x, y)$ a PDE, called the *two-dimensional* **Helmholtz equation**

$$F_{xx} + F_{yy} + v^2F = 0 \tag{8.84}$$

Separation of the Helmholtz equation is achieved if we set $F(x, y) = H(x)Q(y)$. By substitution of this into (8.84) we obtain

$$\frac{d^2H}{dx^2}Q = -\left(H\frac{d^2Q}{dy^2} + v^2HQ\right)$$

To separate the variables, we divide both sides by HQ, finding

$$\frac{1}{H}\frac{d^2H}{dx^2} = -\frac{1}{Q}\left(\frac{d^2Q}{dy^2} + v^2Q\right)$$

Both sides must equal a constant, by the usual argument. This constant must be negative, say, $-k^2$, because only negative values will lead to solutions that satisfy (8.81) without being identically zero. Thus

$$\frac{1}{H}\frac{d^2H}{dx^2} = -\frac{1}{Q}\left(\frac{d^2Q}{dy^2} + v^2Q\right) = -k^2$$

This yields two ODEs for H and Q, namely,

$$\frac{d^2H}{dx^2} + k^2H = 0 \tag{8.85}$$

$$\frac{d^2Q}{dy^2} + p^2Q = 0 \tag{8.86}$$

where $p^2 = v^2 - k^2$.

Step 2: Satisfying the Boundary Condition

General solutions of (8.85) and (8.86) are

$$H(x) = A\cos kx + B\sin kx \qquad \text{and} \qquad Q(y) = C\cos py + D\sin py$$

with constant A, B, C, D. From $u = FG$ and (8.81) it follows that $F = HQ$ must be zero on the boundary, that is, on the edges $x = 0, x = a, y = 0, y = b$; see Figure 6. This gives the conditions

$$H(0) = 0, \qquad H(a) = 0, \qquad Q(0) = 0, \qquad Q(b) = 0$$

Hence $H(0) = A = 0$ and then $H(a) = B\sin ka = 0$. Here we must take $B \neq 0$ since otherwise $H(x) \equiv 0$ and $F(x, y) \equiv 0$. Hence $\sin ka = 0$ or $ka = m\pi$, that is,

$$k = \frac{m\pi}{a} \quad (m \text{ integer})$$

In precisely the same fashion we conclude that $C = 0$ and p must be restricted to the values $p = n\pi/b$ where n is an integer. We thus obtain the solutions $H = H_m$, $Q = Q_n$, where

$$H_m(x) = \sin\frac{m\pi x}{a} \quad \text{and} \quad Q_n(y) = \sin\frac{n\pi y}{b}, \quad m = 1, 2, \ldots;\ n = 1, 2, \ldots$$

As in the case of the vibrating string, it is not necessary to consider $m, n = -1, -2, \ldots$ since the corresponding solutions are essentially the same as for positive m and n, except for a factor -1. Hence the functions

$$F_{mn}(x, y) = H_m(x)Q_n(y) = \sin\frac{m\pi x}{a}\sin\frac{n\pi y}{b}, \quad m = 1, 2, \ldots; \; n = 1, 2, \ldots \tag{8.87}$$

are solutions of the Helmholtz equation (8.84) that are zero on the boundary of our membrane.

Eigenfunctions and Eigenvalues: Having taken care of (8.84), we turn to (8.83). Since $p^2 = v^2 - k^2$ in (8.86) and $\lambda = cv$ in (8.83), we have

$$\lambda = c\sqrt{k^2 + p^2}$$

Hence to $k = m\pi/a$ and $p = n\pi/b$ there corresponds the value

$$\lambda = \lambda_{mn} = c\pi\sqrt{\frac{m^2}{a^2} + \frac{n^2}{b^2}}, \quad m = 1, 2, \ldots; \; n = 1, 2, \ldots \tag{8.88}$$

in the ODE (8.83). A corresponding general solution of (8.83) is

$$G_{mn}(t) = B_{mn}\cos\lambda_{mn}t + B_{mn}^{*}\sin\lambda_{mn}t$$

It follows that the functions $u_{mn}(x, y, t) = F_{mn}(x, y)G_{mn}(t)$, written as

$$u_{mn}(x, y, t) = (B_{mn}\cos\lambda_{mn}t + B_{mn}^{*}\sin\lambda_{mn}t)\sin\frac{m\pi x}{a}\sin\frac{n\pi y}{b} \tag{8.89}$$

with λ_{mn} according to (8.88), are solutions of the wave equation (8.80) that are zero on the boundary of the rectangular membrane in Figure 6. These functions are called the **eigenfunctions** or *characteristic functions,* and the numbers λ_{mn} are called the **eigenvalues** or *characteristic values* of the vibrating membrane. The frequency of u_{mn} is $\lambda_{mn}/2\pi$.

Discussion of Eigenfunctions: It is very interesting that, depending on a and b, several functions F_{mn} may correspond to the same eigenvalue. Physically this means that there may exist vibrations having the same frequency but entirely different **nodal lines** (curves of points on the membrane that do not move). Let us illustrate this with the following example.

Example 15 Eigenvalues and Eigenfunctions of the Square Membrane

Consider the square membrane with $a = b = 1$. From (8.88) we obtain its eigenvalues

$$\lambda_{mn} = c\pi\sqrt{m^2 + n^2} \tag{8.90}$$

Hence $\lambda_{nm} = \lambda_{nm}$, but for $m \neq n$ the corresponding functions

$$F_{mn} = \sin m\pi x \sin n\pi y \quad \text{and} \quad F_{nm} = \sin n\pi x \sin m\pi y$$

are certainly different. For example, to $\lambda_{12} = \lambda_{21} = c\pi\sqrt{5}$ there correspond the two functions

$$F_{12} = \sin \pi x \sin 2\pi y \qquad \text{and} \qquad F_{21} = \sin 2\pi x \sin \pi y$$

Hence the corresponding solutions

$$u_{12} = (B_{12} \cos c\pi\sqrt{5}\, t + B_{12}{*} \sin c\pi\sqrt{5}\, t)F_{12} \quad \text{and} \quad u_{21} = (B_{21} \cos c\pi\sqrt{5}\, t + B_{21}{*} \sin c\pi\sqrt{5}\, t)F_{21}$$

have the nodal lines $y = 1/2$ and $x = 1/2$, respectively (see Figure 7). Taking $B_{12} = 1$ and $B_{12}{*} = B_{21}{*} = 0$ we obtain

$$u_{12} + u_{21} = \cos c\pi\sqrt{5}\, t\,(F_{12} + B_{21}F_{21}) \tag{8.91}$$

which represents another vibration corresponding to the eigenvalue $c\pi\sqrt{5}$. The nodal line of this function is the solution of the equation

$$F_{12} + B_{21}F_{21} = \sin \pi x \sin 2\pi y + B_{21} \sin 2\pi x \sin \pi y = 0$$

or, since $\sin 2\alpha = 2 \sin\alpha \cos\alpha$

$$\sin \pi x \sin \pi y\,(\cos \pi y + B_{21} \cos \pi x) = 0 \tag{8.92}$$

This solution depends on the value of B_{21} (see Figure 8). From (8.90) we see that even more than two functions may correspond to the same numerical value of λ_{mn}. For example, the four functions F_{18}, F_{81}, F_{47}, and F_{74} correspond to the value

$$\lambda_{18} = \lambda_{81} = \lambda_{47} = \lambda_{74} = c\pi\sqrt{65}$$

because

$$1^1 + 8^8 = 4^2 + 7^2 = 65$$

This happens because 65 can be expressed as the sum of two squares of positive integers in several ways. According to a theorem by Gauss, this is the case for every sum of two squares among whose prime factors there are at least two different ones of the form $4n + 1$ where n is a positive integer. In our case we have $65 = 5 \times 13 = (4 + 1)(12 + 1)$.

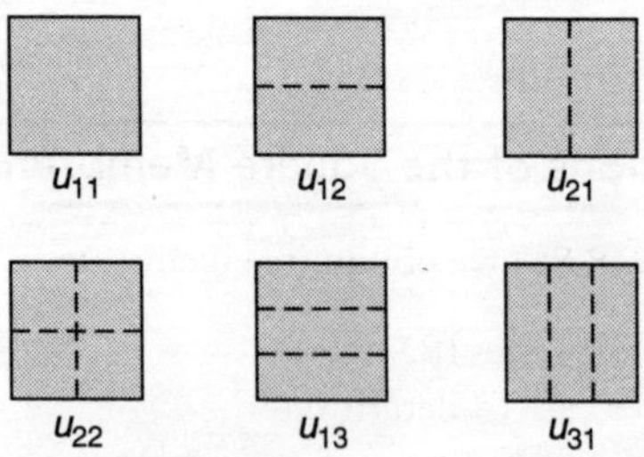

Figure 7 Nodal lines of the solutions u_{11}, u_{12}, u_{21}, u_{22}, u_{13}, u_{31} in the case of the square membrane.

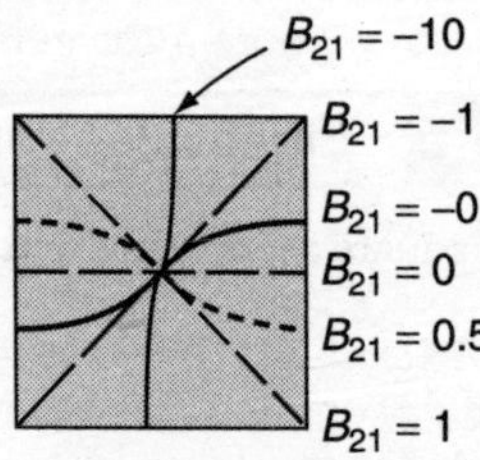

Figure 8 Nodal lines of the solution (8.91) for some values of B_{21}.

Step 3: Solution of the Model (8.80)–(8.82)

So far we have solutions (8.89) satisfying (8.80) and (8.81) only. To obtain the solution that also satisfies (8.82). We consider the double series

$$u(x,y,t)=\sum_{m=1}^{\infty}\sum_{n=1}^{\infty} u_{mn}(x,y,t)$$

$$=\sum_{m=1}^{\infty}\sum_{n=1}^{\infty}(B_{mn}\cos\lambda_{mn}t+B_{mn}{*}\sin\lambda_{mn}t)\sin\frac{m\pi x}{a}\sin\frac{n\pi y}{b} \tag{8.93}$$

(without discussing convergence and uniqueness). From (8.93) and (8.82), setting $t=0$, we have

$$u(x,y,0)=\sum_{m=1}^{\infty}\sum_{n=1}^{\infty}B_{mn}\sin\frac{m\pi x}{a}\sin\frac{n\pi y}{b}=f(x,y) \tag{8.94}$$

Suppose that $f(x, y)$ can be represented by (8.94). (Sufficient for this is the continuity of f, $\partial f/\partial x$, $\partial f/\partial y$, $\partial^2 f/\partial x\partial y$ in R.) Then (8.94) is called the **double Fourier series** of $f(x, y)$. Its coefficients can be determined as follows. Setting

$$K_m(y)=\sum_{n=1}^{\infty}B_{mn}\sin\frac{n\pi y}{b} \tag{8.95}$$

we can write (8.94) in the form

$$f(x,y)=\sum_{m=1}^{\infty}K_m(y)\sin\frac{m\pi x}{a}$$

For fixed y this is the Fourier sine series of $f(x, y)$, considered as a function of x. We see that the coefficients of this expansion are

$$K_m(y)=\frac{2}{a}\int_0^a f(x,y)\sin\frac{m\pi x}{a}\,dx \tag{8.96}$$

Furthermore, (8.95) is the Fourier sine series of $K_m(y)$, and the coefficients are

$$B_{mn}=\frac{2}{b}\int_0^b K_m(y)\sin\frac{n\pi y}{b}\,dy$$

From this and (8.96) we obtain the **generalized Euler formula**

$$B_{mn}=\frac{4}{ab}\int_0^b\int_0^a f(x,y)\sin\frac{m\pi x}{a}\sin\frac{n\pi y}{b}\,dx\,dy \qquad m=1,2,\ldots;\ n=1,2,\ldots \tag{8.97}$$

for the **Fourier coefficients** of $f(x, y)$ in the double Fourier series (8.94).

The B_{mn} in (8.93) are now determined in terms of $f(x, y)$. To determine the $B_{mn}*$, we differentiate (8.93) termwise with respect to t; using (8.82b), we obtain

$$\left.\frac{\partial u}{\partial t}\right|_{t=0}=\sum_{m=1}^{\infty}\sum_{n=1}^{\infty}B_{mn}{*}\,\lambda_{mn}\sin\frac{m\pi x}{a}\sin\frac{n\pi y}{b}=g(x,y)$$

Suppose that $g(x, y)$ can be developed in this double Fourier series. Then, proceeding as before, we find that the coefficients are

$$B_{mn}^* = \frac{4}{ab\lambda_{mn}} \int_0^b \int_0^a g(x, y) \sin \frac{m\pi x}{a} \sin \frac{n\pi y}{b} \, dx \, dy \qquad m = 1, 2, \ldots; \quad n = 1, 2, \ldots \tag{8.98}$$

Result: *If f and g in* (8.82) *are such that u can be represented by* (8.93), *then* (8.93) *with coefficients* (8.97) *and* (8.98) *is the solution of the model* (8.80)–(8.82).

Example 16 Vibration of a Rectangular Membrane

Find the vibrations of a rectangular membrane of sides $a = 4$ ft and $b = 2$ ft (Figure 9) if the tension is 12.5 lb/ft, the density is 2.5 slugs/ft^2 (as for light rubber), the initial velocity is 0, and the initial displacement is

$$f(x, y) = 0.1(4x - x^2)(2y - y^2) \text{ ft} \tag{8.99}$$

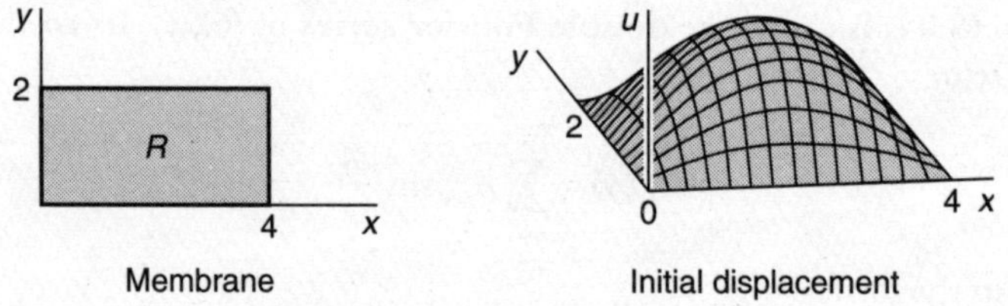

Figure 9 Example 16.

Solution: $c^2 = T/\rho = 12.5/2.5 = 5$ [ft^2/sec^2]. Also, $B_{mn}^* = 0$ from (8.98). From (8.97) and (8.99)

$$B_{mn} = \frac{4}{4 \times 2} \int_0^2 \int_0^4 0.1(4x - x^2)(2y - y^2) \sin \frac{m\pi x}{4} \sin \frac{n\pi y}{2} \, dx \, dy$$

$$= \frac{1}{20} \int_0^4 (4x - x^2) \sin \frac{m\pi x}{4} \, dx \int_0^2 (2y - y^2) \sin \frac{n\pi y}{2} \, dy$$

Two integrations by parts give for the first integral on the right

$$\frac{128}{m^3\pi^3}[1 - (-1)^m] = \frac{256}{m^3\pi^3} \qquad (m \text{ is odd})$$

and for the second integral

$$\frac{16}{n^3\pi^3}[1 - (-1)^n] = \frac{32}{n^3\pi^3} \qquad (n \text{ is odd})$$

For even m or n we get 0. Together with the factor 1/20 we thus have $B_{mn} = 0$ if m or n is even and

$$B_{mn} = \frac{256 \times 32}{20 m^3 n^3 \pi^6} = \frac{0.426050}{m^3 n^3} \qquad (m \text{ and } n \text{ are both odd})$$

From this, (8.88), and (8.93) we obtain the answer

$$u(x, y, t) = 0.426050 \sum_{m,n \text{ odd}} \sum \frac{1}{m^3 n^3} \cos\left(\frac{\sqrt{5}\pi}{4} \sqrt{m^2 + 4n^2}\right) t \sin \frac{m\pi x}{4} \sin \frac{n\pi y}{2}$$

$$= 0.426050\left[\cos\frac{\sqrt{5}\pi\sqrt{5}}{4}t\sin\frac{\pi x}{4}\sin\frac{\pi y}{2} + \frac{1}{27}\cos\frac{\sqrt{5}\pi\sqrt{37}}{4}t\sin\frac{\pi x}{4}\sin\frac{3\pi y}{2}\right.$$

$$\left. + \frac{1}{27}\cos\frac{\sqrt{5}\pi\sqrt{13}}{4}t\sin\frac{3\pi x}{4}\sin\frac{\pi y}{2} + \frac{1}{729}\cos\frac{\sqrt{5}\pi\sqrt{45}}{4}t\sin\frac{3\pi x}{4}\sin\frac{3\pi y}{2} + \cdots\right] \quad (8.100)$$

To discuss this solution, we note that the first term is very similar to the initial shape of the membrane, has no nodal lines, and is by far the dominating term because the coefficients of the next terms are much smaller. The second term has two horizontal nodal lines ($y = 2/3, 4/3$), the third term two vertical ones ($x = 4/3, 8/3$), the fourth term two horizontal and two vertical ones, and so on.

One-Dimensional Heat Equation

Derivation

Consider a heat-conducting homogeneous rod of length L, placed along the x axis with one end of the rod at $x = 0$ (origin) and the other end of the rod at $x = L$ (Figure 10). Assume that the rod has a constant density ρ and a uniform cross-section A. Also assume that the lateral surface of the rod is impenetrable to heat transfer, that is the rod is insulated laterally and therefore heat flows only in the x direction. The rod is sufficiently thin and so the temperature is same at all points of any cross-sectional area of the rod. Let $u(x, t)$ be the temperature of the cross-section at the point x at any time t. The amount of heat $Q(t)$ in a small segment of the rod between the cross-sections at x and $x + \Delta x$ is

$$Q(t) = \int_{x}^{x+\Delta t} (c)(\rho A)u(s, t)dx \quad (8.101)$$

where c is the specific heat of the rod. By Fourier's law of heat conduction, rate of propagation of heat (i.e, the quantity of heat passing through a cross-section at x in unit time) is

$$q = -k\frac{\partial u}{\partial x}A \quad (8.102)$$

where k is the coefficient of thermal conductivity. Since heat flows in the direction of decreasing temperature, a negative sign appears in (8.102). The rate of heat flows at cross-section $x + \Delta x$ is

$$-kA\frac{\partial u}{\partial x}(x + \Delta x, t)$$

The rate of change of heat content in the segment of the rod between x and $x + \Delta x$ must be equal to the net heat flow into this segment of the rod. Thus

$$\frac{\partial Q}{\partial t} = kA\left[\left.\frac{\partial u}{\partial x}\right|_{x+\Delta x} - \left.\frac{\partial u}{\partial x}\right|_{x}\right] \quad (8.103)$$

By mean value theorem

$$\frac{\partial Q}{\partial t} = \frac{\partial}{\partial t}\int_{x}^{x+\Delta x} c\rho Au(s, t)\,dt = \int_{x}^{x+\Delta x} c\rho A\frac{\partial u}{\partial t}ds = c\rho Au_t(\xi, t)\Delta x \quad (8.104)$$

where ξ lies between x and $x + \Delta x$. Replacing the left-hand side of (8.103) by (8.104) we get

$$c\rho A u_t(\xi, t)\Delta x = kA\left[\left.\frac{\partial u}{\partial x}\right|_{x+\Delta x} - \left.\frac{\partial u}{\partial x}\right|_{x}\right]$$

Rewriting this and taking the limit as $\Delta x \to 0$, we obtain

$$\frac{\partial u(x, t)}{\partial t} = a^2 \frac{\partial^2 u}{\partial x^2} \tag{8.105}$$

where $a^2 = k/c\rho$ is known as the diffusivity constant. Equation (8.105) is the one-dimensional heat equation which is second-order, homogeneous and parabolic type.

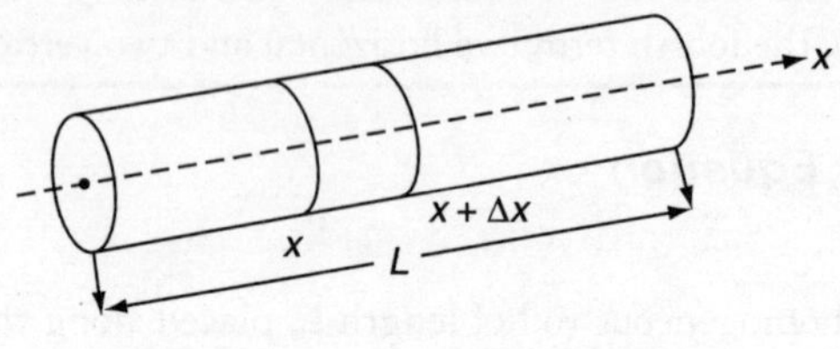

Figure 10 Cross-section of the rod.

Solution of One-Dimensional Heat Equation

Consider a long thin wire or rod or bar of constant cross-section and a homogeneous heat conducting material. Consider a bar of length L oriented along the x axis with one end A coinciding with origin (Figure 11). Suppose that the lateral surface of the bar is perfectly insulated. Then the heat flows in the bar along the x direction only.

Thus the temperature u of the bar depends only on x and t. This phenomenon is described by the initial boundary value problem consisting of one-dimensional heat equation

$$\frac{\partial u}{\partial t} = c^2 \frac{\partial^2 u}{\partial x^2} \tag{8.106}$$

where c^2 is the thermal diffusivity. The boundary conditions (BCs) prescribed at the end points A and B are

$$x(0, t) = 0; \quad x(L, t) = 0 \tag{8.107}$$

that is the end points are assumed to be at zero temperature. The initial temperature distribution in the bar given by

$$u(x, 0) = f(x) \tag{8.108}$$

where $f(x)$ is a given (prescribed) function of x. Solution by the separation of variables technique or the method of separating variables or product method reduces the initial boundary value problem to the solution of two ordinary differential equations as follows.

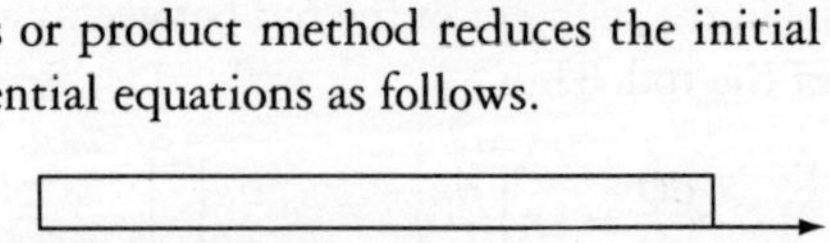

Figure 11 Bar for solution of one-dimensional heat equation.

***Step* 1:** Assume that the solution $u(x, t)$ is separable, that is

$$u(x, t) = X(x)T(t) \tag{8.109}$$

where $X(x)$ is a function of x alone and $T(t)$ is function of t only. Differentiating (8.109) with respect to t and x we get

$$\frac{\partial u}{\partial t} = X\dot{T}; \quad u_x = X'T; \quad u_{xx} = X''T$$

Substituting these in the PDE (8.106) we get

$$X\dot{T} = c^2 X''T$$

$$\frac{X''}{X} = \frac{\dot{T}}{c^2 T}$$

$$\frac{X''}{X} = \frac{\dot{T}}{c^2 T} = k = \text{constant}$$

which results in the following two ordinary differential equations:

$$X'' - kX = 0 \tag{8.110}$$

$$\dot{T} - c^2 kT = 0 \tag{8.111}$$

Using (8.109) the boundary conditions (8.107) reduces to

$$0 = u(0, t) = X(0)T(t) \Rightarrow X(0) = 0 \tag{8.112}$$

$$0 = u(L, t) = X(L)T(t) \Rightarrow X(L) = 0 \tag{8.113}$$

Since $T(t) \neq 0$ (otherwise if $T(t) = 0$ the $u = 0$ for all t, which is a trivial solution)

***Step* 2:**

Case 1: $k = 0, X'' = 0$

The solution is $X(x) = ax + b$. Using (8.112) and (8.113) we have

$$0 = X(0) = a(0) + b \Rightarrow b = 0$$
$$0 = X(L) = a(L) \Rightarrow a = 0$$

Thus $X(x) = 0$ for all x or $u = 0$ for all x, which is a trivial solution.

Case 2: $k = \lambda^2 > 0, \quad X'' - \lambda^2 X = 0$

The solution is $X(x) = Ae^{\lambda x} + Be^{-\lambda x}$

Using (8.112) we get

$$0 = X(0) = A + B \Rightarrow A = -B$$

Using (8.113) we get

$$0 = X(L) = A(e^{\lambda L} - e^{-\lambda L}) \Rightarrow A = 0$$

We finally get

$$A = 0 = B$$

Now $e^{\lambda L} - e^{-\lambda L} = 2\sinh \lambda L = 0$ only when $\lambda = 0$. Thus $X(x) = 0$ for all x or $u = 0$ for all x which is a trivial solution.

Case 3: $k = -\lambda^2$
The equation is $X'' + \lambda^2 X = 0$. The solution is

$$X(x) = A\cos\lambda x + B\sin\lambda x$$

Using (8.112) we get

$$0 = X(0) = A + B(0) \Rightarrow A = 0$$

Using (8.113) we get

$$0 = X(L) - B\sin\lambda L$$

We must take $B \neq 0$ since otherwise $X(x) = 0$. Hence

$$\sin\lambda L = 0$$
$$\lambda L = n\pi$$

Therefore

$$\lambda = \frac{n\pi}{L} \quad (n \text{ is an integer}) \tag{8.114}$$

For $B = 1$, we get infinitely many solutions:

$$X_n(x) = \sin\frac{n\pi x}{L} \quad (n = 1, 2, 3, \ldots) \tag{8.115}$$

Solving (8.111), we get

$$T_n(t) = B_n e^{-\lambda_n{}^2 t} = B_n e^{-n^2\pi^2 t/L^2} \tag{8.116}$$

Hence the eigenfunctions

$$u_n(x,t) = X_n(x)T_n(t)$$
$$= B_n \frac{\sin n\pi x}{L} e^{-\lambda_n{}^2} \quad (n = 1, 2, 3, \ldots) \tag{8.117}$$

satisfy (8.106) and boundary conditions (8.107).

***Step* 3:** In order to satisfy the initial condition (8.108) we invoke the superposition or linearity principle which states that

$$u(x,t) = \sum_{n=1}^{\infty} u_n(x, t) = \sum_{n=1}^{\infty} B_n \sin\frac{n\pi x}{L} e^{-c^2 n^2 \pi^2 t/L^2} \tag{8.118}$$

These satisfy (8.06) and (8.17). To determine the unknown constants B_n use initial condition (8.18) in (8.118). We get

$$f(x) = u(x,0) = \sum_{n=1}^{\infty} B_n \sin\frac{n\pi x}{L} \tag{8.119}$$

Since the RHS of (8.119) is a half-range Fourier sine series expansion of $f(x)$ in the interval $(0, L)$, the Fourier coefficients B_n's are given by

$$B_n = \frac{2}{L}\int_0^L f(x)\sin\frac{n\pi x}{L}\,dx \quad (n = 1, 2, \ldots) \tag{8.120}$$

Thus the temperature distribution in the bar is given by the solution (8.118) with B_n's calculated from (8.120).

***Note*:** Solution (8.118) is transient (negative exponential), that is, u decreases as t increases which is consistent with physical nature of the heat conduction problem.

CONCEPT CHECK

1. Find $u(x, t)$ for the string of length $L = 1$ and $c^2 = 1$ when the initial velocity is zero and the initial deflection with small k (say, 0.01) is as follows. Sketch or graph $u(x, t)$ as in Figure 5.

(a) $k \sin 2\pi x$ (b) $k[\sin \pi x - (\sin 3\pi x\ 3)]$

(c) $kx(1-x)$ (d) $kx(1-x^2)$

(e)

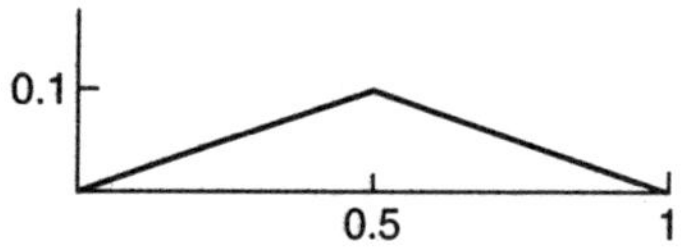

(f)

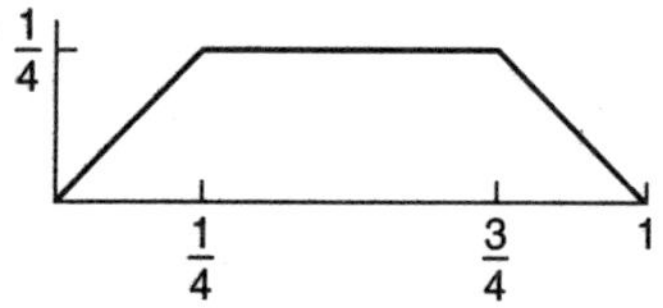

(g)

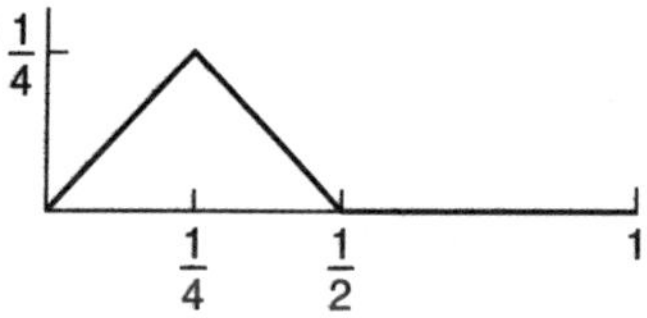

(h)

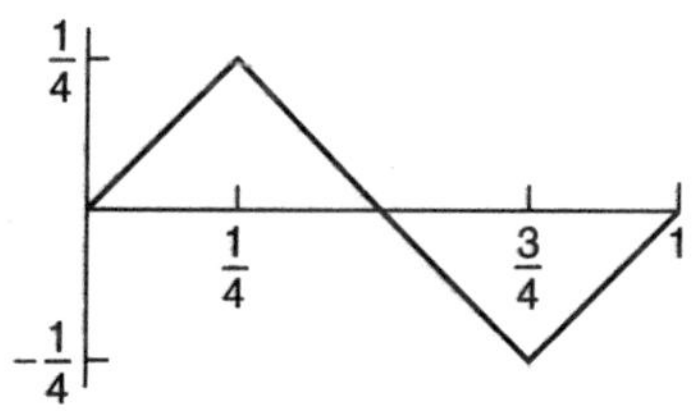

(i)

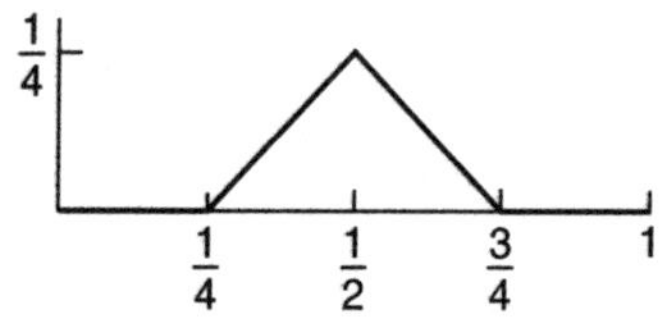

(j)

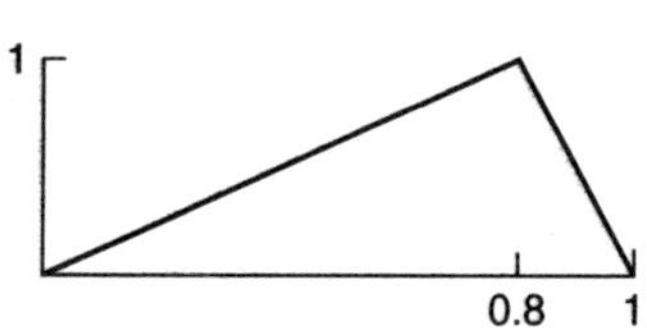

2. **(Frequency)** How does the frequency of the fundamental mode of the vibrating string depend on the length of the string? On the mass per unit length? What happens to the string if we double the tension? Why is a contrabass larger than a violin?

3. **(Nonzero initial velocity)** Find the deflection $u(x, t)$ of the string of length $L = \pi$ and $c^2 = 1$ for zero initial displacement and "triangular" initial velocity $u_t(x, 0) = 0.01x$ if $0 \leqq x \leqq \pi/2$, $u_t(x, 0) = 0.01(\pi - x)$ if $\pi/2 \leqq x \leqq \pi$. (Initial conditions with $u_t(x, 0) \neq 0$ are hard to realize experimentally.)

4. **TEAM PROJECT. Forced Vibrations of an Elastic String.** Show the following.

(a) Substitution of

$$u(x,t) = \sum_{n=1}^{\infty} G_n(t) \sin \frac{n\pi x}{L} \tag{8.121}$$

(L = length of the string) into the wave equation (8.61) governing free vibrations leads to

$$\ddot{G}_n + \lambda_n^2 G = 0, \ \ \lambda_n = \frac{cn\pi}{L} \tag{8.122}$$

(b) Forced vibrations of the string under an external force $P(x, t)$ per unit length acting normal to the string are governed by the PDE

$$u_{tt} = c^2 u_{xx} + \frac{P}{\rho} \tag{8.123}$$

(c) For a sinusoidal force $P = A\rho \sin \omega$ we obtain

$$\frac{P}{\rho} = A \sin \omega = \sum_{n=1}^{\infty} k_n(t) \sin \frac{n\pi x}{L} \tag{8.124}$$

$$k_n(t) = \begin{cases} (4A/n\pi)\sin\omega & (n \text{ odd}) \\ 0 & (n \text{ even}) \end{cases}$$

Substituting (8.121) and (8.124) into (8.123) gives

$$\ddot{G}_n + \lambda_n^2 G_n = \frac{2A}{n\pi}(1 - \cos n\pi)\sin\omega$$

If $\lambda_n^2 \neq \omega^2$, the solution is

$$G_n(t) = B_n \cos \lambda_n t + B_n^* \sin \lambda_n t$$
$$+ \frac{2A(1 - \cos n\pi)}{n\pi(\lambda_n^2 - \omega^2)} \sin \omega$$

Determine B_n and B_n^* so that u satisfies the initial conditions $u(x,0) = f(x), u_t(x,0) = 0$

(d) **(Resonance)** Show that if $\lambda_n = \omega$, then

$$G_n(t) = B_n \cos \omega + B_n^* \sin \omega$$
$$- \frac{A}{n\pi\omega}(1 - \cos n\pi) t \cos \omega$$

(e) **(Reduction of boundary conditions)** Show that prob. 1(a)–1(c) with more complicated boundary conditions $u(0, t) = 0$, $u(L, t) = h(t)$, can be reduced to a problem for a new function v satisfying conditions $v(0, t) = v(L, t) = 0, v(x, 0) = f_1(x), v_t(x, 0) = g_1(x)$ but a nonhomogeneous wave equation. *Hint:* Set $u = v + w$ and determine w suitably.

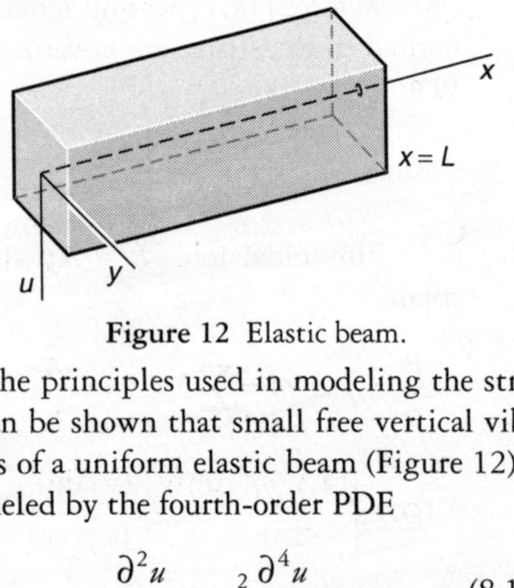

Figure 12 Elastic beam.

5. By the principles used in modeling the string it can be shown that small free vertical vibrations of a uniform elastic beam (Figure 12) are modeled by the fourth-order PDE

$$\frac{\partial^2 u}{\partial t^2} = -c^2 \frac{\partial^4 u}{\partial x^4} \qquad (8.125)$$

where $c^2 = EI/\rho A$ (E = Young's modulus of elasticity, I = moment of intertia of the cross section with respect to the y axis in the figure, ρ = density, A = cross-sectional area).

6. Substituting $u = F(x)G(t)$ into (8.125), show that

$$F^{(4)}/F = -\ddot{G}/c^2 G = \beta^4 = \text{const}$$
$$F(x) = A \cos \beta x + B \sin \beta x + C \cosh \beta x + D \sinh \beta x$$
$$G(t) = a \cos c\beta^2 t + b \sin c\beta^2 t$$

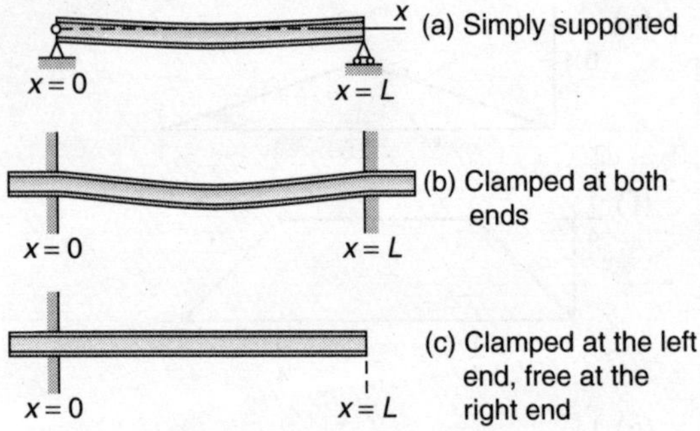

Figure 13 Supports of a beam.

7. **[Simply supported beam in Figure 7(a)]** Find solutions $u_n = F_n(x)G_n(t)$ of (8.125) corresponding to zero initial velocity and satisfying the boundary conditions

$$u(0,t) = 0, u(L,t) = 0$$

(ends simply supported for all times t),

$$u_{xx}(0,t) = 0, u_{xx}(L,t) = 0$$

(zero moments, hence zero curvature, at the ends)

8. Find the solution of (8.125) that satisfies the conditions in Prob. 7 as well as the initial condition

$$u(x,0) = f(x) = x(L - x)$$

9. Compare the results of Probs. 8 and 1(c). What is the basic difference between the frequencies of the normal modes of the vibrating string and the vibrating beam?

10. **[Clamped beam in Figure 13(b)]** What are the boundary conditions for the clamped beam in Figure 13(b)? Show that F in Prob. 6 satisfies these conditions if βL is a solution of the equation

$$\cosh \beta L \cos \beta L = 1 \qquad (8.126)$$

Determine approximate solutions of (8.126), for instance, graphically from the intersections of the curves of cos βL and 1/cosh βL.

11. **[Clamped-free beam in Figure 13(c)]** If the beam is clamped at the left and free at the right [Figure 13(c)], the boundary conditions are

$$u(0, t) = 0, \qquad u_x(0, t) = 0$$

$$u_{xx}(L, t) = 0, \qquad u_{xxx}(L, t) = 0$$

Show that F in Prob. 6 satisfies these conditions if βL is a solution of the equation.

$$\cosh \beta L \cos \beta L = -1$$

Find approximate solutions of Prob. 9.

12. **(Frequency)** How does the frequency of the eigenfunctions of the rectangular membrane change if (a) we double the tension, (b) we take a membrane of half the mass of the original one, (c) we double the sides of the membrane? (Give reason.)
13. Determine and sketch the nodal lines of the eigenfunctions of the square membrane for $m = 1, 2, 3, 4$ and $n = 1, 2, 3, 4$.
14. Represent $f(x, y)$ by a series, where $0 < x < 1$, $0 < y < 1$.
    (a) $f(x, y) = 1$
    (b) $f(x, y) = x$
    (c) $f(x, y) = y$
    (d) $f(x, y) = x + y$
    (e) $f(x, y) = xy$
    (f) $f(x, y) = xy(1 - x)(1 - y)$
15. Find the deflection $u(x, y, t)$ of the square membrane of side π and $c^2 = 1$ if the initial velocity is 0 and the initial deflection is
    (a) $k \sin 2x \sin 5y$
    (b) $0.1 \sin x \sin y$
    (c) $0.1xy(\pi - x)(\pi - y)$
16. Verify the discussion of the terms of (8.100) in Example 16.
17. Repeat the task of Prob. 13 when $a = 4$ and $b = 1$.
18. Verify the calculation of B_{mn} in Example 16 by integration by parts.
19. Find eigenvalues of the rectangular membrane of sides $a = 2$ and $b = 1$ to which there correspond two or more different (independent) eigenfunctions.
20. **(Minimum property)** Show that among all rectangular membranes of the same area $A = ab$ and the same c the square membrane is that for which u_{11} has the lowest frequency.
21. **Double Fourier Series.** Represent $f(x, y)$ $(0 < x < a, 0 < y < b)$ by a double Fourier series (8.94).
    (a) $f(x, y) = k$
    (b) $f(x, y) = 0.25xy$
    (c) $f(x, y) = xy(a^2 - x^2)(b^2 - y^2)$
    (d) $f(x, y) = xy(a - x)(b - y)$
22. **(Deflection)** Find the deflection of the membrane of sides a and b with $c^2 = 1$ for the initial deflection

$$f(x, y) = \sin\left(\frac{3\pi x}{a}\right) \sin\left(\frac{4\pi y}{b}\right)$$

and initial velocity 0.

23. Repeat the task in Prob. 22 with $c^2 = 1$, for $f(x, y)$ as in Prob. 21(d) and initial velocity 0.
24. **(Forced vibrations)** Show that forced vibrations of a membrane are modeled by the PDE $u_{tt} = c^2\nabla^2 u + P/\rho$, where $P(x, y, t)$ is the external force per unit area acting perpendicular to the xy plane.

Solved Problems

Q 1. Form a PDE by eliminating the arbitrary constants a and b from

$$\log(az - 1) = x + ay + b$$ **[JNTU Aug./Sep. 2007, April/May 2007]**

Answer: The given equation is

$$\log(az - 1) = x + ay + b \qquad (8.127)$$

where a and b are arbitrary constants. Differentiating (8.127) partially with respect to x, we get

$$\frac{1}{az-1}ap = 1 \qquad (8.128)$$
$$ap = az - 1$$

where $p = \partial z/\partial x$. Again differentiating (8.127) partially with respect to y, we get

$$\frac{1}{az-1}aq = a \qquad (8.129)$$
$$aq = a(az-1)$$

where $q = \partial z/\partial y$. Dividing (8.129) by (8.128), we get

$$\frac{aq}{ap} = \frac{a(az-1)}{az-1}$$

$$\frac{q}{p} = a \Rightarrow q = ap$$

Substituting the values a and ap in (8.128), we get

$$q = \left(\frac{q}{p}\right)z - 1$$

$$pq = qz - p$$

$$p(q+1) = qz$$

which is the required equation.

Q 2. Form the PDE by eliminating the arbitrary constants a and b from $2z = (x+a)^{1/2} + (y-a)^{1/2} + b$.

[JNTU Aug./Sep. 2007, May/June 2006]

Answer: Differentiating the given equation partially with respect to x, we get

$$\frac{2\partial z}{\partial x} = \frac{1}{2\sqrt{x+a}} + 0$$

$$\frac{4\partial z}{\partial x} = \frac{1}{\sqrt{x+a}}$$

$$4p = \frac{1}{\sqrt{x+a}}$$

$$\sqrt{x+a} = \frac{1}{4p}$$

$$x + a = \frac{1}{16p^2}$$

$$a = \frac{1}{16p^2} - x \qquad (8.129)$$

Now, differentiating the given equation with respect to y, we get

$$\frac{2\partial z}{\partial y} = \frac{1}{2\sqrt{y-a}} + 0$$

$$\frac{4\partial z}{\partial y} = \frac{1}{\sqrt{y-a}}$$

$$4q = \frac{1}{\sqrt{y-a}}$$

$$\sqrt{y-a} = \frac{1}{4q}$$

$$y - a = \frac{1}{16q^2}$$

$$y = \frac{1}{16q^2} + a \tag{8.130}$$

Substituting the value of a from (8.129) to (8.130), we get

$$y = \frac{1}{16q^2} + \frac{1}{16p^2} - x$$

$$y + x = \frac{1}{16p^2} + \frac{1}{16q^2}$$

$$16(x+y) = \frac{1}{p^2} + \frac{1}{q^2}$$

which is the required equation.

Q 3. Form the partial differential equation by eliminating the arbitrary function from

$$z = f(y) + \phi(x+y)$$ [JNTU Nov. 2005, May 2005]

Answer: The given function is

$$z = f(y) + \phi(x+y) \tag{8.131}$$

By differentiating (8.131) partially with respect to x, we get

$$p = \frac{\partial z}{\partial x} = 0 + \phi'(x+y) \times 1 \tag{8.132}$$

By differentiating (8.131) partially with respect to y, we get

$$q = \frac{\partial z}{\partial y} = f'(y) = \phi'(x+y) \times 1 \tag{8.133}$$

Differentiating (8.132) partially with respect to x, we get

$$\frac{\partial^2 z}{\partial x^2} = \phi''(x+y) \times 1$$

$$\frac{\partial^2 z}{\partial x^2} = r = \phi''(x+y) \tag{8.134}$$

Differentiating (8.133) partially with respect to y, we get

$$\frac{\partial^2 z}{\partial y^2} = \phi''(x+y)\times 1$$

$$\frac{\partial^2 z}{\partial x \partial y} = s = \phi''(x+y) \tag{8.135}$$

From (8.134) and (8.135), we have

$$r = s \text{ or } r - s = 0$$

which is the required partial differential equation.

Q 4. Form the partial differential equation by eliminating the arbitrary function from $xyz = f(x^2+y^2+z^2)$.

[JNTU May/June 2006, Nov. 2004]

Answer: The given equation is

$$xyz = f(x^2+y^2+z^2) \tag{8.136}$$

Differentiating (8.136) partially with respect to x, we get

$$y\left(z + x\frac{\partial z}{\partial x}\right) = f'(x^2+y^2+z^2)\left(2x + 2z\frac{\partial z}{\partial x}\right)$$

$$y(z+xp) = f'(x^2+y^2+z^2)(2x+xzp) \tag{8.137}$$

Differentiating (8.136) partially with respect to y, we get

$$x\left(z + y\frac{\partial z}{\partial y}\right) = f'(x^2+y^2+z^2)\left(2y + 2z\frac{\partial z}{\partial y}\right)$$

$$x(z+yq) = f'(x^2+y^2+z^2)(2x+xzq) \tag{8.138}$$

Dividing (8.137) by (8.138), we get

$$\frac{y(z+xp)}{x(z+yp)} = \frac{2x-2zp}{2y-2zp}$$

$$(y+zq)y(z+zp) = (x+zq)x(z+yp)$$

which is the required equation.

Q 5. Solve the partial differential equation $x^2p^2 + y^2q^2 = 1$.

[JNTU Aug./Sep. 2008, May/June 2008]

Answer: The given equation is $x^2p^2 + y^2q^2 = 1$. Let

$$x^2p^2 = 1 - y^2q^2 = a \quad \text{(say)}$$

$$1 - a = y^2q^2$$

$$(yq)^2 = 1 - a$$

$$yq = \sqrt{1-a}$$

$$q = \frac{\sqrt{1-a}}{y}$$

Also,

$$x^2 p^2 = a$$
$$(xp)^2 = a$$
$$xp = \sqrt{a}$$
$$p = \frac{\sqrt{a}}{x}$$

We have,

$$\partial z = p\partial x + q\partial y$$
$$\partial z = \frac{\sqrt{a}}{x}\partial x + \frac{\sqrt{1-a}}{y}\partial y$$

Integrating the above equation, we get

$$\int \partial z = \int \frac{\sqrt{a}}{x}\partial x + \int \frac{\sqrt{1-a}}{y}\partial y$$
$$z = \log x(\sqrt{a}) + \log y(\sqrt{1-a})$$

Q 6. Solve $x^2(z-y)p + y^2(x-z)q = z^2(y-x)$. **[JNTU May/June 2008, Nov. 2005]**

Answer: The given equation can be simplied as

$$x^2(z-y)p + y^2(x-z)q = z^2(y-x)$$
$$-x^2(y-z)p - y^2(z-x)q = -z^2(x-y) \qquad (8.139)$$
$$x^2(y-z)p + y^2(z-x)q = z^2(x-y)$$

Comparing (8.139) with $Pp + Qq = R$, we get

$$P = x^2(y-z);\ Q = y^2(z-x);\ R = z^2(x-y)$$

The Lagrange's auxiliary equations are

$$\frac{dx}{x^2(y-z)} = \frac{dy}{y^2(z-x)} = \frac{dz}{z^2(x-y)} \qquad (8.140)$$

Now, using $1/x, 1/y, 1/z$ as multipliers, we get

$$z = \frac{(1/x)dx + (1/y)dy + (1/z)dz}{x(y-z) + y(z-x) + z(x-y)} = \frac{(1/x)dx + (1/y)dy + (1/z)dz}{xy - xz + yz - yx + zx - zy}$$
$$= \frac{(1/x)\,dx + (1/y)\,dy + (1/z)dz}{0}$$

This implies

$$\frac{1}{x}dx + \frac{1}{y}dy + \frac{1}{z}dz = 0$$

Integrating both sides we get

$$\int \frac{1}{x}dx + \int \frac{1}{y}dy + \int \frac{1}{z}dz = \int 0$$

$$\log x + \log y + \log z = \log k_1$$

$$\log(xyz) = \log k_1$$

$$xyz = k_1 \tag{8.141}$$

where k is an arbitrary constant. Using $(y+z),(z+x),(x+y)$ as multipliers, we get

$$z = \frac{(y-z)dx + (z-x)dy + (x-y)dz}{x^2(y-z)(y-z) + y^2(z-x)(z-x) + z^2(x-y)(x-y)}$$

$$= \frac{(y-z)dx - (z-x)dy - (x-y)dz}{x^2(y^2-z^2) + y^2(z^2-x^2) + z^2(x^2-y^2)}$$

$$= \frac{(y-z)dx + (z-x)dy + (x-y)dz}{x^2y^2 - x^2z^2 + y^2z^2 - y^2x^2 + z^2x^2 - z^2y^2}$$

$$= \frac{(y+z)dx + (z+x)dy + (x+y)dz}{0}$$

This implies

$$(y+z)dx + (z+x)dy + (x+y)dz = 0$$

Integrating we get

$$\int (y+z)dx + \int (z+x)dy + \int (x+y)dz = \int 0$$

$$xy + xz + zy + xy + xz + yz = k_2$$

$$2(xy + yz + zx) = k_2 \tag{8.142}$$

where k_2 is an arbitrary constant. Using (8.141) and (8.142) the general solution is

$$\phi\,(xyz, 2(xy + yz + zx)) = 0$$

where ϕ is an arbitrary function.

Q 7. Solve

$$\frac{x^2}{p} + \frac{y^2}{q} = z \qquad \text{[JNTU 2006, 2003]}$$

Answer: The given equation can be written as

$$(x^{-2}p)^{-1} + (y^{-2}q)^{-1} = z \tag{8.143}$$

This is of the form $f(x^m p, y^n q, z) = 0$ with $m = -2$ and $n = -2$. Substitute $X = x^{1-m} = x^{1-(-2)} = x^3$ and $Y = y^{1-n} = y^3$. Now

$$p = \frac{\partial z}{\partial x} = \frac{\partial z}{\partial X}\frac{\partial X}{\partial x} = P(3x^2)$$

where $P = \partial z / \partial X$. Therefore we have

$$x^{-2} p = 3P$$

Similarly

$$y^{-2} q = 3Q$$

where

$$Q = \frac{\partial z}{\partial Y}$$

Now (8.143) becomes

$$(3P)^{-1} + (3Q)^{-1} = z \tag{8.144}$$

This is of the form $f(z, P, Q) = 0$. Let $z = g(X + aY)$ be the solution of (8.144). Put $u = X + aY$. Then $z = g(u)$. Therefore

$$P = \frac{dz}{du} \quad \text{and} \quad Q = a\frac{dz}{du}$$

Now (8.144) becomes

$$\left(3\frac{dz}{du}\right)^{-1} + \left(3a\frac{dz}{du}\right)^{-1} = z$$

$$\frac{1}{3}\frac{du}{dz} + \frac{1}{3a}\frac{du}{dz} = z$$

$$\frac{1}{3}\frac{du}{dz}\left(1 + \frac{1}{a}\right) = z$$

Separating the variables, we get

$$zdz = \left(\frac{a+1}{3a}\right) du$$

Integrating, we get

$$\frac{z^2}{2} = \frac{a+1}{3a} u + k$$

where k is a constant. Solving the above equation we get

$$\frac{z^2}{2} = \left(\frac{a+1}{3a}\right)(X + aY) + k$$

$$\frac{z^2}{2} = \left(\frac{a+1}{3a}\right)(x^3 + ay^3) + k$$

which is the required solution.

Summary

In this chapter we have discussed the concept of **partial differential equation** (PDE). A PDE is an equation involving one or more partial derivatives of an unknown function that depends on two or more variables. Just as for ordinary differential equations we say that a PDE is **linear** if it is of first degree in the unknown function and its partial derivatives. Otherwise we call it **nonlinear**. We call a linear PDE **homogeneous** if each of its terms contains either the unknown function or one of its partial derivatives. Otherwise we call the equation **nonhomogeneous.**

Next we discuss the various solutions to the PDE: complete integral, particular integral, general solution, and singular integral. Standard forms I, II, III, IV have been clearly indentified and a method for solving each is presented.

We conclude the chapter by discussing the method of separation of variables, one-dimensional and two-dimensional wave equations. Numerous solved examples are provide within the chapter as well as in a separate section.

Review Questions and Problems

1. Form a partial differential equation by eliminating l, m, n from

$$\frac{x^2}{l^2} = \frac{y^2}{m^2} = \frac{z^2}{n^2}$$

2. Form a partial differential equation by eliminating arbitrary constants l and m from $z = lxe^x + (1/2)l^2e^{2y} + m$.
3. Find the differential equation of all planes passing through the origin.
4. Form a differential equation of all planes having equal intercepts on x and y axes.
5. Obtain partial differential equation from $z = f_1(y + 2x) + f_2(y - 3x)$. **[JNTU 2004]**
6. Obtain partial differential equation from $z = f(\sin x + \cos y)$. **[JNTU 2006, 2004]**
7. Form the partial differential equation by eliminating the arbitrary functions from $z = f(y) + \phi(x + y)$. **[JNTU 2008, 2005]**
8. Form the partial differential equation by eliminating the arbitrary functions from $z = xf_1(x + t) + f_2(x + t)$. **[JNTU 2007]**
9. Form the partial differential equation by eliminating the arbitrary functions from $\phi\left(\frac{y}{x}, x^2 + y^2 + z^2\right) = 0$ **[JNTU 2006, 2005]**
10. Form the partial differential equation by eliminating the arbitrary functions from $z = f(x^2 + y^2 + z^2)$. **[JNTU 2007, 2006]**
11. Solve the equation $(x^2 + y^2)(p^2 + q^2) = 1$. **[JNTU 2008]**
12. Solve $q^2 = z^2p^2(1 - p^2)$.
13. Solve $p^3 + q^3 = 8z$.
14. Solve $p^2z^2 + q = p^2q$.
15. Solve $q^2 - p = y - x$.
16. Solve $pqz = p^2(qx + p^2) + q^2(py + q^2)$. **[JNTU 2003, 2002]**
17. Solve $x^2(y - z)p + y^2(z - x)q = z^2(x - y)$. **[JNTU 2008, 2004]**
18. Solve $x^2p^2 + y^2q^2 = z^2$.
19. Solve $z^2 = 1 + p^2 + q^2$.
20. Solve $z^2(p^2x^2 + q^2) = 1$. **[JNTU 2006]**
21. Solve $z^2(p^2 + q^2) = x^2 + y^2$. **[JNTU 2007, 2006, 2003]**
22. Solve the partial differential equation $\frac{p}{x^2} + \frac{q}{y^2} = z$. **[JNTU 2006]**
23. Solve $u_{xt} = e^{-t}\cos x$ with $u(x, 0) = 0$ and $u(0, t) = 0$ by separation of variables.

Frequently Asked Questions in Examinations

1. Express the following system in matrix form and solve by Gauss elimination method:

$$\begin{aligned} 2x_1 + x_2 + 2x_3 + x_4 &= 6 \\ 6x_1 + 6x_2 + 6x_3 + 2x_4 &= 36 \\ 4x_1 + 3x_2 + 3x_3 - 3x_4 &= -1 \\ 2x_1 + 2x_2 - x_3 + x_4 &= 10 \end{aligned}$$

{JNTU May 2008}

2. Prove that the following set of equations are consistent and solve them:

$$\begin{aligned} 3x + 3y + 2z &= 1 \\ x + 2y \quad &= 4 \\ 10y + 3z &= -2 \\ 2x - 3y - z &= 5 \end{aligned}$$

{JNTU May 2008, Aug. 2007}

3. Find the rank of

$$\begin{bmatrix} 2 & -4 & 3 & -1 & 0 \\ 1 & -2 & -1 & -4 & 2 \\ 0 & 1 & -1 & 3 & 1 \\ 4 & -7 & 4 & -4 & 5 \end{bmatrix}$$

{JNTU May 2008}

4. Solve the following system of equations:

$$\begin{aligned} 3x + y + 2z &= 3 \\ 2x - 3y - z &= -3 \\ x + 2y + z &= 4 \end{aligned}$$

{JNTU May 2008}

5. Find the rank of

$$\begin{bmatrix} 1 & 4 & 3 & -2 & 1 \\ -2 & -3 & -1 & 4 & 3 \\ -1 & 6 & 7 & 2 & 9 \\ -3 & 3 & 6 & 6 & 12 \end{bmatrix}$$

{JNTU May 2008}

6. Solve the following system of equations:

$$\begin{aligned} x + y + \quad w &= 0 \\ y + z \quad &= 0 \\ x + y + z + w &= 0 \\ x + y + 2z \quad &= 0 \end{aligned}$$

{JNTU May 2008}

7. Find the rank of the following matrix:

$$\begin{bmatrix} 0 & 1 & -3 & -1 \\ 1 & 0 & 1 & 1 \\ 3 & 1 & 0 & 2 \\ 1 & 1 & -2 & 0 \end{bmatrix}$$

[JNTU May 2008]

8. Solve the following system of equations:

$$\begin{aligned} x + 2y + 3z &= 1 \\ 2x + 3y + 8z &= 2 \\ x + y + z &= 3 \end{aligned}$$

[JNTU May 2008]

9. Determine the rank of the matrix by reducing it to the normal form:

$$A = \begin{bmatrix} 2 & -1 & 3 & 4 \\ 0 & 3 & 4 & 1 \\ 2 & 3 & 7 & 5 \\ 2 & 5 & 11 & 6 \end{bmatrix}$$

[JNTU Aug. 2007]

10. Find whether the following equations are consistent; if so, solve them:

$$\begin{aligned} x + 2y - z &= 3 \\ 3x - y - 2z &= 1 \\ 2x - 2y + 3z &= 4 \\ x - y + z &= -1 \end{aligned}$$

[JNTU Aug. 2007]

11. Find the value of "k" such that the rank of the matrix A is 3, where

$$A = \begin{bmatrix} 1 & 2 & -1 & 3 \\ 4 & 1 & 2 & 1 \\ 3 & -1 & 1 & 2 \\ -1 & 2 & 0 & k \end{bmatrix}$$

[JNTU Aug. 2007]

12. Find whether the system of equations is consistent; if so, solve them:

$$\begin{aligned} x_1 + 2x_2 + x_3 - 2x_4 &= 6 \\ 2x_1 + 3x_2 + 2x_3 - 2x_4 &= 8 \\ 3x_1 + x_2 + 2x_3 - x_4 &= 4 \\ 4x_1 + 2x_2 + 2x_3 - 3x_4 &= 9 \end{aligned}$$

[JNTU Aug. 2007]

13. Solve the tridiagonal system by writing the coefficient matrix as a product of a lower triangular and upper triangular matrices:

$$\begin{aligned} 3x_1 - x_2 \qquad\qquad &= 4 \\ 2x_1 - x_2 + x_3 \qquad &= 6 \\ 2x_2 + 3x_3 + 2x_4 &= 11 \\ x_3 - 2x_4 &= -11 \end{aligned}$$

[JNTU Aug. 2007]

14. Find whether the following equations are consistent; if so solve them:

$$\begin{aligned} x + y + 2z &= 4 \\ 2x - y + 3z &= 9 \\ 3x - y - z &= 2 \end{aligned}$$

[JNTU Aug. 2007]

15. Find the rank of the matrix by reducing to the normal form:

$$\begin{bmatrix} 1 & 2 & 3 & 0 \\ 2 & 4 & 3 & 2 \\ 3 & 2 & 1 & 3 \\ 6 & 8 & 7 & 5 \end{bmatrix}$$

[JNTU Aug. 2007]

16. Determine whether the following equations will have a nontrivial solution; if so, solve them:

$$\begin{aligned} 3x + 4y - z - 6w &= 0 \\ 2x + 3y + 2z - 3w &= 0 \\ 2x + y - 14z - 9w &= 0 \\ x + 3y + 13z + 3w &= 0 \end{aligned}$$

[JNTU Aug. 2007]

17. Find an LU decomposition of the matrix and solve the linear system **AX** = **B**:

$$\begin{bmatrix} -3 & 12 & -6 \\ 1 & -2 & 2 \\ 0 & 1 & 1 \end{bmatrix} \begin{bmatrix} x \\ y \\ z \end{bmatrix} = \begin{bmatrix} -33 \\ 7 \\ -1 \end{bmatrix}$$

[JNTU Aug. 2007]

18. Determine the values of λ for which the following set of equations may possess nontrivial solution and solve them in each case:

$$\begin{aligned} 3x_1 + x_2 - \lambda x_3 &= 0 \\ 4x_1 - 2x_2 - 3x_3 &= 0 \\ 2\lambda x_1 + 4x_2 + \lambda x_3 &= 0 \end{aligned}$$

[JNTU Aug. 2007]

19. Solve the following tridiagonal system by LU decomposition:

$$\begin{aligned} x_1 + 2x_2 \quad\quad &= 7 \\ x_1 + 3x_2 - x_3 &= 4 \\ 4x_2 + 3x_3 &= 5 \end{aligned}$$

[JNTU Aug. 2007]

20. Show that the only real value of λ for which the following equations have nontrivial solution is 6 and solve them:

$$\begin{aligned} 6x + 2y + 3z &= \lambda x \\ 3x + y + 2z &= \lambda y \\ 2x + 3y + z &= \lambda z \end{aligned}$$

[JNTU Aug. 2007]

21. Reduce the matrix A to its normal form and hence find the rank:

$$A = \begin{bmatrix} 0 & 1 & 2 & -2 \\ 4 & 0 & 2 & 6 \\ 2 & 1 & 3 & 1 \end{bmatrix}$$

{JNTU Aug. 2007, Aug. 2006}

22. Define the rank of the matrix and find the rank of the following matrix:

$$\begin{bmatrix} 2 & 1 & 3 & 5 \\ 4 & 2 & 1 & 3 \\ 8 & 4 & 7 & 13 \\ 8 & 4 & -3 & -1 \end{bmatrix}$$

{JNTU Aug. 2007}

23. Determine a, b, c so that A is orthogonal where

$$A = \begin{bmatrix} 0 & 2b & c \\ a & b & -c \\ a & -b & c \end{bmatrix}$$

{JNTU Aug. 2007}

24. Determine whether the following equations will have a nontrivial solution; if so, solve them:

$$\begin{aligned} 4x + 2y + z + 3w &= 0 \\ 6x + 3y + 4z + 7w &= 0 \\ 2x + y + w &= 0 \end{aligned}$$

{JNTU Aug. 2006}

25. Test for consistency the set of equations and solve them if they are consistent:

$$\begin{aligned} x + 2y + 2z &= 2 \\ 3x - 2y - z &= 5 \\ 2x - 5y + 3z &= -4 \\ x + 4y + 6z &= 0 \end{aligned}$$

{JNTU Aug. 2006}

26. Determine whether the following equations will have a nontrivial solution; if so, solve them:

$$\begin{aligned} x + y - 2z + 3w &= 0 \\ x - 2y + z - w &= 0 \\ 4x + y - 5z + 8w &= 0 \\ 5x - 7y + 2z - w &= 0 \end{aligned}$$

{JNTU Aug. 2006}

27. Find the rank of the following matrix:

$$\begin{bmatrix} 2 & 3 & -1 & -1 \\ 1 & -1 & -2 & -4 \\ 3 & 1 & 3 & -2 \\ 6 & 3 & 0 & -7 \end{bmatrix}$$

{JNTU Aug. 2006}

28. By reducing the matrix into normal form, find its rank:

$$\begin{bmatrix} 1 & 2 & 3 & 4 \\ 2 & 1 & 4 & 3 \\ 3 & 0 & 5 & -10 \end{bmatrix}$$

[JNTU Aug. 2006]

29. Solve the following system:

$$\begin{aligned} x - 4y + 7z &= 8 \\ 3x + 8y - 2z &= 6 \\ 7x - 8y + 26z &= 31 \end{aligned}$$

[JNTU Aug. 2006]

30. Find the rank of the following matrix:

$$\begin{bmatrix} 1 & 1 & -3 & 2 \\ 2 & -1 & 2 & -3 \\ 3 & -2 & 1 & -4 \\ -4 & 1 & -3 & 1 \end{bmatrix}$$

[JNTU April 2006]

31. Determine whether the matrix **A** has inverse:

$$\mathbf{A} = \begin{bmatrix} 4 & 0 & 0 & 1 \\ 1 & 3 & 1 & -1 \\ 0 & 0 & 0 & 2 \\ 2 & 4 & 2 & -3 \end{bmatrix}$$

[JNTU April 2004]

32. Explain solution of tridiagonal systems of linear equations. [JNTU April 2004]
33. Explain Gauss elimination method. [JNTU April 2004]
34. Find the inverse of matrix **A** using elementary row operations:

$$\mathbf{A} = \begin{bmatrix} -2 & 1 & 3 \\ 0 & -1 & 1 \\ 1 & 2 & 0 \end{bmatrix}$$

[JNTU April 2004]

35. Solve by Gauss–Seidal method:

$$\begin{aligned} 4x_1 + 3x_2 - x_3 &= 6 \\ 3x_1 + 5x_2 + 3x_3 &= 4 \\ x_1 + x_2 + x_3 &= 1 \end{aligned}$$

[JNTU April 2004]

36. Find the value of x such that matrix **A** is singular:

$$\mathbf{A} = \begin{bmatrix} 3-x & 2 & 2 \\ 2 & 4-x & 1 \\ -2 & -4 & -(1+x) \end{bmatrix}$$

[JNTU Aug. 2003, Aug. 2002]

37. Find the eigenvalues and the corresponding eigenvectors of the matrix:

$$\begin{bmatrix} 2 & 2 & 1 \\ 1 & 3 & 1 \\ 1 & 2 & 2 \end{bmatrix}$$

[JNTU May 2008, Aug. 2007]

38. If

$$A = \begin{bmatrix} 1 & 0 \\ 0 & 3 \end{bmatrix}$$

find A^{50}. [JNTU May 2008]

39. Find the eigenvalues and eigenvectors of the matrix

$$\begin{bmatrix} 1 & -6 & -4 \\ 0 & 4 & 2 \\ 0 & -6 & -3 \end{bmatrix}$$

[JNTU May 2008]

40. Diagonalize the matrix

$$\begin{bmatrix} 8 & -6 & 2 \\ -6 & 7 & -4 \\ 2 & -4 & 3 \end{bmatrix}$$

[JNTU Aug. 2007]

41. Diagonalize the matrix

$$\begin{bmatrix} 8 & -8 & -2 \\ 4 & -3 & -2 \\ 3 & -4 & 1 \end{bmatrix}$$

[JNTU Aug. 2007]

42. If A and B are *n*-rowed square matrices and if A is invertible show that $A^{-1}B$ and BA^{-1} have the same eigenvalues. [JNTU Aug. 2007]

43. Verify Cayley–Hamilton theorem and hence find A^{-1} for

$$A = \begin{bmatrix} 7 & 2 & -2 \\ -6 & -1 & 2 \\ 6 & 2 & -1 \end{bmatrix}$$

[JNTU April 2007]

44. Diagonalize the matrix A and hence find A^4:

$$A = \begin{bmatrix} 1 & 1 & 1 \\ 0 & 2 & 1 \\ -4 & 4 & 3 \end{bmatrix}$$

[JNTU April 2007]

45. Show that the matrix A satisfies its characteristic equation and hence find A^{-1}:

$$A = \begin{bmatrix} 1 & -2 & 2 \\ 1 & 2 & 3 \\ 0 & -1 & 2 \end{bmatrix}$$

[JNTU April 2007]

46. Verify that the sum of eigenvalues is equal to the trace of A for the matrix A and find the corresponding eigenvectors:

$$A = \begin{bmatrix} 3 & -1 & 1 \\ -1 & 5 & -1 \\ 1 & -1 & 3 \end{bmatrix}$$

[JNTU April 2007]

47. Verify Cayley–Hamilton theorem for the matrix A:

$$A = \begin{bmatrix} 8 & -8 & 2 \\ 4 & -3 & -2 \\ 3 & -4 & 1 \end{bmatrix}$$

{JNTU Aug. 2006}

48. Verify Cayley–Hamilton theorem for A and hence find A^{-1} and A^4, where

$$A = \begin{bmatrix} 3 & 1 & 1 \\ -1 & 5 & -1 \\ 1 & -1 & 3 \end{bmatrix}$$

{JNTU Aug. 2006}

49. Verify Cayley–Hamilton theorem for A. Find A^4 and A^{-1} using Cayley–Hamilton theorem.

$$A = \begin{bmatrix} 1 & 2 & -1 \\ 2 & 1 & -2 \\ 2 & -2 & 1 \end{bmatrix}$$

{JNTU Aug. 2006}

50. Find the eigenvalues and the corresponding eigenvectors of

$$\begin{bmatrix} 1 & 1 & 1 \\ 1 & 1 & 1 \\ 1 & 1 & 1 \end{bmatrix}$$

{JNTU May 2006}

51. Prove that the product of the eigenvalues is equal to the determinant of the matrix.

{JNTU May 2006}

52. Diagonalize the matrix A and hence find A^4.

$$A = \begin{bmatrix} 1 & 1 & 1 \\ 0 & 2 & 1 \\ -4 & 4 & 3 \end{bmatrix}$$

{JNTU May 2006}

53. Find the eigenvalues and the corresponding eigenvectors of the matrix

$$\begin{bmatrix} 8 & -6 & 2 \\ -6 & 7 & -4 \\ 2 & -4 & 3 \end{bmatrix}$$

{JNTU May 2006}

54. If λ is an eigenvalue of a non singular matrix A then show that
   a. $1/\lambda$ is an eigenvalue of A^{-1}.
   b. $|A|/\lambda$ is an eigenvalue of the matrix adj A. {JNTU May 2006}

55. Find the eigenvalues and the corresponding eigenvectors of the matrix A:

$$A = \begin{bmatrix} 5 & -2 & 0 \\ -2 & 6 & 2 \\ 0 & 2 & 7 \end{bmatrix}$$

{JNTU May 2006}

56. Prove that the two eigenvectors corresponding to the two different eigenvalues are linearly independent. [JNTU May 2006]

57. Find eigenvalues and eigenvectors of matrix **A**.

$$A = \begin{bmatrix} -2 & 2 & -3 \\ 2 & 1 & -6 \\ -1 & -2 & 0 \end{bmatrix}$$

[JNTU May 2006, May 2001]

58. Diagonalize the matrix **A**.

$$A = \begin{bmatrix} -1 & 1 & 2 \\ 0 & -2 & 1 \\ 0 & 0 & -3 \end{bmatrix}$$

[JNTU May 2001]

59. Show that the given matrix satisfies the matrix equation $A^3 - A^2 - 18A - 30I = 0$:

$$\begin{bmatrix} 1 & 2 & 3 \\ 2 & -1 & 4 \\ 3 & 1 & 1 \end{bmatrix}$$

[JNTU May 2001]

60. Discuss the nature of the quadratic forms and reduce it to canonical form $x^2 4y + 6x + y^2 + 2y + 4z^2$. [JNTU May 2008]

61. Reduce the quadratic form $2x^2 + 2y^2 + 2z^2 - 2xy + 2xz - 2yz$ to the canonical form. [JNTU May 2008, Aug. 2006]

62. Find a matrix **P** which diagonalizes the matrix associated with the quadratic form $3x^2 + 5y^2 + 3z^2 - 2yz + 2zx - 2xy$. [JNTU May 2008, April 2007]

63. Find the transformation which will transform $4x^2 + 3y^2 + z^2 - 8xy - 6yz + 4xz$ into a sum of squares and find the reduced form. [JNTU May 2008]

64. Show that every square matrix can be expressed uniquely as a sum of a Hermitian and skew-Hermitian matrices. [JNTU Aug. 2007]

65. Find a real symmetric matrix **C** of the quadratic form $Q = x_1^2 + 3x_2^2 + 2x_3^2 + 2x_1x_2 + 6x_2x_3$ and find the index and signature. [JNTU Aug. 2007]

66. Find the orthogonal transformation which transforms the quadratic form $x_1^2 + 3x_2^2 + 3x_3^2 - 2x_2x_3$ to canonical form. [JNTU Aug. 2007]

67. Prove that every Hermitian matrix can be written as $\mathbf{A} + i\mathbf{B}$ where **A** is real and symmetric and **B** is real and skew-symmetric. [JNTU Aug. 2007]

68. Prove that the eigenvalues of a real symmetric matrix are real. [JNTU April 2007]

69. Reduce the quadratic form $7x^2 + 6y^2 + 5z^2 - 4xy - 4yz$ to the canonical form. [JNTU April 2007]

70. Define the following:
    a. Hermitian matrix
    b. Skew-Hermitian matrix
    c. Unitary matrix
    d. Orthogonal matrix

    [JNTU April 2007]

71. Show that the eigenvalues of a unitary matrix is of unit modulus. [JNTU April 2007]

72. Define:
   a. Spectral matrix
   b. Quadratic form
   c. Canonical form [JNTU Aug. 2006]

73. Reduce the quadratic form $3x^2 + 5y^2 + 3z^2 - 2yz + 2zx - 2xy$ to the canonical form. [JNTU Aug. 2006]

74. Show that matrix **A** is a skew-Hermitian matrix and also unitary. Find eigenvalues and the corresponding eigenvectors of **A**:

$$\mathbf{A} = \begin{bmatrix} i & 0 & 0 \\ 0 & 0 & i \\ 0 & i & 0 \end{bmatrix}$$

[JNTU Aug. 2006]

75. Show that the matrix **A** is Hermitian. Find its eigenvalues and the corresponding eigenvectors:

$$\mathbf{A} = \begin{bmatrix} 2 & 3+4i \\ 3-4i & 2 \end{bmatrix}$$

[JNTU Aug. 2006]

76. Use Lagrange reduction method to reduce the quadratic form $x_1^{\ 2} + 6x_2^{\ 2} + 18x_3^{\ 2} + 4x_1x_2 + 8x_1x_3 - 4x_2x_3$ to the sum of squares and find the corresponding linear transformation also find the index and signature. [JNTU May 2006]

77. Reduce the quadratic form $x^2 + y^2 + 2z^2 - 2xy + 4xz + 4yz$ to the canonical form. [JNTU May 2006]

78. Reduce the quadratic form $8x^2 + 7y^2 + 3z^2 - 12xy - 8yz + 4xz$ to the canonical form. [JNTU May 2006]

79. Verify that **A*A** is a Hermitian matrix:

$$\mathbf{A} = \begin{bmatrix} 2+i & 3 & -1+3i \\ -5 & i & 4-2i \end{bmatrix}$$

[JNTU May 2006]

80. Reduce $3x^2 + 3z^2 + 4xy + 8xz + 8yz$ into canonical form. [JNTU May 2006]

81. Reduce $3x^2 - 2y^2 - z^2 - 4xy + 8xz + 12yz$ into canonical form by orthogonal transformation and also find its rank, index and signature. [JNTU May 2006]

82. Prove that matrix **A** is unitary:

$$\mathbf{A} = (1/2)\begin{bmatrix} 1+i & -1+i \\ 1+i & 1-i \end{bmatrix}$$

[JNTU May 2002]

83. Prove that eigewnvalues of Hermitian matrix are real. [JNTU 2005, 2004, 2002, 2001]

84. Find a positive root of equation $x^3 - x^2 - 1 = 0$ by bisection method. [JNTU May 2008]

85. Find the interpolating polynomial $f(x)$ from the table:

x	0	1	4	5
$f(x)$	4	3	24	39

[JNTU May 2008]

86. Using Lagrange's interpolation formula, find $y(10)$ from the following table:

x	5	6	9	11
y	12	13	14	16

[JNTU May 2008]

87. Find the second difference of the polynomial $x^4 - 12x^3 + 42x^2 - 30x + 9$ with interval of differencing $h = 2$. [JNTU May 2008]

88. Solve $x^3 = 2x + 5$ for a positive root by iteration method. [JNTU May 2008]

89. Find the parabola passing through the points (0, 1), (1, 3) and (3, 55) using Lagrange's interpolation formula. [JNTU May 2008]

90. If the interval of differencing is unity, prove that

$$\Delta \frac{2^x}{x!} = \frac{2^x(1-x)}{(x+1)!}$$ [JNTU May 2008]

91. If the interval of differencing is unity, prove that

$$\Delta[x(x+1)(x+2)(x+3)] = 4(x+1)(x+2)(x+3)$$ [JNTU May 2008]

92. Find a real root of the equation $f(x) = x + \log x - 2$ using Newton–Raphson method. [JNTU Aug. 2007]

93. Find $f(22)$ from the following data using Newton's backward formula:

x	20	25	30	35	40	45
y	354	332	291	260	231	204

[JNTU Aug. 2007]

94. Find an iterative formula to find the cubic root of a number using Newton–Raphson method. Hence evaluate cubic root of 15. [JNTU Aug. 2007]

95. For x = 20, 25, 32, 49 and cos(x) = 0.939, 0.906, 0.848, 0.656, find cos(43) using Lagrange's formula. [JNTU Aug. 2007]

96. Find a real root of $x^4 - x - 10 = 0$ using bisection method. [JNTU Aug. 2007]

97. Find $f(9)$ by Newton's backward formula given that $f(2) = 94.8$, $f(5) = 87.9$, $f(8) = 81.3$ and $f(11) = 75.1$. [JNTU Aug. 2007]

98. Find a real root of $xe^x - \cos x = 0$ using Newton–Raphson method. [JNTU April 2007]

99. Using Gauss backward difference formula, find $y(8)$ from the following table:

x	0	5	10	15	20	25
y	7	11	14	18	24	32

[JNTU April 2007]

100. Find a real root of the equation $x^3 - x - 11 = 0$ by bisection method. [JNTU April 2007]

101. Construct difference table for the following data:

x	0.1	0.3	0.5	0.7	0.9	1.1	1.3
$f(x)$	0.003	0.067	0.148	0.248	0.370	0.518	0.697

Find $f(0.6)$ using a cubic polynomial that fits at $x = 0.3, 0.5, 0.7$ and 0.9 using Newton's forward formula. [JNTU April 2007]

102. Find a real root of $x + \log x_{10} - 2 = 0$ using Newton–Raphson method. [JNTU April 2007]

103. If y_x is the value of y at x for which the fifth differences are constant and $y_1 + y_7 = -784$, $y_2 + y_6 = 686$, $y_3 + y_5 = 1088$, then find y_4. [JNTU April 2007]

104. Find a real root of $xe^x = 2$ using regula falsi method. [JNTU April 2007]

105. Find $f(22)$ from the following table using Gauss forward formula:

x	20	25	30	35	40	45
$f(x)$	354	332	291	260	231	204

[JNTU April 2007]

106. Find a real root of $\exp(x) \times \sin x = 1$ using regula falsi method. [JNTU Aug. 2006]

107. Find $y(25)$, given that $y_{20} = 24$, $y_{24} = 32$, $y_{28} = 35$, $y_{32} = 40$ using Gauss forward difference formula. [JNTU Aug. 2006]

108. Find a real root of $x \tan x + 1 = 0$ using Newton–Raphson method. [JNTU Aug. 2006]

109. Find the unique polynomial $P(x)$ of degree 2 or less such that $P(1) = 1$, $P(3) = 27$, $P(4) = 64$ using Lagrange interpolation formula and Newton divided difference formula. [JNTU Aug. 2006]

110. Find a root of $\exp(x) \times \sin x = 1$ using Newton–Raphson method. [JNTU Aug. 2006]

111. Find a real root of $xe^x - \cos x = 0$ using Newton–Raphson method. [JNTU Aug. 2006]

112. Find a real root of the equation $x^3 - 6x - 4 = 0$ by bisection method. [JNTU May 2006]

113. Find $f(2.5)$ using Newton's forward formula from the following table:

x	0	1	2	3	4	5	6
y	0	1	16	81	256	625	1296

[JNTU May 2006]

114. Find a real root of $x^3 - 5x + 3 = 0$ using bisection method. [JNTU May 2006]

115. Given that $\sin 45° = 0.7077$, $\sin 50° = 0.766$, $\sin 55° = 0.8192$, $\sin 60° = 0.866$, find $\sin 52°$ using Newton's forward difference formula. [JNTU May 2006]

116. Find a real root of the equation $2x - \log x = 7$ by successive approximation method. [JNTU May 2006]

117. Find $y(1.6)$ using Newton's forward difference formula from the table:

x	1	1.4	1.8	2.2
y	3.49	4.82	5.96	6.5

[JNTU May 2006]

118. Find a real root of $xe^x = 3$ using regula falsi method. [JNTU May 2006]

119. Show that $\Delta^2 f_i = (f_i + f_{i+1})\Delta f_i$. [JNTU May 2006]

120. If $f(x) = u(x)\, v(x)$, show that $f(x_0, x_1) = u(x_0)\, v(x_0, x_1) + u(x_0, x_1)\, v(x_1)$. [JNTU May 2006]

121. The Newton–Raphson method is to be used to find the root of the equation $f(x) = 0$ where x_0 is the initial approximation and f' is the derivate of f, the method converges:
   a. Always
   b. Only if f is a polynomial
   c. Only if $f(x_0) < 0$
   d. None **[JNTU May 2006]**

122. Which of the following statements applies to the bisection method used for finding roots of functions?
   a. Converges within a few iterations
   b. Guaranteed to work for all continuous functions
   c. Is faster than the Newton–Raphson method
   d. Requires that there be no error in determining the sign of the function **[JNTU May 2006]**

123. Find the real root of the equation $x^6 - x - 1 = 0$ with an accuracy of 0.001 by bisection method, false position method. **[JNTU May 2006]**

124. Find the root of the equation $x\log_{10} x = 1.2$. **[JNTU May 2006]**

125. Find the one real root of equation $x^4 - x - 10 = 0$ up to 4 decimal points by successive approximation method. **[JNTU May 2006]**

126. From the following table of values of x and $f(x)$, determine $f(0.29)$:

x	0.20	0.22	0.24	0.26	0.28	0.30
$f(x)$	1.6596	1.6698	1.6804	1.6912	1.7024	1.7139

[JNTU May 2006]

127. Find $f(2.5)$ using the following table:

x	1	2	3	4
$f(x)$	1	8	27	64

[JNTU May 2006]

128. Fit a parabola to the data given below:

x	0	1	4	5
$f(x)$	4	3	24	39

[JNTU May 2008]

129. For the table below, find $f'(1.76)$ and $f'(1.72)$:

x	1.72	1.73	1.74	1.75	1.76
$f(x)$	0.17907	0.17728	0.17552	0.17377	0.17204

[JNTU May 2008]

130. Find a curve $y = ae^{bx}$ to the data:

x	0	2	4
y	5.1	10	31.1

[JNTU May 2008]

131. Using the table below, find $f'(0)$ and $\int_0^9 f(x)dx$:

x	0	2	3	4	7	9
$f(x)$	4	26	58	110	460	920

{JNTU May 2008}

132. Using Simpson's 3/8th rule, evaluate $\int_0^6 dx/(1+x^2)$, by dividing the range into 6 equal parts. {JNTU May 2008}

133. Fit a curve of the form $y = ae^{bx}$ to the data:

x	0	1	2	3
y	1.05	2.10	3.85	8.30

{JNTU May 2008}

134. Fit the least square straight line $y = a + bx$ to the following data:

x	−5	−3	−1	0	1	2	4
$f(x)$	0.4	−0.1	−0.2	−0.3	−0.3	0.1	0.4

{JNTU May 2008}

135. Find $f(2.36)$ from the following table:

x	1.6	1.8	2.0	2.2	2.4	2.6
$f(x)$	4.95	6.05	7.39	9.03	11.02	13.46

{JNTU May 2008}

136. Fit a curve of the form $y = a\exp(bx)$ from the following data:

x	1	2	3	4	5	6
y	1.6	4.5	13.8	40.2	125	300

{JNTU Aug. 2007}

137. Evaluate $\int_0^1 e^{-x^2} dx$ by taking $h = 2$ using:
   a. Simpson's 1/3rd rule
   b. Trapezoidal rule {JNTU Aug. 2007}

138. Fit a parabola of the form $y = a + bx + cx^2$ for the following data:

x	1.0	1.5	2.0	2.5	3.0	3.5	4.0
y	1.1	1.3	1.6	2.0	2.7	3.4	4.1

{JNTU Aug. 2007}

139. By the method of least squares fit a parabola of the form $y = a + bx + cx^2$ for the following data:

x	2	4	6	8	10
y	3.07	12.85	31.47	57.38	91.29

{JNTU Aug. 2007}

140. Use the trapezoidal rule with $n = 4$ to estimate $\int_0^1 dx/(1+x^2)$ correct to four decimal places. {JNTU Aug. 2007}

141. Fit a parabola of the form $y = A_1 e^{\lambda_1 x} + A_2 e^{\lambda_2 x}$ for the following data:

x	1.0	1.1	1.2	1.3	1.4	1.5	1.6	1.7	1.8
y	1.175	1.336	1.51	1.698	1.904	2.129	2.376	2.646	2.942

[JNTU Aug. 2007]

142. Fit a parabola of the form $y = a + bx + cx^2$ to the following data:

x	1	2	3	4	5	6	7
y	23	5.2	9.7	16.5	29.4	35.5	54.4

[JNTU Aug. 2007, Aug. 2006]

143. The table below shows the temperature $f(t)$ as a function of time t, find $f'(1.5)$.

t	1	2	3	4	5	6	7
$f(t)$	81	75	80	83	78	70	60

[JNTU April 2007]

144. Derive normal equations to fit the straight line $y = a + bx$. [JNTU April 2007, May 2006]

145. Fit a curve of the form $y = a + bx + cx^2$ for the following data:

x	10	15	20	25	30	35
y	35.3	32.4	29.2	26.1	23.2	20.5

[JNTU April 2007]

146. Fit a straight line of the form $y = a + bx$ for the following data:

x	0	5	10	15	20	25
y	12	15	17	22	24	30

[JNTU April 2007]

147. Fit a curve of the form $y = ae^{bx}$ to the following data:

x	1	5	7	9	12
y	10	15	12	15	21

[JNTU Aug. 2006]

148. Evaluate $\int_0^1 \sqrt{1+x^2}\, dx$ taking $h = 1$ using

a. Simpson's 1/3rd rule
b. Trapezoidal rule

[JNTU Aug. 2006]

149. Fit the curve of the form $y = ae^{bx}$.

x	77	100	185	239	285
y	2.4	3.4	7.0	11.1	19.6

[JNTU Aug. 2006]

150. Fit a curve of the form $y = ab^x$ for the following data by the method of least squares:

x	2	3	4	5	6
y	8.3	15.4	33.1	64.2	127.4

[JNTU Aug. 2006]

151. Derive the normal equations to fit the parabola $y = a + bx + cx^2$. [JNTU May 2006]

152. Given that

x	4.0	4.2	4.4	4.6	4.8	5.0	5.2
$\log(x)$	1.3863	1.4351	1.4816	1.5261	1.5686	1.6094	1.6487

evaluate $\int_4^{5.2} \log x \, dx$ by Simpson's 3/8th rule. [JNTU May 2006]

153. Evaluate $\int_0^2 e^{-x^2} dx$ using Simpson's rule taking $h = 0.25$. [JNTU May 2006]

154. Fit a straight line $y = a + bx$ from the following data:

x	0	1	2	3	4
y	1	1.8	3.3	4.5	6.3

[JNTU May 2006]

155. A rocket is launched from the ground. Its acceleration measured in every 5 seconds is tabulated below. Find the velocity and the position of the rocket at $t = 40$ seconds using trapezoidal rule.

t	0	5	10	15	20	25	30	35	40
$A(t)$	40.0	42.25	48.50	51.25	54.35	59.48	61.5	64.3	68.7

[JNTU May 2006]

156. From the following table, find Y when $X = 38$.

X	30	35	40	45	50
Y	15.9	14.9	14.1	13.3	12.5

[JNTU May 2006]

157. Fit a straight line through the five points (1, 3), (3, 4), (4, 6), (6, 7) and (7, 8) and estimate the value of y at $x = 5$. [JNTU May 2006]

158. Solve the following using RK fourth-order method $y' = y - x$, $y(0) = 2$, $h = 0.2$. Find $y(0.2)$. [JNTU May 2008]

159. Given $y' = x^2 - y, y(0) = 1,$ find correct to four decimal places the value of $y(0,1)$, by using Euler's method. [JNTU May 2008]

160. Find $y(0.1)$, $y(0.2)$, $z(0.1)$, $z(0.2)$ given $dy/dx = x + z$, $dz/dx = x - y^2$ and $y(0) = 2$, $z(0) = 1$ by using Taylor's series method. [JNTU May 2008]

161. Solve $y' = y - x^2, y(0) = 1$ by Picard's method upto the third approximation. Hence find the value of $y(0.1)$, $y(0.2)$. [JNTU May 2008]

162. Solve $y' = x + y$ given $y(1) = 0$. Find $y(1.1)$ and $y(1.2)$ by Taylor's series method. [JNTU May 2008]

163. Solve $dy/dx = xy$ using RK method at $x = 0.2$ given $y(0) = 1$, $y'(0) = 0$ taking $h = 0.2$. [JNTU May 2008]

164. Solve the equation $dy/dx = x - y^2$ with the conditions $y(0) = 1$, $y'(0) = 1$. Find $y(0.2)$ and $y(0.4)$ using Taylor's series method. [JNTU May 2008]

165. Given $y' = x + \sin y$, $y(0) = 1$, compute $y(0.2)$ and $y(0.4)$ with $h = 0.2$ using Euler's modified method. **[JNTU Aug. 2007]**

166. Find $y(0.5)$, $y(1)$, and $y(1.5)$, given that $y' = 4 - 2x$, $y(0) = 2$, with $h = 0.5$ using modified Euler method. **[JNTU Aug. 2007]**

167. Find the solution of $dy/dx = x - y$ at $x = 0.1, 0.2, 0.3, 0.4$, and 0.5 using modified Euler method where $y(0) = 1$. **[JNTU Aug. 2007, May 2006]**

168. Find $y(0.1)$, $y(0.2)$, and $y(0.3)$ using Taylor's series method where $dy/dx = 1 - y$, $y(0) = 0$. **[JNTU Aug. 2007]**

169. Find $y(0.2)$ using Picard's method given that $dy/dx = xy$, $y(0) = 1$, by taking $h = 0.1$. **[JNTU April 2007]**

170. Given $y' = x + \sin y$, $y(0) = 1$, compute $y(0.2)$ and $y(0.4)$ with $h = 0.2$ using Euler's modified method. **[JNTU April 2007]**

171. Use Runge–Kutta fourth-order method to evaluate $y(0.1)$ and $y(0.2)$, given that $dy/dx = x + y$, $y(0) = 1$. **[JNTU April 2007]**

172. Find the solution of $dy/dx = x - y$ at $x = 0.4$ subject to the condition $y = 1$, at $x = 0$ and $h = 0.1$ using Milne's method. Use Euler's modified method to evaluate $y(0.1)$, $y(2)$, and $y(0.3)$. **[JNTU April 2007]**

173. If $dy/dx = 2xy$, $y(0) = 2$, find $y(0.4)$ using Adam's predictor–corrector formula by calculating $y(0.1)$, $y(0.2)$, and $y(0.3)$ using Euler's modified formula. **[JNTU Aug. 2006]**

174. Find $y(0.1)$ and $y(0.2)$ using Taylor's series method given that $y' = y^2 + x$ and $y(0) = 1$. **[JNTU Aug. 2006]**

175. Find $y(0.1)$ and $y(0.2)$ using Runge–Kutta fourth-order formula given that $y' = x^2 - y$ and $y(0) = 1$. **[JNTU Aug. 2006]**

176. Given that $dy/dx = 1 + xy$ and $y(0) = 1$, compute $y(0.1)$ and $y(0.2)$ using Picard's method. **[JNTU May 2006]**

177. Tabulate $y(0.1)$, $y(0.2)$, and $y(0.3)$ using Taylor's series method given that $y' = y^2 + x$ and $y(0) = 1$. **[JNTU May 2006]**

178. Find $y(0.1)$ and $y(0.2)$ using Euler's modified formula given that $dy/dx = x^2 - y$, $y(0) = 1$. **[JNTU May 2006]**

179. Solve the differential equation, $dy/dx = 2xy$, $y(0) = 0.5$ for $1 > x > 0$ by Runge–Kutta second-order method. **[JNTU May 2006]**

180. Solve $y' = 4 - 2x$, $y(0) = 2$, with $h = 0.5$ using the following methods:
    a. Euler method
    b. Modified Euler method
    and compare the results with the theoretical values. **[JNTU May 2006]**

181. Using Euler's method solve $y' = x + 2y$, $y(0) = 0$, $0 \leq x \leq 1$ with step size $h = 0.25$. **[JNTU May 2006]**

182. Find $dy/dx + xy = 0$ from $x = 0$ to $x = 0.25$, $y(0) = 1$, $h = 0.05$ using Euler's method. **[JNTU May 2006]**

183. Use Runge–Kutta fourth-order method to solve $dy/dx = -2xy^2$, $y(0) = 1$, $h = 0.2$ in the interval [0, 1]. [JNTU May 2005]

184. Solve the following differential equation by Runge–Kutta fourth-order method: $dy/dx = x^2y^2 - xy$ with $y(0) = 1$ at $x = 0$ and $h = 0.1$ in the range $0 \leq x \leq 1$. [JNTU May 2005]

185. Use Milne's predictor–corrector method to solve $dy/dx = -2xy^2$, $y(0) = 0.1$, $h = 0.2$ in the interval [0, 1]. [JNTU May 2005]

186. Use Milne's predictor–corrector method to compute the solution of $dy/dx = -2xy$, $y(0) = 0.5$, $h = 0.1$ in the interval [0, 1]. [JNTU May 2005]

187. Obtain the Fourier series expansion of $f(x)$ given that $f(x) = (\pi - x)^2$ in $0 < x < 2\pi$ and deduce the following:

$$\frac{1}{1^2} + \frac{1}{2^2} + \frac{1}{3^2} + \cdots = \frac{\pi^2}{6}$$

[JNTU May 2008]

188. Find the half-range Fourier cosine series for $f(x) = x$ in $0 < x < 2\pi$. [JNTU May 2008]

189. Find the half-range cosine series for $f(x) = \sin kx$ for k not an integer. [JNTU May 2008]

190. Expand $f(x) = x^2, 0 < x < 2\pi$ as a Fourier series. [JNTU May 2008]

191. Using Parsevals identity, evaluate

$$\int_0^\infty \frac{x^2 dx}{(a^2 + x^2)^2}, \ a > 0$$

[JNTU May 2008]

192. Evaluate

$$\int_0^\infty \frac{dx}{(a^2 + x^2)(b^2 + x^2)}$$

using transforms. [JNTU May 2008]

193. Obtain the Fourier series expansion of $f(x)$ given that $f(x) = kx(\pi - x)$ in $0 < x < 2\pi$ where k is a constant. [JNTU May 2008]

194. Find the Fourier series of periodicity 3 for $f(x) = 2x - x^2$, in $0 < x < 3$. [JNTU May 2008]

195. Find the half-range cosine series for the function $f(x) = (x - 1)^2$ in the interval $0 < x < 1$. Hence show that

$$\sum_{n=1}^\infty \frac{1}{(2n-1)^2} = \frac{x^2}{8}$$

[JNTU Aug. 2007]

196. State and prove Fourier integral theorem. [JNTU Aug. 2007]

197. Obtain a Fourier expansion for $\sqrt{1 - \cos x}$ in the interval $-\pi < x < \pi$. [JNTU Aug. 2007]

198. Write the Dirichlet's conditions for the existence of Fourier series of a function $f(x)$ in the interval $(\alpha, \alpha + 2\pi)$. Find the Fourier series representation for $f(x) = x$ in $0 < x < 2\pi$. [JNTU Aug. 2007]

199. Find a half-range sine series for $f(x) = ax + b$ in $0 < x < 1$. [JNTU Aug. 2007]

200. Find Fourier cosine transform of

$$f(x) = \begin{cases} \cos x & 0 < x < a \\ 0 & x \geq a \end{cases}$$

[JNTU Aug. 2007]

201. Find the Fourier series of $f(x) = x^2 - 2$, when $-2 < x < 2$. [JNTU April 2007]

202. Find the Fourier cosine transform of $5e^{-2x} + 2e^{-5x}$. [JNTU April 2007]

203. Represent the following function by a Fourier sine series:

$$f(t) = \begin{cases} t, & 0 < t \le \dfrac{\pi}{2} \\ \dfrac{\pi}{2}, & \dfrac{\pi}{2} < t \le \pi \end{cases}$$ [JNTU April 2007]

204. Find the finite cosine transform of

$$f(x) = \frac{\pi}{3} - x + \frac{x^2}{2\pi}$$ [JNTU Aug. 2006]

205. Find the Fourier series to represent the function $f(x) = |\sin x|, -\pi < x < \pi$. [JNTU Aug. 2006]

206. Find Fourier sine transform of xe^{-ax}. [JNTU Aug. 2006]

207. Obtain the Fourier expansion of $x \sin x$ as cosine series in $(0, \pi)$ and show that

$$\frac{1}{1\times 3} - \frac{1}{3\times 5} + \frac{1}{5\times 7} - \frac{1}{7\times 9} + \cdots = \frac{\pi - 2}{4}$$ [JNTU May 2006]

208. Find the Fourier series corresponding to the following $f(x)$ defined in $(-2, 2)$:

$$f(x) = \begin{cases} 2, & -2 \le x \le 0 \\ x, & 0 < x < 2 \end{cases}$$ [JNTU May 2006]

209. Find the finite Fourier cosine transform of $f(x) = x^2$ in $(0, l)$. [JNTU May 2006]

210. Expand $f(x) = x \sin x$, $0 < x < 2\pi$ as a Fourier series. [JNTU May 2006]

211. Using Fourier integral, show that

$$\int_0^\infty \frac{1-\cos \pi\lambda}{\lambda} \sin x\lambda \, d\lambda = \frac{\pi}{2}, \quad 0 < x < \pi$$ [JNTU May 2006]

212. Define a periodic function. Find the Fourier expansion for the function $f(x) = x - x^2, -1 < x < 1$.
[JNTU May 2006]

213. If $f(x) = [(\pi - x)/2]^2$ in the interval $(0, 2\pi)$ then show that

$$f(x) = \frac{\pi^2}{12} + \sum_{n=1}^{\infty} \frac{\cos nx}{n^2}$$ [JNTU May 2005]

214. Prove that in the range $-p < x < p$,

$$\cosh ax = \frac{2a}{\pi} \sinh a\pi \left[\frac{1}{2a^2} + \sum_{n=1}^{\infty} \frac{-1^n}{n^2 + a^2} \cos nx \right]$$ [JNTU May 2005]

215. Find the Fourier series for the function $f(x) = x^2$, $0 \le x \le p$

$= -x^2, -p \le x \le 0$ [JNTU May 2005]

216. Find the Fourier expansion for $f(x) = px$ in $(-c, c)$. [JNTU May 2005]

217. Find the half-range sine series for

$$f(x) = 1/4 - x, \quad 0 < x < 1/2$$
$$= x - 3/4, \quad 1/2 < x < 1$$

[JNTU May 2005]

218. Find the Fourier sine and cosine transforms of

$$f(x) = x, \quad 0 < x < 1$$
$$= 2 - x, \quad 1 < x < 2$$
$$= 0, \quad x > 2$$

[JNTU May 2005]

219. Solve the partial differential equation $x^2 p^2 + y^2 q^2 = 1$. [JNTU May 2008]

220. Solve the difference equation, using Z-transform $y(k + 2) - 5y(k + 1) + 6y(k) = 5^n$, given $y(0) = 0$, $y(1) = 0$. [JNTU May 2008]

221. Solve $(x + y)p + (y + z)q = (z + x)$. [JNTU May 2008]

222. Solve the difference equation, using Z-transform $y(k + 2) - 2 \cos \alpha \times y(k + 1) + y(k) = 0$, given $y(0) = 1, y(1) = 1$. [JNTU May 2008]

223. Solve the difference equation, using Z-transform $y(n + 2) + 3y(n + 1) + 2y(n) = 0$, given $y(0) = 0, y(1) = 1$. [JNTU May 2008]

224. Solve $x^2 (z- y)p+ y^2(x- z)q = z^2 (y- x)$. [JNTU May 2008]

225. Solve $(x^2+ y^2 + z^2)p - 2xyq = -2xz$. [JNTU May 2008]

226. Solve the difference equation, using Z-transform $y(k + 2) - 4y(k + 1) + 4y(k) = \pi$, given $x(0) = 0$, $x(1) = 0$. [JNTU May 2008]

227. Form the partial differential equation by eliminating the arbitrary constants from

$$(x - a)^2 + (y - b)^2 + z^2 = r^2$$

[JNTU Aug. 2007]

228. Solve the partial differential equation $z^2(p^2 + q^2) = x^2 + y^2$. [JNTU Aug. 2007]

229. Form the partial differential equation by eliminating the arbitrary function $z = f(x^2 + y^2 + z^2)$. [JNTU Aug. 2007]

230. Solve the partial differential equation $p^2x + q^2y = z$. [JNTU Aug. 2007]

231. Solve the partial differential equation $(p/x^2) + (q/y^2) = z$. [JNTU Aug. 2006]

232. Form the partial differential equation by eliminating the arbitrary constants.

$$\log (az - 1) = x + ay + b$$

[JNTU Aug. 2007]

233. Solve the partial differential equation $px(y^2 + z) - qy(x^2 + z) = z(x^2 - y^2)$. [JNTU Aug. 2007]

234. Form the partial differential equation by eliminating the arbitrary constants a, b from

$$2z = (x + a)^{1/2} + (y - a)^{1/2} + b$$

[JNTU Aug. 2007]

235. Solve the partial differential equation, $z^4p^2 + z^4q^2 = x + y^2$. [JNTU Aug. 2007]

236. Form the partial differential equation by eliminating the arbitrary functions from $z = xf_1(x+t) + f_2(x+t)$ [JNTU April 2007]

237. Solve the partial differential equation $p^2x^4 + y^2zq = 2z^2$. [JNTU April 2007]

238. Form the partial differential equation by eliminating the arbitrary function from $z = yf(x^2 + z^2)$ [JNTU April 2006, Aug. 2007]

239. Solve the partial differential equation $x(y-z)p + y(z-x)q = z(x-y)$. [JNTU April 2007]

240. Form the partial differential equation by eliminating the arbitrary constants $z = ax^3 + by^3$. [JNTU April 2007]

241. Solve the partial differential equation $z(x-y) = px^2 - qy^2$. [JNTU April 2007]

242. Solve the difference equation, using Z-transforms $u_{n+2} - 3u_{n+1} + 2u_n = 0$ given that $u_0 = 0, u_1 = 1$. [JNTU April 2007]

243. Solve the partial differential equation $p^2x + q^2y = z$. [JNTU Aug. 2006]

244. Form the partial differential equation by eliminating the arbitrary function from $z = f(x^2 - y^2)$. [JNTU Aug. 2006]

245. Solve the partial differential equation $(y+z)p + (z+x)q = x + y$. [JNTU Aug. 2006]

246. Find $Z[1/(n+2)(n+3)]$. [JNTU Aug. 2006]

247. Find the Z-transform of $\sin \alpha k, k \geq 0$. [JNTU May 2006]

248. Form the partial differential equation by eliminating arbitrary constants.

$$2z = [\sqrt{x+a} + \sqrt{y-a}] + b$$ [JNTU May 2006]

249. Find $Z(2 \times 3^n + 5 \times n)$ and deduce $Z[2 \times 3^{n+3} + 5(n+3)]$ using shifting theorem. [JNTU May 2006]

250. Form the partial differential equation by eliminating the arbitrary function from $xyz = f(x^2 + y^2 + z^2)$ [JNTU May 2006]

251. Find $z^{-1}[(z^2+z)/(z-1)^2]$. [JNTU May 2006]

252. Find $z[1/(n+2)(n-1)]$. [JNTU May 2006]

253. Find $z^{-1}[1/(z-1/2)(z-1/3)]$ if $1/3 < |z| < 1/2$. [JNTU May 2006]

254. Form partial differential equation by eliminating arbitrary function from $F(x^2 + y^2, z - xy) = 0$. [JNTU May 2006]

255. Solve $x^2(y-z)p + y^2(z-x)q = z^2(x-y)$. [JNTU May 2006]

256. Solve $z^2(p^2 + q^2 + 1) = 1$. [JNTU May 2006]

257. Solve $z(p^2 - q^2) = x - y$. [JNTU May 2006]

258. Solve the difference equation, using Z-transforms $u_{n+2} - u_n = 2^n$ where $u_0 = 0, u_1 = 1$. [JNTU Aug. 2007]